AN INTEGRATED APPROACH

Fresh grapes taste sweet because they contain glucose. Glucose, a carbohydrate also known as dextrose, is a simple sugar. But how can a simple sugar be used to fuel a complex system like your body? How does excess carbo-  hydrate from a big meal of pasta turn into fat and get stored in the body? The answers can be found in chemistry. We developed an integrated approach to teaching General, Organic, and Biological

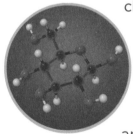

chemistry so that you will learn the basics of chem- istry and the applications behind them. For example in Chapter 5, we apply general, organic and biological chemistry concepts to glucose, and later bring all of them together to understand how your body uses glucose to produce energy.

Using similar strategies to weave concepts together instead of presenting them as stand-  alone principles, we integrate the topics of gen- eral, organic, and biological chemistry to make learning chemistry relevant and fun.

ATOMIC MASSES OF THE ELEMENTS

Name	Symbol	Atomic Number	Atomic Mass[a]	Name	Symbol	Atomic Number	Atomic Mass[a]
Actinium	Ac	89	(227)	Neodymium	Nd	60	144.2
Aluminum	Al	13	26.98	Neon	Ne	10	20.18
Americium	Am	95	(243)	Neptunium	Np	93	(237)
Antimony	Sb	51	121.8	Nickel	Ni	28	58.69
Argon	Ar	18	39.95	Niobium	Nb	41	92.91
Arsenic	As	33	74.92	Nitrogen	N	7	14.01
Astatine	At	85	(210)	Nobelium	No	102	(259)
Barium	Ba	56	137.3	Osmium	Os	76	190.2
Berkelium	Bk	97	(247)	Oxygen	O	8	16.00
Beryllium	Be	4	9.012	Palladium	Pd	46	106.4
Bismuth	Bi	83	209.0	Phosphorus	P	15	30.97
Bohrium	Bh	107	(264)	Platinum	Pt	78	195.1
Boron	B	5	10.81	Plutonium	Pu	94	(244)
Bromine	Br	35	79.90	Polonium	Po	84	(209)
Cadmium	Cd	48	112.4	Potassium	K	19	39.10
Calcium	Ca	20	40.08	Praseodymium	Pr	59	140.9
Californium	Cf	98	(251)	Promethium	Pm	61	(145)
Carbon	C	6	12.01	Protactinium	Pa	91	231.0
Cerium	Ce	58	140.1	Radium	Ra	88	(226)
Cesium	Cs	55	132.9	Radon	Rn	86	(222)
Chlorine	Cl	17	35.45	Rhenium	Re	75	186.2
Chromium	Cr	24	52.00	Rhodium	Rh	45	102.9
Cobalt	Co	27	58.93	Roentgenium	Rg	111	(272)
Copper	Cu	29	63.55	Rubidium	Rb	37	85.47
Curium	Cm	96	(247)	Ruthenium	Ru	44	101.1
Darmstadtium	Ds	110	(271)	Rutherfordium	Rf	104	(261)
Dubnium	Db	105	(262)	Samarium	Sm	62	150.4
Dysprosium	Dy	66	162.5	Scandium	Sc	21	44.96
Einsteinium	Es	99	(252)	Seaborgium	Sg	106	(266)
Erbium	Er	68	167.3	Selenium	Se	34	78.96
Europium	Eu	63	152.0	Silicon	Si	14	28.09
Fermium	Fm	100	(257)	Silver	Ag	47	107.9
Fluorine	F	9	19.00	Sodium	Na	11	22.99
Francium	Fr	87	(223)	Strontium	Sr	38	87.62
Gadolinium	Gd	64	157.3	Sulfur	S	16	32.07
Gallium	Ga	31	69.72	Tantalum	Ta	73	180.9
Germanium	Ge	32	72.64	Technetium	Tc	43	(98)
Gold	Au	79	197.0	Tellurium	Te	52	127.6
Hafnium	Hf	72	178.5	Terbium	Tb	65	158.9
Hassium	Hs	108	(269)	Thallium	Tl	81	204.4
Helium	He	2	4.003	Thorium	Th	90	232.0
Holmium	Ho	67	164.9	Thulium	Tm	69	168.9
Hydrogen	H	1	1.008	Tin	Sn	50	118.7
Indium	In	49	114.8	Titanium	Ti	22	47.87
Iodine	I	53	126.9	Tungsten	W	74	183.8
Iridium	Ir	77	192.2	Uranium	U	92	238.0
Iron	Fe	26	55.85	Vanadium	V	23	50.94
Krypton	Kr	36	83.80	Xenon	Xe	54	131.3
Lanthanum	La	57	138.9	Ytterbium	Yb	70	173.0
Lawrencium	Lr	103	(260)	Yttrium	Y	39	88.91
Lead	Pb	82	207.2	Zinc	Zn	30	65.41
Lithium	Li	3	6.941	Zirconium	Zr	40	91.22
Lutetium	Lu	71	175.0	—	—	112	(285)
Magnesium	Mg	12	24.31	—	—	113	(284)
Manganese	Mn	25	54.94	—	—	114	(289)
Meitnerium	Mt	109	(268)	—	—	115	(288)
Mendelevium	Md	101	(258)	—	—	116	(292)
Mercury	Hg	80	200.6	—	—	118	(293)
Molybdenum	Mo	42	95.94				

[a]Values in parentheses are the mass number of the most stable isotope.

GENERAL, ORGANIC, AND BIOLOGICAL CHEMISTRY

AN INTEGRATED APPROACH

LAURA FROST
Georgia Southern University

TODD DEAL
Georgia Southern University

KAREN C. TIMBERLAKE
Los Angeles Valley College

Prentice Hall

Boston Columbus Indianapolis New York San Francisco Upper Saddle River
Amsterdam Cape Town Dubai London Madrid Milan Munich Paris Montréal Toronto
Delhi Mexico City São Paulo Sydney Hong Kong Seoul Singapore Taipei Tokyo

Library of Congress Cataloging-in-Publication Data
Frost, Laura D.
    General, organic, and biological chemistry : an integrated approach / Laura D. Frost, S. Todd Deal, Karen C. Timberlake.
       p.   cm.
    ISBN 978-0-8053-8178-8
    1. Chemistry—Textbooks   I. Deal, S. Todd. II. Timberlake, Karen. III. Title.

QD251.3.F76 2011
540—dc22   2009043873

Acquisitions Editor: Dawn Giovanniello
Editor in Chief, Chemistry and Geosciences: Nicole Folchetti
Marketing Manager: Erin Gardner
Assistant Editor: Laurie Hoffman
Editorial Assistant: Lisa Tarabokjia
Marketing Assistant: Nicola Houston
VP/Executive Director, Development: Carol Trueheart
Development Editor: Ray Mullaney
Managing Editor, Chemistry and Geoscience: Gina Cheselka
Project Manager, Production: Shari Toron
Copy Editor: Martha Williams
Proofreader: Marne Evans
Senior Manufacturing and Operations Manager: Nick Sklitsis
Operations Specialist: Maura Zaldivar
Composition/Full-Service: Preparé Inc.

Production Editor, Full Service: Rebecca Dunn
Senior Technical Art Specialist: Connie Long
Art Studio: Precision Graphics
Project Manager: John C. Morgan
Design Director/Cover Designer: Mark Ong
Interior Design: Maureen Eide
Photo Research Manager: Elaine Soares
Photo Researcher: Eric Schrader
Senior Media Producer: Angela Bernhardt
Media Producer: Kristin Mayo
Senior Media Production Supervisor: Liz Winer
Media Production Coordinator: Shannon Kong
Cover Photo Credit: Grapes and Pasta: JupiterUnlimited; Dancer:
       Shutterstock; Molecular art: Precision Graphics

ISBN 10: 0-8053-8178-3; ISBN 13: 978-0-8053-8178-8 (Student Edition)

Prentice Hall
is an imprint of

www.pearsonhighered.com

About the Authors

LAURA D. FROST is an Associate Professor of Chemistry at Georgia Southern University where she has taught chemistry to allied health students since 2000. She received her bachelor's degree in chemistry from Kutztown University and a Ph.D. in chemistry with a biophysical focus from the University of Pennsylvania.

Professor Frost is actively engaged in the teaching and learning of chemistry and uses a guided inquiry approach in her classes. She is very involved in the scholarship of teaching and learning and has demonstrated that the use of inquiry-based activities increases student learning in her one-semester allied health chemistry course.

Dr. Frost is a member of the American Chemical Society and its Chemical Education division and the Biophysical Society. In 2007, she was honored with the Regent's Award for the Scholarship of Teaching and Learning by the University System of Georgia and was inducted into the Regent's Hall of Fame for Teaching Excellence. She is an advocate for increased student learning in science, technology, engineering, and mathematics (STEM) disciplines using guided inquiry and has spoken at numerous conferences and workshops on this topic.

When not teaching or writing, Dr. Frost enjoys running, disc golf, hiking, camping, and most importantly her family—her husband Baxter and their two elementary school-age children, Iris and Little Baxter.

TODD DEAL received his B.S. degree in chemistry in 1986 from Georgia Southern College (now University) in Statesboro, Georgia, and his Ph.D. in chemistry in 1990 from The Ohio State University. He joined the faculty of his undergraduate alma mater in 1992 where he currently serves as Director of the Office of Student Leadership and Civic Engagement. During his tenure at Georgia Southern, Professor Deal has also served as Associate Dean of the Allen E. Paulson College of Science and Technology.

Professor Deal has taught chemistry to allied health and preprofessional students for 20 years. In 1994, he was selected Professor of the Year by the students at Georgia Southern University. Professor Deal is also the recipient of the Allen E. Paulson College of Science and Technology's Award for Excellence in Teaching (2003), the Georgia Southern University Award for Excellence in Contributions to Instruction (2003), and the Allen E. Paulson College of Science and Technology's Award for Excellence in Service (2006).

Professor Deal is a member of American Chemical Society and its Organic, Carbohydrate, and Chemical Education divisions. In 1996, he was named to the Project Kaleidoscope Faculty for the 21st Century in recognition of his innovative teaching in the sciences. He is also a member of the Omicron Delta Kappa National Leadership Honor Society.

When he is not teaching, Professor Deal enjoys spending as much time as possible with his wife Karen and their daughters, Abbie and Anna. He is an avid cyclist, a decent kayaker, a huge fan of Starbucks coffee, and is the "Voice of the Eagles" as public address announcer for Georgia Southern's football team. Professor Deal also enjoys interacting with his students outside of class in his role as mentor and advisor for Georgia Southern's Baptist Collegiate Ministries.

KAREN TIMBERLAKE is Professor Emerita of chemistry at Los Angeles Valley College, where she taught chemistry for allied health and preparatory chemistry for 36 years. She received her bachelor's degree in chemistry from the University of Washington and her master's degree in biochemistry from the University of California at Los Angeles.

Professor Timberlake has been writing chemistry textbooks for 33 years. During that time, her name has become associated with the strategic use of pedagogical tools that promote student success in chemistry and the application of chemistry to real-life situations. More than one million students have learned chemistry using texts, laboratory manuals, and study guides written by Karen Timberlake. In addition to *General, Organic, and Biological Chemistry: Structures of Life*, third edition, she is also the author of *Basic Chemistry*, third edition, and *Chemistry: An Introduction to General, Organic, and Biological Chemistry*, tenth edition with the accompanying *Study Guide, Selected Solutions Manual, Laboratory Manual*, and *Essential Laboratory Manual*.

Professor Timberlake belongs to numerous science and educational organizations including the American Chemical Society (ACS) and the National Science Teachers Association (NSTA). She was the Western Regional Winner of Excellence in College Chemistry Teaching Award given by Chemical Manufacturers Association. In 2004, she received the McGuffey Award in Physical Sciences from the Text and Academic Authors Association for her textbook *Chemistry: An Introduction to General, Organic, and Biological Chemistry*, eighth edition, which has demonstrated excellence over time. In 2006, she received the Textbook Excellence Award for the first edition of *Basic Chemistry*. She has participated in education grants for science teaching including the Los Angeles Collaborative for Teaching Excellence (LACTE) and a Title III grant at her college. She speaks at conferences and educational meetings on the use of student-centered teaching methods in chemistry to promote learning success of students.

Dedication

To my husband Baxter for his unwavering support.

Laura Frost

I dedicate this book to my beautiful wife and bff, Karen. Thank you for taking this journey with me and for always believing in me. And to my amazing daughters, Abbie and Anna, you inspire me everyday.

Todd Deal

I dedicate this book to:
My husband for his patience, loving support, and preparation of late meals. My son, John, daughter-in-law, Cindy, grandson, Daniel, and granddaughter, Emily, for the precious things in life.
The wonderful students over many years whose hard work and commitment always motivated me and put purpose in my writing.

Karen Timberlake

Brief Contents

Contents

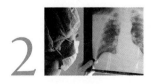

6
Intermolecular Forces
State Changes, Solubility, and Cell Membranes   211

Looking Ahead—Guided Inquiry Activities for Chapter 7   244

7
Solution Chemistry
How Sweet Is Your Tea?   247

Looking Ahead—Guided Inquiry Activities for Chapter 8   284

8
Acids, Bases, and Buffers in the Body   287

12

Food as Fuel—
A Metabolic Overview   421

Preface

TO THE STUDENT

You are about to embark on an incredible journey into the world of chemistry. *General, Organic, and Biological Chemistry: An Integrated Approach* was written especially for the student interested in pursuing a health science career like nursing, nutrition, dental hygiene, or respiratory therapy, but it has applications for any student interested in applying the concepts of chemistry to real-world situations. Wherever possible we have integrated the concepts of general, organic, and biological chemistry to create a seamless framework to relate chemistry to your everyday life.

One of our goals in writing this book is to help you become better problem solvers so that you can critically assess situations within your workplace, the popular press, and your world in general. As you explore the pages of this book you will encounter materials that

- apply chemistry to your life.
- apply chemistry to health careers that interest you.
- encourage you to develop problem-solving skills, allowing you to be successful in this chemistry course and other courses.

As we share this book with you, we hope that our enthusiasm for teaching chemistry helps to engage you in the learning of chemistry, a subject near and dear to our hearts. Chemistry truly is everywhere!

TEXT FEATURES

You may often wonder, "What does chemistry have to do with me?" This book aims to answer that question. Many sections of this book begin with a familiar everyday example as a way to introduce a chemical topic that may seem unfamiliar. In this way we hope to connect chemistry to your career and life. No two students learn the same way, so this text offers a variety of artwork, activities, and problems to introduce, drill, and challenge you on topics and "make them stick." The features of this text are discussed here.

Guided Inquiry Activities

Guided Inquiry activities precede each chapter. These activities offer an alternative approach to learning the material in the noted chapter section and are intended to engage you during class. They are best worked in groups. All the information to answer the questions in these activities can be found in previous sections, or directly in the activity itself in the form of information, a data set, or a model. The questions are carefully crafted to guide you through an exploration of the given information to develop chemical concepts that you can then apply to further examples. This approach offers a more student-centered environment in the classroom, allowing the instructor to be less of a pure knowledge source and serve more as a learning facilitator.

Math Matters

Some areas of chemistry are more math intensive than others. The Math Matters feature offers optional tutorial information for students who may not be comfortable with certain math skills such as unit conversion, significant figures, percentages, and logarithms.

Questions and Problems in Every Section

At the end of most sections, a set of Practice Problems allows you to apply problem solving to the concepts in that section. By working the problems after each section, you immediately reinforce newly learned concepts rather than waiting until you get to the end of the chapter. We encourage you to be an active learner by working problems as you progress through each chapter.

Matched-Pair Problems

Many problems in the text are matched with a similar type of question or problem. For each pair, the answer for the odd-numbered problem is located at the end of each chapter rather than at the end of the textbook. Checking your answers gives you immediate feedback on problem-solving success. The answers to the even-numbered problems are not in the text, so your instructor may use these problems for homework and/or quiz questions.

End-of-Chapter Problems

At the end of every chapter, more problems encourage you to think about the concepts you have learned. The Additional Problems provide in-depth questions that integrate the topics from the entire chapter to improve your understanding and critical thinking. Challenge Problems provide in-depth study and may also be used in cooperative learning environments. The *Study Guide* contains additional learning exercises for every section and worked-out solutions for all the odd-numbered problems. The *Instructor's Manual* has the solutions to all the guided inquiry activities in the text.

Macro-to-Micro Art Illustrates Molecular Organization

Macro-to-micro illustrations show how the function and behavior of everyday objects are organized at the atomic and molecular level. This helps put the molecular world into context, making connections between the atomic world we cannot see and the macroscopic world we can see.

Clear Illustrations Allow Students to Visualize Chemistry

The art program is not only beautifully rendered, but also pedagogically effective.

CHAPTER ORGANIZATION

Throughout the text the general, organic, and biological chemistry topics are integrated using relevant examples to apply concepts and principles. This text intentionally contains only 12 chapters, allowing all chapters to be covered in a one-semester course. The first 10 chapters must be covered in sequence as material in later chapters builds upon material in earlier chapters. If time begins to run short, Chapters 11 and 12 can stand alone more than other chapters. A general description of the contents in each chapter follows.

Chapter 1

Chapter 1 serves as a basic introductory chapter highlighting the states, properties, and types of matter. The initial focus is on the motion of the particles of which matter is composed, which leads to an introduction to the gas laws. The classification of matter, in turn, leads to a presentation of the periodic table in the context of the elements. The chapter concludes by distinguishing between physical changes and chemical reactions in matter and introducing students to the basic scaffolding of a chemical reaction. This chapter also features the first two Math Matters sections—Metric Units, Prefixes, and Conversion Factors and Significant Figures and Rounding Calculator Results.

Chapter 2

Chapter 2 is an introduction to the atom. The material builds from the subatomic particles to the nucleus, electron cloud, electron energy levels, and valence electrons. These concepts refocus the student's attention on the periodic table through the introduction of atomic numbers and atomic mass. The concept of the mole is introduced in terms of number of atoms and atomic mass for completeness and to re-emphasize unit conversions. From this foundation, the chapter builds on its discussion of isotopes to lead into radioisotopes and the elementary concepts of nuclear chemistry. Chapter 2 also introduces students to scientific notation in a Math Matters section.

Chapter 3

Building on the introduction to atoms and subatomic particles from Chapter 2, Chapter 3 begins with an introduction to the octet rule, leading to a discussion of bonding and compound formation. The initial focus is on ion formation and ionic compounds. Students get their first taste of chemical nomenclature as they learn to name ionic compounds. The discussion moves to atoms that do not form ions and their compounds. The concepts of covalent bonding are introduced, and students learn to construct Lewis structures. The focus then moves to using the Lewis structures to determine molecular shape. The concepts of VSEPR are developed with a focus on carbon-containing compounds. The chapter concludes with an introduction to the concepts of electronegativity and bond polarity and the application of these concepts to the determination of molecular polarity.

Chapter 4

Chapter 4 utilizes the structural concepts developed in Chapter 3 to introduce students to organic compounds. The foundation of this chapter is structural and the chapter includes representations of organic compounds. The chapter includes a brief introduction to organic functional groups, focusing on unsaturated hydrocarbons. Further discussion of functional groups is integrated into later chapters as they appear in biomolecules. Students are also introduced to their first biomolecules—fatty acids—in the context of the structural and polarity concepts with which they are now familiar. Treatment of several types of isomerism—including structural isomers, cis/trans stereoisomers, and enantiomers—begins here.

Chapter 5

Chapter 5 integrates the functional groups alcohol, aldehyde, and ketone into the context of carbohydrates to illustrate relevant structure, bonding, and chirality. Students are introduced to Fischer projections and diastereomers using carbohydrates as examples. Several reactions are introduced in this chapter including oxidation–reduction, hemiacetal formation, and condensation/hydrolysis. The ring formation of carbohydrates serves as the focus of hemiacetal formation reactions and leads nicely to a consideration of the Haworth formulas. Glycoside formation as a condensation reaction and the naming of glycosidic bonds introduce important disaccharides. The chapter continues with some important polysaccharides and concludes with treatment of relevant blood carbohydrates, the ABO blood groups, and heparin.

Chapter 6

Building on the concepts of molecular polarity and the structure of organic compounds, Chapter 6 is an integrated treatment of intermolecular forces. After a brief introduction to the types of intermolecular forces, the chapter focuses on applications of these forces to solubility, state changes, and cell membranes. Each of the topics is developed around organic and/or biochemical compounds. Reactivity is included in a discussion of hydrogenation reactions and their use in the production of margarine from vegetable oils.

Chapter 7

This chapter expands on the solubility concepts from Chapter 6 to introduce properties of solutions and the solubility of ionic compounds. Section 7.3 reintroduces students to the brief introduction of balancing chemical equations by balancing solvation equations. The focus then shifts to concentration and the determination of solute concentration in a solution. Because equivalents are often used in medical applications, this concept is introduced and developed for ionic solutions. Percent concentrations are emphasized in the solution calculations because of their widespread use in health services. A Math Matters segment discussing percent is included in this chapter for students needing extra instruction in this area. The chapter concludes with a discussion of osmosis and diffusion and its application to cells and membrane transport.

Chapter 8

Chapter 8 naturally follows solution chemistry as an introduction to acid–base chemistry. The reactivity of organic and biochemical compounds that act as acids and bases are included in this discussion. After an introduction to strong acids, strong bases, and neutralization reactions, equilibrium and LeChâtelier's principle are introduced as a prelude to weak acids and bases. This is followed by a thorough treatment of pH and pK_a. Amino acids are introduced as examples of biochemical acids and bases. The chapter concludes with a thorough discussion of biological buffers, highlighting the bicarbonate buffer, acidosis, and alkalosis.

Chapter 9

Chapter 9 is a synthesis of many of the concepts developed in previous chapters. The discussion begins with a further consideration of amino acids (Chapter 8), focusing on the chirality of amino acids (Chapter 4), and the structure and polarity of side chains (Chapter 6). Remaining functional groups (Chapter 4) not discussed in previous chapters appear here in amino acid side chains. The discussion moves to the condensation reaction of amino acids forming peptides (amide functional group) and thus proteins. The ensuing material is developed around the theme of protein structure with a focus on intermolecular forces (Chapter 6) at all levels of structure. In addition to peptide formation, reactivity is considered through disulfide formation and denaturation. The chapter concludes by introducing some important proteins found in the human body and highlighting their structure–function relationship.

Chapter 10

Chapter 10 integrates concepts of elementary thermodynamics with enzymatic catalysis. The material draws on the understanding of protein structure from Chapter 9 to develop the concepts of enzymatic catalysis. After a brief introduction to some enzyme terminology, a basic treatment of thermodynamics is then applied to enzyme catalysis. This is followed up with a description of enzyme activity and the factors affecting it.

Chapter 11

Chapter 11 provides the student with basic information regarding the structure and important functions nucleic acids play in living systems. The chapter builds a nucleotide from the simple biological molecules already introduced in previous chapters and strings these together to make a nucleic acid. Emphasis is placed on the chemical reactions that govern the building of nucleotides and nucleic acids (condensation reactions). The discovery and structure of DNA are discussed in the familiar terms of primary, secondary, and tertiary structures. Sections discussing mutations and viruses are included as applications of the concepts in the chapter. The chapter ends with a section discussing recombinant DNA technology, providing further relevance and understanding of the information presented earlier in the chapter and the distinction between gene cloning and organism cloning.

Chapter 12

Chapter 12 provides the student with a basic overview of the metabolism of carbohydrates, proteins, and lipids. This includes their digestion, catabolism, and use in energy production. The focus is the catabolic oxidation of glucose and ATP production. The chapter concludes with the alternative fuel sources of fatty acids and amino acids and discusses how these feed into catabolic oxidation and metabolism in general.

SUPPLEMENTS

For the Instructor

MasteringChemistry® (www.masteringchemistry.com) MasteringChemistry® is an online homework and tutorial system. Instructors can create online assignments for their students by choosing from a wide range of items, including end-of-chapter problems and enhanced tutorials. Assignments are automatically graded with up-to-date diagnostic information, helping instructors recognize areas in which students are having difficulty either individually or as a class.

Instructor Solutions Manual (0-321-67705-6) includes answers and solutions for all questions and problems in the text.

Printed Test Bank (0-321-67706-4) includes more than 600 multiple choice questions.

Instructor Resource DVD (0-321-68260-2) This DVD includes all the art and tables from the book in high-resolution format for use in classroom projection or when creating study materials and tests. In addition, the Instructor can access PowerPoint™ lecture outlines or create their own easily with the art provided in PPT format. Also available are downloadable files of the Instructor Solutions Manual and Test Bank.

Blackboard (0-321-67704-8) and **WebCT** (0-321-67497-9) provide powerful course management capability.

For the Student

MasteringChemistry® (www.masteringchemistry.com) MasteringChemistry® provides students with an extensive self-study diagnostic with an interactive eText. In addition, MasteringChemistry® offers a chemistry homework and tutorial system if the instructor chooses to give online homework assignments.

Study Guide with Selected Solutions (0-805-38186-4) Prepared by Jason W. Ribblett of Ball State University, this manual for students contains complete solutions to the selected odd-numbered end-of-chapter problems in the book.

Pearson eText: The integration of eText within MasteringChemistry® gives students, with new books, easy access to the electronic text when they are logged into MasteringChemistry®. Pearson eText pages look exactly like the printed text, offering powerful new functionality for students and instructors. Users can create notes, highlight text in different colors, create bookmarks, zoom, view in single-page or two-page view, etc.

ACKNOWLEDGMENTS

The preparation of a first edition for new authors is a considerable undertaking. We are thankful for the support, encouragement, and dedication of the many people whom we have encountered along our journey. After our first meeting with co-author Karen Timberlake, we knew having her involved was the best way to introduce our integrated approach to a one-semester general, organic, and biological chemistry course. Without her experience, we may have become lost in the details of creating a first edition. She was always lighting the way and offering endless encouragement and necessary editing to help make this book a quality product. Her expertise has been invaluable to this project. One of us (LF) would also like to acknowledge James Spencer, one of the granddaddys of process-oriented guided inquiry learning (POGIL), who opened her eyes to a new style of teaching when she heard his Pimentel address at the 2005 spring American Chemical Society meeting.

We would like to express special appreciation to Jim Smith and his team at Benjamin-Cummings for believing so earnestly in the original project. Jim's vision for this project provided a foundation for us as an author team. The editorial staff at Pearson has been exceptional. We are grateful for the stamina of Ray Mullaney, editor in chief of science book development, who answered every question we ever asked, no matter how obvious to him, and for listening to our frustrations as we navigated the world of publishing. We also want to acknowledge the enthusiasm of Dawn Giovanniello, acquisitions editor, who has believed in this project from day one and has been very patient as she coordinated all the extras behind a quality textbook including the book's design palette. We also want to recognize copyeditor Martha Williams for catching grammatical and spelling errors and for introducing us to editing and style sheets when we had never heard of such a thing. We also appreciate the patience of Eric Schrader, photo researcher, as he sought to find the perfect photo to help students visualize chemistry. We want to thank the production team including Shari Toron, project manager, Connie Long, art manager, and Rebecca Dunn, who have been very patient with us as we embarked on the production process. We appreciate the contributions of Laurie Hoffman and Lisa Tarabokjia, who encouraged us to meet our deadlines and organized and managed the review process.

This text reflects the contributions of many professors who took the time to review and edit the manuscript and provided outstanding comments, help, and suggestions; we are grateful for your contributions.

In addition, we could not have completed this textbook without the support of our exceptional colleagues in the Department of Chemistry at Georgia Southern University, whom we feel are the best aggregate of chemistry teachers on the planet. They allowed us to explore this approach in our classrooms, were always willing to discuss seemingly random chemistry topics, and continue to be just a fun bunch of coworkers. We also want to thank the many faculty who have waited patiently for this project to become a reality.

If you would like to share your experience using this textbook, either as a student or faculty member, or you have questions regarding its content, we would love to hear from you.

Laura Frost
ldelong@georgiasouthern.edu

Todd Deal
stdeal@georgiasouthern.edu

LIST OF REVIEWERS

Eric Arnoys, *Calvin College*

George Bandik, *University of Pittsburgh*

Allesandra L. Barrera, *Georgia Gwinnett College*

Shay Bean, *Chattanooga State Technical Community College*

Joe C. Burnell, *The University of Indianapolis*

Kathy Carrigan, *Portland Community College*

Kent Chambers, *Hardin Simmons University*

Paul Chamerlain, *George Fox University*

Ron Paul Choppi, *Chaffey College*

Jeannie T. B. Collins, *University of Southern Indiana*

Christa L. Colyer, *Wake Forest University*

William Davis, *University of Texas at Brownville*

Brahmadeo Dewprashad, *Borough of Manhattan Community College*

P. K. Duggal, *Maple Woods Community College*

Stephen U. Dunham, *Moravian College*

George C. Flowers, *Darton College*

Hao Fong, *South Dakota School of Mines and Technology*

Don Fujito, *LaRoche College*

Emily Halvorson, *Pima Community College*

Kirk Hunter, *Texas State Technical College*

Louis Giacinti, *Milwaukee Area Technical College*

Christina Goode, *CSU Fullerton*

Matthew A. Johnston, *Lewis-Clark State University*

John R. Kiser, *Isothermal Community College*

Andrea D. Leonard, *University of Louisiana, Lafayette*

Charles F. Marth, *Western Carolina University*

Janice J. O'Donnell, *Henderson State University*

Rebecca O'Malley, *University of South Florida*

Julie Peyton, *Portland State University*

Jennifer Powers, *Kennesaw State University*

Deboleena Roy, *American River College*

Gillian E. A. Rudd, *Northwestern State University*

Michael Russell, *Mt. Hood Community College*

Karen Sanchez, *Florida State College, Jacksonville*

John Singer, *Jackson Community College*

Dan Stasko, *University of Southern Maine, Lewiston-Auburn College*

James D. Stickler, *Allegheny College of Maryland*

Koni Stone, *CSU Stanislaus*

Everett Shane Talbott, *Somerset Community College*

Kwok-tuen Tse, *Harold Washington College*

Maria Vogt, *Bloomfield College*

James Zubricky, *University of Toledo*

GENERAL, ORGANIC, AND BIOLOGICAL CHEMISTRY

AN INTEGRATED APPROACH

Guided Inquiry Activities **FOR CHAPTER 1**

EXERCISE 1 Classifying Matter

Information

Chemistry is the study of matter and its changes. There are two main types of matter: *pure substances* and *mixtures* of substances. A mixture that is mostly water is called an aqueous solution. Matter can exist in several different phases or *states*, the three most common being solid, liquid, or gas.

Types of Matter Flow Chart with Examples, Their Formula(s), and State(s)

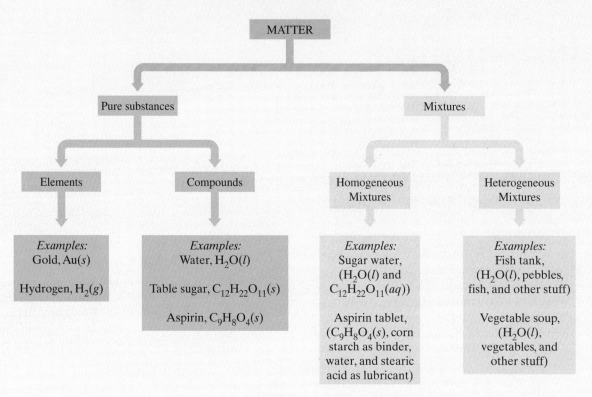

MATTER

Pure substances

Mixtures

Elements

Compounds

Homogeneous Mixtures

Heterogeneous Mixtures

Examples:
Gold, Au(s)

Hydrogen, H$_2(g)$

Examples:
Water, H$_2$O(l)

Table sugar, C$_{12}$H$_{22}$O$_{11}(s)$

Aspirin, C$_9$H$_8$O$_4(s)$

Examples:
Sugar water, (H$_2$O(l) and C$_{12}$H$_{22}$O$_{11}(aq)$)

Aspirin tablet, (C$_9$H$_8$O$_4(s)$, corn starch as binder, water, and stearic acid as lubricant)

Examples:
Fish tank, (H$_2$O(l), pebbles, fish, and other stuff)

Vegetable soup, (H$_2$O(l), vegetables, and other stuff)

Elements can exist as individual *atoms* (neon, Ne) or as pairs of atoms (hydrogen, H$_2$) forming *molecules*.

Questions

1. Consider the examples in the flow chart. How does the formula of an element differ from that of a compound?

2. Can an element be a pure substance? Can a compound be a pure substance?

3. Using the information given, how might you define a pure substance? How does a pure substance differ from a mixture?

4. Based on the flow chart, would you classify the following substances as (a) an element or compound, (b) atom or molecule?

Matter	Element or Compound	Atom or Molecule
He		
N_2		
CH_2O (formaldehyde)		
CH_3COOH (vinegar)		

5. As a group, devise a definition for a compound.

6. In your own words, describe the difference between a homogeneous and heterogeneous mixture.

7. Would you classify the following matter as element, compound, or mixture? If you classify it as a mixture, classify it as homogeneous or heterogeneous.
   a. table salt, (NaCl)   b. nickel (Ni)
   c. chocolate chip cookie dough   d. air

8. What do you think the labels (s), (l), (g), and (aq) on the formulas in the data set mean?

SECTION 1.4

EXERCISE 2   The Periodic Table

Information

All *elements* are listed individually on the *periodic table of the elements*. There is a periodic table on the inside front cover of your textbook, and an alphabetical listing of all the elements on the facing page. Refer to this as you answer the questions in this section.

The rows on the periodic table are referred to as *periods*, and the columns are referred to as *groups*. The table is organized with metals on the left and nonmetals on the right, with a staircase dividing line between the two. Find this staircase on the periodic table in your textbook.

Chemical formulas show the type and number of each element present in a compound. For example, water's chemical formula is H_2O. It contains two hydrogen atoms and one oxygen atom.

Questions

1. In which group on the periodic table are the following elements found?
   a. sodium   b. oxygen   c. calcium   d. carbon

2. In which period on the periodic table are the following elements found?
   a. hydrogen   b. nitrogen   c. sulfur   d. phosphorus

3. Provide the names of the following elements and identify them as metals or nonmetals.
   a. Cu   b. Na   c. Cl   d. C   e. K   f. P

4. Look at the periodic table of the elements. About how many elements are there in nature?

5. Identify the number and name of each element in the following formulas.
   a. $C_6H_{12}O_6$ (dextrose)   b. NaOH (lye, found in drain cleaners)
   c. $NaHCO_3$ (baking soda)   d. $C_{15}H_{21}NO_2$ (demerol, a painkiller)

6. Are most of elements on the periodic table metals or nonmetals? Considering that most of Earth is made up of silicon (the main component in sand) and that Earth's biomass is made up mostly of carbon, does this seem surprising?

Whether we are vacationing at the beach or studying for an exam, everything that we touch is made up of matter. Some of it can even seem invisible, like the air we breathe. Chapter 1 begins by exploring some of the properties of matter.

Chemistry
IT'S ALL ABOUT "STUFF"

Did you know that everything you do every day involves chemistry? Yes, everything. From the water and shampoo in your shower, to the food you eat for breakfast, to the gas that powers your car, to the medicines you take, to the sunscreen lotion that protects your skin, even the clothes that you wear—all of these somehow involve chemistry. You are about to embark on an incredible and exciting journey. Learning chemistry is really learning about everyday life and how chemistry impacts our lives and even provides for life itself. To understand concepts as diverse as how our bodies function and the wide variety of conveniences that make our lives easier, you need to understand chemistry. So, come journey with us. It will be challenging, but fun, we promise!

1.1 "Stuff" Is Matter

All the "stuff" that we just mentioned is composed of something that chemists call matter. **Matter** can be defined as anything that takes up space. From the smallest pill dispensed by a pharmacist to the shampoo in a bottle to the air in a balloon, each of these takes up some amount of space and is a form of matter.

Matter: Volume

Saying that something "takes up space" is another way to say that it has **volume**. You have probably heard a large bottle of a soft drink referred to as a "2-liter" bottle. This tells the volume of the soft drink in the bottle—2 liters (1 liter = 1.057 quart). You are also aware that as you blow up a balloon, it gets bigger. This is simply an increase in the volume of air in the balloon. Thus, volume is a measure of the amount of space occupied by a substance. Volume is a three-dimensional measurement.

Some of the commonly encountered units of volume are milliliter (mL), cubic centimeter (cc; same volume as a mL), and liter (L). A teaspoon of cough syrup contains 5 mL or 5 cc, and, as we have stated, soft drinks are routinely sold in bottles containing 2 L.

Matter: Mass

Now consider that anything that takes up space can also be placed on a scale and weighed, that is, it has mass. So, matter can be more completely defined as anything that takes up space and has mass. **Mass** is the amount of matter in a substance and remains the same regardless of the location of the substance. In science, health care, and many other applications, the most common unit used to measure the mass of a substance is the gram (g). A raisin or a paper clip has a mass of about 1 g. (See the Math Matters box.)

The term *weight* may be more familiar to you than the term *mass*. These terms are related, but they do not mean the same thing. The weight of an object is determined by the pull of gravity on the object, and that force changes based on location. For instance, astronauts weigh less on the moon than on Earth. This is because the gravitational pull of the moon is less than that of Earth. The astronauts' mass did not change, but their weight did.

So, how are mass and weight related to each other? When you step on the bathroom scale, are you determining your mass or your weight?

A scale or balance is used to measure mass. When scales are manufactured, they are set at the factory so that a 1-kilogram (1000-g) object has the same mass on all the scales. This is known as "calibrating" the scales and it accounts for the gravitational pull of Earth; therefore, the mass and the weight of the 1-kg object are the same. If the object and scale were transported to the moon, the scale would have to be adjusted in order to account for the lesser gravitational pull of the moon.

Thus, as long as an object is weighed in roughly the same location on Earth's surface, its mass and weight will have the same measured value. In order to be less confusing, we will always refer to weighable quantities using the term *mass*.

MATH MATTERS

Metric Units, Prefixes, and Conversion Factors

In science, health care, and business, we often find it necessary to measure or report the mass or volume of something. In order for such measurements to be consistent and easily compared, we need to have a defined set of standards, that is, a measurement system. Most scientists and health care professionals throughout the world use the **metric system**. The metric system is part of the *Système Internationale* (SI) or International System of Units understood and accepted around the world. Since we will use the metric system throughout this text, it is important that you become familiar with its units and understand how it works.

The metric system is based on a series of standard units for each type of measurable quantity like mass and volume. The standard metric unit for mass is the **gram** (g). The standard metric unit for volume is the **liter** (L), and the standard metric unit for length is the **meter** (m).

If a paperclip has a mass of about 1 g, imagine how many paperclips it would take to equal your body mass! That would be a huge number that would be awkward to work with. Or think of trying to get the mass of one human hair by comparing it to a paperclip. It would take a very large number of hairs to equal the mass of a paperclip. In order to deal efficiently with quantities that are substantially larger or smaller than the standard unit, the metric system employs a series of prefixes that represent fractions or multiples of a gram. (Table 1.1 lists the common metric prefixes, as well as the abbreviation and value for each.) For example, since the average person has a mass much greater than a paperclip, we apply the prefix *kilo* to the unit "gram." *Kilo* means "1000 times greater than," so a *kilo*gram is 1000 g. Thus, an average person's mass would be expressed as 70 kg, which is a much more manageable number than 70,000 g.

We can use Table 1.1 to obtain equivalency relationships between different units. The numbers in the right-hand column provide the relationship between any unit and the standard unit. For example, when moving from decigrams to grams, we should divide by 10; therefore, the equivalency is 1 dg = 0.1 g. To eliminate the decimal point, we can say that 10 dg = 1 g.

Such equivalencies can be used as **conversion factors** to allow us to convert one unit to another using one or more of these factors. For example, when the equivalency 10 dg = 1 g is written as a conversion factor, it takes the following form:

$$\frac{10 \text{ dg}}{1 \text{ g}} \quad \text{or} \quad \frac{1 \text{ g}}{10 \text{ dg}}$$

These conversion factors are read as, "There are 10 decigrams in 1 gram" or "1 gram contains 10 decigrams," respectively. Conversion factors are intended to be used as tools to allow you to convert a quantity in one unit to the equivalent quantity in a larger or smaller unit. The following example illustrates the use of conversion factors derived from Table 1.1.

How many grams of vitamin C are in a tablet containing 100 mg? The equivalency relationship from the table shows us that 1 mg = 0.001 g or 1000 mg = 1 g, which gives us the following conversion factors:

$$\frac{1000 \text{ mg}}{1 \text{ g}} \quad \text{or} \quad \frac{1 \text{ g}}{1000 \text{ mg}}$$

Which conversion factor should we use? Notice that the question gives us a number of milligrams and asks us for the number of grams. We want to *convert* from milligrams to grams. In order to do this, we set up an equation using a conversion factor to cancel the given unit, leaving the desired unit in the answer. The form of this equation is

$$\text{Given unit} \times \left(\frac{\text{Desired unit}}{\text{Given unit}} \right)$$

$$\uparrow$$

This is the conversion factor

Notice that when the calculation is carried out, the "given unit" in the denominator of the conversion factor will cancel the "given unit" from the problem.

Thus, for our example, the equation is

$$100 \text{ \cancel{mg}} \times \left(\frac{1 \text{ g}}{1000 \text{ \cancel{mg}}} \right) = 0.1 \text{ g}$$

TABLE 1.1	METRIC PREFIXES	
Prefix	**Abbreviation**	**Relationship to Standard Unit**
kilo-	k	1000 ×
base unit (has no prefix)		1 × (gram, liter, meter)
deci-	d	÷ 10
centi-	c	÷ 100
milli-	m	÷ 1000
micro-	μ	÷ 1,000,000

SAMPLE PROBLEM 1.1

Unit Conversions

Using the appropriate conversion factor, convert each of the following quantities to the indicated unit.

a. 10,000 cg = _____ g   **b.** 0.005 L = _____ mL

SOLUTION

a. From Table 1.1, we determine that 100 cg = 1 g. Since we want our answer in grams, that unit needs to be in the numerator of our conversion factor. The equation is set up as follows:

$$10,000 \text{ \cancel{cg}} \times \left(\frac{1 \text{ g}}{100 \text{ \cancel{cg}}} \right) = 100 \text{ g}$$

b. The equivalency is 1000 mL = 1 L, and because we want the answer in mL we need to make sure that mL is in the numerator of the conversion factor. The correct equation is

$$0.005 \, \cancel{L} \times \left(\frac{1000 \text{ mL}}{1 \, \cancel{L}} \right) = 5 \text{ mL}$$

States of Matter

In order to better understand this "stuff" called matter and how it behaves, we need to consider its composition. At the most basic level, matter is a collection or assembly of particles too small to be seen. In Chapters 2 and 3, we will discover more about these particles. What we can see or touch in our everyday lives is the result of many of these individual particles gathered together. The look, feel, and behavior of matter that we can see and touch depend on the composition and behavior of these particles.

You can see and hold a brick or a glass of water and are aware that each has mass and volume. Based on our balloon example earlier in the chapter, air (and other gases) have volume (they have mass as well) and are, therefore, matter, too.

The brick, water, and air represent forms of matter, but they are physically different. The brick has a definite, unchanging shape and volume, whereas water has a definite volume (amount), but its shape changes depending on its container. However, because you cannot "see" air, you can only witness the changes that it causes. For example, it fills up and takes the shape of a container, like a balloon, into which it is placed. Each of these examples represents a different state of matter. A **state of matter** is the physical form in which the matter exists. The three most common states of matter are solid (the brick), liquid (water), and gas (air).

Let's take a look at the submicroscopic particles that make up substances like the brick, water, and air, as shown in Figure 1.1.

The particles in a solid, for example, a brick, are arranged in an orderly manner and are tightly packed together. They are moving, but only very slightly, and not significantly enough to affect the shape of the solid. Therefore, **solids** have a definite shape and a definite volume.

The particles in a liquid, for example, water, are somewhat less orderly and are only loosely associated with each other. They move freely, colliding with and sliding over each other. These behaviors of its particles mean that a **liquid** has a definite volume, but it takes the shape of its container.

The particles in a gas, for example, air, have no orderly arrangement and are far apart from each other. They move at high rates of speed and often collide with each

▶ **FIGURE 1.1 Particles of a solid, liquid, and gas.** The particles in a solid are tightly packed and barely moving. The particles in a liquid are less orderly than those in a solid and are freely moving. The particles in a gas are disordered and rapidly moving.

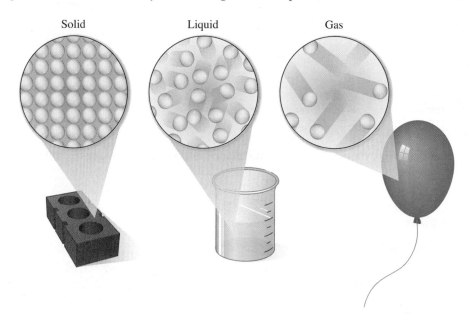

Solid   Liquid   Gas

other and with the walls of their container. Therefore, a **gas** has no definite shape or volume; instead, it expands to fill the container in which it is placed.

Thus, the motion and association of the particles in a given substance determine whether it is a solid, liquid, or gas. A substance's properties, such as shape and volume, are also greatly influenced by the motion of its particles. To develop our understanding of the behavior of matter and its relationship to particle motion, let's take an in-depth look at the particle motion of gases and how that motion is affected by factors such as pressure and temperature.

PRACTICE PROBLEMS

1.1 Contrast the movement of particles in a liquid with that of a gas.

1.2 Contrast the movement of particles in a liquid with that of a solid.

1.3 Indicate whether each of the following describes a solid, liquid, or gas:

a. This substance has no definite volume or shape

b. The particles in this substance are rigidly held in place.

1.4 Indicate whether each of the following describes a solid, liquid, or gas:

a. This substance has a definite volume but takes the shape of the container.

b. The particles in this substance are so far apart they do not interact strongly with each other.

1.2 A Closer Look at Gases—An Introduction to the Behavior of Matter

Because their behavior is easily manipulated and observed, gases have long been used to study the behavior of matter. Gases are noticeably affected by changes in their environment such as increases or decreases in temperature or pressure. Both pressure and temperature are external (environmental) factors that cause changes in the motion of the particles of a gas. In this section, we define pressure and temperature and consider how their changes affect the volume of a gas.

Gases and Pressure

When dealing with gases, one of the most important quantities to consider is pressure. To illustrate the concept of pressure, imagine an empty syringe (with no needle attached), similar to those used to dispense medicine orally to an infant.

If the plunger of the syringe is drawn all the way out, the syringe will fill with air. If you place your finger firmly over the open tip of the syringe and depress the plunger, what happens? The plunger moves in, and the sample of air is "squeezed." The particles of a gas are usually far apart (see Figure 1.1) and not closely associated with each other. In other words, a sample of gas is mostly empty space. When the air in the syringe is squeezed, the space between the particles is decreased, and the particles of the air are forced closer together and have less room to move about. By depressing the plunger, you are compressing the gas by applying pressure to it. **Pressure** is simply a force exerted against a given area.

Pressure measurements are expressed in a variety of different units, but for our discussion we will use two of the more common and familiar units, pounds per square inch (psi) and millimeters of mercury (mmHg). The latter unit is the one used when measuring blood pressure.

Suppose you placed a 5-pound bag of sugar on the first digit of your thumb. Your thumb would definitely feel some pressure! In fact, the pressure exerted by the bag of sugar on your thumb, which is about 1 square inch, would be 5 pounds per square inch or 5 psi. The unit of **pounds per square inch (psi)** is a measure of force (measured in

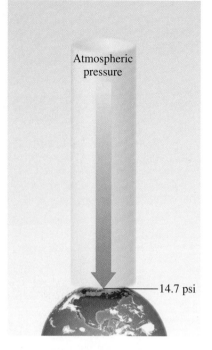

▲ **FIGURE 1.2 Pressure of the atmosphere on the surface of Earth.** At sea level, the atmosphere exerts an average pressure of 14.7 pounds per square inch on Earth's surface.

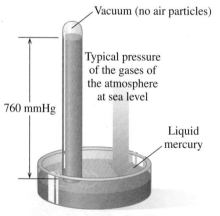

▲ **FIGURE 1.3 The mercury barometer.** The pressure of the gases of Earth's atmosphere is able to support a column of mercury 760 mm high.

pounds) applied to an area of 1 square inch (1 in^2). The pressure of the atmosphere at sea level is about 14.7 pounds per square inch, 14.7 psi as shown in Figure 1.2.

The unit **millimeters of mercury (mmHg)** is also a measure of pressure derived from the force exerted by the atmosphere surrounding Earth.

The pressure of the atmosphere can be measured using a device known as a mercury barometer, which is constructed by filling a long glass tube that is sealed at one end with liquid mercury and inverting the tube into a dish of mercury without letting air into the tube (see Figure 1.3). The force of the atmosphere pushing down on the mercury in the dish prevents most of the mercury in the tube from draining out. In fact, assuming the tube is long enough, a column of mercury 760 mm high (29.92 in) will remain inside the tube at sea level. Because the pressure exerted by the atmosphere determines the height of the column of mercury remaining in the tube, we can measure the pressure by measuring the height of the mercury column in millimeters, thus giving us pressure measurements in the unit mmHg.

As stated previously, the typical pressure exerted by the atmosphere at sea level supports a column of mercury 760 mm high. This pressure is equivalent to 14.7 psi but varies with altitude and changes in the weather.

Pressure and Volume—Boyle's Law

In the example of the syringe, as the plunger was depressed, the applied pressure increased and the volume of the air decreased (see Figure 1.4).

In the mid 1600s, British chemist Robert Boyle noted these changes and began to experiment with the effect of pressure on the volume of a gas.

In his experiments with gases, Boyle discovered that the volume of a gas and the pressure applied to it are related if the temperature and amount of the gas are not allowed to change. He found that when the pressure on a gas was doubled, the volume of the gas was reduced to half of its initial volume. Similarly, if he tripled the pressure on the gas, the volume was reduced to one-third of its initial volume. Figure 1.5 shows how these changes in pressure affect the volume of a gas.

Boyle found, and his law states, that *the volume of a fixed amount of gas at constant temperature is inversely proportional to the pressure.* Simply stated, Boyle's law says that an increase in the pressure on a gas will result in a decrease of its volume and vice versa.

Mathematically, Boyle's law allows us to determine what will happen to a sample of gas of known volume and pressure if we make changes to either the volume or the pressure while the temperature stays the same. The following equation applying Boyle's law gives the formula for such a determination:

$$P_i \times V_i = P_f \times V_f$$

where P_i = the initial or starting pressure

V_i = the initial or starting volume

P_f = the final or ending pressure

V_f = the final or ending volume

Let's see how this equation can be used to illustrate Boyle's law. Assume you have a syringe with one end sealed containing 10 cc of air at a pressure of 14.7 psi. The unit cc stands for cubic centimeter, a common unit found on medical syringes. One cc is equal to one milliliter. If you double the pressure to 29.4 psi—by pressing on the plunger—what will the volume of the air be? (You may be able to think this one through without the equation: Push twice as hard on the plunger of the syringe and the volume of the air should decrease by one-half to 5 cc.) Based on the information given, we see that $P_i = 14.7 \text{ psi}$, $V_i = 10 \text{ cc}$, and $P_f = 29.4 \text{ psi}$. First, we need to solve the Boyle's law equation for the missing variable, V_f:

$$P_i \times V_i = P_f \times V_f$$

Begin by dividing both sides by P_f:

$$\frac{P_i \times V_i}{P_f} = \frac{\cancel{P_f} \times V_f}{\cancel{P_f}}$$

This gives us

$$\frac{P_i \times V_i}{P_f} = V_f$$

Substituting the given values (numbers and units) into the equation and solving gives us

$$\frac{14.7 \cancel{psi} \times 10 \text{ cc}}{29.4 \cancel{psi}} = V_f$$

$$V_f = 5 \text{ cc}$$

When you complete the calculation (taking care to cancel the units on your numbers), you find that the final volume is 5 cc, just as you may have predicted.

Boyle's Law and Breathing

Do you know that you use Boyle's law every day? Breathing is a very practical application of Boyle's law. When you breathe in, you are actually not forcefully drawing air into your lungs; instead, the muscles of your rib cage and your diaphragm contract to cause the volume of your chest cavity to increase (see Figure 1.6). When this happens, the air pressure inside your lungs decreases (volume increase results in pressure decrease), and the pressure of the atmosphere causes air to rush into your lungs to equalize the internal and external pressures. When the muscles relax, the volume of your chest cavity decreases, thereby increasing the pressure in your lungs relative to the outside, and air flows out to the lower pressure (atmosphere).

a.
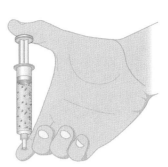

b.

▲ **FIGURE 1.4 Effect of pressure on the volume of a gas.** As the plunger is depressed (going from a to b), the pressure on the gas increases and the volume of the gas decreases.

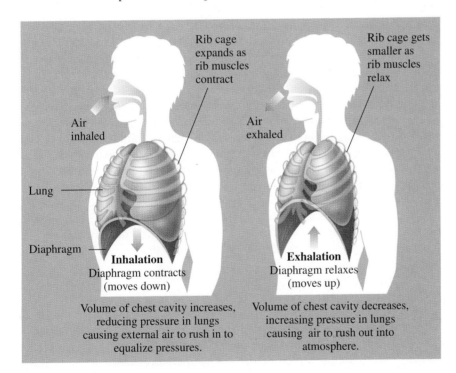

▲ **FIGURE 1.6 Boyle's law and breathing.** Breathing is controlled by contraction and relaxation of the diaphragm muscles.

Rib cage expands as rib muscles contract

Rib cage gets smaller as rib muscles relax

Air inhaled

Air exhaled

Lung

Diaphragm

Inhalation Diaphragm contracts (moves down)

Volume of chest cavity increases, reducing pressure in lungs causing external air to rush in to equalize pressures.

Exhalation Diaphragm relaxes (moves up)

Volume of chest cavity decreases, increasing pressure in lungs causing air to rush out into atmosphere.

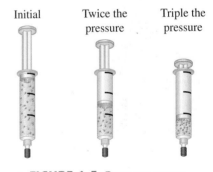

Initial   Twice the pressure   Triple the pressure

▲ **FIGURE 1.5 Pressure versus volume of a gas.** When the pressure on the gas is doubled, the volume is halved. When the pressure is tripled, the volume decreases to one-third of the initial volume.

SAMPLE PROBLEM 1.2

Boyle's Law Calculations

The lungs of a normal adult can hold about 5 L of air under typical atmospheric pressure (760 mmHg). If a cliff diver dives to a depth where the water pressure is 950 mmHg, what will the volume of his lungs be?

SOLUTION

Changes in the volume and pressure of a gas are related using Boyle's law. First, check the problem to determine what is given and what information is requested.

We are given

P_i = 760 mmHg.

V_i = 5 L

P_f = 950 mmHg

We need to find V_f

Solve the Boyle's law equation for the missing variable to get

$$\frac{P_i \times V_i}{P_f} = V_f$$

Substituting the appropriate values into the equation and canceling the units gives us

$$\frac{760 \; \cancel{\text{mmHg}} \times 5 \text{ L}}{950 \; \cancel{\text{mmHg}}} = V_f$$

Do the math to find that V_f = 4 L.

MATH MATTERS

Significant Figures

Look at the man in the picture. How old would you say he is?

Would you agree that he looks to be about 50 years old? Does that mean that you know he is *exactly* 50? Unless you know him personally and also know that this picture was taken on his fiftieth birthday, you cannot know that. So, you've *guessed* that he is about 50 years old.

In estimating this person's age, you encounter the concept of uncertainty in a measurement. All measurements have some level of uncertainty. We can understand the level of uncertainty in any measurement by the number of significant figures used in the number. In any measurement, the **significant figures** are the digits known with certainty plus the estimated digit. When we estimate that the man in the picture is 50 years old, we feel confident that his age is somewhere between 40 and 60 years. This means that our determination has one significant figure. In other words, we are confident that he is not younger than 40 or older than 60, but we cannot say for sure that he is, for example, 48 or 54. We can only estimate the first digit of his age. The zero serves to hold the "ones" place in the number to give the correct magnitude. We do not know the actual value of either number with absolute certainty.

If you were told that the man in the picture was born in 1952 and that this picture was taken in 2004, you could then say with relative certainty that he is 52 years old in the photo. Since you do not know the month of his birth or the month when the picture was taken, you know that he is older than 51, but not yet 53 years old. You know with absolute certainty the first digit of his age, and the second digit is an estimate. Thus, you now know his age to two significant figures.

In any type of measurement (age, mass, volume, length, and so forth), there will be some uncertainty. Working with significant figures allows us to represent and convey to others the uncertainty in a given measurement. In any measurement, all nonzero numbers are considered significant. Zeros tend to be the "troublemakers" because their significance depends on their position in a number. For example, when an object's weight is reported as 20 g, the number has only one significant figure, so its weight is between 10 g and 30 g. However, if an object's weight is reported as 21.0 g, the last zero is significant and tells us that the object's weight is between 20.9 g and 21.1 g. The rules for counting significant figures in measurements are summarized in Table 1.2.

TABLE 1.2 COUNTING SIGNIFICANT FIGURES IN MEASUREMENTS

Rule	Measurement	Number of Significant Figures
1. A digit is significant if it is		
a. not a zero.	41 g	2
	15.3 m	3
b. a zero between nonzero digits.	101 L	3
	6.071 kg	4
c. a zero at the end of a decimal number.	20. g	2
	9.800 °C	4
2. A zero is not significant if it is		
a. at the beginning of a decimal number.	0.03 L	1
	0.00024 kg	2
b. a placeholder in a large number. without a decimal point.	12,000 km	2
	3,450,000 m	3

SAMPLE PROBLEM 1.3

Counting Significant Figures

Give the number of significant figures in each of the following measurements.

a. 380 min   b. 10,800 m

c. 9.317 kg   d. 0.0580 L

SOLUTION

a. Two significant figures—see rule 2b in Table 1.2.

b. Three significant figures—the zero between the 1 and 8 is significant, but the other two aren't. See rules 1b and 2b in Table 1.2.

c. Four significant figures—see rule 1a in Table 1.2.

d. Three significant figures—the last zero is significant, but the first two are not. See rules 1c and 2b in Table 1.2.

MATH MATTERS

Calculator Numbers and Rounding

When measurements are used in calculations, such as the Boyle's law calculations performed earlier, each measurement has some uncertainty. Adding, subtracting, multiplying, or dividing numbers that are uncertain can result in numbers that appear to be more certain than they actually are. Let's go back to the age of the man in the picture introduced earlier. Before we knew his birth date, we estimated that he was 50 years old. If we wanted to know how many days he has been alive, we would simply multiply his age in years by 365.25 days/year. Multiply these numbers in your calculator, and the number you will get is 18,262.5 days. Let's think about that. We know the man's age within ± 10 years, but simply by multiplying we now know it within one-half day! Somehow that doesn't seem logical. In fact, using measurements (which inherently have some uncertainty) in arithmetic cannot *increase* their certainty. That is, multiplying two uncertain numbers cannot lead to a number that is more certain.

Given that fact, how many of the resulting numbers from our calculation of the man's age are *meaningful*? Keeping all the numbers results in an answer with six significant figures, when we started with a number (50) that has only one! A general rule of thumb for such calculations is that *your answer can be no more certain than the least certain number you started with*. For multiplication and division, this means that the number of significant digits in the answer will be the same as the number of significant digits in the least significant starting number. For our previous calculation, the answer can have only one significant digit, so 18,262.5 days must be rounded to give one significant figure. We will use the following rules for rounding a number.

- If the first digit to be dropped (the one following your last retained significant digit) is 4 or less, simply remove it and the remaining digits.
- If the first digit to be dropped is 5 or greater, increase the last retained digit by 1 and remove all other digits.
- In both cases, zeros are substituted as placeholders where appropriate.

Thus, our result of 18,262.5 days is rounded to 20,000 days. (Our estimate of his age in years had only one significant figure; thus the calculation of his age in days can be no more certain.) When adding and subtracting, the answer is rounded to give the same number of decimal places as the number with the fewest decimal places. Thus, when calculating a sum such as 2.5 + 3.25, the answer is not 5.75, but it is rounded to 5.8.

SAMPLE PROBLEM 1.4

Rounding Calculated Numbers

Complete each of the following calculations and give the answer with the correct number of significant figures.

a. $75 - 21.3 =$ _____
b. $8.24 \times 3.0 =$ _____
c. $0.005 + 1.23 =$ _____
d. $12.4 \div 0.06 =$ _____

SOLUTION

a. Answer is 54. The calculator result of 53.7 is rounded to 54 so that the answer has no digits to the right of the decimal, as is true for the starting number 75.
b. Answer is 25. The calculator result of 24.72 is rounded to 25, the same number of significant digits (2) as the less significant of the two starting numbers.
c. Answer is 1.24. The calculator result of 1.235 is rounded to 1.24 to give two places to the right of the decimal.
d. Answer is 200. The calculator result of 206.66666 is rounded to 200 to one significant digit, which is the number of significant digits in the less significant starting number (0.06).

Gases and Temperature

Changing the temperature of a gas directly affects the motion of the particles. When the temperature of a gas is increased, energy in the form of heat is added to the gas, and the motion of the particles increases. Likewise, removing heat energy decreases the temperature of the gas and causes the gas particles to slow down. In each case, the change in the **temperature** results in a change in the motion of the particles of the gas. Before we consider the effects of temperature on gases, let's consider how temperature is measured.

In different parts of the world, temperature is measured using different scales. In the United States, we continue to use the Fahrenheit scale, while most of the rest of the world uses the Celsius scale. However, scientists use yet another scale called the absolute, or Kelvin, scale where the temperature unit is kelvins.

The most straightforward way to compare the temperature scales is to compare temperatures that we are most familiar with and observe their values on each scale. Consider the thermometers in Figure 1.7.

Pay particular attention to the freezing point and boiling point of water on the three scales. Notice that the Celsius and Kelvin scales have the same "degree" unit (the size of a degree is the same), but the two scales are offset by 273 degrees. Mathematically, this relationship is expressed as:

$$\text{Kelvin (K)} = \text{Celsius (°C)} + 273$$

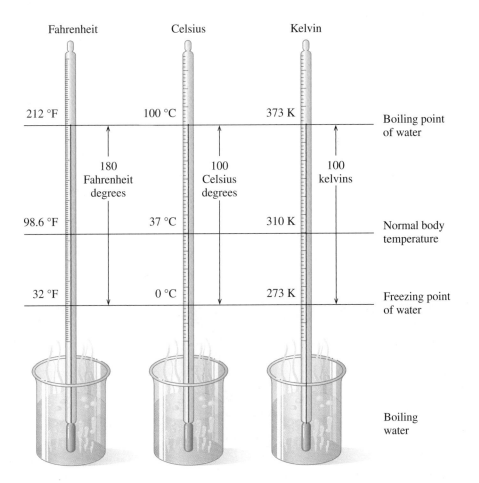

◄ **FIGURE 1.7 Comparison of temperature scales.** The zero point (freezing of water) is different for each of the scales. Comparing the boiling point to the freezing point of water on each scale, we see that the "size" of each degree is the same on the Celsius and Kelvin scales, but smaller on the Fahrenheit scale.

The relationship between the Fahrenheit scale and the others is not as straightforward. A unit of 1 degree (1°) on the Celsius scale is equal to 1.8 degrees (1.8°) on the Fahrenheit scale, and the zero points differ by 32 degrees. The relationship between °F and °C is shown:

$$°C = \frac{(°F - 32)}{1.8}$$

$$°F = (1.8 \times °C) + 32$$

After converting a Fahrenheit temperature to degrees Celsius, you can convert the Celsius temperature to kelvins using the Celsius to Kelvin scale equation, so a conversion equation to convert degrees Fahrenheit to kelvins is not needed.

SAMPLE PROBLEM 1.5

Temperature Conversion

On a hot summer day in Georgia, the thermometer often rises as high as 102 °F. What is this temperature in °C and kelvins?

SOLUTION

Using the Fahrenheit to Celsius equation and substituting the appropriate number, we get

$$°C = \frac{(102 \ °F - 32)}{1.8}$$

$$°C = 39$$

Now that we have the temperature in °C, we can convert it to the Kelvin scale using:

$$K = 39\,°C + 273$$

$$K = 312$$

Temperature and Volume—Charles's Law

To discover how temperature affects the volume of a gas, consider the photos in Figure 1.8. In the first picture, you see a fully inflated balloon just before it was placed into a cooler at very low temperature. In the second picture, you see the same balloon just after it was taken out of the cooler. Notice the difference? After being in the cooler, the balloon is smaller—its volume has decreased. This relationship between the temperature of a gas and its volume was discovered in the late 1700s by French scientist Jacques Charles.

▶ **FIGURE 1.8 Investigating a balloon's volume versus temperature.** The balloon is considerably smaller after being cooled in a cooler (b). This demonstrates that as the temperature of a gas decreases, so does its volume.

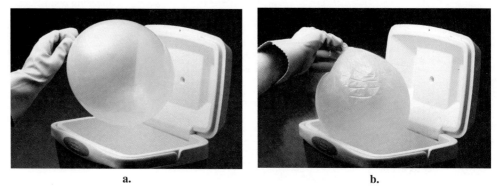

a.   **b.**

In his experiments with gases, Charles found that if the pressure and amount of a gas are not allowed to change, the volume of the gas is directly proportional to its absolute temperature. (We'll discover why he used absolute temperature a little later.) Charles discovered that when the absolute temperature of a gas was doubled, the volume of the gas also doubled (Figure 1.9).

Charles's law states: *The volume of a fixed amount of gas at constant pressure is directly proportional to its temperature.* In other words, an increase in the temperature of the gas will result in an increase in its volume, while a decrease in temperature will result in a decrease in its volume.

We can use Charles's law to determine what will happen to a sample of gas of known volume and temperature if we make changes to either its volume or its temperature. Charles' law can be applied as shown for making such determinations:

$$\frac{V_i}{T_i} = \frac{V_f}{T_f}$$

where T = the absolute temperature, V = volume, and the subscripts i and f stand for "initial" and "final," respectively (as they did in the Boyle's law discussion).

Let's do an example to see how this works. If you remove a balloon containing 105 mL of air from a freezer at 0 °C and allow the balloon to warm to room temperature (25 °C), what would its final volume be? $V_i = 105\,mL$, $T_i = 0\,°C$, and $T_f = 25\,°C$. Solving the Charles' law equation for the missing variable gives

$$\frac{V_i \times T_f}{T_i} = V_f$$

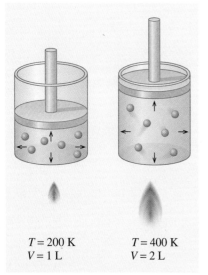

$T = 200\text{ K}$ $T = 400\text{ K}$
$V = 1\text{ L}$ $V = 2\text{ L}$

▲ **FIGURE 1.9 Effect of temperature on the volume of a gas.** When the temperature of the gas in an expandable container is doubled, its volume is also doubled. The increased motion of the gas particles causes the volume to increase.

Substituting the appropriate values into the rearranged equation results in the following:

$$\frac{105 \text{ mL} \times 25\,^\circ\text{C}}{0\,^\circ\text{C}} = V_f$$

Do you notice a problem with the equation? Look at the bottom of the fraction. There is a zero in the denominator! You may remember from your math classes that this is not allowed. So, how is this handled? (*Hint*: Remember the Kelvin scale?) Temperatures used in gas law equations must be converted to the Kelvin scale. On the Kelvin scale, the absence of all heat is called absolute zero, and this temperature has never been reached. So for all practical purposes, using the Kelvin scale prevents us from ever having a temperature of zero.

Converting each of the temperatures in our previous equation to the Kelvin scale and placing these appropriately into the equation, we get

$$\frac{105 \text{ mL} \times 298\,K}{273\,K} = V_f$$

$$115 \text{ mL} = V_f$$

Doing the math (taking care to cancel units) results in the answer $V_f = 115$ mL. Does the final answer make sense? Should the volume increase when the balloon is removed from the freezer? Yes. That is exactly what Charles's law predicts.

SAMPLE PROBLEM 1.6

Using Charles's Law

Determine the volume of the balloon we just discussed (105 mL in the freezer) if it is removed from the freezer (0 °C) and warmed to 145 °F.

SOLUTION

First, we cannot complete our calculation with the temperature in degrees Fahrenheit, so we need to convert the temperature to the Kelvin scale. We convert the temperature first to °C and then to K as follows:

$$^\circ\text{C} = \frac{(145\,^\circ\text{F} - 32)}{1.8}$$

$$^\circ\text{C} = 63$$

$$K = 63 + 273 = 336 \text{ K}$$

So, our final temperature (T_f) is 336 K. From the information in the problem, we know that $V_i = 105$ mL and $T_i = 273$ K (after converting from Celsius). Plugging our numbers into the Charles's law equation and rearranging gives us

$$\frac{105 \text{ mL} \times 336\,K}{273\,K} = V_f$$

Complete the calculation to find that $V_f = 129$ mL.

PRACTICE PROBLEMS

In the following problems, you may assume that the temperature and amount of the gas remain unchanged.

1.5 A balloon is filled with helium gas, which is lighter than air. When the following changes are made at a constant temperature, which of the diagrams on the next page (A, B, or C) show the new volume of the balloon?

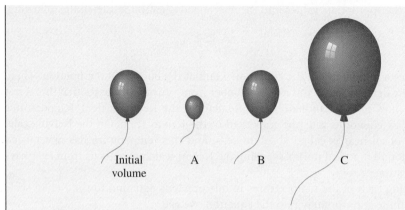

Initial volume   A   B   C

a. The balloon floats to a higher altitude where the outside pressure is lower.
b. The balloon is taken inside a house, but the atmospheric pressure remains the same.
c. The balloon is placed in a hyperbaric chamber where the pressure is increased.

1.6 When you swim underwater, the pressure of the water pushes against your chest cavity and subsequently reduces the size of your lungs. At 10 m below the water surface, the pressure exerted by the water is 14.7 psi. If a swimmer has a lung volume of 6 L at sea level, what would the volume of her lungs be when she is at the bottom of a pool that is 5 m deep?

1.7 A child's balloon containing 8.5 L of helium gas at sea level is released and floats away into the atmosphere. What will the volume of the helium be when the balloon rises to the point where the atmospheric pressure is 380 mmHg?

1.8 Use the Fahrenheit to Celsius equation to convert normal body temperature (98.6 °F) to °C. Then convert the Celsius value to Kelvin.

1.9 To keep a room comfortable, the air is heated or cooled to remain at approximately 75 °F. This is often referred to as ambient or room temperature. What is room temperature in °C?

1.10 On the Kelvin scale, the lowest possible temperature is zero—a temperature so cold, it has never been achieved. Zero kelvin is known as absolute zero. What is this temperature in °C? In °F?

In the following problems, you may assume that amount of the gas remains unchanged.

1.11 Select the diagram that shows the new volume of a balloon when the following changes are made at constant pressure.

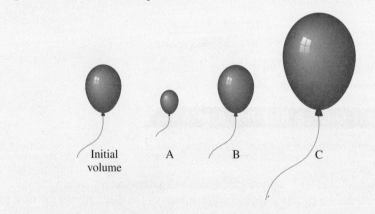

Initial volume   A   B   C

a. If the temperature is changed from 150 K to 300 K.
b. If the balloon is placed in a freezer.
c. If the balloon is warmed and then cooled to its initial temperature.

1.12 Assuming no air can move in or out, what happens to the volume of a hot air balloon as the air is heated?

1.13 A gas has a volume of 10 L at 0 °C. What is the final temperature of the gas (in °C) if its volume increases to 25 L?

1.3 Classifying Matter: Mixture or Pure Substance

Sometime during your education, you may have had a course in biology where you classified an organism taxonomically—according to its family, genus, and species as well as several other larger groupings. For example, the genus and species name for humans is *Homo sapiens*. In chemistry, we classify matter, but not quite to the extent that biologists classify organisms. Unlike the system used in biology, the classification of matter is not intended to result in a unique scientific name; instead, the system used in chemistry provides information about the characteristics of a sample of matter.

We have already encountered the *states of matter*—solid, liquid, and gas—and while the state in which matter exists may be thought of as a classification of matter, we usually use much broader terms when classifying matter. The flowchart shown in Figure 1.10 gives you an idea of how the classification of matter works. Use this flowchart as a guide as we work to develop and understand the scheme that chemists use to classify matter.

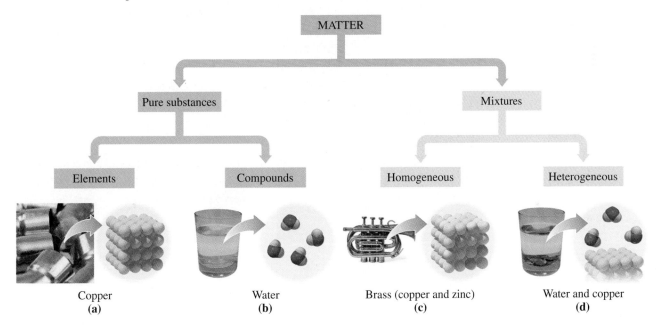

▲ **FIGURE 1.10 Flowchart used for classifying matter.** Matter can be broadly classified as a mixture or a pure substance. Mixtures are classified as homogeneous or heterogeneous and pure substances as elements or compounds.

As two examples of matter, consider blood and a diamond. If you are a blood donor, you may have heard blood discussed in terms of "whole blood," "plasma," "white cells," and "red cells." These terms give you a hint that blood is a mixture containing many components. In contrast, if you have ever bought a diamond, you know that one of the measures used to grade these gemstones is clarity, which is really a measure of purity. The clearer (more pure) a diamond is, the more it costs. So, how do the complexity of a sample of blood and the purity of a diamond contribute to how they are classified?

Take a look at the flowchart. The first thing you notice is that all matter is divided into two large categories or classifications—mixtures and pure substances. Let's consider these one at a time.

Mixtures

A **mixture** is a combination of two or more substances. Blood contains red blood cells, white blood cells, plasma (which is mostly water), and several other things. One of the defining characteristics of a mixture is that it can be separated into its different components. Therefore, blood is a mixture. Blood banks often separate blood into red cells and plasma for storage and for use in different medical procedures.

Mixtures can be further classified as homogeneous or heterogeneous. A **homogeneous mixture** is one whose composition is the same throughout. The air that you breathe is a homogeneous mixture; it contains nitrogen, oxygen, carbon dioxide, argon, and other gases. Every time you take a breath, you get the same substances in the same amount as anyone else who is breathing the same air. In contrast, grandma's chunky garden-style spaghetti sauce is a **heterogeneous mixture** because its composition is not uniform but varies throughout. One spoonful of spaghetti sauce may contain onions, mushrooms, and large chunks of tomato, while another spoonful contains celery, onions, and large chunks of green pepper. No two samples contain the same substances in the same amount.

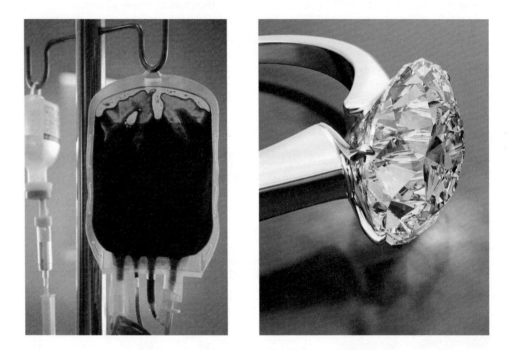

Pure Substances

Much of the matter we encounter in our everyday lives is some sort of mixture. From the air we breathe to the food we eat to the concrete we walk on, all are mixtures. However, because of their varied composition, mixtures are difficult to study. While chemists do routinely work with and study mixtures, they much prefer to handle pure substances.

A **pure substance** is made up of only one type of matter. Diamonds, are composed only of carbon. Because a diamond is composed of a single substance, it is not a mixture but is a pure substance.

Note from the flowchart that pure substances can be one of two types—elements or compounds. An **element** is the simplest type of matter because it is made up of only one type of atom. (An **atom** is the smallest unit of matter that can exist and keep its chemically unique characteristics.) A **compound** is a pure substance that is made of two or more elements that are chemically joined together.

Classifying Mixtures

Classify each of the following mixtures as homogeneous or heterogeneous. Briefly justify your answer.

a. olive oil   **b.** chocolate chip cookie dough

S O L U T I O N

a. Olive oil is a homogeneous mixture. It has the same composition throughout.

b. Chocolate chip cookie dough is a heterogeneous mixture. Each spoonful contains different amounts of chocolate chips and dough.

Mixture Versus Pure Substance

Classify each of the substances below as a mixture or a pure substance.

a. cake batter   **b.** helium gas inside a balloon

S O L U T I O N

a. Cake batter is a mixture of flour, butter, sugar, and other ingredients.

b. The helium gas is a pure substance.

1.14 Classify each of the following mixtures as homogeneous or heterogeneous.

a. a bowl of vegetable soup   **b.** mouthwash

c. an unopened can of cola   **d.** a dinner salad

1.15 Classify each of the following mixtures as homogeneous or heterogeneous.

a. a bottle of sports drink   **b.** a blueberry pancake

c. gasoline   **d.** tapioca pudding

1.16 Classify each of the following substances as a pure substance or a mixture.

a. copper   **b.** ice cream   **c.** table sugar

1.17 Classify each of the following substances as a pure substance or a mixture.

a. salt water   **b.** distilled water   **c.** concrete

1.4 Elements, Compounds, and the Periodic Table

If a pure substance can be an element or a compound, how do we distinguish which is which? Actually, we have a tremendous tool to guide us in this process: the periodic table of the elements (sometimes called the periodic chart). At its most basic level, the periodic table is a "listing" of all the elements found on Earth.

Somewhere along your educational journey, you may have encountered the periodic table, most likely hanging on the wall in a chemistry classroom. But have you ever tried to decipher it? This strangely shaped chart contains an amazing collection of data, both in the way that it is organized and in the actual information listed. We will make extensive use of the periodic table and the information it contains; therefore, we need to understand its basic layout and parts. Take a look at the periodic table in Figure 1.11.

Periodic Table of Elements

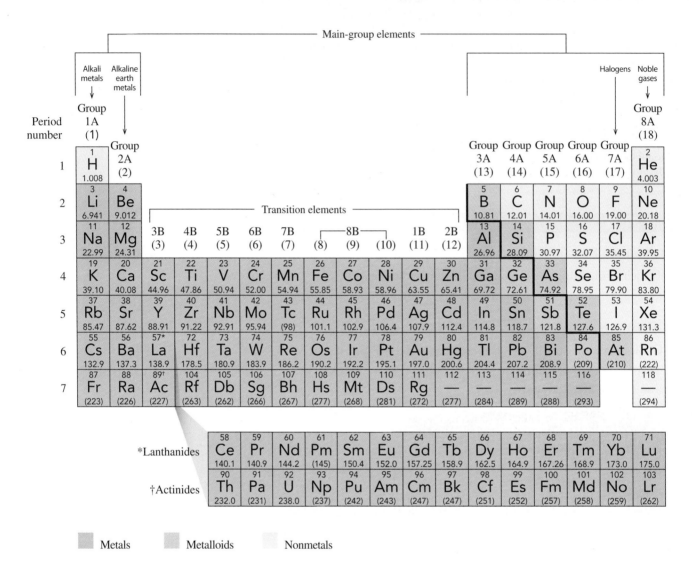

Metals   Metalloids   Nonmetals

▲ **FIGURE 1.11 The periodic table of the elements.** Each element has its own block on the table. The bold staircase-shaped line, which begins at the element boron and runs diagonally down and to the right, separates the metals from the nonmetals. Elements bordered by the line, with the exception of aluminum, are metalloids.

The periodic table consists of many small blocks. Each has a letter or two in its center and numbers above and below these letters. The letters are known as the elemental symbol and represent the name of each element. For many of the elements, the symbols are derived directly from the name of the element, for example, lithium = Li, carbon = C, hydrogen = H, oxygen = O, and so on. A few of the elements have symbols that do not match the first few letters of their name, for example, sodium = Na, gold = Au. Their elemental symbols are derived from the Latin names for the element. Since the elements form the basic "vocabulary" of chemistry, you need to know the names and symbols of the more common elements, especially those that are found in living things (carbon, hydrogen, oxygen, nitrogen, phosphorus, and sulfur).

The blocks on the periodic table are organized in specific arrangements. A vertical column of blocks is known as a **group** of elements. The elements in a group have similar chemical behaviors. Each group has a number and letter designation. The groups with A designations (1A through 8A) are known as the main-group elements, and the groups with B designations are the transition elements. This system of numbering the groups is common in North America. A system using the numbers 1 through 18 for the columns has been recommended by the International Union of Pure and Applied Chemistry (IUPAC) and is also used. Several of the groups have special names to designate the members of the group as shown at the top of Figure 1.11.

A horizontal row of blocks is known as a **period**. The periods are numbered from 1 to 7 with sections of Periods 6 and 7 set apart at the bottom of the periodic table. As we continue on our journey, we will see that each of the groups and periods has special characteristics and helps us better understand the chemistry of the elements.

In the periodic table of the elements, each element has its own block. The bold, staircase line, which begins at the element boron and runs diagonally down and to the right, separates the elements that are nonmetals (to the right of and above the line) from the elements that are metals (to the left of and below the line). The elements that lie *on* the line, with the exception of aluminum, are known as metalloids.

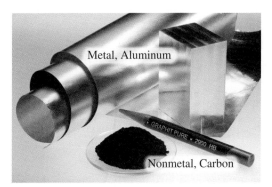

Compounds

How do we classify substances like water (H_2O) or table salt (NaCl) that contain more than one element from the periodic table? Recall that we said a pure substance containing two or more elements chemically combined is classified as a compound; therefore, both water and table salt are compounds.

Before we move on, let's take a quick look at the representations we use for compounds. Why does the representation for water contain the subscript 2, but the one for table salt doesn't? These representations, known as **chemical formulas**, show that water is composed of two particles of hydrogen and one particle of oxygen—the absence of a subscript on oxygen is understood to mean *one* particle—and table salt is composed of 1 particle of sodium and 1 particle of chlorine. A compound's chemical formula identifies both the type and number of particles of each of the elements in a compound.

SAMPLE PROBLEM 1.9

Classifying Pure Substances

Classify each of the following as an element or compound and explain your classification.

a. carbon dioxide (a gas we exhale)

b. helium (a gas sometimes used to inflate balloons)

SOLUTION

a. Carbon dioxide is a compound. From its name, you can deduce that this substance contains carbon and "something else" (even if you did not know that oxide is a name often used for oxygen when it is in compounds). Because carbon dioxide contains two different elements, it is a compound.

b. Helium is an element. Look for helium on the periodic table. You will find it on the upper right. The gas used to inflate a balloon so that it will "float" is pure helium—an element that is lighter than air.

SAMPLE PROBLEM 1.10

Using the Periodic Table

Use the periodic table on the inside front cover of this book to answer the following questions.

a. What is the elemental symbol for oxygen?

b. To what group does oxygen belong?

c. What period is oxygen in?

SOLUTION

a. Oxygen's symbol is O.

b. Oxygen is located in Group 6A.

c. Oxygen is in Period 2.

SAMPLE PROBLEM 1.11

Particles in Compounds

Identify and give the number of atoms of each element in the following compounds.

a. $C_2H_4O_2$, acetic acid, a component of vinegar

b. H_3PO_4, phosphoric acid

SOLUTION

a. Acetic acid contains 2 atoms of carbon, 4 atoms of hydrogen, and 2 atoms of oxygen.

b. Phosphoric acid contains 3 atoms of hydrogen, 1 atom of phosphorus, and 4 atoms of oxygen.

PRACTICE PROBLEMS

1.18 Classify each of the following as an element or compound and explain your classification.

a. aluminum, used in soft drink cans and foil

b. sodium hypochlorite, found in bleach

c. hydrogen, fuel of the sun

d. potassium chloride, a table salt substitute

1.19 Classify each of the following as an element or a compound.

a. Fe   b. $CaCl_2$ c. Si   d. KI

1.20 Use the periodic table to supply the missing information in the following chart.

Name	Elemental Symbol	Group	Period	Metal or Nonmetal
Fluorine				
		1A	2	
	Cl			
		4A		Nonmetal

1.21 Identify and give the number of particles of each element in the following compounds.

a. $NaNO_2$, sodium nitrite, a preservative

b. $C_{12}H_{22}O_{11}$, table sugar

c. $CaCO_3$, calcium carbonate

1.22 Identify and give the number of particles of each element in the following compounds.

a. Na_2SO_4, sodium sulfate, a drying agent

b. CO, carbon monoxide

c. NH_3, ammonia

1.5 How Matter Changes

On a hot summer day, you add several ice cubes to your glass of tea and come back later to find that the ice has disappeared. Now your glass seems to have two liquids—the tea at the bottom and a new, clear liquid on top. You may have seen someone at a science fair pour a clear, vinegar-smelling liquid into a papier-mâché volcano and then watched a foamy "lava" flow out. Each of these instances represents a change in matter, but each is a different type of change: One is a physical change and the other is a chemical reaction.

Physical Change

The ice in the tea is water in the solid state. As it melts, the ice undergoes a physical change and becomes a liquid (the clear liquid on top of your tea). In changing from solid to liquid, the identity of the water did not change—both ice and liquid are still H_2O (see Figure 1.12). In fact, if you add enough heat to the liquid water, it will eventually become steam—a gas—but it is still water, H_2O. A change in the state of matter—from a solid to a liquid or a liquid to a gas, for example—represents a physical change of matter. In a **physical change**, the form of the matter is changed, but its identity remains the same.

Water (H_2O)
as solid water (ice)

Water (H_2O)
as liquid water

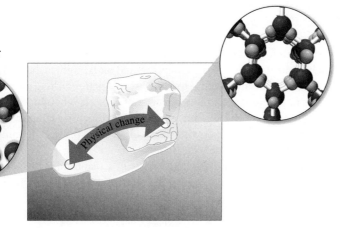
Physical change

◀ **FIGURE 1.12 Physical change of water from solid to liquid to gas.** As water changes from solid to liquid to gas, the arrangement of the particles becomes less orderly and their motion increases.

Chemical Reaction

The volcano in our example was most likely filled with baking soda before the vinegar was poured into it. When vinegar and baking soda combine, water and carbon dioxide bubbles form. Chemically, neither vinegar nor baking soda is the same as water or carbon dioxide. So when these two substances combined, new substances were formed. Unlike a physical change, a chemical change results in a change in the chemical

TABLE 1.3	COMPARING PHYSICAL CHANGES AND CHEMICAL REACTIONS
Physical Change	**Chemical Reaction**
Sawing a log in half	Burning a log in your fireplace
An ice cube melting	Hydrogen and oxygen combining to form water
Molten iron solidifying to form a nail	A nail rusting
Mixing oil, water, and eggs to make mayonnaise	Butter becoming rancid
Mixing cooked spaghetti and sauce to make a meal	A spaghetti supper being metabolized to provide energy for a morning run

identity of the substance (or substances) involved. When a substance (or substances) undergoes such a change, we more commonly refer to it as a **chemical reaction**. Table 1.3 gives some examples of physical changes and chemical reactions.

Chemical Equations

Do you enjoy a thick, juicy hamburger that has been grilled over a charcoal fire? Have you ever wondered how the big pile of charcoal that you start with is reduced to a small pile of gray dust when the fire finally dies out? When charcoal burns, a chemical reaction is occurring. To show what is happening to the element carbon, the main component in charcoal as it burns, we use a **chemical equation** like the one that follows:

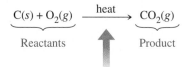

$$\underbrace{C(s) + O_2(g)}_{\text{Reactants}} \xrightarrow{\text{heat}} \underbrace{CO_2(g)}_{\text{Product}}$$

Reaction arrow

The parts of a chemical equation

A chemical equation is the chemist's way of writing a sentence about what happens in a chemical reaction. It is meant to be "read" from left to right. The chemical equation shown explains that "solid carbon and oxygen gas when heated react to form carbon dioxide gas." In this equation, carbon and oxygen are the **reactants**, and carbon dioxide is the **product**. The reaction arrow means "react to form." Special reaction conditions are often written as words or symbols above or below the reaction arrow. (Typical examples include heat, which is often represented as Δ; light, which is often represented as $h\nu$; or the name of an enzyme or other substance required for the reaction to occur.) The labels in parentheses after each substance (state labels) indicate its physical state— (s)olid, (l)iquid, or (g)as. A fourth label, (aq), meaning aqueous or dissolved in water, will be encountered later. So, when solid charcoal is burned, it combines with oxygen in the air and forms carbon dioxide, a gas.

Although this example gives the parts of a reaction, it does not represent all possible reaction scenarios. In other words, a chemical reaction may have one reactant and one product, one reactant and multiple products, multiple reactants and one product, or multiple reactants and multiple products.

SAMPLE PROBLEM 1.12

Physical Change or Chemical Reaction

Determine whether each of the following is a physical change or a chemical reaction.

a. a plant using carbon dioxide and water to make sugar

b. water vapor condensing on a cool evening to form dew on your grass

SOLUTION

a. This is a chemical reaction. Carbon dioxide and water are chemically different substances from sugar.

b. This is a physical change. Water vapor is a gas that condenses to form dew, which is simply liquid water.

PRACTICE PROBLEMS

1.23 Determine whether each of the following is a physical change or a chemical reaction.

a. A copper penny turns green.

b. Sugar is melted to make candy.

c. An antacid tablet is dropped into water and bubbles of carbon dioxide form.

1.24 Determine whether each of the following is a physical change or a chemical reaction:

a. Cream, sugar, and flavorings are mixed and frozen to make ice cream.

b. Milk spoils and becomes sour.

c. A puddle of spilled nail polish remover "disappears."

SUMMARY

1.1 "Stuff" Is Matter

Chemistry is the study of the "stuff" that chemists call matter. Matter is anything that takes up space and has mass. Volume is a measure of the amount of space occupied by a substance, and mass is a measure of the amount of matter in that substance. Mass and weight are similar, but not identical. Mass is always the same, but weight, which is a measure of the force of gravity on an object, changes with location. Most matter exists in one of three different states: solid, liquid, or gas. The three states of matter differ in the motion and association of the particles of which they are composed.

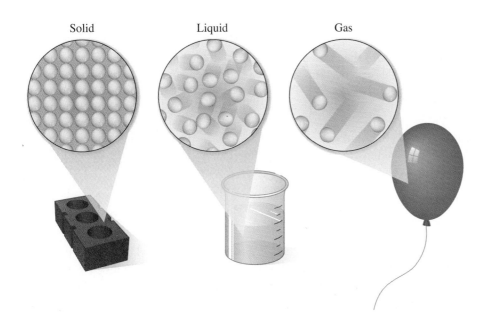

Solid   Liquid   Gas

chapter one

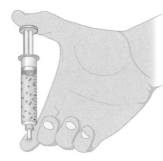

1.2 A Closer Look at Gases—An Introduction to the Behavior of Matter

The motion of the particles in a gas is affected by changes in both pressure and temperature. Pressure is a force exerted on a given area and is typically measured in units of psi or mmHg. Temperature is a measure of the hotness or coldness of a substance and is expressed in units of °F, °C, or kelvins. Changing the pressure or temperature of a gas causes a change in the volume of the gas. Boyle's law and Charles's law allow us to determine how these changes affect the volume of a gas.

1.3 Classifying Matter: Mixture or Pure Substance

In addition to classifying matter as solid, liquid, or gas, chemists use a broader classification of mixture or pure substance. Mixtures can be separated into their component parts and are further classified as homogeneous or heterogeneous. Pure substances are made up of a single component and are classified as elements or compounds.

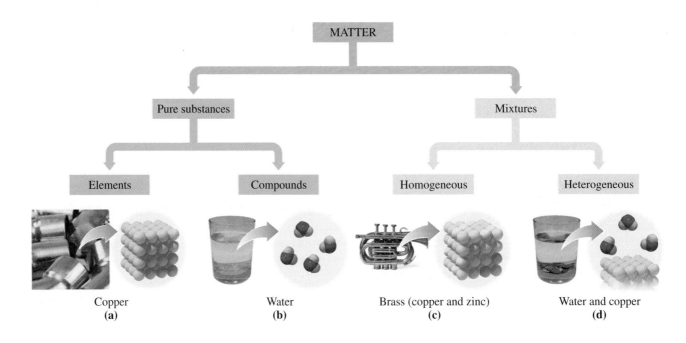

Copper	Water	Brass (copper and zinc)	Water and copper
(a)	(b)	(c)	(d)

1.4 Elements, Compounds, and the Periodic Table

The periodic table of the elements is a useful tool for cataloging all the elements. Each block on the table contains the symbol of a single element along with some other useful information about that element. The blocks are arranged in columns known as groups and rows known as periods. The elements in Groups 1A through 8A are known as the main-group elements and those in the B groups are known as the transition elements. A staircase line on the right side of the periodic table separates the metallic and nonmetallic elements. Compounds are chemical combinations of elements represented by chemical formulas, which provide the identity and number of particles of each element in the compound.

1.5 How Matter Changes

Matter can undergo two types of change: a physical change and a chemical reaction. In a physical change, the substance involved changes states, but its identity remains the same. In a chemical reaction, the identity of the reacting substance (or substances) is changed. A chemical reaction is represented using a chemical equation, which provides the identity of the reacting substance(s) and the product(s), the physical state of all substances in the reaction, and any conditions necessary for the reaction to occur.

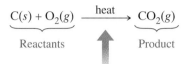

$$\underbrace{C(s) + O_2(g)}_{\text{Reactants}} \xrightarrow{\text{heat}} \underbrace{CO_2(g)}_{\text{Product}}$$

Reaction arrow

The parts of a chemical equation

KEY TERMS

atom—The smallest unit of matter that can exist and keep its unique characteristics.

chemical equation—A chemist's shorthand for representing a chemical reaction. It shows reactants and products, and often reaction conditions.

chemical formula—A representation showing both the identity and number of elements in a compound.

chemical reaction—A change involving the reaction of one or more substances to form one or more new substances; identity of substances is changed.

compound—A pure substance made up of two or more elements that are chemically joined together.

conversion factor—An equivalency written in the form of a fraction showing the relationship between two units.

element—The simplest type of matter; a pure substance composed of only one type of atom.

gas—A state of matter composed of particles that are not associated with each other and are rapidly moving, giving the substance no definite volume or shape.

gram—The base metric unit for measuring mass; roughly equivalent to the mass of a paperclip or raisin.

group—One of the vertical columns of elements on the periodic table.

heterogeneous mixture—A mixture whose composition varies throughout.

homogeneous mixture—A mixture whose composition is the same throughout.

liquid—A state of matter composed of particles that are loosely associated and freely moving, giving the substance a definite volume, but no a definite shape.

liter—The standard metric unit for measuring volume; a liter is about 6% larger than a quart.

mass—The measured amount of matter in a substance.

matter—Anything that takes up space and has mass.

meter—The standard metric unit for measuring length; a meter is approximately 10% larger than a yard.

metric system—The most common measurement system used worldwide in health care, business, and scientific applications.

millimeters of mercury (mmHg)—A unit of pressure derived from the pressure exerted by Earth's atmosphere; this unit is commonly employed in blood pressure measurements.

mixture—A combination of two or more substances.

period—One of the horizontal rows of elements on the periodic table.

physical change—A change in the state of matter, but not its identity.

pounds per square inch (psi)—A measure of pressure expressed as force (pounds) on a defined area (a square inch).

pressure—The amount of force exerted against a given area.

product—The new substance(s) produced by a chemical reaction; shown on the right side of a chemical equation.

pure substance—Contains only one type of matter.

reactant—The substance(s) that react to form different substance(s) in a chemical reaction; shown on the left side of a chemical equation.

significant figures—The digits in any measurement that are known with certainty plus the first estimated digit.

solid—A state of matter composed of particles that are arranged orderly and have very little motion, giving the substance a definite shape and a definite volume.

state of matter—The physical form in which a substance exists, either solid, liquid, or gas.

temperature—A measure of the hotness or coldness of a substance.

volume—The amount of space occupied by a substance.

IMPORTANT EQUATIONS

Boyle's Law

$$P_i \times V_i = P_f \times V_f$$

Temperature Conversions

$$\text{Kelvin (K)} = \text{Celsius (°C)} + 273$$

$$°C = \frac{(°F - 32)}{1.8}$$

$$°F = (1.8 \times °C) + 32$$

Charles's Law

$$\frac{V_i}{T_i} = \frac{V_f}{T_f}$$

ADDITIONAL PROBLEMS

1.25 When completing each of the following conversions, would you expect the number to get larger or smaller?
  a. 50 mg to g   **b.** 0.23 g to μg
  c. 5.2 m to km   **d.** 800 kL to dL

1.26 When completing each of the following conversions, would you expect the number to get larger or smaller?
  a. 789 cm to m   **b.** 0.45 kg to mg
  c. 694 mL to μL   **d.** 800 dm to m

1.27 Supply the missing information in each of the following conversion factors:

  a. $\dfrac{?\ mL}{1\ L}$ **b.** $\dfrac{?\ kg}{1000\ g}$ **c.** $\dfrac{100\ cm}{?\ m}$

  d. $\dfrac{1\ L}{?\ dL}$ **e.** $\dfrac{?\ μg}{1\ g}$

1.28 Supply the missing information in each of the following conversion factors:

  a. $\dfrac{?\ g}{1000\ mg}$ **b.** $\dfrac{10\ dg}{?\ g}$ **c.** $\dfrac{?\ g}{1,000,000\ μg}$

  d. $\dfrac{?\ m}{1\ km}$ **e.** $\dfrac{1\ L}{?\ cL}$

1.29 The typical runner's stride is about 1 meter. If a runner completes a 5K (5-kilometer) race, how many strides did he take? (Assume all his strides were the same length.)

1.30 If a prescription calls for 2000 mg of aspirin per dose, how many grams of aspirin will a patient take if she takes one dose?

1.31 Soft drinks are sold in plastic bottles that will hold 1 L or 2 L. How many mL of soft drink are contained in each size bottle?

1.32 Over-the-counter cough syrups often have doses listed as a number of teaspoons. A teaspoon contains 5.0 mL of cough syrup.
  a. Write a conversion factor relating number of teaspoons and number of mL.
  b. How many mL of cough syrup are in 3 teaspoons?

1.33 A raisin has a mass of approximately 1 g.
  a. Write a conversion factor relating number of raisins and number of grams.
  b. If a box contains 4500 cg of raisins, how many 1-g raisins are present in the box?

1.34 In this chapter, we discussed that the average pressure of the atmosphere is 14.7 psi. We also showed that the pressure of the atmosphere would support a column of mercury 760 mm high. Write a conversion factor relating atmospheric pressure in psi to mmHg.

1.35 In a high-pressure chamber, the pressure can easily rise to 1140 mmHg. Convert this pressure to psi.

1.36 How does the arrangement of particles in a liquid differ from that of a gas?

1.37 In this chapter, you learned that the volume of a gas changes with the applied pressure. The same is not true for a solid. In terms of particle arrangement, explain why solids are not compressible.

1.38 Standing on a boat in your scuba gear, you fill a balloon with helium.
  a. If you take the balloon with you as you dive to a depth of 40 m, what, if anything, will happen to the size of the balloon (assume temperature stays the same)?
  b. If the balloon slips out of your hand and floats upward, before you dive, what, if anything, will happen to the size of the balloon (assume temperature stays the same)?

1.39 A sealed syringe contains 5 cc of oxygen gas under 14.7 psi of pressure. If the pressure on the syringe is increased to 36.8 psi, but the temperature remains constant, what is the volume of the oxygen gas in the syringe?

1.40 Scuba divers wear tanks of compressed air to help them breathe underwater. Based on your knowledge of Boyle's law, explain why the air in a scuba tank must be pressurized in order for a deep-sea diver to breathe while underwater.

1.41 The lowest recorded atmospheric pressure occurred during a typhoon in the Pacific Ocean in 1979 and measured approximately 652 mmHg. The highest atmospheric pressure on record occurred in Mongolia in 2001 and measured approximately 814 mmHg. If a balloon were inflated with air to a volume of 10. L under the conditions in Mongolia and then rapidly transported to the conditions described for the typhoon in the Pacific, what would the volume of the balloon be under these new conditions, assuming no change in temperature?

1.42 If the balloon in Problem 1.38 had a volume of 2 L at a depth of 40 m, what was the original volume of the balloon if we assume the pressure at the surface of the water is 14.7 psi?

HINT: *Remember that underwater pressure increases by 14.7 psi for every 10 m.*

1.43 Give the number of significant figures in each of the following measurements:
- **a.** 25 minutes
- **b.** 44.30 °C
- **c.** 0.037010 m
- **d.** 800. L

1.44 Give the number of significant figures in each of the following measurements:
- **a.** 0.000068 g
- **b.** 100 °F
- **c.** 9,237,200 years
- **d.** 25.00 m

1.45 Give the number of significant figures in each of the following measurements:
- **a.** 200 °C
- **b.** 0.080 m
- **c.** 10,581,000 kg
- **d.** 1.40 L

1.46 Go back and check your answers to Problems 1.39, 1.41, and 1.42. Are your answers recorded with the correct number of significant figures?

1.47 Complete each of the following calculations and give your answer to the correct number of significant figures.
- **a.** $100 \times 23 =$ _____
- **b.** $97.5 - 43.02 =$ _____
- **c.** $8,064/0.0360 =$ _____
- **d.** $54.00 + 78 =$ _____

1.48 Complete each of the following calculations and give your answer to the correct number of significant figures.
- **a.** $340/96 =$ _____
- **b.** $0.305 + 43.0 =$ _____
- **c.** $0.0065 \times 11 =$ _____
- **d.** $19.029 - 0.00801 =$ _____

1.49 A temperature of −40 °F is dangerously cold. What is this temperature on the Celsius scale? On the Kelvin scale?

1.50 The coldest temperature ever recorded on Earth was in Vostok, Antarctica, and was measured at −128.6 °F. Convert this temperature to Celsius.

1.51 The hottest temperature ever recorded in the U.S. was 56.7 °C in Death Valley, California. Convert this temperature to Kelvin. What is this temperature on the Fahrenheit scale?

1.52 Use the formula given in the chapter to show that the boiling point of water on the Fahrenheit scale (212 °F) is equal to 100 °C.

1.53 When working with Charles's law, what variable must be assumed to remain constant?

1.54 The night before a big January "polar bear" bike ride, you take your bike inside to clean it up and pump up the tires. The next morning, after spending the night in your cold garage, the tires on your bike feel "squishy." Assuming they did not leak overnight, why do the tires feel less firm this morning than they did last night?

1.55 A beach ball is filled with 10.0 L of air in a hotel room where the temperature is 65 °F. What is the volume of the air in the beach ball out on the beach where the temperature is 38 °C?

1.56 Liquid nitrogen is an extremely cold liquid (−196 °C) used in chemistry "magic" shows to "shrink" balloons (and do other fun things). If a balloon is inflated to 2 L at 26 °C and then cooled in liquid nitrogen, what is the final volume of the balloon?

1.57 If a hot air balloon has an initial volume of 1000 L at 50 °C, what is the temperature (in °C) of the air inside the balloon if the volume expands to 2500 L?

1.58 Explain the difference between a mixture and a compound.

1.59 Classify each of the following as a mixture or a pure substance:
- **a.** ink in a pen
- **b.** wood
- **c.** blood
- **d.** ammonia (NH_3)
- **e.** silicon

1.60 For each of the substances that you classified as a mixture in Problem 1.59, determine whether it is a homogeneous or heterogeneous mixture.

1.61 For each of the substances that you classified as a pure substance in Problem 1.59, determine whether it is an element or a compound.

1.62 Classify each of the following as a mixture or a pure substance:
- **a.** paint
- **b.** $CaCO_3$ (calcium carbonate)
- **c.** tin
- **d.** milk
- **e.** table sugar ($C_{12}H_{22}O_{11}$)

1.63 For each of the substances that you classified as a mixture in Problem 1.62, determine whether it is a homogeneous or heterogeneous mixture.

1.64 For each of the substances that you classified as a pure substance in Problem 1.62, determine whether it is an element or a compound.

1.65 Use the periodic table to supply the missing information in the following chart.

Name	Elemental Symbol	Group	Period	Metal or Nonmetal
Lithium				
		3B	4	
	S			
		4A		Nonmetal

1.66 Give the name and number of particles of each element in the following compounds:
  a. H_2O_2, hydrogen peroxide, a disinfectant
  b. N_2H_4, hydrazine, a rocket fuel
  c. $MgSO_4$, magnesium sulfate, Epsom salts
  d. $C_6H_{12}O_6$, glucose
  e. $C_{14}H_{18}N_2O_5$, aspartame, an artificial sweetener

1.67 Give the name and number of particles of each element in the following compounds:
  a. TiO_2, titanium dioxide, white pigment in paint
  b. NH_3, ammonia, a household cleaner
  c. Na_2CO_3, sodium carbonate, baking powder
  d. N_2O, nitrous oxide, laughing gas
  e. KOH, potassium hydroxide, used in some drain cleaners

1.68 Identify each of the following as a physical change or a chemical reaction:
  a. water in a cloud forming into snow
  b. gasoline burning in a car's engine
  c. chewing a stick of gum

  d. an exploding hand grenade
  e. breaking a china plate

1.69 Identify each of the following as a physical change or a chemical reaction:
  a. bleaching your hair
  b. cutting your finger
  c. a cut healing
  d. nitric acid dissolving copper metal to form copper nitrate
  e. striking a match

1.70 Label the indicated parts of the following chemical equation:

$$CaCO_3(s) \xrightarrow{\Delta} CaO(s) + CO_2(g)$$

  a. Label the reactant(s).
  b. Label the product(s).
  c. Label the reaction arrow
  d. What does the Δ symbol in the equation mean?
  e. Based on the state labels, what would you expect to observe as this reaction occurs?

CHALLENGE PROBLEMS

1.71 The brakes in your car work on a hydraulic system. When you depress the brake pedal, fluid in your brake line is compressed. In turn, this fluid presses against the brake pads, which then press against the wheels causing them to stop. If air gets into your brake lines, the brakes will work but require you to step on the brake pedal harder to make your car stop. Using your knowledge of the difference in the molecular arrangement in liquids and gases and your understanding of gas behavior, explain why air in your brake lines causes your brakes to work less efficiently.

1.72 The following equation shows the reaction of magnesium metal and chlorine to form magnesium chloride which is used as a de-icer for roads and airport runways:

$$Mg(s) + Cl_2(g) \longrightarrow MgCl_2(s)$$

  a. Is magnesium metal an element or a compound? How do you know?
  b. Is chlorine an element or a compound? How do you know?
  c. What do the labels (s) and (g) represent?

  d. What do you notice about the number of atoms on either side of the chemical equation?
  e. Is this a physical change or a chemical reaction?

1.73 Table sugar ($C_{12}H_{22}O_{11}$) is dissolved in water to make sugar water.

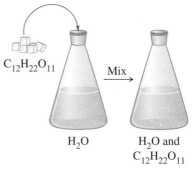

$C_{12}H_{22}O_{11}$ Mix

H_2O H_2O and $C_{12}H_{22}O_{11}$

  a. Is water an element or a compound? How do you know?
  b. Is sugar water a homogeneous or heterogeneous mixture? How do you know?
  c. When sugar is dissolved in water, is it a physical change or a chemical reaction?

ANSWERS TO ODD-NUMBERED PROBLEMS

Practice Problems

1.1 The particles in a liquid are close together but disordered, the particles in a gas are more disordered than a liquid and are far apart.

1.3 a. gas   b. solid

1.5 a. It would look like balloon C.
        b. It would look like balloon B.
        c. It would look like balloon A.

1.7 17 L

1.9 24 °C

1.11 a. It would look most like balloon C.
         b. It would look most like balloon A.
         c. It would look most like balloon B

1.13 410. °C

1.15 a. homogeneous   **b.** heterogeneous
    c. homogeneous   **d.** heterogeneous

1.17 a. mixture   **b.** pure substance
    c. mixture

1.19 a. element   **b.** compound
    c. element   **d.** compound

Additional Problems

1.25 a. smaller   **b.** larger
    c. smaller   **d.** larger

1.27 a. 1000   **b.** 1   **c.** 1
    d. 10   **e.** 1,000,000

1.29 5000 strides

1.31 1 L = 1000 mL and 2 L = 2000 mL

1.33 a. $\dfrac{1 \text{ raisin}}{1 \text{ g}}$ **b.** 45 raisins

1.35 22.1 psi

1.37 The particles of a solid are packed very tightly. Putting pressure on the solid cannot push the particles any closer together.

1.39 2 cc

1.41 12 L

1.43 a. 2   **b.** 4   **c.** 5   **d.** 3

1.45 a. 1   **b.** 2   **c.** 5   **d.** 3

1.47 a. 2000   **b.** 54.5   **c.** 224,000   **d.** 132

1.49 −40 °C and 233 K

1.51 330. K; 134 °F

1.53 pressure and the amount of gas present

1.55 10.7 L

1.57 535 °C

1.59 a. mixture   **b.** mixture   **c.** mixture
    d. pure substance   **e.** pure substance

1.61 d. compound   **e.** element

1.63 a. homogeneous   **d.** homogeneous

1.21 a. 1 sodium, 1 nitrogen, 2 oxygens
    b. 12 carbons, 22 hydrogens, 11 oxygens
    c. 1 calcium, 1 carbon, 3 oxygens

1.23 a. chemical reaction   **b.** physical change
    c. chemical reaction

1.65

Name	Elemental Symbol	Group	Period	Metal or Nonmetal
Lithium	Li	1A	2	Metal
Scandium	Sc	3B	4	Metal
Sulfur	S	6A	3	Nonmetal
Carbon	C	4A	2	Nonmetal

1.67 a. 1 titanium, 2 oxygens
    b. 1 nitrogen, 3 hydrogens
    c. 2 sodiums, 1 carbon, 3 oxygens
    d. 2 nitrogens, 1 oxygen
    e. 1 potassium, 1 oxygen, 1 hydrogen

1.69 a. chemical reaction   **b.** physical change
    c. chemical reaction   **d.** chemical reaction
    e. chemical reaction

1.71 Stepping on the brake pedal when no air is present causes the particles of the liquid to be compressed (move closer together). Since the particles in a liquid are so close together, very little pressure on the brake pedal is required to make the car stop. With air in the brake lines, depressing the pedal first compresses the gas particles, which have much more space between them. The air must be compressed with maximum pressure before the liquid is compressed to stop the car.

1.73 a. compound. It contains more than one element in its formula (H_2O).
    b. homogeneous. The sugar and water particles are evenly distributed when dissolved.
    c. physical change.

Guided Inquiry Activities FOR CHAPTER 2

EXERCISE 1 Isotopes

Information

Symbolic Notation for Isotopes

$${}^{A}_{Z} X$$

A = Mass Number

Z = Atomic Number

X = Atomic symbol

Isotopes of Calcium and the Number of Particles in Each

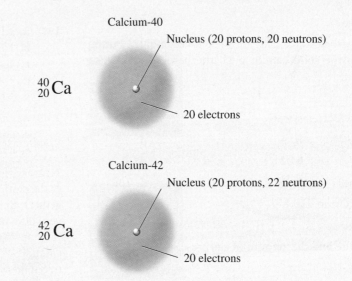

Calcium-40

Nucleus (20 protons, 20 neutrons)

$${}^{40}_{20} Ca$$

20 electrons

Calcium-42

Nucleus (20 protons, 22 neutrons)

$${}^{42}_{20} Ca$$

20 electrons

Questions

1. How many protons are found in calcium-42 and calcium-40?
2. How many neutrons are found in calcium-42 and calcium-40?
3. What is the relationship between the number of protons and the number of electrons in an atom?
4. Based on the information, what is common to all calcium atoms?
5. What does the atomic number Z indicate?
6. How is the mass number A determined?
7. How would you define *isotope*?
8. Where are the protons, neutrons, and electrons located in an atom?
9. What do calcium *isotopes* have in common? How do they differ?
10. Where is most of the mass of an atom?
11. For the following isotopes, determine the number of protons, neutrons, and electrons.

$${}^{27}_{13}Al$$

Protons = _____

Neutrons = _____

Electrons = _____

$${}^{37}_{17}Cl$$

Protons = _____

Neutrons = _____

Electrons = _____

SECTION 2.6

EXERCISE 2 Radioactivity

Information

TYPES OF IONIZING RADIATION PRODUCED IN NUCLEAR REACTIONS			
Emission	Symbol	Charge	Mass Number
Alpha	α or ^4_2He	2+	4
Beta	β or $^0_{-1}\text{e}$	1−	0
Gamma	$^0_0\gamma$	0	0
Positron	^0_1e	1+	0
Neutron	^1_0n	0	1

Nuclear Equation

$$\underset{\text{Reactant}}{^{238}_{92}\text{U}} \longrightarrow \underset{\text{Products}}{^{234}_{90}\text{Th} + {}^4_2\text{He}}$$

Questions

1. Using the table provided, what type of ionizing radiation (radioactivity) is produced in the nuclear equation shown?

2. How does each of the following in the reactant compare with the total number of each found in the products?
   a. number of protons
   b. number of neutrons
   c. mass number

3. Fill in the blanks to balance the mass numbers and the atomic numbers and provide the correct atomic symbol to complete the following equation:

$$^{131}_{53}\text{I} \longrightarrow \underline{\hspace{2cm}} + {}^0_{-1}\text{e}$$

4. Can you provide a nuclear equation for the following statement? Iron-59 emits a beta particle to form cobalt-59.

▶ **FIGURE 2.2 The location of subatomic particles in an atom.** In an atom, the protons and neutrons are packed into the center (nucleus), and the electrons move in the electron cloud surrounding the nucleus. An atom contains a lot of empty space between the nucleus and the electron cloud.

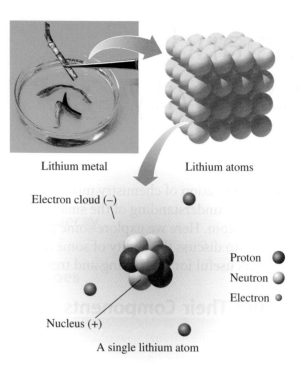

Lithium metal   Lithium atoms

Electron cloud (−)

Proton
Neutron
Electron

Nucleus (+)

A single lithium atom

the electrons are constantly moving in this space (see Figure 2.2). The remaining balloon interior would represent empty space. In fact, most of an atom consists of empty space.

Even though atoms are small and their subatomic particles are even smaller, both still qualify as matter (they have mass and take up space). Most of the mass of an atom lies in the nucleus. The electrons contribute very little to an atom's mass since they weigh about 2000 times less than a proton or neutron. For this reason we will ignore the electron's contribution to the mass of an atom. Chemists use a unit called the **atomic mass unit**, or **amu**, when discussing the mass of atoms. Because a proton and a neutron have about the same mass, each is defined as weighing about 1 amu. Today an amu is defined as one-twelfth of a carbon atom containing six protons and six neutrons. Therefore, the mass of a carbon atom would be 12 amu.

2.2 Atomic Number and Mass Number

All atoms of the same element always have the same number of protons. This feature distinguishes atoms of one element from atoms of all other elements.

Atomic Number

The number of protons in an atom of any given element can be determined from the periodic table. (There is a periodic table on the inside front cover of this text.) The sets of numbers appearing in numerical order above each element block are the **atomic numbers**. *The atomic number indicates the number of protons present in an atom of each element.* The number of protons gives an atom its unique properties. For example, a carbon atom, atomic number 6, contains six protons. *All* atoms of carbon have six protons. Lithium, atomic number 3, contains three protons. *All* atoms of lithium have three protons. Since atoms are neutral (no charge), the number of electrons in an atom is equal to the number of protons. Carbon must contain six electrons and lithium must contain three electrons to be neutral atoms.

Mass Number

How many neutrons are present in an atom of carbon? Remember that virtually all of an atom's mass is in the protons and neutrons and these particles weigh about the same. Suppose the mass of a carbon atom is given as 12. The number of neutrons present must be six since the number of protons in a carbon atom is six. The number of neutrons in an

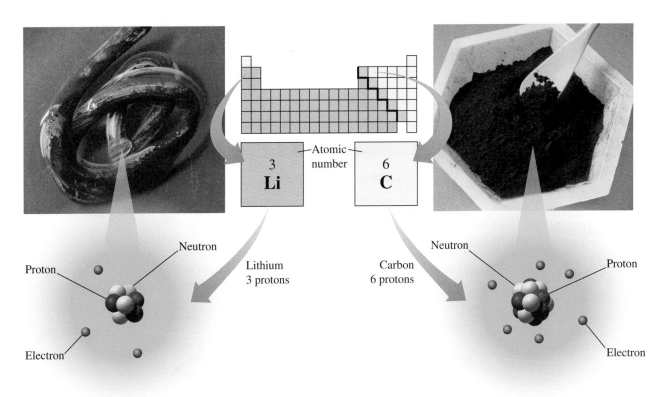

Neutron

Proton

Electron

Lithium
3 protons

Carbon
6 protons

Neutron

Proton

Electron

atom can be found from an atom's **mass number**, which is simply the number of protons plus the number of neutrons. This information is summarized in Table 2.2.

$$\text{Mass number} = \text{number of protons} + \text{number of neutrons}$$

Once the atomic number and the mass number for a given atom are known, you can determine the number of subatomic particles present. We can represent an atom with its mass number and atomic number in *symbolic notation*:

Mass number $\longrightarrow$ $^{12}_{6}C$ $\longleftarrow$ Atomic
Atomic number $\longrightarrow$ symbol

TABLE 2.2   DETERMINING THE NUMBER OF SUBATOMIC PARTICLES IN AN ATOM

Subatomic Particle	Number of Subatomic Particles in Atom
Protons	Atomic number *(found above element in periodic table)*
Electrons	Equal to number of protons in an atom
Neutrons	Mass number minus number of protons *(mass number designated for a given atom)*

SAMPLE PROBLEM 2.1

Determining the Number of Subatomic Particles

Determine the number of protons, electrons, and neutrons present in a fluorine atom that has a mass number of 19.

SOLUTION

Find fluorine on the periodic table (its atomic symbol is F).

The atomic number = the number of protons, so there are nine protons in this fluorine atom.

The number of electrons = the number of protons, so there are nine electrons.

The number of neutrons = mass number − number of protons.

The mass number is given as 19; therefore the number of neutrons = 19 − 9 or 10.

SAMPLE PROBLEM 2.2

Symbolic Notation

How would you represent the fluorine atom and its subatomic particles described in Sample Problem 2.1 in symbolic notation?

SOLUTION

The atomic symbol for fluorine is F. The mass number appears superscripted to the left of the symbol, and the atomic number appears subscripted to the left of the symbol:

$$^{19}_{9}\text{F}$$

PRACTICE PROBLEMS

2.1 How can you determine the following?

a. the number of protons present in an atom

b. the number of neutrons present in an atom

c. the number of electrons present in an atom

2.2 What can be determined from the following?

a. the mass number

b. the atomic number

c. the mass number minus the atomic number

2.3 Provide the name and atomic symbol of the element that has the following atomic number:

a. 8   b. 12   c. 10   d. 29   e. 47

2.4 Provide the name and atomic symbol of the element that has the following atomic number:

a. 79   b. 19   c. 82   d. 33   e. 11

2.5 What is the mass number for a nitrogen atom that contains seven neutrons?

2.6 What is the mass number for a sulfur atom that contains 17 neutrons?

2.7 Determine the number of protons, neutrons, and electrons in the following atoms:

a. a bromine atom that has a mass number of 80

b. a sodium atom that has a mass number of 23

2.8 Determine the number of protons, neutrons, and electrons in the following atoms:

a. a hydrogen atom that has mass number of 1

b. a sulfur atom that has a mass number of 32

2.9 Provide the symbolic notation for atoms with

a. 2 protons and 2 neutrons.   b. 17 protons and 18 neutrons.

c. 16 protons and 16 neutrons.   d. 55 protons and 78 neutrons.

2.10 Provide the symbolic notation for atoms with

a. 3 protons and 4 neutrons.   b. 7 protons and 10 neutrons.

c. 15 protons and 16 neutrons.   d. 46 protons and 60 neutrons.

2.11 Determine the number of protons, neutrons, and electrons for each of the following atoms:

a. $^{18}_{8}O$ b. $^{40}_{20}Ca$ c. $^{108}_{47}Ag$ d. $^{207}_{82}Pb$

2.12 Determine the number of protons, neutrons, and electrons for each of the following atoms:

a. $^{2}_{1}H$ b. $^{27}_{13}Al$ c. $^{23}_{11}Na$ d. $^{19}_{9}F$

2.3 Isotopes and Atomic Mass

Among the carbon atoms found in nature, most have a mass number of 12, a few have a mass number of 13, and fewer still have a mass number of 14. These atoms must have the same number of protons to be considered carbon atoms, so what is different about them? It must be the number of neutrons. Atoms of the same element *can* have different numbers of neutrons, so not all atoms of the same element have the same mass number. Atoms of the same element with different mass numbers are called **isotopes**. Isotopes have the same number of protons (therefore the same atomic number), but different numbers of neutrons (different mass numbers). We can indicate isotopes in two ways. The isotopes of carbon with mass numbers 12, 13, and 14 can be written in symbolic notation as

$$^{12}_{6}C, \ ^{13}_{6}C, \ \text{and} \ ^{14}_{6}C$$

Or, we can simply state the mass numbers after the name of the element — carbon-12, carbon-13, or carbon-14 — since carbon atoms always have six protons.

SAMPLE PROBLEM 2.3

Subatomic Particles in Isotopes

Determine the number of protons, neutrons, and electrons present in the following isotopes of nitrogen:

a. $^{15}_{7}N$ b. $^{14}_{7}N$ c. $^{16}_{7}N$

SOLUTION

Nitrogen's atomic number is 7, so all have seven protons. A nitrogen atom has no overall charge, and therefore they all also have seven electrons. The number of neutrons is different for each of the isotopes.

Number of neutrons = mass number − number of protons

a. seven protons, seven electrons, eight neutrons

b. seven protons, seven electrons, seven neutrons

c. seven protons, seven electrons, nine neutrons

Atomic Mass

To explore isotopes further, consider the large number of carbon atoms present in the half-carat diamond shown earlier. All the atoms will have six protons and six electrons, but the number of neutrons will differ for different isotopes. Carbon has three main isotopes, carbon-12, carbon-13, and carbon-14, which have six, seven, and eight neutrons, respectively. How do we know how common each isotope is? The periodic table gives us a clue. The number below each element on the periodic table shows the *average* atomic mass for that element. Carbon's average atomic mass is 12.01 amu. If the most common isotopes were the heavier isotopes, we would expect the average mass of carbon given in the periodic table to be closer to 13 or 14 amu. The atomic mass of carbon depends on the proportion of each isotope in a sample of carbon. Therefore, carbon's atomic mass of

▶ **FIGURE 2.3 The periodic table and atomic mass.** (a) Each symbol on the periodic table gives information regarding the number of protons and the average atomic mass of the element. (b) Because the atomic mass of carbon depends on the proportion of each isotope in a sample of carbon, the atomic mass of 12.01 amu is closest to the most abundant isotope ^{12}C. So for 10,000 atoms, there would be 9899 carbon-12 atoms, 100 carbon-13 atoms, and 1 carbon-14 atom.

a.

Isotope	^{12}C	^{13}C	^{14}C
Atomic number	6	6	6
Protons	6	6	6
Neutrons	6	7	8
Abundance	Most	1/100	1/10,000

b. Atomic mass = 12.01 amu

12.01 amu indicates that the most abundant isotope is carbon-12. In our half-carat diamond, there are mostly carbon-12 atoms; however, there are a few carbon-13 atoms (1 in every 100) and very few carbon-14 atoms (1 in every 10,000). Because the carbon-13 and carbon-14 isotopes have more neutrons, the *average* atomic mass of the carbon atoms in a sample of pure carbon such as diamond is 12.01, slightly more than 12.00. The **atomic mass** shown below each of the elements on the periodic table is the average atomic mass weighted for all the isotopes of that element found naturally (see Figure 2.3).

SAMPLE PROBLEM 2.4

Isotopes and Atomic Mass

Copper has two naturally occurring isotopes: copper-63 and copper-65. Based on the atomic mass from the periodic table, which of these is more prevalent?

SOLUTION

The atomic mass of copper (element number 29) is 63.55 amu. Since the average atomic mass is closer to 63 than 65, there must be more copper-63 atoms present in the natural world than copper-65 atoms.

PRACTICE PROBLEMS

2.13 There are three naturally occurring isotopes of magnesium: magnesium-24, magnesium-25, and magnesium-26.

a. How many neutrons are present in each isotope?

b. Write complete symbolic notation for each isotope.

c. Based on the average atomic mass given in the periodic table, which isotope of magnesium is the most abundant?

2.14 Some of the naturally occurring isotopes of tin are tin-118, tin-119, tin-120, and tin-124.

a. How many neutrons are present in each isotope?

b. Write complete symbolic notation for each isotope.

2.4 Counting Atoms: The Mole

If atoms are so tiny, how do we know how many of them are being weighed on a balance? It must be a pretty large number. Chemists use a unit called the **mole** to relate the mass of an element in grams to the number of atoms it contains. The mole is a unit for counting atoms not unlike a *dozen* is a unit for counting things like eggs. The number of eggs in 1 dozen is 12. Suppose each egg weighs approximately 50 grams (g). We could determine how many eggs are present in a closed egg carton by weighing it. For example, if a carton of eggs weighs 600 g, we could calculate that there were 12 eggs or 1 dozen eggs inside (assuming the container weighs nothing) even without being able to actually see the eggs inside the carton. We have used the mass of the eggs to count the number of eggs.

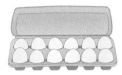

600 g = 12 eggs = 1 dozen

We cannot actually see atoms, but if we know the mass of a sample of atoms, we can then relate the atomic mass to the number of atoms using the mole unit. On the periodic table, carbon has an atomic mass of 12.01 amu. One mole of any element has a **molar mass** in grams numerically equal to the atomic mass of that element. The molar mass tells us the mass of 1 mole of a substance. So, similar to the fact that a dozen eggs in our example has a mass of 600 g, 1 mole of carbon atoms has a molar mass of 12.01 g, or we can say the molar mass of carbon is 12.01 grams per mole of carbon atoms (12.01 g/mole).

Just as we know that there are 12 objects in the unit called a dozen, the number of atoms present in the unit called the mole is defined as 602,000,000,000,000,000,000,000 atoms. Wow! That is a very large number. Huge numbers like this are awkward to handle. It is more convenient to express such a large number in a kind of shorthand called scientific notation. The number of atoms in a mole can be represented in scientific notation as 6.02×10^{23} atoms.

MATH MATTERS

Scientific Notation

To better understand scientific notation, consider the number 70,000. One way to arrive at this number is to multiply $7 \times 10 \times 10 \times 10 \times 10$ (try it on your calculator). In scientific notation we can express the four times we multiplied by 10 as an exponent, showing it as 10^4. The scientific notation for 70,000 would be written as 7×10^4. A number smaller than 1, like 0.005, can be determined by dividing 5 by $10 \times 10 \times 10$ or $5/10^3$. Dividing by a number is equivalent to multipying by its reciprocal. So, in exponential form, $5/10^3$ can be written as 5×10^{-3}. The general form for scientific notation is

$$C \times 10^n$$

where C is called the coefficient and is a number between 1 and 9 and n is the exponent telling us the number of tens places applying to the coefficient. A positive exponent tells us the actual number is greater than 1, and a negative exponent tells us the number is between 0 and 1. In scientific notation, only significant figures are shown in the coefficient (all placeholding zeros are not part of the notation). For example, the scientific notation of the number 0.00650 would be 6.50×10^{-3} since the zeros between the decimal point and the 6 are placeholders and the zero after the 5 is significant. Table 2.3 shows how numbers and scientific notation are related for a wide range of numbers.

TABLE 2.3	RELATING NUMBERS TO SCIENTIFIC NOTATION	
Number	**Meaning**	**Scientific Notation**
1,000,000	$10 \times 10 \times 10 \times 10 \times 10 \times 10$	1×10^6
100,000	$10 \times 10 \times 10 \times 10 \times 10$	1×10^5
10,000	$10 \times 10 \times 10 \times 10$	1×10^4
1000	$10 \times 10 \times 10$	1×10^3
100	10×10	1×10^2
10	10	1×10^1
1	1	1×10^0
0.1	1/10	1×10^{-1}
0.01	$1/(10 \times 10)$	1×10^{-2}
0.001	$1/(10 \times 10 \times 10)$	1×10^{-3}
0.0001	$1/(10 \times 10 \times 10 \times 10)$	1×10^{-4}
0.00001	$1/(10 \times 10 \times 10 \times 10 \times 10)$	1×10^{-5}
0.000001	$1/(10 \times 10 \times 10 \times 10 \times 10 \times 10)$	1×10^{-6}

SAMPLE PROBLEM 2.5

Expressing Numbers in Scientific Notation

Express the following numbers in scientific notation:

a. 820,000 b. 0.00096

SOLUTION

a. To determine the correct scientific notation for a number greater than 1, count the number of tens places to the right of the first number, 8. In this case there are 5 tens places to the right of 8. This becomes the exponent on the number 10 in scientific notation. Any other nonzero numbers go after a decimal point. The answer is 8.2×10^5.

b. To determine the correct scientific notation for a number less than 1, count the number of tens places between the decimal point and the first number, 9, and add 1. In this case there are 3 tens places to the left of the 9 and adding 1 would give a total of 4. The negative of this number becomes the exponent on the tens place. Any other nonzero numbers go after the decimal point. The answer is 9.6×10^{-4}.

SAMPLE PROBLEM 2.6

Changing Numbers from Scientific Notation to Decimal Format

Express the following numbers in decimal format:

a. 6.5×10^4 b. 8.34×10^{-6}

SOLUTION

a. The number in the exponent tells you how many tens places are beyond the decimal place in the scientific notation when converting from scientific notation to decimal format. In this problem, the number 5 will occupy one of the tens places, so three more zeros after the 5 are necessary. The answer is 65,000.

b. Negative exponents indicate that this number is smaller than 1, but greater than zero. The number in the exponent tells you how many tens places should appear to the left of the decimal place when converting from scientific to decimal format. The number 8 will occupy one of these places, so 5 more zeros should appear in the final answer to the right of the decimal place. All digits to the right of the decimal place in the scientific notation should appear in the final answer. The answer is 0.00000834.

Avogadro's Number

The large number of atoms present in 1 mole is called **Avogadro's number** (N) after Amedeo Avogadro, an Italian physicist who first distinguished atoms and molecules. The relationship between moles and number of atoms was later determined.

$$6.02 \times 10^{23} \text{ atoms} = 1 \text{ mole of atoms}$$

In general, for any substance Avogadro's number can be defined as

$$N = 6.02 \times 10^{23} \text{ particles/mole}$$

We have used the mass in grams for the carbon atoms to count the number of carbon atoms:

$$12.01\text{g C} = 1 \text{ mole of C} = 6.02 \times 10^{23} \text{ atoms of C}$$

Just as a dozen eggs, a dozen paper clips, and a dozen bowling balls have different masses, a mole of copper atoms and a mole of carbon atoms will not have the same mass. Since atoms have different atomic masses, the same number of atoms of two different elements have different masses.

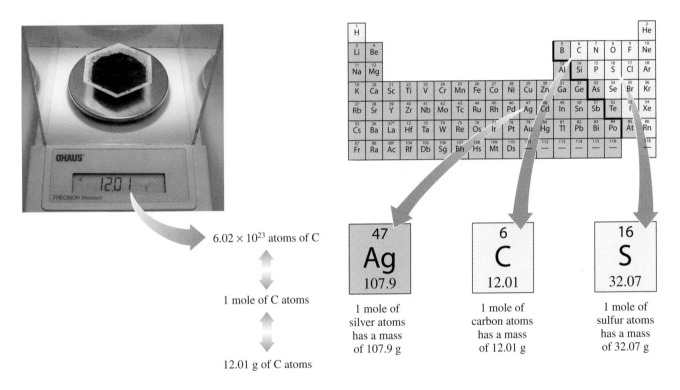

6.02×10^{23} atoms of C

1 mole of C atoms

12.01 g of C atoms

1 mole of silver atoms has a mass of 107.9 g

1 mole of carbon atoms has a mass of 12.01 g

1 mole of sulfur atoms has a mass of 32.07 g

Let's solve a problem to see how atoms and moles are related to each other.

Two quantities like atoms and moles that can be related to each other with an equal sign are referred to as **equivalent units**.

From our definition of Avogadro's number we know the equivalent unit relating atoms and moles is

$$6.02 \times 10^{23} \text{ atoms} = 1 \text{ mole of atoms}$$

This equivalent unit can be rearranged to

$$\frac{6.02 \times 10^{23} \text{ atoms}}{1 \text{ mole}}$$

and can be used as a conversion factor when we solve a problem.

How many atoms are present in 2.00 moles of carbon?

To determine the answer to a problem like this, the problem can be set up using the following guidelines:

- Examine the entire problem first, identifying the information you have (in this case moles) and the information you are trying to determine (in this case atoms).
- The quantity you are trying to determine (atoms) must appear in the numerator so your answer has those units.
- Equivalent units should have a unit in the numerator and a unit in the denominator, arranged so that the units in the problem cancel each other out so your answer is in the desired units.
- Set up the relationship to cancel the given units (moles) and to provide the needed unit in the answer (atoms). The equation can be set up as follows:

$$2.00 \ \cancel{\text{mole}} \ C \times \frac{6.02 \times 10^{23} \text{ atoms C}}{1 \ \cancel{\text{mole}} \ C} = 1.20 \times 10^{24} \text{ atoms C}$$

- Ask yourself if your answer makes sense. When solving a problem for the number of atoms, your answer should be a large number and have a large positive exponent.

SAMPLE PROBLEM 2.7

Converting from Moles to Grams

What is the mass of 4.00 moles of carbon in grams?

SOLUTION

The unit for the final answer (g) must appear in the numerator and other units (moles) must cancel out. In this problem, we are looking for the number of grams. The equivalent unit that we have been working with that contains grams is the atomic mass where

$$12.01 \text{ g C} = 1 \text{ mole C}$$

Two possible conversion factors are

$$\frac{12.01 \text{ g C}}{1 \text{ mole C}} \quad \text{or} \quad \frac{1 \text{ mole C}}{12.01 \text{ g C}}$$

The unit for the final answer (g) must appear in the numerator and the other unit (moles) must cancel out. Using the conversion factor that cancels the given unit (mole), we can set up the problem as follows:

$$4.00 \ \cancel{\text{mole}} \ C \times \frac{12.01 \text{ g C}}{1 \ \cancel{\text{mole}} \ C} = 48.0 \text{ g C}$$

If 1 mole of carbon weighs about 12 g, then 4 moles should weigh four times that amount, so your answer makes sense.

SAMPLE PROBLEM 2.8

Converting from Grams to Number of Atoms

a. How many atoms of aluminum are present in 1.00 g of aluminum?

b. How does this compare to the number of atoms in 1.00 g of lead?

SOLUTION

To solve these two problems we use two conversion factors, the molar mass for the element from the periodic table (located beneath the atomic symbol with units of g/mole) and the conversion factor for Avogadro's number, to relate the mass to the number of atoms. The units on Avogadro's number will be atoms per mole.

a. How many atoms of aluminum?

$$1.00 \text{ g Al} \times \frac{1 \text{ mole Al}}{26.98 \text{ g Al}} \times \frac{6.02 \times 10^{23} \text{ atoms of Al}}{1 \text{ mole Al}} = 2.23 \times 10^{22} \text{ atoms of Al}$$

Information given   Molar mass (conversion factor)   Avogadro's number

b. How many atoms of lead?

$$1.00 \text{ g Pb} \times \frac{1 \text{ mole Pb}}{207.2 \text{ g Pb}} \times \frac{6.02 \times 10^{23} \text{ atoms of Pb}}{1 \text{ mole Pb}} = 2.91 \times 10^{21} \text{ atoms of Pb}$$

Information given   Molar mass (conversion factor)   Avogadro's number

The exponents in the two answers show that there are about 10 times more atoms in 1.00 g of aluminum than in 1.00 g of lead. Since aluminum atoms are lighter, it takes more aluminum atoms to obtain the same mass.

PRACTICE PROBLEMS

2.15 Compare (a) the number of atoms and (b) the number of grams present in 1.00 mole of silver (Ag) and 1.00 mole of gold (Au).

2.16 Compare (a) the number of atoms and (b) the number of grams present in 1.00 mole of helium (He) and 1.00 mole of argon (Ar).

2.17 Calculate the following:

a. the number of Na atoms in 0.250 mole of Na

b. the number of moles in 4.0 g of Pb

c. the number of atoms in 12.0 g of Si

2.18 Calculate the following:

a. the number of S atoms in 0.660 mole of S

b. the number of moles in 15.0 g of Fe

c. the number of atoms in 66.0 g of Cu

2.5 Electron Arrangements

In Section 2.1 we showed that the electrons occupy a large space outside the nucleus of an atom, forming what chemists call an electron cloud. The exact location of any given electron in this cloud is difficult to determine because the electrons are moving around very quickly in this space. Because electrons are charged and in constant motion, they possess energy. The electrons in an atom are actually found in distinct energy levels based on the amount of energy the electrons possess. Think of the energy levels as an uneven staircase (see Figure 2.4) where the height of the steps in the staircase decreases as you ascend.

In atoms, the electrons can exist only at distinct energy levels (steps on the staircase), not in between. In general, electrons will occupy the lowest energy step first. This lowest energy level is found closest to the nucleus. Unlike a real staircase, the energy levels get closer together the farther you get from the nucleus. The maximum number of electrons that can be found in any given energy level can be calculated by the formula $2n^2$ where n is the number of the energy level. In the first energy level ($n = 1$), the maximum number of electrons present is 2, in the second energy level ($n = 2$), the maximum number of electrons is 8, in the third energy level, 18, and so on.

Table 2.4 shows the electron arrangements for the first 20 elements. In the table, look at element 19, potassium (K), and element 20, calcium (Ca). They have electrons in the fourth energy level even though they do not have the maximum 18 electrons filling up the third level. An energy level does not have to be full before the next energy level begins to fill with electrons.

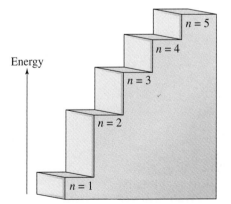

▲ **FIGURE 2.4 An energy staircase.** Electrons in an atom occupy the lower energy levels first. Electrons are only found at energy levels represented by each step, not between the steps. The number of the energy level is represented by n.

TABLE 2.4	ENERGY-LEVEL ARRANGEMENT OF ELECTRONS FOR THE FIRST TWENTY ELEMENTS*		Number of Electrons in Energy Level			
Element	Group Number	Total Number of Electrons	$n = 1$	$n = 2$	$n = 3$	$n = 4$
H	1A	1	1			
He	8A	2	2			
Li	1A	3	2	1		
Be	2A	4	2	2		
B	3A	5	2	3		
C	4A	6	2	4		
N	5A	7	2	5		
O	6A	8	2	6		
F	7A	9	2	7		
Ne	8A	10	2	8		
Na	1A	11	2	8	1	
Mg	2A	12	2	8	2	
Al	3A	13	2	8	3	
Si	4A	14	2	8	4	
P	5A	15	2	8	5	
S	6A	16	2	8	6	
Cl	7A	17	2	8	7	
Ar	8A	18	2	8	8	
K	1A	19	2	8	8	1
Ca	2A	20	2	8	8	2

*Valence electrons in blue.

SAMPLE PROBLEM 2.9

Electrons and Energy Levels

How many electrons are in each energy level of the following elements?

a. O   b. Ca

SOLUTION

a. Oxygen has an atomic number of 8, so it has 8 electrons total. Two are located in the first energy level while 6 are in the second energy level.

n	Number of Electrons
1	2
2	6

b. Calcium has an atomic number of 20, so it has 20 electrons total. Two are located in the first energy level, 8 in the second, 8 in the third, and 2 in the fourth.

n	Number of Electrons
1	2
2	8
3	8
4	2

Examining the energy levels and number of electrons present in those energy levels provides insight into how the elements in a group or period are related to each other. As shown in Table 2.4, the elements with the same number of electrons in their highest energy level are in the same group. For example, boron (B) and aluminum (Al) have three electrons in their highest energy level and they are both in Group 3A. The highest energy level that contains electrons is referred to as the valence level or **valence shell**, and the electrons residing in that energy level are referred to as the **valence electrons**. The valence electrons are farthest from the nucleus and are important since these are the electrons responsible for chemical reactions.

The number and energy levels of valence electrons correspond to the element group numbers and periods on the periodic table. The group (column) number represents the number of valence electrons for the main-group elements. Group 1A elements have one valence electron, Group 2A elements have two, and so on across the periodic table. The period (row) represents the outermost energy level or shell containing electrons. The electrons in the elements H and He (Period 1) are located in energy level 1, which can hold a maximum of only two electrons. The electrons in the elements in Period 2, lithium (Li) through neon (Ne), hold their valence electrons in energy level 2 which can hold a maximum of eight electrons. Valence electrons for the main-group elements in Periods 3 and higher are similarly found in energy level 3 and higher.

SAMPLE PROBLEM 2.10

Valence Electrons

How many valence electrons are in the following elements?

a. N   b. Si

SOLUTION

The number of valence electrons can be determined by the group number for the main-group elements.

a. Nitrogen is in Group 5A; therefore the number of valence electrons is five.

b. Silicon is in Group 4A; therefore the number of valence electrons is four.

PRACTICE PROBLEMS

2.19 How many electrons constitute a full energy level $n = 1$? Energy level $n = 2$

2.20 Which electrons have higher energy, those in energy level $n = 1$ or those in energy level $n = 2$?

2.21 How many electrons are in each energy level of the following elements?

a. He   b. C   c. Na   d. Ne

2.22 How many electrons are in each energy level of the following elements?

a. H   b. F   c. Ar   d. K

2.23 How many valence electrons are present in the following atoms?

a. O   b. C   c. P   d. Na

2.24 How many valence electrons are present in the following atoms?

a. S   b. Mg   c. Be   d. F

2.6 Radioactivity and Radioisotopes

Do you recognize the following symbol?

This symbol is called the *trefoil* (or tri-foil) and is the international symbol for radiation. What is radiation? Energy given off spontaneously from the nucleus of an atom is called **nuclear radiation**. Nuclear radiation is something that we are encouraged to avoid in everyday life because it can sometimes be life threatening. Elements that emit radiation are said to be radioactive. What is radioactivity? What elements possess this property and why? If radiation is so dangerous, why do we see these stickers in different areas in a hospital? These are all good questions that will be answered in the next few sections.

Radiation is a type of energy that comes in many forms, and we get it from both natural and human-made sources. Light and heat from the sun are both natural sources of radiation. Microwaves used in ovens and radio waves used for communication are examples of human-made radiation. Similarly, the nucleus of some atoms can also "radiate." Radiating nuclei were first detected in the late 1800s with materials that are still used today in professional film photography.

In 1896, Antoine Becquerel got an exposure on a photographic plate (in those days they had plates, not film or digital images), not from interaction with visible light, but by exposing the plate to a rock in the dark! Elements that spontaneously emit radiation from their nucleus are said to be **radioactive** (see Figure 2.5). The rock that Bequerel used in his experiment contained the element uranium, which we know today to be a radioactive element.

Nuclear radiation can be in the form of high-energy particles or in the form of high-energy rays. Like the uranium rock that left a pattern on Bequerel's photographic plate, some radioactive isotopes of elements, or **radioisotopes**, are useful in medical imaging. Certain radioisotopes concentrate in particular tissues. The emitted radiation

▶ **FIGURE 2.5 Radioactive decay.** A radioactive element such as the uranium rock pictured here in a light-tight container will emit radiation, chemically altering photographic film.

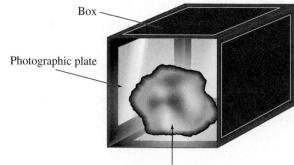

Box

Photographic plate

Radioactive rock containing uranium

TABLE 2.5	SOME RADIOISOTOPES AND THEIR USE IN DIAGNOSING AND TREATING DISEASE
Radioisotope	**Clinical Use**
Chromium-51	Red blood cell labeling for determining blood volume
Cobalt-60	Cancer therapy
Gallium-67	Soft tissue tumor (lymphoma) detection
Iodine-123, iodine-131	Thyroid imaging and uptake
Technetium-99m	Bone imaging, renal imaging, breast tumor imaging, heart perfusion imaging
Xenon-133	Pulmonary ventilation imaging

Source: nuclearpharmacy.uams.edu/RPLIST.html

from those locations can create an image on a photographic plate or detector. Some actual images are shown in Figure 2.6. Some elements used in imaging are shown in Table 2.5.

Are all atomic nuclei radioactive? No. Most naturally occurring isotopes actually have a stable nucleus and these isotopes are therefore *not* radioactive. Isotopes that are not stable become stable by spontaneously emitting radiation from their nuclei. This process is called **radioactive decay**. Several naturally occurring radioisotopes can be mined from the earth; many more can be prepared in scientific laboratories. All the isotopes of elements with atomic number 84 and higher are radioactive. Some of the lower elements also have radioisotopes. The symbolic notation used to distinguish isotopes (Section 2.2) is also used to distinguish radioisotopes. For example, the most abundant isotope of gallium is $^{69}_{31}Ga$, which is not radioactive. The radioisotope $^{67}_{31}Ga$ is used in medicine to image tumor cells.

Types of Radiation

The first three types of nuclear radiation that were discovered are the positively charged alpha (α) particle, the negatively charged beta (β) particle, and the neutral gamma (γ) ray. There are two other less common forms; the positron (+ charge) and the subatomic particle the neutron. The properties of these five types are summarized in Table 2.6.

An **alpha particle** is represented by the Greek letter α or as $^{4}_{2}He$, a helium nucleus containing two protons and two neutrons. Because it is a helium atom without electrons, it carries a 2+ charge from its two protons.

In some atoms with an unstable nucleus, one of the neutrons may eject a high-energy electron. This electron is emitted as a radioactive beta particle. The neutron that emits the beta particle becomes a proton as a result. A **beta particle** is a high-energy electron and carries a charge of 1−. It is represented by the Greek letter β, or symbolically as $^{0}_{-1}e$ since e⁻ is the symbol for an electron.

▲ **FIGURE 2.6 Chest image taken using a radioactive element.** Radioactive elements like xenon-133 can be used to image lung tissue. Notice the large dark spot on the right side indicating a tumor in this patient who was diagnosed with lung cancer.

TABLE 2.6	TYPES AND PROPERTIES OF NUCLEAR RADIATION	
Emission	**Symbol**	**Charge**
Alpha	α or $^{4}_{2}He$	2+
Beta	β or $^{0}_{-1}e$	1−
Gamma	$^{0}_{0}\gamma$	0
Positron	$^{0}_{1}e$	1+
Neutron	$^{1}_{0}n$	0

Gamma rays are not high-energy particles, but simply high-energy radiation. Gamma rays are released when an unstable nucleus rearranges to a more stable state. Neither the charge nor the mass of the isotope will change, so the gamma ray is simply represented by the Greek letter γ. Gamma rays are very similar to X-rays, which are also high-energy radiation. X-rays are not a form of nuclear radiation; however, they are included in this section because they are commonly used in the clinical setting and share many properties with gamma rays.

BIOLOGICAL EFFECTS OF RADIATION

Why is radiation dangerous to living organisms? Radioactive emissions like the ones just described contain a lot of energy, so when emitted, they will interact with any atoms they come into contact with, whether in a living organism or not. Alpha and beta particles, neutrons, and gamma rays and X-rays are also known as **ionizing radiation** because when they interact with another atom they have the effect of bumping off one of that atom's electrons (remember, electrons are on the outside of the atom, far from the nucleus), making the atom more reactive and less stable. The loss of too many electrons from too many atoms over long periods of time in living cells can affect a cell's chemistry and genetic material. In humans, this can cause a number of health problems, the most common of which is cancer. Not all ionizing radiation has the same amount of energy. Radiation of higher energy can penetrate farther into a tissue, affecting cells located deeper in the body. The penetrating power for different forms of ionizing radiation is shown in Figure 2.7. Persons who work with radioactive materials—for example, radiologists—must take special precautions to protect themselves from radiation exposure by wearing a heavy lab coat, lab glasses, gloves, and, depending on the type of radiation with which they are working, they may have to stand behind a plastic or lead shield (Table 2.7). People who routinely work with radioactive materials or X-rays usually wear a film badge to monitor their total exposure to radiation over a given time period (see Figure 2.8).

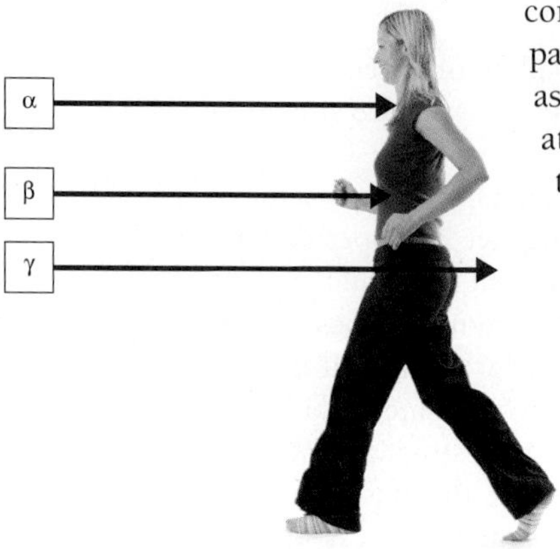

▲ **FIGURE 2.7 Pentration of radiation.** Different types of ionizing radiation have different energies and are therefore able to penetrate different materials to different extents.

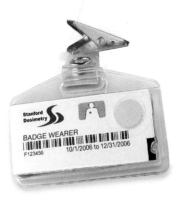

▲ **FIGURE 2.8 Film badge.** A film badge can be used to monitor exposure to nuclear radiation or X-rays over a given time period.

TABLE 2.7	PROPERTIES OF COMMON IONIZING RADIATION		
	Travel Distance Through Air	Tissue Penetration	Protective Shielding
Alpha (α)	A few centimeters	Stops at the skin surface; only dangerous if inhaled or eaten	Paper, clothing
Beta (β)	A few meters	Will not penetrate past skin layer	Heavy clothing, plastic, aluminum foil, gloves
Gamma (γ)	Several hundred meters	Fully penetrates body	Thick lead, concrete, layer of water
X-ray	Several meters	Penetrates tissues, but not bone	Lead apron, concrete barrier

SAMPLE PROBLEM 2.11

Properties of Radiation

Based on its penetrating power into tissue, which form of ionizing radiation is the most dangerous?

SOLUTION

An examination of Table 2.7 indicates that gamma radiation is the most penetrating. It will pass through a person's whole body.

PRACTICE PROBLEMS

2.25 If the symbol for an alpha particle is ^4_2He, explain how an alpha particle is different from a helium atom.

2.26 What is the charge on a beta particle? An alpha particle? A gamma ray?

2.27 Which type of nuclear radiation has the greatest penetrating power?

2.28 Which type of nuclear radiation is most similar to X-ray radiation? Why?

2.29 Complete the following table:

Isotope Name	Symbolic Notation	Number of Protons	Number of Neutrons	Mass Number	Medical Use
Thallium-201					Tumor imaging
	$^{123}_{53}I$	53			Thyroid imaging
		54		133	Pulmonary ventilation imaging
Fluorine-18					Positron emission tomography (PET) imaging

2.7 Nuclear Equations and Radioactive Decay

Uranium-238 is a radioactive isotope that emits an alpha particle when it undergoes radioactive decay. We can represent this process with a special type of chemical equation, called a nuclear decay equation:

$$^{238}_{92}U \longrightarrow ^{234}_{90}Th + ^4_2He$$

Reactant   Products

Notice that the mass number of the reactant (238 for uranium) is equal to the total mass number of the products (234 + 4: thorium and an alpha particle) and the atomic number of the reactants (92) is equal to the sum of the atomic numbers of the products (90 + 2). The symbol changed from uranium to thorium because the number of protons present changed (92 to 90). Atoms with different numbers of protons are different elements.

The general form for a **nuclear decay equation** can be described as follows:

Radioactive nucleus undergoing decay $\longrightarrow$ *new nucleus formed + radiation emitted*

Remember, the number of protons and neutrons on both sides of the nuclear decay equation will be the same whether they are all in the same atom, or divided up due to the decay. Let's practice writing nuclear decay equations with an example for each of the three most common forms of ionizing radiation, α, β, and γ.

Alpha Decay

Try to fill in the missing isotope in the following nuclear decay equation representing an alpha decay:

$$^{224}_{88}\text{Ra} \longrightarrow \boxed{} + {}^{4}_{2}\text{He}$$

Remember that the mass and atomic numbers must be equal on both sides of the equation. If radium (Ra) has a mass number of 224 and it loses 4 particles in an alpha decay, the resulting atom will have a mass number of 220 (224 minus 4). During alpha decay 2 protons are lost, so the resulting atom will have two fewer protons than Ra, or 86 (88 minus 2). The new symbol for the element will be the element that has an atomic number of 86, radon. The symbol of the resulting atom is $^{220}_{86}\text{Rn}$

$$^{224}_{88}\text{Ra} \longrightarrow \boxed{^{220}_{86}\text{Rn}} + {}^{4}_{2}\text{He}$$

Beta Decay

In beta decay, a high-energy electron is emitted from the nucleus and in the process, the neutron becomes a proton. The total mass of the isotope does not change. Try to determine the missing isotope for the following beta decay equation:

$$^{56}_{25}\text{Mn} \longrightarrow \boxed{} + {}^{0}_{-1}\text{e}$$

Remember, the total mass (56) does not change, but one more proton is present. The number of protons increases on the right side by one so the resulting isotope has a total of 26 protons. The symbol for the element that has 26 protons is Fe, iron. The symbol of the resulting atom is $^{56}_{26}\text{Fe}$.

$$^{56}_{25}\text{Mn} \longrightarrow \boxed{^{56}_{26}\text{Fe}} + {}^{0}_{-1}\text{e}$$

Gamma Decay

Since gamma rays are only energy, an isotope that is a pure gamma emitter will not change its atomic number or mass number upon decay. Radioactive technetium (atomic symbol Tc) is one of the few pure gamma emitters. It is used extensively in medical imaging since most (~96%) of the radioactivity decays from the metastable (semi-stable) state containing the gamma ray within 24 hours of injection. Short decay times mean that there is less radioactivity in the patient for a shorter time period, and this is a desirable feature for medical radioisotopes. This radioactive isotope is designated in its metastable state with the symbol m, technetium-99m ($^{99m}_{43}\text{Tc}$). By emitting a gamma ray, the technetium becomes more stable. The equation for such a gamma decay is

$$^{99m}_{43}\text{Tc} \longrightarrow {}^{99}_{43}\text{Tc} + {}^{0}_{0}\gamma$$

SAMPLE PROBLEM 2.12

Writing Nuclear Decay Equations

Cobalt-60 is used in cancer therapy. Provide a nuclear decay equation for the beta decay of cobalt-60.

SOLUTION

Set up a nuclear decay equation with cobalt-60 on the reactant side and a beta particle on the product side.

$$^{60}_{27}\text{Co} \longrightarrow ? + {}^{0}_{-1}\text{e}$$

The mass number of the product isotope is still 60 since release of a beta particle does not change the overall mass number. The atomic number (number of protons) increases to 28 since one neutron has become a proton. Element number 28 is nickel. Therefore the complete equation is

$$^{60}_{27}\text{Co} \longrightarrow {}^{60}_{28}\text{Ni} + {}^{0}_{-1}\text{e}$$

Producing Radioactive Isotopes

Although some radioisotopes occur in nature, many more are prepared in chemical laboratories. Radioisotopes can be prepared by bombarding stable isotopes with fast-moving alpha particles, protons, or neutrons. The source for technetium-99m is the radioisotope molybdenum-99, which is produced in a nuclear reactor by bombarding the stable isotope molybdenum-98 with neutrons.

$$\ce{^{98}_{42}Mo + ^{1}_{0}n \longrightarrow ^{99}_{42}Mo}$$

Here, the particle appears on the reactants side of the equation since a radioisotope is being produced. The nuclear equation is still balanced (the mass numbers and the atomic numbers on both sides of the equation are equal).

Since technetium-99m is used so much in nuclear medicine, many labs keep a supply of molybdenum-99 on hand, which beta decays to give the technetium-99m.

$$\ce{^{99}_{42}Mo \longrightarrow ^{99m}_{43}Tc + ^{0}_{-1}e}$$

SAMPLE PROBLEM 2.13

Writing Nuclear Decay Equations

Cobalt-60 is produced by bombarding naturally occurring cobalt-59 with neutrons. Provide the nuclear equation for this reaction.

SOLUTION

The symbol for a neutron is $\ce{^{1}_{0}n}$ (Table 2.4). A neutron is added to cobalt-59 and cobalt-60 is produced:

$$\ce{^{59}_{27}Co + ^{1}_{0}n \longrightarrow ^{60}_{27}Co}$$

PRACTICE PROBLEMS

2.30 Provide a balanced nuclear equation for the decay of each of the following:

a. thorium-232 undergoing alpha decay

b. strontium-92 undergoing beta decay

c. aluminum-28 undergoing beta decay

d. californium-251 undergoing alpha decay

2.31 Provide a balanced nuclear equation for the decay of each of the following:

a. carbon-14 undergoing beta decay

b. polonium-212 undergoing alpha decay

c. copper-66 undergoing beta decay

d. americium-241 undergoing alpha decay

2.8 Radiation Units and Half-Lives

Radioactivity Units

The activity of a radioactive sample is measured as the number of radioactive emissions or particles, often called disintegrations, that occur in a second. For example, the loss of one alpha particle from a nucleus is one disintegration. The unit for measuring disintegrations is called the **curie (Ci)**. The curie was named for the Polish scientist Marie Curie, who was very active in the study of radioactivity at the turn of the twentieth century. The *activity* of a radioactive isotope defines how quickly (or slowly) it emits radiation. A curie is a unit of activity equal to 3.7×10^{10} disintegrations per second.

TABLE 2.8	UNITS FOR RADIATION ACTIVITY
Common Unit	**Relationship to Other Units**
curie (Ci)	1 Ci = 3.7×10^{10} disintegrations per second
millicurie (mCi)	1 Ci = 1000 mCi
microcurie (μCi)	1 Ci = 1,000,000 μCi

One curie is a dangerously high level of radiation, so a fraction of a curie like a millicurie (mCi—one-thousandth of a curie) or a microcurie (μCi—one-millionth of a curie) is often used in medical applications (see Table 2.8).

Half-Life

Every radioactive isotope emits its radiation at a different rate. In other words, some isotopes are more unstable than others and emit radiation more rapidly. This can be measured as the **half-life**, which is the time it takes for one-half (50%) of the atoms in a radioactive sample to decay (emit radiation). Those radioisotopes that are more unstable than others emit radiation more rapidly and have shorter half-lives.

Naturally occurring radioisotopes tend to have long half-lives, disintegrating slowly over a number of years. Radioisotopes used in medicine tend to have much shorter half-lives, decaying rapidly (in hours or days); this allows radioactivity to be quickly eliminated from the body. The half-lives of several radioactive elements are shown in Table 2.9.

TABLE 2.9	HALF-LIVES OF SOME RADIOISOTOPES	
Radioisotope	**Symbol**	**Half-Life**
Naturally occurring radioisotopes		
Hydrogen-3 (tritium)	^{3}H	12.3 years
Carbon-14	^{14}C	5730 years
Radium-226	^{226}Ra	1600 years
Uranium-238	^{238}U	4.5 billion years
Radioisotopes used in medicine		
Chromium-51	^{51}Cr	28 days
Iron-59	^{59}Fe	46 days
Phosphorus-32	^{32}P	14.3 days
Technitium-99m	^{99m}Tc	6 hours
Iodine-123	^{123}I	13.3 hours
Iodine-131	^{131}I	8 days

Knowing the half-life allows us to determine how much radioactivity is left after a given amount of time has passed. Let's look at an example.

Radioactive iodine-131 has a half-life of 8 days. If a dose with an activity of 200 μCi is given to a patient today, about how much of the radioactivity will still remain active after 32 days?

First, ask yourself, how many half-lives are in 32 days? Since one half-life is 8 days, there are 4 half-lives in 32 days. Mathematically, you could arrive at this answer this way:

$$32 \text{ days} \times \frac{1 \text{ half-life}}{8 \text{ days}} = 4 \text{ half-lives}$$

Then, how much of the dose will remain active after 4 half-lives? During each half-life, half of what remained at the end of the last half-life is lost.

This problem can be solved by diagramming the solution as follows:

$$200\mu\text{Ci} \xrightarrow[\text{1 half-life}]{\text{8 days}} 100\mu\text{Ci} \xrightarrow[\text{2 half-lives}]{\text{16 days}} 50\mu\text{Ci} \xrightarrow[\text{3 half-lives}]{\text{24 days}} 25\mu\text{Ci} \xrightarrow[\text{4 half-lives}]{\text{32 days}} 12.5\mu\text{Ci}$$

Half-Life

Iron-59 has a half-life of 46 days. If 96.0 g of a radioactive iron (^{59}Fe) is received in the lab today, how many grams of the iron will still be radioactive after 138 days?

SOLUTION

Begin by determining how many half-lives will be in 138 days:

$$138 \text{ days} \times \frac{1 \text{ half-life}}{46 \text{ days}} = 3 \text{ half-lives}$$

Then, diagram the solution using the number of half-lives you determined in the first step:

$$96 \text{ g} \xrightarrow[\substack{\text{1 half-life}}]{\substack{46 \text{ days}}} 48 \text{ g} \xrightarrow[\substack{\text{2 half-lives}}]{\substack{92 \text{ days}}} 24 \text{ g} \xrightarrow[\substack{\text{3 half-lives}}]{\substack{138 \text{ days}}} 12 \text{ g}$$

2.32 Define half-life in your own words.

2.33 How do the radioisotopes used in medical imaging differ from naturally occurring radioisotopes?

2.34 Technetium-99m is very useful in diagnostic imaging since it has a short half-life of 6 hours. If a patient receives a dose with an activity of 25 mCi of technetium-99m for cardiac imaging, how much radioactivity will be left in the patient's body 48 hours after injection?

2.35 Radioactive iodine-131 has a half-life of 8 days. If a dose with an activity of exactly 400 μCi of ^{131}I is given to a patient on October 1, how much of the ^{131}I will still be active on November 1 (32 days later)?

2.9 Medical Applications for Radioisotopes

Because it is important to expose patients to the smallest possible dose of radiation for the shortest time period, radioisotopes with short half-lives are selected for use in nuclear medicine. Specific radioisotopes can provide images of specific body tissues. For example, iodine is used in the body only by the thyroid gland so any iodine—radioactive or nonradioactive—put into the body will accumulate in the thyroid (see Figure 2.9). Radioisotopes used in medicine are prepared in a laboratory (unlike naturally occurring isotopes). A number of radioisotopes and their medical uses were shown earlier in Table 2.5.

The two main uses of medical radioisotopes are (1) in diagnosing diseased states and (2) in therapeutically treating diseased tissues.

When diagnosing a diseased state, a minimum amount of radioisotope is administered because the isotope is used for detection only and thus should have minimal or no effect on body tissue. A radioisotope used in this way is called a **tracer**. As an example, iodine-123 can be used to diagnose proper thyroid function. It is a gamma emitter and has a half-life of 13 hours. Gamma emitters are useful for diagnosis since gamma radiation is highly penetrating and thus can more easily exit the body. When injected, the iodine-123 will concentrate in the thyroid. The ionizing radiation (gamma rays) emitted from the iodine-123 is detected by a scanner that produces an image of the thyroid (see Figure 2.9a). If the thyroid is functioning normally, the radioisotope will be evenly distributed throughout

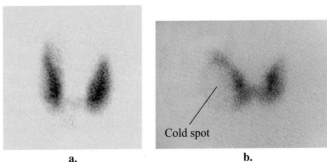

a.   b.

▲ **FIGURE 2.9 The thyroid gland has a butterfly shape.** Nuclear radiation scans of (a) a normal thyroid and (b) a thyroid gland with a cold spot on the left lobe. Can you see a difference?

the organ. If there is a nonfunctioning area in the thyroid, the iodine will not be distributed in this area and it shows up as a "cold" spot (absence of tracer) on the scan (see Figure 2.9b). Unusually high areas of activity, like an area with rapidly dividing cancer cells, would show up as a "hot" spot (more tracer than normal).

Radioisotopes and Cancer Treatment

Radioisotopes can be used in therapy to destroy diseased and cancerous tissues. In the case of thyroid cancer, iodine-131, a beta emitter, is administered in a dose approximately one thousand times higher than that of the iodine-123 tracer used to detect the disease. The radioactive iodine will be absorbed only by the thyroid and the beta emissions will destroy cells in that specific location. Cells that are rapidly dividing (cancer cells) are more susceptible to ionizing radiation damage because their genetic material is exposed more often during cell division. These cells are destroyed at a higher rate.

The emissions from radioisotopes can also be used therapeutically without actually injecting the patient with the radioisotope. For example, the radioisotope cobalt-60 is used in this way in the treatment of some cancers. A beam of gamma radiation that is emitted from cobalt-60 can be aimed directly at a tumor area to destroy the tissue. The radiation is applied externally (no injection) and affects only the diseased area.

SAMPLE PROBLEM 2.15

Radioisotopes in Medicine

A patient suspected of having thyroid cancer can be diagnosed and, if necessary, treated with radioactive iodine. Compare the dose and radioisotope that would be given for (a) diagnosis and (b) treatment?

SOLUTION

a. For diagnosis, a patient would be receiving only a low dose or trace amount (typically in the µCi range) of radioactive iodine—preferably a gamma emitter since this radiation travels farther—to capture a scan of the thyroid gland. Iodine-123, a gamma emitter, works well for this purpose.

b. For treating cancer, a higher dose (typically in the mCi range) and less penetrating radioisotope is more suitable. Iodine-131, a beta emitter, works well for treatment.

SAMPLE PROBLEM 2.16

Medical Isotopes and Half-Life

Gold-198 is a beta emitter used in the treatment of leukemia. It has a half-life of 2.7 days. How long would it take for a dose with an activity of 96 mCi to decay to an activity of 3.0 mCi?

SOLUTION

This problem is most easily done by diagramming the problem to determine the number of half-lives that will have passed. After the passing of each half-life, one-half of the previous amount is still present.

$$96 \text{ mCi} \xrightarrow[\text{1 half-life}]{2.7 \text{ days}} 48 \text{ mCi} \xrightarrow[\text{2 half-lives}]{5.4 \text{ days}} 24 \text{ mCi} \xrightarrow[\text{3 half-lives}]{8.1 \text{ days}} 12 \text{ mCi} \xrightarrow[\text{4 half-lives}]{10.8 \text{ days}} 6.0 \text{ mCi} \xrightarrow[\text{5 half-lives}]{13.5 \text{ days}} 3.0 \text{ mCi}$$

Therefore, about 14 days.

PRACTICE PROBLEMS

2.36 What is a diagnostic tracer? What are the preferable characteristics?

2.37 Because calcium is an element in bone, why do you think it might be useful to use the radioisotope calcium-47 in the diagnosis and treatment of bone diseases?

2.38 Iodine-123 is used for thyroid imaging and has a half-life of 13.3 hours. How much would be left after 40 hours (3 half-lives) if a dose with an activity of 300 µCi was initially administered?

2.39 Chromium-31 is used in imaging red blood cells and has a half-life of 28 days. If a dose with an activity of 50 µCi is given to a patient, how long will it take for the patient to have less than 5 µCi (one-tenth the original dose) present?

SUMMARY

2.1 Atoms and Their Components

An atom consists of three subatomic particles: protons, neutrons, and electrons. Protons have a positive charge, neutrons have no charge, and electrons have a negative charge. Most of the mass of an atom comes from the protons and neutrons located in the center, or nucleus, of an atom. The unit for the mass of an atom is the atomic mass unit (amu); each proton and neutron present in an atom weighs approximately 1 amu.

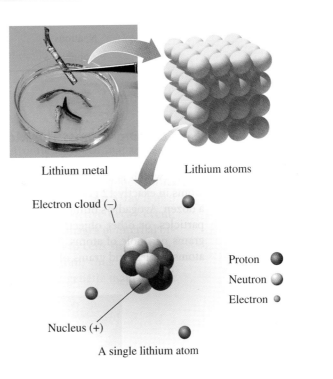

Lithium metal   Lithium atoms

Electron cloud (–)

Proton
Neutron
Electron

Nucleus (+)

A single lithium atom

2.2 Atomic Number and Mass Number

The atomic number of an atom defines the number of protons present in an atom. All atoms of a given element have the same number of protons. The mass number of an atom is the total number of protons and neutrons present in a given atom of an element.

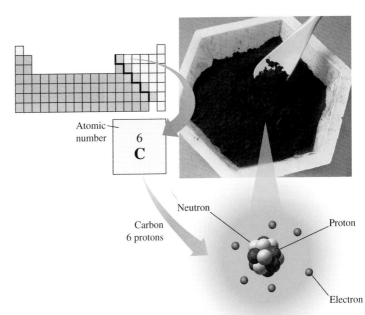

Atomic number

6
C

Carbon
6 protons

Neutron
Proton
Electron

2.7 Nuclear Equations and Radioactive Decay

Radioactive decay of a radioisotope can be represented symbolically in the form of a nuclear decay equation. The number of protons and the mass number found in the reactant (the decaying radioisotope) is equal to the sum of the number of protons and the mass numbers found in the products (the stable isotope and the radioactive particle).

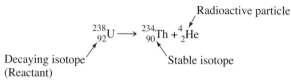

Radioactive particle

$$^{238}_{92}\text{U} \longrightarrow {}^{234}_{90}\text{Th} + {}^{4}_{2}\text{He}$$

Decaying isotope
(Reactant)

Stable isotope

2.8 Radiation Units and Half-Lives

Radioactive decay is measured as the number of decay events, or disintegrations, that occur in 1 second. The standard unit for measuring radioactive decay in a radioisotope is the curie (Ci). For smaller quantities used in medical applications, the microcurie (μCi) is often used. A curie is defined as 3.7×10^{10} disintegrations per second. The half-life of a radioisotope is the amount of time it takes for one-half of the radiation in a given sample to decay. Most radioisotopes used in medicine have short half-lives, allowing the radioactivity to be more quickly eliminated from the body.

2.9 Medical Applications for Radioisotopes

Certain elements concentrate in particular organs of the body. If a radioisotope of this element can be made, this area of the body can be imaged using that radioisotope. A patient can be injected with a trace amount of a radioisotope to diagnose a diseased state. Radioisotopes can also be applied directly to tumors in high doses to eliminate the cancerous cells.

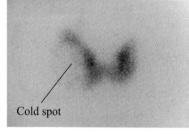

Cold spot

KEY TERMS

alpha particle—A form of nuclear radiation consisting of two protons and two neutrons (a helium nucleus).

atomic mass—The weighted average mass of all the naturally occurring isotopes of an atom.

atomic mass unit (amu)—The small unit of mass used to quantify the mass of very small particles; 1 amu is equal to one-twelfth of a carbon atom containing six protons and six neutrons.

atomic number—The number above the symbol of the elements on the periodic table. It is equal to the number of protons in the element.

Avogadro's number—The number of particles present in a mole of a substance, 6.02×10^{23}.

beta particle—A form of nuclear radiation where an unstable neutron emits a high-energy electron, the beta particle, changing the neutron into a proton.

curie (Ci)—A unit of radiation equal to 3.7×10^{10} disintegrations per second.

electron—A negatively charged subatomic particle, symbol e^-.

electron cloud—The area of an atom outside of the nucleus where the electrons can be found.

equivalent units—Two quantities that can be related to each other with an equal sign.

gamma ray—High-energy nuclear radiation emitted to stabilize a radioactive nucleus.

half-life—The time it takes for one-half of the atoms in a radioactive sample to decay.

ionizing radiation—A general name for high-energy radiation of any kind (for example, nuclear, X-ray).

isotope—Atoms that have the same number of protons but different numbers of neutrons.

mass number—The total number of protons and neutrons in an atom.

molar mass—The mass in grams of 1 mole of a substance. For an element this is numerically equivalent to atomic mass.

mole—A conversion unit used to count small objects like atoms and molecules. The number of objects in a mole is 6.02×10^{23}.

nuclear decay equation—A chemical equation where the reactant is a radioactive isotope and the products are a radioactive particle and a second radioisotope.

nuclear radiation—Energy emitted spontaneously from the nucleus of an atom.

neutron—A subatomic particle with no charge, symbol n.

nucleus—The central part of an atom containing protons and neutrons.

proton—A positively charged subatomic particle, symbol p.

radioactive—Elements that spontaneously emit nuclear radiation.

radioactive decay—The process of a nucleus spontaneously emitting radiation.

radioisotope—A radioactive isotope of an element.

subatomic particles—The small parts that organize to form atoms: protons, neutrons, and electrons.

tracer—A minimum amount of a radioactive substance used for detection purposes only.

valence electrons—Electrons found in the highest occupied energy level.

valence shell—The highest occupied energy level of an atom containing electrons.

ADDITIONAL PROBLEMS

2.40 Complete the following statements:
 a. The mass number is the sum of the number of _____ and _____.
 b. Most of the mass of an atom is found in the _____.
 c. In an atom, the number of _____ equals the number of _____.

2.41 Complete the following statements:
 a. A _____ is a subatomic particle that has a neutral charge.
 b. The atomic number on the periodic table equals the number of _____ in the atom.
 c. Two atoms that have the same number of protons and electrons but a different number of neutrons are called _____.

2.42 Provide the number of protons and electrons in the following atoms:
 a. chlorine   **b.** element number 11   **c.** S

2.43 Provide the number of protons and electrons in the following atoms:
 a. gold   **b.** Zn   **c.** element number 29

2.44 Determine the number of protons, neutrons and electrons in the following atoms:
 a. $^{39}_{19}K$ **b.** $^{16}_{8}O$ **c.** $^{81}_{35}Br$ **d.** $^{4}_{2}He$

2.45 Determine the number of protons, neutrons and electrons in the following atoms:
 a. $^{55}_{26}Fe$ **b.** $^{15}_{7}N$ **c.** $^{52}_{24}Cr$ **d.** $^{137}_{56}Ba$

2.46 Complete the following table:

Symbol	Number of Protons	Number of Neutrons	Number of Electrons	Mass Number	Name
$^{14}_{6}C$				14	
	13	14	13		
$^{19}_{9}F$			9		

2.47 Complete the following table:

Symbol	Number of Protons	Number of Neutrons	Number of Electrons	Mass Number	Name
$^{1}_{1}H$					
	12	12	12		
$^{9}_{4}Be$		5			

2.48 What are the units on Avogadro's number?

2.49 What is the mass of 4.00 moles of each of the following elements?
 a. H   **b.** Li   **c.** Na   **d.** Cl

2.50 How many atoms are in 5 moles of each of the following elements?
 a. O   **b.** N   **c.** F   **d.** Ne

2.51 How many atoms are present in a 1.00-kg block of gold?

2.52 A pencil mark (made with graphite carbon) on paper has a mass of about 0.000005 g (5×10^{-6} g). How many atoms of carbon graphite are present in the pencil mark?

2.53 Provide scientific notation for the following numbers:
 a. 203,000,000   **b.** 12.4   **c.** 0.0000000278

2.54 Provide scientific notation for the following numbers:
 a. 1,400,000   **b.** 0.079   **c.** 0.00000354

2.55 Express the following numbers in a decimal format:
 a. 1.56×10^{-5} **b.** 2.8×10^{5} **c.** 9.0×10^{-2}

2.56 Express the following numbers in a decimal format:
 a. 7.4×10^{-2} **b.** 3.75×10^{3} **c.** 1.19×10^{-8}

2.57 What is the number of electrons in the outer energy level and the group number for each of the following elements?

Example: Fluorine—seven electrons, Group 7A
 a. magnesium   **b.** chlorine   **c.** oxygen
 d. nitrogen   **e.** barium   **f.** bromine

2.58 What is the number of electrons in the outer energy level and the group number for each of the following elements?

Example: Fluorine—seven electrons, Group 7A
   a. lithium   **b.** silicon   **c.** neon
   d. argon   **e.** tin   **f.** cesium

2.59 How many valence electrons are present in the following atoms?
   a. C   **b.** I   **c.** K   **d.** B

2.60 How many valence electrons are present in the following atoms?
   a. N   **b.** Mg   **c.** S   **d.** Ne

2.61 Identify the most penetrating radiation associated with the following:
   a. particle that has the same number of protons and neutrons as a helium nucleus
   b. an electron that is emitted from an atom's nucleus
   c. has no mass

2.62 Identify the most penetrating radiation associated with the following:
   a. can be stopped by paper
   b. the highest energy (most penetrating) emission
   c. can be stopped by plastic or aluminum foil

2.63 Provide the symbolic notation for the following radioactive isotopes:
   a. phosphorus-32   **b.** cobalt with 33 neutrons
   c. element number 24 containing 27 neutrons.

2.64 Provide the symbolic notation for the following radioactive isotopes:
   a. uranium-238   **b.** xenon with 89 neutrons
   c. element number 53 containing 78 neutrons.

2.65 Complete the following nuclear equations:
   a. $^{15}_{8}O \longrightarrow ^{15}_{9}F + ?$ **b.** $^{46}_{23}V \longrightarrow ? + ^{0}_{-1}e$
   c. $^{234}_{92}U \longrightarrow ? + ^{4}_{2}He$ **d.** $^{8}_{4}Be \longrightarrow ? + ^{4}_{2}He$

2.66 Complete the following nuclear equations:
   a. $^{214}_{82}Pb \longrightarrow ? + ^{0}_{-1}e$ **b.** $^{18}_{8}O + ^{0}_{-1}e \longrightarrow ? + ^{1}_{0}n$
   c. $^{218}_{84}Po \longrightarrow ^{214}_{82}Pb + ?$ **d.** $^{60}_{27}Co \longrightarrow ? + ^{0}_{0}\gamma$

2.67 Write balanced nuclear decay equations for each of the following emitters:
   a. copper-66 (β)   **b.** platinum-192 (α)
   c. tin-126 (β)   **d.** gallium-72 produces germanium-72

2.68 Write balanced nuclear decay equations for each of the following emitters:
   a. thorium-225 (α)   **b.** bismuth-210 (α)
   c. cesium-137 (β)   **d.** sulfur-35 (β)

2.69 When boron-10 is bombarded with alpha particles, nitrogen-12 and neutrons are produced. Provide a nuclear equation for this reaction.

2.70 A 120-mg sample of technetium-99m is used for a diagnostic test. If technetium-99m has a half-life of 6.0 hours, how much of the technetium-99m remains 24 hours after the test?

2.71 Fluorine-18, which has a half-life of 110 minutes, is used in tomography scans. If exactly 100 mg of fluorine-18 is shipped at 8:00 A.M, how many milligrams of the radioisotope are still active if the sample arrives at the nuclear medicine laboratory at 1:30 P.M the same day?

2.72 If the amount of radioactive phosphorus-32 in a sample decreases from 1.2 g to 0.30 g in 28 days, what is the half-life of phosphorus-32?

2.73 If the amount of radioactive iodine-123 in a sample decreases from 0.400 g to 0.100 g in 26.2 hours, what is the half-life of iodine-123?

2.74 What is the importance of a cold spot when using a radioactive tracer? A hot spot?

CHALLENGE PROBLEMS

2.75 Precious metals are commonly measured out in troy ounces. One troy ounce of platinum is equivalent to 31.10 g of platinum. How many atoms of platinum are in 1.00 troy ounce of platinum?

2.76 On an archaeological dig, a wooden canoe is unearthed and analyzed for carbon-14. About 25% of the carbon-14 initially present remains. What is the approximate age of the canoe? The half-life of carbon-14 is 5730 years.

2.77 Iron-59 has a half-life of 46 days. If 168 g of a radioactive iron (^{59}Fe) is received in the lab today, what percentage of the original is left after 276 days?

ANSWERS TO ODD-NUMBERED PROBLEMS

Practice Problems

2.1 **a.** The number of protons is the atomic number.
   b. The number of neutrons is the mass number minus the atomic number.
   c. The number of electrons is the same as the number of protons in an atom.

2.3 **a.** oxygen, O   **b.** magnesium, Mg   **c.** neon, Ne
   d. copper, Cu   **e.** silver, Ag

2.5 14

2.7 **a.** 35 protons, 45 neutrons, 35 electrons
   b. 11 protons, 12 neutrons, 11 electrons

2.9 **a.** ^4_2He **b.** $^{35}_{17}\text{Cl}$ **c.** $^{32}_{16}\text{S}$ **d.** $^{133}_{55}\text{Cs}$

2.11 **a.** 8 protons, 10 neutrons, 8 electrons
    b. 20 protons, 20 neutrons, 20 electrons
    c. 47 protons, 61 neutrons, 47 electrons
    d. 82 protons, 125 neutrons, 82 electrons

2.13 **a.** Magnesium-24 has 12 neutrons, magnesium-25 has 13 neutrons, magnesium-26 has 14 neutrons.
    b. $^{24}_{12}\text{Mg}$, $^{25}_{12}\text{Mg}$, $^{26}_{12}\text{Mg}$ **c.** magnesium-24

2.15 **a.** They are equal; 6.02×10^{23} atoms.
    b. Gold weighs more, 197.0g versus 107.9g.

2.17 **a.** 1.51×10^{23} atoms   **b.** 0.019 mole
    c. 2.57×10^{23} atoms

2.19 two electrons, eight electrons

2.21 **a.** 2 e^- in first energy level
    b. 2 e^- in first energy level, 4 e^- in second energy level
    c. 2 e^- in first energy level, 8 e^- in second energy level, 1 e^- in third energy level
    d. 2 e^- in first energy level, 8 e^- in second energy level

2.23 **a.** 6   **b.** 4   **c.** 5   **d.** 1

2.25 An alpha particle is a helium nucleus. There are no electrons in an alpha particle.

2.27 gamma

2.29

Isotope Name	Symbolic Notation	Number of Protons	Number of Neutrons	Mass Number	Medical Use
Thallium-201	$^{201}_{81}\text{Tl}$	81	120	201	Tumor imaging
Iodine-123	$^{123}_{53}\text{I}$	53	70	123	Thyroid imaging
Xenon-133	$^{133}_{54}\text{Xe}$	54	79	133	Pulmonary ventilation imaging
Fluorine-18	$^{18}_{9}\text{F}$	9	9	18	Positron emission tomography (PET) imaging

2.31 **a.** $^{14}_{6}\text{C} \longrightarrow ^{14}_{7}\text{N} + ^{0}_{-1}\text{e}$ **b.** $^{212}_{84}\text{Po} \longrightarrow ^{208}_{82}\text{Pb} + ^{4}_{2}\text{He}$
    c. $^{66}_{29}\text{Cu} \longrightarrow ^{66}_{30}\text{Zn} + ^{0}_{-1}\text{e}$ **d.** $^{241}_{95}\text{Am} \longrightarrow ^{237}_{93}\text{Np} + ^{4}_{2}\text{He}$

2.33 The half-lives are shorter, and they are prepared in the lab.

2.35 25 μCi

2.37 Ca-47 will concentrate in bone.

2.39 After four half lives (112 days) the patient will have less than one-tenth the dose present.

Additional Problems

2.41 **a.** neutron   **b.** protons   **c.** isotopes

2.43 **a.** 79 protons, 79 electrons
    b. 30 protons, 30 electrons
    c. 29 protons, 29 electrons

2.45 **a.** 26 protons, 29 neutrons, 26 electrons
    b. 7 protons, 8 neutrons, 7 electrons
    c. 24 protons, 28 neutrons, 24 electrons
    d. 56 protons, 81 neutrons, 56 electrons

2.47

Symbol	Number of Protons	Number of Neutrons	Number of Electrons	Mass Number	Name
^1_1H	1	0	1	1	Hydrogen-1
$^{24}_{12}\text{Mg}$	12	12	12	24	Magnesium-12
^9_4Be	4	5	4	9	Beryllium-9

2.49 **a.** 4.03 g   **b.** 27.8 g   **c.** 92.0 g   **d.** 142 g

2.51 3.06×10^{24} atoms Au

2.53 **a.** 2.03×10^8 **b.** 1.24×10^1 **c.** 2.78×10^{-8}

2.55 **a.** 0.0000156   **b.** 280,000   **c.** 0.090

2.57 **a.** two electrons, Group 2A
    b. seven electrons, Group 7A
    c. six electrons, Group 6A
    d. five electrons, Group 5A
    e. two electrons, Group 2A
    f. seven electrons, Group 7A

2.59 **a.** 4 e^- **b.** 7 e^- **c.** 1 e^- **d.** 3 e^-

2.61 **a.** alpha particle   **b.** beta particle   **c.** gamma ray

2.63 **a.** $^{32}_{15}\text{P}$ **b.** $^{60}_{27}\text{Co}$ **c.** $^{51}_{24}\text{Cr}$

2.65 **a.** $^{15}_{8}\text{O} \longrightarrow ^{15}_{9}\text{F} + ^{0}_{-1}\text{e}$ **b.** $^{46}_{23}\text{V} \longrightarrow ^{46}_{24}\text{Cr} + ^{0}_{-1}\text{e}$
    c. $^{234}_{92}\text{U} \longrightarrow ^{230}_{90}\text{Th} + ^{4}_{2}\text{He}$ **d.** $^{8}_{4}\text{Be} \longrightarrow ^{4}_{2}\text{He} + ^{4}_{2}\text{He}$

2.67 **a.** $^{66}_{29}\text{Cu} \longrightarrow ^{66}_{30}\text{Zn} + ^{0}_{-1}\text{e}$
    b. $^{192}_{78}\text{Pt} \longrightarrow ^{188}_{76}\text{Os} + ^{4}_{2}\text{He}$
    c. $^{126}_{50}\text{Sn} \longrightarrow ^{126}_{51}\text{Sb} + ^{0}_{-1}\text{e}$
    d. $^{72}_{31}\text{Ga} \longrightarrow ^{72}_{32}\text{Ge} + ^{0}_{-1}\text{e}$

2.69 $^{10}_{5}\text{B} + ^{4}_{2}\text{He} \longrightarrow ^{12}_{7}\text{N} + 2^{1}_{0}\text{n}$

2.71 12.5 mg

2.73 13.1 hours

2.75 9.60×10^{22} atoms of Pt

2.77 1.56%

Guided Inquiry Activities **FOR CHAPTER 3**

EXERCISE 1 Ionic Compounds

Part 1. Information

Ionic compounds form between elements that can give or take electrons (exist as cations or anions) to yield a full valence shell. In an ionic compound, the cation gives its electron(s) to the anion and a strong attraction of opposites is formed between the two. This type of chemical bond is called an *ionic bond*.

Examine the following data set. Notice that none of the compounds have a net charge (+ and − charges add up to zero).

DATA SET	
Formula	**Name**
NaI	Sodium iodide
K_2S	Potassium sulfide
CaF_2	Calcium fluoride
Mg_3N_2	Magnesium nitride
$CuBr_2$	Copper(II) bromide
Fe_2O_3	Iron(III) oxide
FeO	Iron(II) oxide
AlP	Aluminum phosphide

Questions

1. In naming ionic compounds, which goes first in the name, the *metal* or the *nonmetal*? The *anion* or the *cation*?

2. a. In potassium sulfide, the charge on the potassium ion is _____ and the charge on the sulfide is _____.
   b. Why does the formula for potassium sulfide have two potassium ions and one sulfide ion?

3. Some of the names of ionic compounds have Roman numerals after the metals and others do not. Consider their position on the periodic table. What is different about these two groups of metals?

4. a. The charge on one Fe in Fe_2O_3 is _____.
   b. The charge on the Fe in FeO is _____.

5. a. Based on your answer to question 4, what does the Roman numeral found in some of the names in the data set represent?
   b. Based on your answer to question 3, when is the Roman numeral used?

6. How does the name of the nonmetal (anionic) part of the compound differ from that of its original element?

Part 2. Information

Polyatomic ions are groups of nonmetal atoms that together act like an anion or cation. When combined with a cation or anion respectively, they form ionic compounds.

Formula	Name	Formula	Name
NH_4^+	Ammonium	OH^-	Hydroxide
$C_2H_3O_2^-$	Acetate	HPO_4^{2-}	Hydrogen phosphate
HCO_3^-	Bicarbonate	NO_3^-	Nitrate
CO_3^{2-}	Carbonate	SO_4^{2-}	Sulfate

7. Complete the following table:

Formula	Name
KCl	
	Sodium sulfate
Fe(OH)$_3$	
NaHCO$_3$	
	Ammonium nitrate
	Silver acetate
CaHPO$_4$	
	Mercury(II) chloride

SECTION 3.5

EXERCISE 2 Molecular Shape

Information

Electron pairs joining atoms (whether in single, double, or triple bonds) will get as *far away from each other as possible* when forming bonds in a molecule. Electrons in bonds will also repel lone pair electrons. This is known as the valence-shell electron-pair repulsion theory, or *VSEPR*. VSEPR allows us to predict the three-dimensional shape (also called molecular geometry) that a molecule will adopt.

Lone pairs repel each other more than they repel electrons involved in bonds.

Activity

Using the toothpicks for bonds, a half toothpick for lone pairs on the central atom(s), large marshmallows for C, N, and O, and small marshmallows for H's build the five structures outlined below. You will need 24 toothpicks, 8 large marshmallows, and 13 small marshmallows. Keep in mind the bonding patterns outlined in Section 3.4.

a. Make a model of CO_2 being sure to get all bonds (toothpicks) as far away from each other as possible. What is the bond angle between OCO?

b. Make a model of H_2CCH_2 being sure to get all bonds (toothpicks) as far away from each other as possible. What is the bond angle between HCH?

c. Make a model of CH_4 being sure to get all bonds (toothpicks) as far away from each other as possible. Can you estimate the bond angle between HCH?

d. Make a model of NH_3 being sure to get all bonds and lone pairs (toothpicks) as far away from each other as possible. Is the HNH bond angle greater or less than the HCH bond angle estimated for CH_4? Why do you think that is so?

e. Make a model of H_2O, being sure to get all bonds and lone pairs (toothpicks) as far away from each other as possible. Is the HOH bond angle greater or less than the HCH bond angle estimated for CH_4?

Keep your structures for reference.

Questions

You have built the following molecular shapes with your toothpicks and marshmallows: tetrahedral, trigonal planar, linear, pyramidal, and bent.

1. Based on the names of the molecules shapes and the molecules that you built, assign each of the following molecules a shape:

        a. CO_2 _____

        b. H_2CCH_2 _____

        c. CH_4 _____

        d. NH_3 _____

        e. H_2O _____

2. Complete the following table:

Molecular Formula	Lewis Structure	Molecular Shape (geometry) [from question 1]	Number of Atoms Bonded to Carbon or Central Atom	Number of Lone Pairs on Carbon or Central Atom
CO_2				
H_2CCH_2				
CH_4				
NH_3				
H_2O				

3. Complete the following table:

Molecular Formula	Lewis Structure	Molecular Shape (geometry)	Number of Atoms Bonded to Carbon or Central Atom	Number of Lone Pairs on Carbon or Central Atom
CH_3F				
HCCH				
CH_2CHCl				
NH_4^+				
CH_3CH_3				

4. Based on the tables in questions 2 and 3, does the number of atoms attached to a central carbon atom affect its shape? List the shapes that carbon atoms can adopt and the number of atoms attached in each case.

SECTION 3.6

EXERCISE 3 Bond Polarity

Information

A *covalent bond* is formed when two atoms share electrons between them. If the two atoms are different, they will not share the electrons equally, and a *polar covalent bond* is formed. In this case, one of the atoms will have the electrons *more* of the time and be partially negatively charged (symbol δ^-) and the other atoms will have the electrons *less* of the time and be partially positively charged (symbol δ^+). How much charge is present can actually be measured as a quantity called a *dipole moment*. This is represented by an arrow with a hash mark on the tail $\longmapsto$. The head of the arrow points toward the more negative atom.

Electronegativity

Fluorine is the most electronegative element. The closer an element is to fluorine on the periodic table, the more electronegative it is. For example, oxygen is more electronegative than nitrogen which is more electronegative than carbon.

▼ **FIGURE 1A** A nonpolar covalent bond.

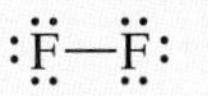

:F̈—F̈:   In a nonpolar covalent bond the electrons are shared equally.
This occurs if the atoms in the bond are identical.
One exception: C—H bonds are also nonpolar covalent.

▼ **FIGURE 2A** A polar covalent bond.

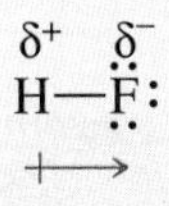

δ^+ δ^-
H—F̈:   In a polar covalent bond the electrons are not shared equally. The
more electronegative element is the negative pole of the dipole. This
⊢—→   polarity can be represented with delta symbols (above) or the dipole
moment arrow (below).

Questions

Bond Polarity

1. Are the following bonds considered polar covalent or nonpolar covalent? Explain your reasoning.
   a. C—N
   b. C—Cl
   c. O—H
   d. H—H
   e. C—H
   f. N—H

2. For the polar covalent bonds in question 1, indicate the polarity using the partial symbols (δ^+/δ^-) and the dipole moment arrow.

Molecule Polarity

3. Draw the Lewis structure for ammonia, NH_3. Show the polarity of the bonds between N and H on the structure you drew using the dipole moment arrow. What is the molecular shape (geometry) around the N in ammonia? Draw the overall dipole to the right of your structure using the dipole moment arrow with the head pointing toward the negative side of your Lewis structure and the tail pointing toward the positive side.

4. Draw the Lewis structure for H_2O. Show the polarity of the bonds on the structure you drew using the dipole moment arrow. Considering that the shape of H_2O is bent, do you think that H_2O has a dipole? (In other words, is there a side of the molecule that is more negative and a side more positive?) If so, indicate this to the right of your structure using the dipole moment arrow.

5. Draw the Lewis structure for carbon dioxide. Although CO_2 has polar bonds, it is a nonpolar molecule. Provide an explanation for this. [**HINT:** *Consider its shape.*]

6. Why do some molecules have polar bonds and yet are nonpolar molecules whereas other molecules like water and ammonia have polar bonds and are polar molecules?

7. Using a dipole moment arrow (↔) indicate the overall molecular polarity for the following molecules (you may have to draw out Lewis structures to do this). If the molecule is nonpolar, state no dipole.
   a. HBr
   b. OCS
   c. CH_3F

   d.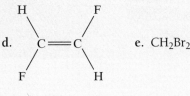
   e. CH_2Br_2

Why is the formula for water H_2O and not H_3O or H_4O? Why do most elements exist in combination with others as compounds? What is so special about the arrangement of the periodic table? Read on to explore these ideas in Chapter 3.

Compounds
PUTTING PARTICLES TOGETHER

In the previous two chapters, you were introduced to matter and its component parts—atoms. Interestingly, the majority of elements do not occur naturally by themselves or in uncombined states, with the notable exception of the members of Group 8A that appear on the far right-hand column of the periodic table and a few others. In other words, most elements do not exist as individual atoms; instead atoms tend to combine with other atoms to form compounds. For example, the formula for water is H_2O, which means that water is the unique combination of two hydrogen atoms with one oxygen atom.

What advantage does an atom gain through combining with other atoms? What is special or unique about the elements that do exist as single atoms?

As we explore why most elements exist in combination with other elements, we will see how this is related to the element's position on the periodic table. We will also see that metals and nonmetals combine differently than do nonmetals with other nonmetals.

3.1 The Octet Rule

In order to understand what "drives" most elements to form compounds, let's first consider the exceptions. All the Group 8A elements, also called the noble gases, are chemically unreactive under all but extreme conditions, meaning that the atoms of these gases are highly stable. Chemists have long understood that if an element is more stable, it is less likely to react with another element, and if an element is less stable, it is more likely to react. In other words, the behaviors of *stability and reactivity are inversely related*.

As we will see throughout this chapter and the remainder of this book, the reactivity of an atom is determined by the arrangement of the electrons, specifically the valence electrons, in its electron cloud. While we saw in Chapter 2 that nuclei can be reactive (radioactive), in this chapter we will be less concerned with the reactivity of nuclei than we are with that of the electron cloud. You could say that chemists live with their heads in the clouds!

What can we learn about the electron clouds of the noble gas elements that will explain why they are resistant to combining with other atoms? What makes these atoms so much more stable and less reactive than other atoms? In order to answer these questions, let's take a look at the valence electrons of the noble gases.

Recall that the group number of the main-group elements on the periodic table gives the number of valence electrons for any element in that group. Therefore, each of the noble gases has eight valence electrons (except for helium, which has only two valence electrons). While we do not completely understand why an electron arrangement of eight valence electrons is unusually stable, chemists have long observed this fact to be experimentally true. Most atoms in living things react with other atoms to achieve a total of eight electrons in their valence shell. This is known as the **octet rule**.

In order to achieve a valence octet, some atoms give away electrons, others will accept or take electrons, and still others share electrons with other atoms. These different modes of achieving a valence octet lead to different kinds of compounds.

3.2 In Search of an Octet Part 1: Ion Formation

Let's first take a look at atoms that gain or lose electrons in order to achieve a valence octet. In Section 2.1, we saw that an atom of a given element contains an equal number of protons and electrons, which means that an atom is electrically neutral—it has no charge. If an atom gains or loses electrons in order to achieve an octet, this newly formed particle would have an *unequal* number of protons and electrons and, therefore, would have a net charge. We refer to these charged atoms as **ions**.

As our first example, let's consider chlorine, element number 17 on the periodic table. As a member of Group 7A, a chlorine atom has seven valence electrons. In order to achieve an octet in its valence shell, the chlorine atom needs to gain one electron. By gaining that one electron, the chlorine atom would then have an extra electron in its electron cloud, which means it has more negative charges than positive charges and, therefore, a net negative charge. Ions formed when atoms gain electrons and become negatively charged are referred to as **anions**. When chlorine gains the one electron necessary to complete its octet, the newly formed ion has a net charge of 1−. As Figure 3.1 shows, adding one electron to the chlorine atom means that it now has 18 electrons, but still only 17 protons, resulting in a net charge of 1−.

▶ **FIGURE 3.1 Chlorine forming an anion.** A chlorine atom gains an electron when forming an anion.

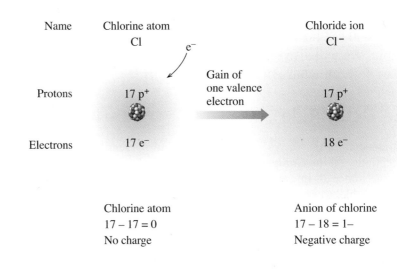

Name	Chlorine atom Cl		Chloride ion Cl⁻
Protons	17 p⁺	Gain of one valence electron	17 p⁺
Electrons	17 e⁻		18 e⁻

Chlorine atom
17 − 17 = 0
No charge

Anion of chlorine
17 − 18 = 1−
Negative charge

Notice that when chlorine gains an electron to complete the octet in its valence shell, it achieves not only the same number of valence electrons as argon but also has the same number of total electrons as argon. The ion formed by chlorine is **isoelectronic** (*iso*, a prefix meaning "same") with the argon atom. The chlorine ion is more stable than the chlorine atom because it has satisfied the octet rule.

As a second example, let's consider sodium, element number 11 on the periodic table. The sodium atom is in Group 1A and has one valence electron. In order to complete its octet, as the chlorine atom did in the previous example, the sodium atom would need to gain seven more electrons. Such a process takes more energy than the atom has available for attracting electrons and, therefore, is not favorable. However, take a look at the possibility shown in Figure 3.2. If the sodium atom gives up one electron, it becomes isoelectronic with neon. This means that the ion formed by the loss of that one electron would have a complete octet in its valence shell—exactly like neon. After giving up the electron, the ion would have only 10 electrons, but still 11 protons. Thus, it would have a net charge of 1+ (11 protons − 10 electrons = 1+). Ions formed when atoms give up electrons and become positively charged are referred to as **cations**.

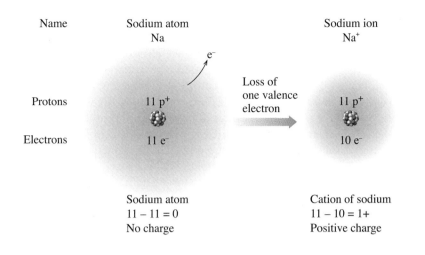

Name	Sodium atom Na		Sodium ion Na$^+$
Protons	11 p$^+$	Loss of one valence electron →	11 p$^+$
Electrons	11 e$^-$		10 e$^-$

Sodium atom
11 − 11 = 0
No charge

Cation of sodium
11 − 10 = 1+
Positive charge

Trends in Ion Formation

How can we determine which atoms will gain electrons to form anions and which atoms will give up electrons to form cations? The periodic table is a useful tool for this task.

Recall from Chapter 1 that we can use the periodic table to distinguish the elements that are nonmetals from those that are metals (see Figure 3.3). As a rule, when elements form ions, those that are metals give up electrons to form cations, while elements that are nonmetals gain electrons to form anions. So for the main-group elements, those in Groups 1A, 2A, and 3A form cations. On the other hand, the nonmetallic elements in Groups 5A, 6A, and 7A form anions. (The elements of Group 4A tend not to form ions but gain their stability in ways we will discover later in this chapter.)

We now know how to determine which atoms form anions and which atoms form cations, but as we will see, ions of different elements have different charges. The main-group metals form cations with charges from 1+ up to 3+, and the main-group nonmetals in Groups 1A–3A form anions with charges from 1− up to 3−. We need to be able to identify the charge of an ion formed by each element. Once again, the periodic table is a useful tool for helping us to do this.

Group 1A (1)	Group 2A (2)	3B (3)	4B (4)	5B (5)	6B (6)	7B (7)	8B (8)	8B (9)	8B (10)	1B (11)	2B (12)	Group 3A (13)	Group 4A (14)	Group 5A (15)	Group 6A (16)	Group 7A (17)	Group 8A (18)
H$^+$																	
Li$^+$														N^{3-}	O^{2-}	F$^-$	
Na$^+$	Mg^{2+}											Al^{3+}		P^{3-}	S^{2-}	Cl$^-$	
K$^+$	Ca^{2+}				Cr^{2+} Cr^{3+}		Fe^{2+} Fe^{3+}			Cu$^+$ Cu^{2+}	Zn^{2+}					Br$^-$	
Rb$^+$	Sr^{2+}									Ag$^+$	Cd^{2+}		Sn^{2+} Sn^{4+}			I$^-$	
Cs$^+$	Ba^{2+}									Au$^+$ Au^{3+}			Pb^{2+} Pb^{4+}				

◻ Metals　　◻ Metalloids　　◻ Nonmetals　　☐ Main-group elements

▲ **FIGURE 3.3 Trends in ion formation.** Positive ions are produced from metals and negative ions are formed from nonmetals. Common ions formed are shown.

In our previous discussion, we saw that a sodium atom gives up one electron to form a cation with a 1+ charge. In fact, the same is true of all the elements of Group 1A. Since all members of that group have one valence electron, they all behave the same way, giving up that electron and forming a cation with a charge of 1+. The elements of Group 2A all have two valence electrons. During ion formation, these elements give up those two electrons to form cations with a 2+ charge. Similarly, the metallic elements of group 3A give up three electrons to form cations with a 3+ charge. Notice that there is a pattern here. The metallic elements in Groups 1A to 3A form cations with a charge that is the same as the number of the group to which the element belongs. In each case the cation formed is isoelectronic with one of the noble gases.

Recall that the nonmetals gain electrons to form anions. Previously, we used the example of chlorine gaining one electron to complete its octet and form an anion with a 1− charge. The other members of Group 7A behave the same way, each gaining one electron to form an anion with a 1− charge. Now take a look at the elements in Group 6A. Recall that each of these elements has six valence electrons. In order for an element of Group 6A to complete its octet, it needs to gain *two* electrons. In so doing, the element forms an anion with two more electrons than the atom from which it originated and therefore has a charge of 2−. Following this trend, the nonmetallic elements of Group 5A (nitrogen and phosphorus) form anions with a charge of 3−. The charge on the anions formed by nonmetallic main-group elements is equal to the group number of the element minus eight.

The rules for predicting charges on ions formed by the main-group elements can be summarized as follows:

Predicting Charge on an Ion Formed by a Main-Group Element

Cations: Charge = Group Number;

for example, $K = 1+, Mg = 2+, Al = 3+$

Anions: Charge = Group Number − 8;

for example, $Br (7 - 8) = 1-, O (6 - 8) = 2-, N (5 - 8) = 3-$

For elements other than the main-group elements, the charge on the ions they form is not as simple to predict. In fact, the transition elements or transition metals form more than one ion. For example, both copper and iron can form two cations each. Copper forms the ions Cu^+ and Cu^{2+}, while iron forms Fe^{2+} and Fe^{3+}. Unlike the main-group metals, it is not possible to determine the charge of the ions formed by the transition metals from their group number. Later, we will show how it is possible to predict the charges on these metal ions, but for now common charges on some of the transition metal ions are shown in Figure 3.3.

Another group of common ions whose charges cannot be determined from the periodic table are the **polyatomic ions**. These are a special group of ions consisting of a group of nonmetals that interact with each other to form an ion. Examples include HCO_3^- (bicarbonate) and NH_4^+ (ammonium). Notice that each ion contains more than one type of nonmetal, but they interact to form an ion with a single charge. The charge shown is for the entire group of atoms together. See Table 3.1 for a listing of some common polyatomic ions.

Naming Ions

In order to distinguish an ion and the atom from which it was formed, we give the ion a different name. For metal ions, this simply involves adding the word *ion* to the name of the metal, so Na^+ is called sodium *ion*. For the transition metals, which typically form more than one ion, a Roman numeral in parentheses following the name of the metal indicates the charge on the ion. So, Fe^{2+} is the iron(II) ion, and the Fe^{3+} ion is the iron(III) ion.

For the nonmetals, the suffix *ide* replaces the last few letters of the name of the element. For example, the anion formed from fluorine is fluor*ide* and that of oxygen is ox*ide*. There is no definitive rule for how many letters to remove from the end of the name of a nonmetal before adding the *ide* suffix, but with some practice, you will catch on.

Most of the common polyatomic ions end in *ate* (see Table 3.1). The *ite* ending is used for the names of related ions that have one less oxygen atom. By recognizing these endings you will be able to identify when a polyatomic ion is present in a compound. The hydroxide (OH^-), hydronium (H_3O^+), and cyanide ions (CN^-) are exceptions to this naming pattern. As when learning vocabulary words, you may have to memorize the number of oxygens and charges associated with some of these ions. By memorizing the formulas in the boxes, the other related ions can be derived. For example, the sulfate ion is SO_4^{2-}. The formula for the sulfite ion with one less oxygen is therefore SO_3^{2-}.

Important Ions in the Body

A number of ions found in the fluids and cells of the human body perform important functions. The main cations found in the body include Na^+, K^+, Ca^{2+}, and Mg^{2+}, which are important in maintaining solution concentrations inside and outside cells. The main anions are Cl^-, HCO_3^-, and HPO_4^{2-}.

These ions help maintain the charge neutrality between the blood and fluids inside cells. In contrast to ions in solution, tooth enamel contains a hard mineral called hydroxyapaptite containing calcium (Ca^{2+}), phosphate (PO_4^{3-}), and hydroxide (OH^-) ions. Table 3.2 shows a listing of the main ions in bodily fluids and their functions.

TABLE 3.1	COMMON POLYATOMIC ION NAMES AND FORMULAS	
Nonmetal	**Formula of Ion[a]**	**Name of Ion**
Hydrogen	OH^-	Hydroxide
Nitrogen	NH_4^+	Ammonium
	NO_3^-	**Nitrate**
	NO_2^-	Nitrite
Chlorine	ClO_3^-	**Chlorate**
	ClO_2^-	Chlorite
Carbon	CO_3^{2-}	**Carbonate**
	HCO_3^-	Hydrogen carbonate (or bicarbonate)
	CN^-	Cyanide
	$C_2H_3O_2^-$	Acetate
Sulfur	SO_4^{2-}	**Sulfate**
	HSO_4^-	Hydrogen sulfate (or bisulfate)
	SO_3^{2-}	Sulfite
	HSO_3^-	Hydrogen sulfite (or bisulfite)
Phosphorus	PO_4^{3-}	**Phosphate**
	HPO_4^{2-}	Hydrogen phosphate
	$H_2PO_4^-$	Dihydrogen phosphate
	PO_3^{3-}	Phosphite

[a]Boxed formulas are the most common polyatomic ion for that element.

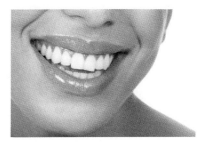

TABLE 3.2	BIOLOGICALLY IMPORTANT IONS	
Ion	**Function**	**Sources**
Cations		
Na^+	Regulates fluids outside cells	Table salt, seafood
K^+	Maintains ion concentration in cells, induces heartbeat	Dairy, bananas, meat
Ca^{2+}	Found outside cells; involved in muscle contraction, formation of bones and teeth, regulates heartbeat	Dairy, whole grains, leafy vegetables
Mg^{2+}	Found inside cells, involved in nerve impulse transmission	Nuts, seafood, leafy vegetables
Fe^{2+}	Found in the protein hemoglobin which is responsible for oxygen transport from lungs to tissue	Liver, red meat, leafy vegetables
Anions		
Cl^-	Found in gastric juice and outside cells; involved in fluid balance in cells	Table salt, seafood
HCO_3^-	Controls acid-base balance in blood	Body produces own supply through breathing and food breakdown
HPO_4^{2-}	Controls acid–base balance in cells	Fish, poultry, dairy

SAMPLE PROBLEM 3.1

Determining Protons and Electrons in Ions

How many protons and electrons are in the following ions?

a. Mg^{2+} b. O^{2-}

SOLUTION

Recall, the number of protons is the atomic number of the element as shown on the periodic table.

a. Protons = 12. An ion with a 2+ charge has two fewer electrons than protons, or in this example, 10 electrons.

b. Protons = 8. An ion with a 2− charge has two more electrons than protons, or in this example, 10 electrons.

SAMPLE PROBLEM 3.2

Naming Ions

Name the following ions.

a. Ca^{2+} b. O^{2-} c. Cr^{3+} d. NO_3^-

SOLUTION

a. Calcium is a main-group metal, so the name will be the metal name plus the word *ion*. The name is calcium ion.

b. Oxygen is a nonmetal, so the ending *ide* will replace the current ending. The name is oxide.

c. Chromium is a transition metal, so it is necessary to identify the charge using a Roman numeral after the metal name. The name is chromium(III) ion.

d. This is a polyatomic ion. It has the highest number of oxygens of any polyatomic nitrogen ion. The name is nitrate.

PRACTICE PROBLEMS

3.1 What is the difference between an atom and an ion of the same element?

3.2 How many protons and electrons are present in the following ions?

a. Al^{3+} b. Br^- c. Hg^+ d. N^{3-}

3.3 How many protons and electrons are present in the following ions?

a. Ca^{2+} b. I^- c. S^{2-} d. Zn^{2+}

3.4 Name the ions in Problem 3.2.

3.5 Name the ions in Problem 3.3.

3.6 Give the name and symbol of the ion with the following number of protons and electrons.

a. 11 protons, 10 electrons   b. 33 protons, 36 electrons

3.7 Give the name and symbol of the ion with the following number of protons and electrons.

a. 9 protons, 10 electrons   b. 24 protons, 21 electrons

3.8 Name the following ions.

a. Cu^{2+} b. SO_4^{2-} c. HPO_4^{2-}

3.9 Name the following ions.

a. NH_4^+ b. $C_2H_3O_2^-$ c. CN^-

3.3 Ionic Compounds—Electron Give and Take

In the previous section, we discovered that metal atoms lose electrons, forming cations, in order to achieve an octet, whereas nonmetal atoms gain electrons, forming anions, for the same reason. However, the process of ion formation does not occur in isolation. In other words, cations do not form unless anions are also formed and vice versa. In fact, when a metal and a nonmetal combine, electrons are transferred between the atoms, thus forming oppositely charged ions as shown in Figure 3.4.

Because opposite charges attract, the newly formed cation and anion are strongly attracted to each other. This attraction of cation and anion is called an **ionic bond**. The resulting combination of cation and anion held together by this strong attraction is called an **ionic compound**. Figure 3.5 demonstrates this process for the formation of common table salt, sodium chloride.

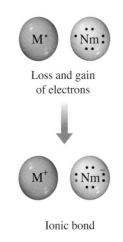

Loss and gain
of electrons

Ionic bond

M is a metal
Nm is a nonmetal

▲ **FIGURE 3.4 Ion formation.** When ions form, metals transfer electrons to nonmetals.

Electon transfer   Ions formed   Ionic bond

Sodium and
chlorine atoms

Sodium and
chloride ions

Sodium chloride, NaCl

▲ **FIGURE 3.5 Formation of the ionic compound sodium chloride.** Both sodium and chlorine are able to complete their octet by giving or taking electrons, respectively.

Formulas of Ionic Compounds

In the example shown in Figure 3.5, a sodium atom gives up one electron to a chlorine atom to form a sodium ion (charge = 1+) and a chloride ion (charge = 1−). The oppositely charged ions, both of which now have completed octets of valence electrons, are attracted to each other and form an ionic bond. The result is a compound whose overall charge is neutral (zero) as shown:

One
sodium ion

One
chloride ion

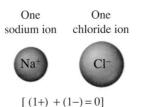

$[(1+) + (1-) = 0]$

Although we recognize sodium chloride, NaCl, as an ionic compound because it contains a metal combined with a nonmetal, the formula does not indicate the charges present on the ions. The compound is written with no charges shown. For sodium chloride, only one sodium ion (Na^+) and one chloride ion (Cl^-) need to combine to form a neutral compound. When writing formulas for chemical compounds, we represent the number of each particle (ions in this case) in the compound with a subscript. Just as a charge of 1+ on an ion is represented as +, the "1" is understood in the formula for the compound and is written as NaCl instead of Na_1Cl_1.

How would we write a formula if one of the ions has a greater charge (although still opposite sign) than the other? The number of ions that combine to form an ionic compound is determined by the charge of the cation and the charge of the anion. Cations and anions combine so that the charges cancel each other out when combined, providing a formula with no net charge.

For example, we saw that calcium forms ions with a 2+ charge, and fluorine, like chlorine, forms ions with a 1− charge. When calcium and fluorine react to form an ionic compound, calcium must give away two electrons, but fluorine can accept only one. Therefore, a calcium atom reacts with *two* fluorine atoms, giving up one electron to each of the fluorines to form a calcium ion (Ca^{2+}) ion and two fluoride (F^-) ions.

Bonds forming

Ionic bonds formed

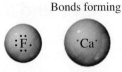 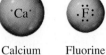 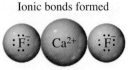

FIGURE 3.6 The formation of an ionic compound from calcium and fluorine. Calcium gives up two electrons (and each fluorine takes one of these electrons), to become isoelectronic with argon.

Fluorine atom   Calcium atom   Fluorine atom   Calcium fluoride, CaF_2

The resulting mineral, called fluorite, forms a hard transparent crystal used in telescope and camera lenses. Figure 3.6 demonstrates the formation of the ionic compound from calcium and fluorine.

Once again the charges on the ions that were formed and the resulting ionic bond formation result in a compound whose net (overall) charge is zero as shown by the following equation:

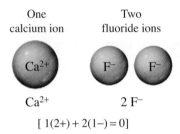

One calcium ion

Two fluoride ions

Ca^{2+} $2\ F^-$

$[\ 1(2+) + 2(1-) = 0\ ]$

Since the neutral ionic compound has 1 calcium ion and 2 fluoride ions, the correct formula is CaF_2.

If we examine each formula more closely, we see that the subscript on each ion is actually the magnitude of the charge on the other ion. Using CaF_2 as an example we see that a calcium ion has a charge of 2 (ignore the sign for the moment). Notice the subscript on fluorine. It is a 2. Similarly, fluoride ion has a charge of 1, and the subscript on calcium is also 1.

In many cases, this pattern can be used to determine the correct formula for an ionic compound. This method, called the *crossover approach*, is a straightforward way of getting the correct formula for an ionic compound. As demonstrated in Figure 3.7, the number representing the magnitude of the charge on the cation becomes the subscript on the anion and vice versa. Note that in the case where the charges on the ions are equal and opposite, for example $+2$ and -2, the formula is reduced to the smallest common ratio. (see Sample Problem 3.3b).

FIGURE 3.7 The crossover approach for writing formulas of ionic compounds.

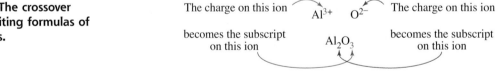

The charge on this ion Al^{3+} O^{2-} The charge on this ion

becomes the subscript on this ion Al_2O_3 becomes the subscript on this ion

In order to form ionic compounds correctly, you must know the charge of the ions formed by the main-group elements on the periodic table. If you don't remember how to determine which atoms form ions and what the charge will be, review Section 3.2 before continuing.

SAMPLE PROBLEM 3.3

Writing Formulas for Ionic Compounds

Determine the correct formula of the ionic compound formed from the following combination of ions.

a. sodium ions and sulfide ions

b. magnesium ions and oxide ions

c. calcium ions and hydroxide ions

SOLUTION

a. Sodium ions have a 1+ charge, Na^+, and sulfide ions have a 2− charge, S^{2-}. We can ignore the signs of the charges and use the magnitudes (the numbers) of the charges. Then, as shown, cross the numbers over so that the superscript on sodium becomes the subscript on sulfide and the superscript on sulfide becomes the subscript on sodium.

$$\left(Na^+ \; S^{2-}\right) = Na_2S_1 \; \blacktriangleright \; Na_2S$$

b. Magnesium ions have a charge of 2+, Mg^{2+}, and oxide ions have a charge of 2−, O^{2-}. Employing the same method used in part a, gives:

$$\left(Mg^{2+} \; O^{2-}\right) = Mg_2O_2 \; \blacktriangleright \; MgO$$

This formula reminds us of an important point. When we use the crossover approach in a case where the charges on the ions are equal, but opposite, the correct formula should contain the smallest common ratio (here 1 : 1). The correct formula for this compound combines one Mg^{2+} ion and one O^{2-} ion in a neutral compound with the formula MgO.

c. Calcium ions have a 2+ charge, Ca^{2+}, and the polyatomic ion, hydroxide, has a 1− charge, OH^-. Use the magnitudes (the numbers) of the charges in the crossover approach. Cross the numbers over so that the superscript on calcium becomes the subscript on hydroxide and the superscript on hydroxide becomes the subscript on calcium. Parentheses are used when more than one polyatomic ion is present. The subscript 2 outside the parentheses indicates that two hydroxides combine with one calcium ion.

$$\left(Ca^{2+} \; OH^-\right) = Ca_1(OH)_2 \; \rightarrow \; Ca(OH)_2$$

SAMPLE PROBLEM 3.4

Interpreting Formulas of Ionic Compounds

Determine the number of each type of ion present in each of the following formulas.

a. KBr, used to treat epilepsy in dogs
b. $MgCl_2$, used as a coagulant in tofu making
c. $Al_2(SO_4)_3$, used in water purification

SOLUTION

a. The ionic compound of potassium and bromide has one potassium ion and one bromide ion.

b. The ionic compound of magnesium and chloride has one magnesium ion and two chloride ions.

c. Note that this compound contains sulfate, a polyatomic ion (see Section 3.2). This group of atoms functions together as a single ion; therefore, this compound contains two aluminum ions and three sulfate ions.

Naming Ionic Compounds

Now that we can write formulas for ionic compounds, the next step is to name them. We can already name cations and anions. To name an ionic compound, put the names of the two ions together, always remembering to list the cation first. So, NaCl is sodium chloride (notice you do not include the word "ion"—sodium ion—in the name) and

MgO is magnesium oxide. Remember that when naming the cation of a transition metal, a Roman numeral is used to designate the charge on the transition metal since it can vary in different ionic compounds.

If the compound contains one of the polyatomic ions, the name of the ion remains unchanged in the name of the compound. For example, $Ca_3(PO_4)_2$ is named calcium phosphate.

SAMPLE PROBLEM 3.5

Naming Ionic Compounds

Name the ionic compounds represented by the formulas given.

a. LiF, used in some batteries

b. CuO, used as a pigment in ceramic glazes

c. $Mg(NO_3)_2$, used as a pesticide

d. NH_4Cl, used to treat electrolyte imbalances

SOLUTION

a. This compound is lithium fluoride.

b. This compound contains the transition metal copper. Before naming this compound we must determine the charge on the copper ion. Because an oxygen ion (oxide) always has a 2− charge, and it combines one-to-one with copper, the copper ion in this case has a 2+ charge. (This gives the compound no net charge.) This compound is copper(II) oxide.

c. Remembering the names of the polyatomic ions, this compound is magnesium nitrate.

d. Remembering the names of the polyatomic ions, this compound is ammonium chloride. Note that this is an example of an ionic compound that does not contain a metal combined with a nonmetal. Ammonium compounds are ionic compounds that do not contain a metal ion.

PRACTICE PROBLEMS

3.10 Give the formula and name for the ionic compounds formed from the following pairs of ions.

a. potassium and chloride, the main ingredients in salt substitutes used by people on low-sodium diets

b. magnesium and oxide, used in heartburn medicine

c. copper(I) and sulfur, a common copper mineral mined for copper metal

3.11 Give the formula and name for the ionic compounds formed from the following pairs of elements.

a. gold(II) and chloride, an antimicrobial

b. calcium and carbonate, the mineral limestone

c. magnesium and hydroxide, the main compound in milk of magnesia

3.12 Give the number of each type of ion present in each of the ionic compounds in Problem 3.10.

3.13 Give the number of each type of ion present in each of the ionic compounds in Problem 3.11.

3.14 Give the formula and name for the ionic compound formed from the combination of the indicated metal and carbonate, CO_3^{2-}, a polyatomic ion.

a. Na   **b.** Ba   **c.** Al

3.15 Repeat Problem 3.14 using the acetate, $C_2H_3O_2^-$, ion.

3.16 Give the number of each type of ion present in each of the ionic compounds in Problem 3.14.

3.17 Give the number of each type of ion present in each of the ionic compounds in Problem 3.15.

3.4 In Search of an Octet Part 2: Covalent Bonding

In the previous sections, we discovered that ionic compounds are formed from the chemical combination of cations and anions. Cations are usually formed from metals, and anions are formed from nonmetals. The ionic bond that results from this give-and-take of electrons is due to the strong attraction of the oppositely charged ions. While ionic compounds are abundant in nature in nonliving things like rocks and minerals, compounds containing mostly nonmetal components are found in living things.

Consider carbon dioxide, CO_2. This compound is a by-product of our cellular energy production process and is used by plants to manufacture their food source. Carbon dioxide is everywhere. Notice that CO_2 contains one atom of carbon (a nonmetal) and two atoms of oxygen (also a nonmetal). This combination means that it is *not* an ionic compound.

Earlier in this chapter, we stated that carbon does not form ions because of the tremendous energy involved in losing or gaining four electrons to complete its valence shell. However, given the example of CO_2, carbon *does* form compounds with other elements. How does it form compounds without forming ions?

While some atoms *transfer* electrons (form ions) to achieve an octet in their valence shell, many atoms chemically combine with other atoms by *sharing* valence electrons in order to achieve an octet. This sharing of electrons, which occurs between nonmetal atoms, results in the formation of a **covalent bond**. Unlike an ionic bond, no electrons are transferred between the atoms in a covalent bond; instead, electrons are shared, thus making both atoms more stable. In other words, the valence electrons in the bond belong to both of the atoms. This may sound strange, but it is true!

When atoms share electrons to form covalent bonds, the resulting new compound is known as a **covalent compound**. The smallest (or fundamental) unit of a covalent compound is a **molecule**. So, carbon dioxide, CO_2 is a molecule and contains atoms held together by covalent bonds.

Nonliving mineral
Mostly ionic and metallic compounds

Living frog and plant
Mostly nonmetal compounds

SAMPLE PROBLEM 3.6

Distinguishing Covalent and Ionic Compounds

Determine whether each of the following is a covalent or ionic compound.

a. NaBr   b. CCl_4 c. NH_4CN

SOLUTION

a. NaBr contains a metal (sodium) and a nonmetal (bromine). Compounds formed from the combination of a metal and a nonmetal are ionic.

b. CCl_4 is made up of carbon (a nonmetal) and chlorine (also a nonmetal). Compounds formed from the combination of nonmetals are covalent compounds.

c. Initial inspection of the formula of NH_4CN shows that all the atoms involved are nonmetals (N, H, and C). However, in Section 3.2, we saw that both NH_4^+ and CN^- are polyatomic ions. Despite the fact that this compound is composed entirely of nonmetals, it is an ionic compound because it is composed of polyatomic ions.

Covalent Bond Formation

Atoms share electrons with other atoms, forming covalent bonds to complete an octet, but how many covalent bonds does an atom form? A simple rule of thumb is this: *The number of covalent bonds that an atom will form equals the number of electrons that it needs to complete its octet.*

We can see how this works with atoms by using dots to represent electrons. The use of electron dot symbols was first developed by chemist G. N. Lewis to show how atoms share electrons forming covalent bonds. The electron dot symbol for any atom consists of the elemental symbol plus a dot for each valence electron. The electron dot symbol for chlorine is

$$:\ddot{\text{Cl}}\cdot$$

We know that chlorine, a member of Group 7A, has seven valence electrons. We construct its electron dot symbol by placing one electron on each side of the elemental symbol (top, bottom, left, and right) and then adding the remaining electrons to form pairs until all the valence electrons are represented. After the first four electrons are in place, each of the remaining electrons can be placed with any one of the existing electrons to form pairs. So, these two electron dot symbols for chlorine are equivalent:

$$:\ddot{\text{Cl}}\cdot \;=\; \cdot\ddot{\text{Cl}}:$$

Because chlorine needs only one electron to complete its octet, it shares its unpaired electron with another nonmetal atom, here a chlorine atom, when forming a covalent bond. A shared pair of electrons, known as a **bonding pair** or simply a bond, is represented as a dash or line connecting the two electron dot symbols (see Figure 3.8). Note that each of the atoms in the newly formed molecule now has an octet—three **lone pairs** of electrons, which are not shared, and the shared bonding pair.

▶ **FIGURE 3.8 Formation of a covalent bond between two chlorine atoms.**

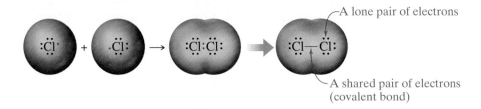

A glance at the properly constructed electron dot symbols of the main-group non-metals in the second period of the periodic table helps to determine the preferred covalent bonding patterns for these elements.

Let's look at the element on which all life on Earth is based, carbon. With its four unpaired valence electrons, carbon needs to form four covalent bonds to complete its octet. To accomplish this, carbon can share one electron each with four other atoms to form four **single bonds**. Or it can form **double bonds** by sharing two of its electrons with one atom. Carbon can also share three of its electrons with one atom to form a **triple bond**. The most common covalent bonding patterns observed for carbon are shown in Figure 3.9. Note that in each case carbon has a complete octet.

▶ **FIGURE 3.9 Preferred bonding patterns used by carbon to complete its octet.**

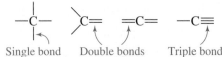

Returning to the electron dot structures, we see that nitrogen forms three covalent bonds, oxygen forms two covalent bonds, and fluorine, like its group member chlorine, forms one covalent bond. Note that hydrogen is an exception to the octet rule. As a member of Period 1, hydrogen needs only two electrons to complete its valence shell; therefore, hydrogen forms just one covalent bond.

Table 3.3 shows the common bonding patterns for the main-group nonmetals that are encountered frequently in living systems.

TABLE 3.3	PREFERRED COVALENT BONDING PATTERNS FOR MAIN-GROUP ELEMENTS			
Group 1A	Group 4A	Group 5A	Group 6A	Group 7A

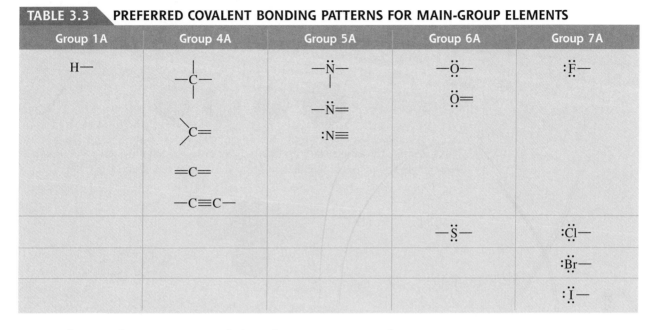

Formulas and Structures of Covalent Compounds

Like the chemical formulas for ionic compounds, the **molecular formula** for a covalent compound completely identifies *all* the components in a molecule. Glucose, the sugar we use for energy, is a covalent compound whose formula is $C_6H_{12}O_6$. This tells us that a molecule of glucose has 6 carbon atoms, 12 hydrogen atoms, and 6 oxygen atoms. Molecular formulas do not reduce to the smallest whole number ratio like the chemical formulas for ionic compounds.

While the molecular formula does tell us the number of atoms in a molecule, it does not tell us *how* the atoms are joined together; it does not describe the structure. The electron dot symbols come in handy for this.

Consider the natural gas called methane, which has the molecular formula CH_4. This is a covalent compound (all nonmetals). Its molecular formula indicates that methane has one carbon atom and four hydrogen atoms. To determine how the atoms are connected, start by drawing the electron dot symbols for all the atoms involved in the molecule. Because carbon has four unpaired valence electrons, it needs to make four bonds to complete its octet. Each of the hydrogen atoms has one valence electron and needs to have two; therefore, each hydrogen atom will make one bond. So, carbon needs to form four bonds and each of the *four* hydrogens needs to form one bond. Connecting the atoms as shown (recall that a line connecting two atoms represents a bonding pair of electrons) fulfills these requirements.

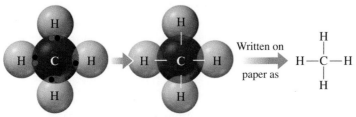

The drawing shows the number and type of each atom in the molecule and their connectivity. This representation is called a **Lewis structure**.

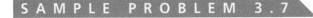

SAMPLE PROBLEM 3.7

Drawing Lewis Structures for Covalent Compounds

Draw the correct Lewis structure for each of the following covalent compounds.

a. NH_3 **b.** C_2H_4

SOLUTION

a. Start by drawing the correct electron dot symbol for each of the atoms involved in the formula.

Next, arrange the atoms so that unpaired electrons are next to each other, much as you line up the pieces of a puzzle in order to fit them together.

To complete the Lewis structure, replace the pairs of dots between atoms with lines representing bonds connecting the two atoms.

Finally, compare the structure with those in Table 3.3, which shows that a structure with three single bonds and one lone pair is one of nitrogen's preferred bonding patterns.

b. Once again, start by drawing the correct electron dot symbol for each of the atoms involved in the formula.

In this case there are two carbon atoms present, and since each hydrogen atom can form only one bond, the two carbons must bond to each other. Arrange the atoms as presented in part a using just single bonds and you would get

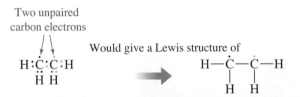

Note that each of the carbon atoms in this structure still has an unpaired electron *and* each has only seven valence electrons (three bonding pairs = 6 electrons +1 unpaired electron = 7 electrons). In this book, carbon will always form four

bonds to complete its octet. So, to complete their octets in this case, the carbons share their unpaired electrons to give

Finally, when we compare the structure with Table 3.3, we find that the arrangement with two single bonds and one double bond is one of carbon's preferred bonding patterns.

Naming Covalent Compounds

As we saw in the last section, naming ionic compounds is a relatively straightforward process—name the cation and then name the anion. Naming covalent compounds is a little different. Here we focus on naming the simplest covalent compounds and look at a few common exceptions.

Covalent compounds composed of only two elements are known as **binary compounds**. These compounds can be named by the following three-step procedure:

STEP 1. Name the first element in the formula.

STEP 2. Name the second element in the formula and change the ending to *ide*.

STEP 3. Designate the number of each element present using one of the Greek prefixes shown in Table 3.4.

Indicating the number of each element present is important when naming covalent compounds because nonmetals can combine with each other in multiple ways. For example, sulfur and oxygen can combine to form both SO_2 and SO_3.

Applying the three-step procedure to the compounds of sulfur and oxygen, the correct name for SO_2 is sulfur dioxide and the correct name for SO_3 is sulfur trioxide. Notice that in both cases, the *mono* prefix for sulfur was understood and was not included in the name. This exception (for *mono*) holds true for only the first element in the name.

TABLE 3.4

GREEK PREFIXES USED WHEN NAMING BINARY COVALENT COMPOUNDS

Prefix	Meaning
mono-	1
di-	2
tri-	3
tetra-	4
penta-	5
hexa-	6
hepta-	7

SAMPLE PROBLEM 3.8

Naming Binary Covalent Compounds

Give the correct name for each of the following binary covalent compounds.
a. N_2O_4, used as a rocket fuel
b. CO_2, produced in respiring cells
c. Cl_2O, a degradation product found in the ozone layer

SOLUTION

a. dinitrogen tetroxide   b. carbon dioxide   c. dichlorine monoxide

SAMPLE PROBLEM 3.9

Writing Formulas of Binary Covalent Compounds

Give the correct formula for each of the following binary covalent compounds.
a. carbon monoxide, a poisonous gas
b. phosphorus trichloride, used in the manufacture of flame retardants
c. dinitrogen pentoxide, a greenhouse gas

SOLUTION

a. CO   b. PCl_3 c. N_2O_5

One of the most notable exceptions to the rules for naming binary covalent compounds is H_2O. According to the rules, this compound is dihydrogen monoxide, but of course it is recognized as water. Ammonia—a component of smelling salts, many household cleaners, and agricultural fertilizers—has a molecular formula of NH_3 and is also an exception to the naming rules. Both of these compounds are so prevalent in our world that their long-standing traditional names have never been replaced by the rules-derived names.

PRACTICE PROBLEMS

3.18 Determine whether each of the following is a covalent or ionic compound.

a. Li_2O b. CS_2 c. $AlCl_3$ d. XeF_2

e. CrO_3 f. $Ca_3(PO_3)_2$ g. ICl   h. NO_2

i. NH_4NO_3 j. PH_3

3.19 Determine whether each of the following is a covalent or ionic compound.

a. NCl_3 b. Fe_2O_3 c. Cs_2CO_3 d. PBr_3

e. C_5H_{12} f. NH_4OH g. SeO_2 h. BaS

i. CBr_4 j. OF_2

3.20 Draw the correct Lewis structure for each of the following covalent compounds.

a. H_2O b. C_3H_8, propane, a fuel used for heating

c. N_2, nitrogen gas   d. SiF_4, used in circuits

3.21 Draw the correct Lewis structure for each of the following covalent compounds.

a. $CHCl_3$, chloroform   b. H_2S, smell of rotting eggs

c. CO_2, dry ice   d. C_2H_2, acetylene, a fuel used in welding

3.22 Using Table 3.3, decide whether molecules with the following formulas are likely to exist (form Lewis structures that obey the octet rule). Briefly explain your answer.

a. CCl_3 b. HF   c. OF_4 d. C_2H_6

3.23 Using Table 3.3, decide whether molecules with the following formulas are likely to exist (form Lewis structures that obey the octet rule). Briefly explain your answer.

a. CH_2 b. NH_2 c. BrCl   d. H_2S

3.24 Complete the table by supplying the missing name or formula for each covalent compound.

Formula	Name
CF_4	?
?	Nitrogen dioxide
PCl_5	?
?	Carbon disulfide
N_2O_5	?

3.25 Complete the table by supplying the missing name or formula for each covalent compound.

Formula	Name
?	Sulfur trioxide
P_2O_5	?
?	Selenium tetrafluoride
CO	?
?	Dinitrogen trioxide

3.5 Getting Covalent Compounds into Shape

Just as a road map allows us to see how towns and cities are connected by roads but gives very little information about the topography, Lewis structures tell us how atoms are connected in molecules but do not tell us anything about a molecule's three-dimensional shape. Our world is three-dimensional, so to understand molecules it is important to identify the three-dimensional shapes they can adopt.

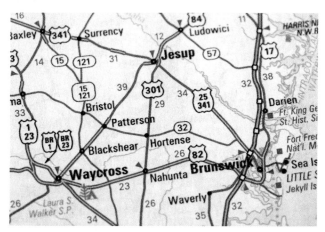

For example, the Lewis structure for methane appears flat or two dimensional on paper. If this were how methane was actually arranged in three dimensions, the angle between two adjacent pairs of electrons would be about 90°.

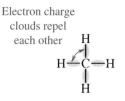

To understand how methane looks in three dimensions, first remember that electrons are negatively charged and that like charges repel each other. So the four bonding pairs of electrons around the carbon atom in CH_4 form four clouds of negative charge that want to get *as far away from each other as possible*.

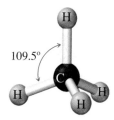

▲ FIGURE 3.10 **The tetrahedral shape of methane, CH_4.**

If, however, the electrons are allowed to rearrange in three-dimensional space to get as far apart as possible, the result is an arrangement of the atoms of CH_4 that achieves an angle of 109.5° between the negatively charged clouds of electrons. The molecule is no longer two dimensional; instead, it has adopted a three-dimensional shape that we call **tetrahedral**, as shown in Figure 3.10.

This reasoning describes the **valence-shell electron-pair repulsion** model that is often referred to simply as **VSEPR**. This model is used to predict the shape of a molecule based on the number of electron **charge clouds** on a given atom. According to VSEPR, in the valence shell of an atom the charge clouds formed by groups of electrons will arrange themselves to be as far away from each other as possible in order to reduce repulsions.

Determining the Shape of a Molecule

To determine the shape (also called geometry) of a molecule using VSEPR, a molecule is assigned a VSEPR form. The VSEPR form is a tool used to relate the number and type of charge clouds on an atom to the shape around that atom. To determine the VSEPR form and, from that, the shape of a molecule, first determine the atom around which you want to know the shape. For small molecules, this is the central atom. (We will discuss larger molecules later.) The central atom is represented with a capital *A* in the VSEPR form.

Next, determine the number of charge clouds around the central atom. When counting charge clouds, keep in mind that a single bond, a double bond, and a triple bond are counted as one charge cloud. Charge clouds in bonds are called *bonding clouds* and are represented with a capital *B* in the VSEPR form.

A lone pair of electrons is also counted as one charge cloud, but as we will see, these charge clouds affect the shape of the molecule in a slightly different way. Charge clouds from lone pairs of electrons are called *nonbonding clouds* and are represented with a capital N in the VSEPR form.

Here are some examples to guide you as you learn to count charge clouds and determine a molecule's VSEPR form.

Formaldehyde has three charge clouds (all bonding clouds) around the central atom, carbon

Formaldehyde, a preservative

The VSEPR form for formaldehyde is AB_3—a central atom surrounded by three bonding charge clouds.

Ammonia has three bonding charge clouds and one nonbonding charge cloud (its lone pair) around the central atom, nitrogen

Ammonia

The VSEPR form for ammonia is AB_3N—a central atom surrounded by three bonding charge clouds and one nonbonding charge cloud.

The molecular shapes associated with the VSEPR forms are shown in Table 3.5 for the common forms containing the elements C, N, and O found in living things.

TABLE 3.5	PREDICTING MOLECULAR SHAPE USING VSEPR		
VSEPR Form	**Molecular Shape**	**Bond Angle**	**Example**
AB_4	Tetrahedral	109.5°	
AB_3N	Pyramidal	<109.5°	
AB_2N_2	Bent	<109.5°	
AB_3	Trigonal planar	120°	
AB_2	Linear	180°	

SAMPLE PROBLEM 3.10

Determining a Molecule's Shape

From the following Lewis structures, assign the VSEPR form for each molecule and, using Table 3.5, determine the shape of each molecule.

a. $:\ddot{O}=C=\ddot{O}:$ b. $H—\ddot{S}—H$ c. $:\ddot{B}r—P—\ddot{B}r:$
$\qquad\qquad\qquad\qquad\qquad\qquad\qquad\qquad\qquad\qquad\qquad\qquad\qquad\quad|$
$\qquad\qquad\qquad\qquad\qquad\qquad\qquad\qquad\qquad\qquad\qquad\qquad\quad:\ddot{B}r:$

SOLUTION

a. Linear. The central atom in the molecule is carbon. Carbon is participating in two double bonds, indicating it has two bonding charge clouds in its valence shell. So this molecule has the VSEPR form AB_2. This VSEPR form corresponds to the molecular shape of linear (Table 3.5). Note that we did not concern ourselves with the oxygen atoms in this molecule. Because oxygen is not the central atom, its arrangement of charge clouds is not included when we assign the VSEPR form.

b. Bent. Sulfur is the central atom in this molecule. Careful examination shows that sulfur is participating in two single bonds and has two nonbonding pairs (also called lone pairs) of electrons. Therefore, the sulfur is surrounded by four charge clouds with a VSEPR form of AB_2N_2. This VSEPR form corresponds to the molecular shape of bent (Table 3.5). Note that the Lewis structures of CO_2 and H_2S look fairly similar—a central atom connected to two other atoms. Both Lewis structures can be drawn with all the atoms in a straight line, but only CO_2 actually has a linear shape. It is important to remember that *Lewis structures show only the connectivity of the atoms in a molecule, not the molecule's shape.*

c. Pyramidal. Phosphorus is the central atom in this molecule; it is participating in three single bonds and has one nonbonding pair of electrons. The VSEPR form of this molecule is AB_3N. This VSEPR form corresponds to the molecular shape of pyramidal (Table 3.5).

Nonbonding Pairs and Their Effect on Molecular Shape

The first three entries in Table 3.5 have the VSEPR forms AB_4, AB_3N, and AB_2N_2. Each has a total of four charge clouds around a central atom, but each has a different shape. How can this be? Isn't there one optimal arrangement of four charge clouds around a central atom? To explore this, let's look at the molecules methane (CH_4), ammonia (NH_3), and water (H_2O) that represent these three VSEPR forms.

Nonbonding pairs do affect the shape of a molecule. The tetrahedral shape of methane, which has four atoms around a central atom, changes to pyramidal for ammonia because a nonbonding pair occupies the fourth space (see Figure 3.11). Because nonbonding pairs of electrons take up space but are invisible in the molecule's shape, the VSEPR forms are different. A molecule's shape is determined by the relative positions of the *atoms* in the molecule.

Nonbonded pairs have a subtle effect on the shape of a molecule. In the Bond Angle column of Table 3.5, notice that the presence of nonbonded pairs on the central

Lone pair

Tetrahedral   Pyramidal

◀ **FIGURE 3.11 Methane and ammonia.** A nonbonded pair changes a tetrahedral shape to a pyramidal shape.

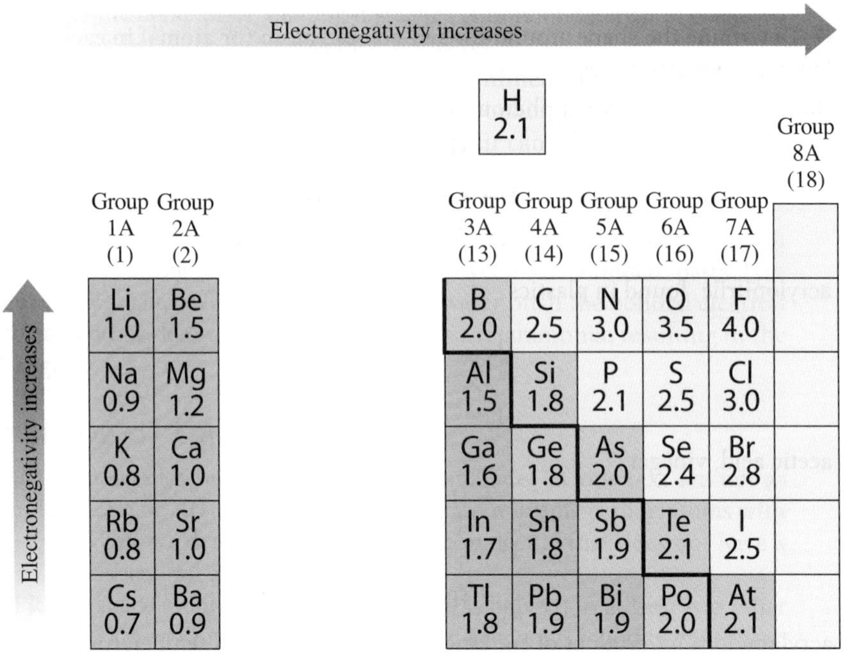

▶ **FIGURE 3.13 Electro-negativities of the main-group elements.** Fluorine is the most electronegative element. The noble gases have no electronegativity since they are not reactive.

Electronegativity increases

				H 2.1			Group 8A (18)

Electronegativity increases

Group 1A (1)	Group 2A (2)		Group 3A (13)	Group 4A (14)	Group 5A (15)	Group 6A (16)	Group 7A (17)
Li 1.0	Be 1.5		B 2.0	C 2.5	N 3.0	O 3.5	F 4.0
Na 0.9	Mg 1.2		Al 1.5	Si 1.8	P 2.1	S 2.5	Cl 3.0
K 0.8	Ca 1.0		Ga 1.6	Ge 1.8	As 2.0	Se 2.4	Br 2.8
Rb 0.8	Sr 1.0		In 1.7	Sn 1.8	Sb 1.9	Te 2.1	I 2.5
Cs 0.7	Ba 0.9		Tl 1.8	Pb 1.9	Bi 1.9	Po 2.0	At 2.1

We can denote the uneven sharing or distribution of electrons in a polar covalent bond using the symbol δ (lowercase Greek delta, meaning "partial"). Figure 3.14 demonstrates how this notation is used in the case of HCl. We can also use an arrow with a hash mark (↦) to represent bond polarity. This arrow is commonly referred to by chemists as the "dipole moment" arrow. Both notations are shown in Figure 3.14. (Note that the arrow points toward the more electronegative element and the plus on the tail of the arrow is at the partial positive end.)

▶ **FIGURE 3.14 In the nonpolar covalent bond of H_2, electrons are shared equally.** In the polar covalent bond of HCl, electrons are shared unequally.

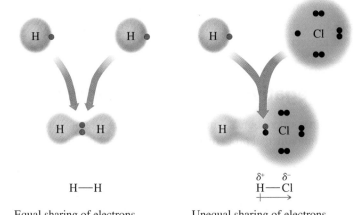

H—H

Equal sharing of electrons in a nonpolar covalent bond

δ^+ δ^-
H—Cl

Unequal sharing of electrons in a polar covalent bond (dipole)

Like the opposite poles on a magnet, a covalent bond that does not share electrons equally has two distinct poles or ends—one that is partially negative and one that is partially positive. Because of this we refer to these bonds as *polar*. Of course, a bond that is "nonpolar" (sharing equally) has no ends or poles.

We can predict the type of bond likely to form by taking the difference between the two element's electronegativities as shown in the table in Figure 3.15. Generally, elements with an electronegativity difference of 1.8 or more will form an ionic bond resulting from the *transfer* of electrons from the less electronegative element to the more electronegative element. For example, NaCl, which we know as an ionic compound, has an electronegativity difference of 2.1 [3.0 (Cl) − 0.9 (Na)] between chlorine and sodium.

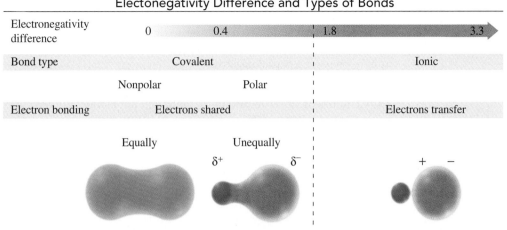

Electonegativity Difference and Types of Bonds

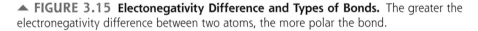

▲ FIGURE 3.15 **Electonegativity Difference and Types of Bonds.** The greater the electronegativity difference between two atoms, the more polar the bond.

In turn, two elements with an electronegativity difference of less than 1.8 will most likely *share* electrons to form a covalent bond. When two atoms are involved in a covalent bond, the greater the difference in electronegativity of the two elements, the more polar the bond.

While it is possible to use the information in Figures 3.13 and 3.15 to calculate an electronegativity difference for any bond and determine where along the bond-type continuum it lies, we simply need to be able to distinguish ionic bonds from covalent bonds and determine if a covalent bond is polar or nonpolar. In Section 3.4, we saw that ionic compounds contain metals and nonmetals, and covalent compounds are made up of only nonmetals. Now we need to distinguish between nonpolar bonds, slightly polar bonds, and strongly polar bonds.

We can do this without numbers by remembering two facts: (1) Fluorine is the most electronegative of all the elements and (2) electronegativities increase as you move toward fluorine on the periodic table. With this in mind, let's consider the following carbon-containing bonds. Which of these bonds is the most polar?

$$C—N \qquad C—O \qquad C—F$$

Since each of the three bonds contains a carbon atom, we can compare the electronegativity of each of the other atoms relative to carbon. Looking at the periodic table, we see that carbon and nitrogen are side by side, which means that their electronegativities are more similar than those of carbon and oxygen or carbon and fluorine. Therefore, the C—N bond, with the smallest difference in electronegativities, is the least polar. Applying the same reasoning, we can determine that the C—O bond would fall next in the order and the C—F bond would be the most polar. When we want to determine relative polarities of bonds, we can do it without electronegativity numbers. In fact, our example illustrates a rule of thumb we can use for comparing the polarity of covalent bonds: *The farther apart two nonmetals are on the periodic table, the greater the polarity of the bond between them.* One major exception to this rule, which we will explore further in Chapter 4, is that a C—H bond is considered to be nonpolar since the electronegativities of carbon and hydrogen are not that different.

SAMPLE PROBLEM 3.11

Polarity of Covalent Bonds

For each of the following covalent bonds, label the less electronegative atom with a δ^+ and the more electronegative atom with a δ^-. Then, label the bond using the dipole moment arrow (↔).

a. N—C   **b.** N—Cl   **c.** P—F

SOLUTION

a. $\overset{\delta^-}{N}\!\!-\!\!\overset{\delta^+}{C}$ On the periodic table, nitrogen is closer to fluorine than carbon, so nitro-
   $\longleftrightarrow$ gen is more electronegative and attracts the bonding electrons more
   strongly, giving it a partial negative charge.

b. $\overset{\delta^+}{N}\!\!-\!\!\overset{\delta^-}{Cl}$ Despite the fact that nitrogen and chlorine are not in the same period or
   $\longleftrightarrow$ group, chlorine is closer in proximity (although not atomic number) to
   fluorine on the periodic table. As the more electronegative element, chlorine attracts the bonding electrons more strongly, giving it a partial negative charge.

c. $\overset{\delta^+}{P}\!\!-\!\!\overset{\delta^-}{F}$ Fluorine is the most electronegative of all elements, so it attracts the elec-
   $\longleftrightarrow$ trons in this bond more strongly, giving it a partial negative charge.

SAMPLE PROBLEM 3.12

Comparing Covalent Bonds

In each of the following bond pairs, circle the pair with the more polar bond.

a. C—N and C—Cl   b. N—Cl and S—N   c. Si—F and Si—O

SOLUTION

In general, the farther apart two atoms are on the periodic table, the greater the difference in their electronegativities and, therefore, the more polar the bond between
them.

a. C—Cl is the more polar bond.

b. N—Cl is the more polar bond.

c. Si—F is the more polar bond.

Molecular Polarity

Molecules, like the bonds that hold them together, can be polar or nonpolar. A **polar
molecule** is a molecule in which one end or area of the molecule is more negatively
charged than the rest of the molecule. In other words, the electrons in a polar molecule
are unevenly distributed over the molecule. In contrast, a **nonpolar molecule** has an
even distribution of electrons over the entire molecule.

For simple molecules containing only two atoms connected by a covalent bond, we
can determine the polarity of the molecule simply by determining the polarity of the
bond. In our previous examples, the bond between the chlorines in Cl_2 is nonpolar, so
Cl_2 is a nonpolar molecule. Likewise, since the bond in HCl is a polar bond, HCl is a
polar molecule.

What happens when a molecule is composed of three or more atoms connected by
several bonds? How do we determine the polarity of a molecule if it has more than one
bond dipole? We *can* determine the polarity of larger molecules, but when we do, we
must consider both the electronegativity of the atoms involved *and* the shape of the
molecule.

As a first example, consider carbon dioxide and water, whose Lewis structures are
shown.

$$:\ddot{O}\!=\!C\!=\!\ddot{O}: \qquad H\!-\!\ddot{O}\!-\!H$$

First, notate the bond polarity for each bond in each molecule:

$$\overset{\delta^-}{:\ddot{O}}\!\!=\!\!C\!\!=\!\!\overset{\delta^-}{\ddot{O}:} \qquad \overset{\delta^+}{H}\!\!-\!\!\overset{\delta^-}{\ddot{O}}\!\!-\!\!H$$
$$\underset{\delta^+ \quad \delta^-}{} \qquad \underset{\delta^- \quad \delta^+}{}$$

Each molecule has two polar bonds. Despite this apparent similarity, we know from experimental evidence that carbon dioxide is a nonpolar molecule whereas water is a polar molecule. How can two molecules that seem so similar actually be so different? Remember, these two molecules have different shapes.

Applying the principles of VSEPR from Section 3.5, we know that carbon dioxide is a linear molecule, but water is bent. To see how shape affects the polarities of these molecules, let's use the dipole moment arrow to represent the bond dipoles and draw the molecules to more closely represent their true shape.

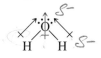

Notice that in CO_2, the bonding electrons are being pulled equally toward each of the oxygen atoms, but in opposite directions. So, like a tug of war that is equally matched, the electrons that are being pulled away from carbon by one of the oxygen atoms are also being pulled away from the carbon atom in the opposite direction by the other oxygen atom. In other words, the bond dipoles cancel each other and the net result is that the electrons are spread out equally over the entire molecule.

Because of the bent shape of the water molecule, the bond dipoles, instead of canceling each other as they do in carbon dioxide, actually add together to give an uneven distribution of electrons. The area around the oxygen atom where the lone pairs of electrons are positioned has more negative charge than the rest of the molecule. This results in an overall molecular dipole for water, shown as a red arrow.

The interaction of electronegativity and molecular shape is the basis for determining whether a molecule is polar or nonpolar. However, the process can be somewhat tedious and time consuming. In most cases, we can use a few shortcuts or rules of thumb to help us quickly determine if a molecule is polar or nonpolar.

When determining molecular polarity,

- if the central atom (or one of the main atoms) has one or more pairs of nonbonding electrons, the molecule is polar.

- if the central atom has no pairs of nonbonding electrons and all the atoms connected to the central atom are identical, the molecule is nonpolar.

- if the central atom has no pairs of nonbonding electrons and at least one of the atoms connected to the central atom is different from the others that are attached, the molecule is polar.

These shortcuts generally apply to small molecules but can be adapted (as we will see in later chapters) to help us determine the polarity of much larger molecules and to understand their behavior.

SAMPLE PROBLEM 3.13

Determining the Polarity of Molecules

Use the shortcuts discussed above to determine whether each of the following molecules is polar or nonpolar. If polar, show the direction of the molecular dipole using a dipole moment arrow.

a. CCl_4 b. NH_3 c. CH_3F

SOLUTION

In order to see all the bonds in the molecule, drawing the correct Lewis structure is critical to determining if the central atom has lone pairs to apply the shortcuts.

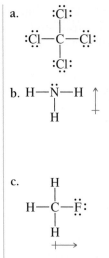

a. The nonbonding pairs are not on the central atom (carbon) and all the atoms connected to carbon are identical (they are all chlorines). Despite the many nonbonding pairs on the chlorines in this molecule, the compound is *nonpolar*.

b. We see from the Lewis structure that ammonia has a nonbonding pair of electrons on the central atom (nitrogen): therefore, using the shortcuts, we can determine that ammonia is a *polar* compound. The negative end of the molecule is the nitrogen end as indicated by the molecule's dipole moment arrow (red).

c. This molecule has four atoms bonded to the central atom (carbon), but fluorine is more electronegative than the three hydrogens; therefore, the compound is *polar*. The negative end of the molecule is the fluorine side as indicated by the molecule's dipole moment arrow (red).

PRACTICE PROBLEMS

3.29 For each of the following molecules, (1) draw the correct Lewis structure; (2) label each polar covalent bond with a dipole moment arrow; and (3) determine if the molecule is polar or nonpolar and draw the dipole moment arrow for the molecule.

a. HCl

b. CH_4

c. NCl_3

d. CSO (C is the central atom)

e. H_2S

f. CH_2Br_2 (C is the central atom)

3.30 For each of the following molecules, (1) draw the correct Lewis structure; (2) label each polar covalent bond with a dipole moment arrow; and (3) determine if the molecule is polar or nonpolar and draw the dipole moment arrow for the molecule.

a. CH_3Br (C is the central atom)

b. HF

c. CH_2O (C is the central atom)

d. SiO_2

e. PBr_3

f. OF_2

SUMMARY

3.1 The Octet Rule

Atoms react with other atoms in order to become more stable. The noble gases (Group 8A) are stable atoms and each has an octet of electrons in its valence shell. An octet is eight electrons for all elements except hydrogen and helium, where a full valence shell is two electrons. Atoms, other than the noble gases, attain a stable valence octet by reacting with each other.

3.2 In Search of an Octet Part 1: Ion Formation

Gaining or losing electrons to form ions is one way that atoms can attain a valence octet. An atom that gains electrons becomes a negatively charged anion, and an atom that loses electrons becomes a positively charged cation. Nonmetal atoms form anions and metal atoms form cations. During ion formation, atoms gain or lose electrons to attain the same number of electrons as the nearest noble gas. For the main-group elements, the atom's position on the periodic table helps determine what its charge will be if it forms an ion. Polyatomic ions are groups of nonmetals that together form ions.

chapter three

Cations have the same name as the element with the word *ion* added to the end, but the name of an anion is derived by changing the ending of the elemental name to *ide*.

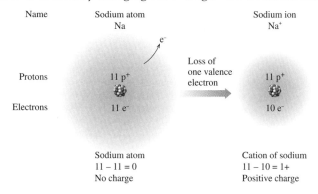

3.3 Ionic Compounds—Electron Give and Take

When an atom loses electrons to form a cation, those electrons are gained by a different atom to form an anion. These oppositely charged ions are attracted to each other and form an ionic bond. The new compound created by the bond between these ions is an ionic compound. Ions combine to form neutral compounds. In other words, the total charge (number of ions times the charge of each) of the cations must equal the total charge of the anions in the compound. To name an ionic compound, first name the cation (dropping the word *ion*) and then name the anion. If a transition metal is present, a Roman numeral is placed in parentheses following the name of the metal to designate its charge. Polyatomic ions are groups of nonmetals that together act as ions. In writing ionic formulas, the metal ion always goes first. If more than one polyatomic ion is present, the number of ions present is represented outside a set of parentheses.

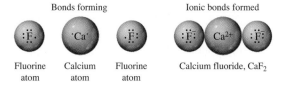

3.4 In Search of an Octet Part 2: Covalent Bonding

Sharing electrons is another way that atoms can achieve a valence octet. This sharing of electrons between two atoms is known as a covalent bond and typically occurs between two or more nonmetal atoms. The number of covalent bonds that an atom will form is determined by the number of electrons that it needs to complete its valence octet and leads to the atom's preferred covalent bonding pattern. The molecular formula of a covalent compound gives the number of each type of atom in a molecule and the Lewis structure shows how those atoms are bonded together. Covalent compounds containing only two elements can be named by naming the first element as the element, naming the second element and changing the ending to *ide*, and using Greek prefixes to indicate the number of atoms of each in the compound.

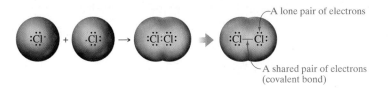

3.5 Getting Covalent Compounds into Shape

Molecules of covalent compounds have a three-dimensional shape that is determined by the arrangement of the electrons in the molecule. A Lewis structure does not imply the three-dimensional shape of a molecule. Valence-shell electron-pair repulsion theory (VSEPR) explains that the electrons around any given atom arrange themselves to get as far away from each other as possible. For carbon atoms, this leads to shapes such as tetrahedral, trigonal planar, and linear. Each shape gives rise to a specific bond angle. The presence of nonbonding pairs on an atom reduces the bond angles around that atom and alters the shape of the molecule.

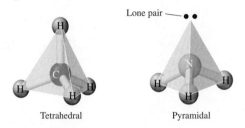

Tetrahedral   Pyramidal

3.6 Electronegativity and Molecular Polarity

The equity in sharing electrons in covalent bonds is determined by comparing the electronegativities of the atoms involved. Nonpolar covalent bonds are formed when atoms share electrons equally (with C—H bonds being a notable exception), and polar covalent bonds are formed when the sharing of electrons is unequal. The presence of polar covalent bonds in a molecule may result in a molecule that is polar, meaning that the electron density is greater in some areas of the molecule than it is in others. To determine whether a molecule is polar or not, we must consider the electronegativity of its atoms as well as the shape of the molecule. Polarity can be represented by the delta (δ) symbol or the dipole moment arrow (↔).

KEY TERMS

anion—An ion with a net negative charge; formed when an atom gains electrons.

binary compounds—Covalent compounds that are composed of only two elements.

bonding pair—A pair of electrons shared by two atoms to form a covalent bond.

cation—An ion with a net positive charge; formed when an atom gives up electrons.

charge clouds—Groupings of electrons (one, two, or three pairs) in a molecule.

covalent bond—The result of atoms sharing electrons in order to achieve an octet in their valence shells.

covalent compound—A type of compound made up of atoms held together by covalent bonds.

double bond—Covalent bonds formed when two atoms share two pairs of electrons.

electronegativity—A measure of the ability of an atom in a covalent bond to attract the bonding electrons to itself.

ion—An atom with a charge due to an unequal number of protons and electrons.

ionic bond—The strong attraction between an anion and a cation.

ionic compound—The result of two or more ions attracted by ionic bonds to form a new electrically neutral species.

isoelectronic—Having the same number of electrons.

Lewis structure—A chemical formula used for covalent compounds to show the number, type, and connectivity of atoms in a molecule.

lone pair—A pair of electrons belonging to only one atom. A lone pair is also called a nonbonded pair.

molecular formula—The chemical formula of a molecular compound that represents all atoms in a molecule.

molecule—The smallest unit of a covalent compound; formed when two or more nonmetal atoms are joined by covalent bonds.

nonbonded pair—A pair of electrons belonging to only one atom. It is also referred to as a lone pair.

nonpolar covalent bond—A bond in which the electrons are shared equally by the two atoms.

nonpolar molecule—A molecule that has an even distribution of electrons over the entire molecule.

octet rule—Atoms react with other atoms to obtain eight electrons in their outer shell.

polar covalent bond—A bond in which the electrons are more strongly attracted by one atom than by the other.

polar molecule—A molecule that has an area that is more negatively charged than the rest of the molecule.

polyatomic ion—A group of atoms that interact to form an ion.

single bond—A covalent bond formed when two atoms share one pair of electrons.

tetrahedral—A molecular shape with four atoms arranged around a central atom to give an angle of 109.5°.

triple bond—Covalent bonds formed when two atoms share three pairs of electrons.

valence-shell electron-pair repulsion model (VSEPR)—A method for determining molecular shape based on the repulsion of groups of electrons in the valence shell of the central atom.

ADDITIONAL PROBLEMS

3.31 How many valence electrons are present in the following atoms?
   a. C   **b.** Cl   **c.** K   **d.** Al

3.32 How many valence electrons are present in the following atoms?
   a. N   **b.** Mg   **c.** S   **d.** Si

3.33 For each of the atoms in Problem 3.31, determine the minimum number of electrons the atom needs to gain, lose, or share to become stable (isoelectronic with a noble gas).

3.34 For each of the atoms in Problem 3.32, determine the minimum number of electrons the atom needs to gain, lose, or share to become stable (isoelectronic with a noble gas).

3.35 Complete the following statements:
   a. An anion has more/fewer (*circle one*) protons than electrons.
   b. The electrons found in the outermost shell of an atom are called _____ electrons.
   c. Ions formed from metals have a positive/negative (*circle one*) charge.

3.36 Complete the following statements:
   a. An ion is charged because the numbers of _____ and electrons are not the same.
   b. An ion that has a positive charge is called a _____.
   c. The ion formed from a sulfur atom is called _____.

3.37 Determine the number of protons and electrons in the following ions:
   a. chloride   **b.** Fe^{3+}
   c. Cr^{6+} **d.** N^{3-}
   e. sodium ion   **f.** H^+

3.38 Determine the number of protons and electrons in the following ions:
   a. oxide   **b.** Ag^+
   c. copper(I) ion   **d.** H^-
   e. zinc(II) ion   **f.** I^-

3.39 Give the name of each of the unnamed ions in Problem 3.37.

3.40 Give the name of each of the unnamed ions in Problem 3.38.

3.41 Give the name and symbol of the ion that has the following number of protons and electrons:
   a. 3 protons, 2 electrons
   b. 35 protons, 36 electrons

3.42 Give the name and symbol of the ion that has the following number of protons and electrons:
   a. 9 protons, 10 electrons
   b. 38 protons, 36 electrons

3.43 Each of the following ions is isoelectronic with a noble gas. Give the symbol of the noble gas.
   a. Rb^+ **b.** F^- **c.** S^{2-} **d.** Ca^{2+}

3.44 Each of the following ions is isoelectronic with a noble gas. Give the symbol of the noble gas.
   a. Li^+ **b.** Mg^{2+} **c.** Al^{3+} **d.** O^{2-}

3.45 Complete the table of polyatomic ions below by supplying the missing information.

Name	Formula
a. acetate	
b. bicarbonate	
c.	NO_3^-
d.	CN^-

3.46 Complete the table of polyatomic ions by supplying the missing information.

a.	NH_4^+
b.	CO_3^{2-}
c. phosphate	
d. hydroxide	

3.47 Give the formula for the ionic compound formed by the ammonium ion and each of the following anions:
   a. sulfide   **b.** chloride
   c. sulfate   **d.** hydroxide

3.48 Give the formula for the ionic compound formed by the phosphate ion and each of the following cations:
   a. sodium   **b.** aluminum
   c. magnesium   **d.** iron(III)

3.49 Give the formula for each of the following ionic compounds:
   a. sodium hydroxide, drain cleaner
   b. aluminum oxide, used as an abrasive
   c. potassium nitrate, an ingredient in gunpowder

3.50 Give the formula for each of the following ionic compounds:
   a. potassium chlorate, used in fireworks displays
   b. magnesium sulfate, Epsom salts
   c. silver(I) iodide, used to seed clouds in rainmaking

3.51 Name the following ionic compounds:
   a. Na_2O **b.** $BaSO_4$
   c. $CuCl_2$ **d.** $Mg(NO_3)_2$
   e. Fe_2O_3 **f.** KF

3.52 Name the following ionic compounds:
- **a.** $AlCl_3$
- **b.** $NH_4C_2H_3O_2$
- **c.** AgBr
- **d.** K_3PO_4
- **e.** CsF
- **f.** MnO

3.53 Supply the missing information in each of the following formulas:
- **a.** $CaCl_?$
- **b.** $Sr_?(PO_4)_?$
- **c.** $Rb_?CO_3$
- **d.** $K_?S$

3.54 Supply the missing information in each of the following formulas:
- **a.** $Sr(C_2H_3O_2)_?$
- **b.** $Li_?SO_4$
- **c.** $Al_?S_?$
- **d.** $Ba(CN)_?$

3.55 Magnesium salicylate is sold as an over-the-counter drug to treat pain and inflammation. If there are two salicylate ions for each magnesium ion in the compound, what is the charge on the salicylate ion?

3.56 Explain the difference between an ionic bond and a covalent bond.

3.57 How many covalent bonds does each of the following atoms prefer to form?
- **a.** C
- **b.** F
- **c.** P
- **d.** O

3.58 How many covalent bonds does each of the following atoms prefer to form?
- **a.** Cl
- **b.** Si
- **c.** S
- **d.** N

3.59 For each of the following compounds, indicate whether it is covalent or ionic and give its correct name:
- **a.** CBr_4
- **b.** SiO_2
- **c.** $MgBr_2$
- **d.** NCl_3
- **e.** $CrCl_3$

3.60 For each of the following compounds, indicate whether it is covalent or ionic and give its correct name:
- **a.** Cu_2O
- **b.** $SiCl_4$
- **c.** H_2O
- **d.** BaS
- **e.** CS_2

3.61 Explain the difference between an ionic formula and a molecular formula.

3.62 Explain the difference between a Lewis structure and a molecular formula.

3.63 Draw a Lewis structure for each of the following covalent compounds:
- **a.** HCN
- **b.** CS_2
- **c.** PCl_3
- **d.** CH_5N

3.64 Draw a Lewis structure for each of the following covalent compounds:
- **a.** CH_4O
- **b.** N_2H_4
- **c.** Br_2
- **d.** C_2H_3Cl

3.65 Give the name of each of the following covalent compounds:
- **a.** SeO_2
- **b.** SiF_4
- **c.** P_4S_3
- **d.** OF_2

3.66 Give the name of each of the following covalent compounds:
- **a.** SO_2
- **b.** CI_4
- **c.** SiS_2
- **d.** SCl_2

3.67 Aspartic acid, a naturally occurring amino acid, is a component of the structure of aspartame (Nutrasweet™), an artificial sweetener widely used in diet soft drinks.
- **a.** Determine the shape around each red-colored atom in the Lewis structure of aspartic acid.
- **b.** Give the bond angle around each of the red-colored carbon atoms.

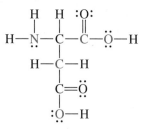

3.68 Cyanoacrylic acid is one of the compounds used to make Super Glue®. Determine the shape of the molecule around each red-colored atom in its Lewis structure.
- **a.** Determine the shape around each red-colored atom in the Lewis structure of cyanoacrylic acid.
- **b.** Give the bond angle of each of the red-colored carbon atoms.

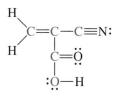

3.69 Explain why the bond angle between H—O—H in water is smaller than the bond angle between H—N—H in ammonia when O and N obey the octet rule.

3.70 Vinyl acetate is used in the production of safety glass for the automotive industry. Determine the shape around each of the carbon atoms in the Lewis structure of vinyl acetate.

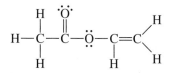

3.71 Methyl isocyanate is used in the manufacturing of pesticides and was the deadly gas released from a chemical plant in the 1984 disaster in Bhopal, India, that killed hundreds. Determine the shape around each of the carbon and nitrogen atoms in the Lewis structure of methyl isocyanate.

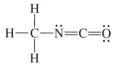

3.72 Identify the more electronegative atom in each of the following pairs:
- **a.** H and Cl
- **b.** N and O
- **c.** C and N
- **d.** F and Br

3.73 Identify the more electronegative atom in each of the following pairs:

a. S and Se   b. P and N

c. C and Cl   d. N and I

3.74 Using the Lewis structures provided,

a. label each polar covalent bond with a dipole moment arrow.

b. determine if the molecule is polar or nonpolar and draw the molecular dipole.

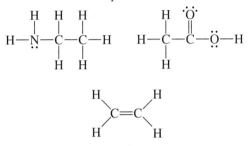

3.75 Using the Lewis structures provided,

a. label each polar covalent bond with a dipole moment arrow.

b. determine if the molecule is polar or nonpolar and draw the molecular dipole,

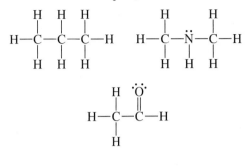

CHALLENGE PROBLEMS

3.76 Two elements, E and X, are combined and react to form a compound containing E^{2+} and X^{2-} ions. Answer the following questions about E and X based on this information.

a. Which of the elements, E or X, is more likely to be a metal?

b. Which of the elements, E or X, is more likely to be a nonmetal?

c. What is the formula of the compound formed from E and X?

d. To what group on the periodic table does each element belong?

3.77 Tooth enamel contains a hard mineral called hydroxyapatite composed of calcium, phosphate, and hydroxide ions. If the formula for the compound contains six phosphate ions and two hydroxide ions, provide the complete formula for the ionic compound.

3.78 Vinyl chloride, C_2H_3Cl, is used in the production of polyvinyl chloride, PVC, a plastic used to make tubing and pipes for a variety of medical applications.

a. Draw the Lewis structure of vinyl chloride.

b. Determine the shape and the bond angle around each of the carbon atoms.

c. Label each of the polar covalent bonds with a dipole moment arrow.

d. Determine if this compound is polar or nonpolar.

3.79 One of the most common compounds used in commercial dry cleaning is tetrachloroethylene, which has the molecular formula C_2Cl_4.

a. Draw the Lewis structure of tetrachloroethylene.

b. Determine the shape and the bond angle around each of the carbon atoms.

c. Label each of the polar covalent bonds with a dipole moment arrow.

d. Determine if this compound is polar or nonpolar.

ANSWERS TO ODD-NUMBERED PROBLEMS

Practice Problems

3.1 In atoms, the number of protons and electrons are the same and there is no net charge. In ions, the number of protons and electrons are different, giving the ion a charge.

3.3 a. 20 p, 18 e^- b. 53 p, 54 e^-

c. 16 p, 18 e^- d. 30 p, 28 e^-

3.5 a. calcium ion   b. iodide

c. sulfide   d. zinc ion

3.7 a. fluoride, F^- b. chromium(III) ion, Cr^{3+}

3.9 a. ammonium ion   b. acetate   c. cyanide

3.11 a. $AuCl_2$ b. $CaCO_3$ c. $Mg(OH)_2$

3.13 a. one Au^{2+}, two Cl^- b. one Ca^{2+}, one CO_3^{2-}

c. one Mg^{2+}, two OH^-

3.15 a. sodium acetate, $NaC_2H_3O_2$

b. barium acetate, $Ba(C_2H_3O_2)_2$

c. aluminum acetate, $Al(C_2H_3O_2)_3$

3.17 a. one Na^+, one $C_2H_3O_2^-$

b. one Ba^{2+}, two $C_2H_3O_2^-$

c. one Al^{3+}, three $C_2H_3O_2^-$

3.19 a. covalent   b. ionic   c. ionic

d. covalent   e. covalent   f. ionic

g. covalent   h. ionic   i. covalent

j. covalent

3.21 a.

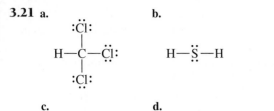

b. H—S̈—H

c. Ö=C=Ö

d. H—C≡C—H

3.23 a. Carbon prefers four bonds, hydrogen only one each, so CH_2 is unlikely to exist.

b. Nitrogen prefers three bonds, hydrogen only one each, so NH_2 is unlikely to exist.

c. These halogens could form one covalent bond like Cl_2, so BrCl could exist.

d. Sulfur will form two bonds, hydrogen one each, so H_2S could exist.

3.25 SO_3, diphosphorus pentoxide, SeF_4, carbon monoxide, N_2O_3

3.27 a.

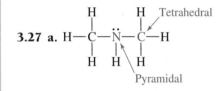

b.

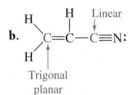

c.

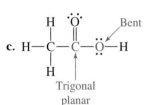

d.

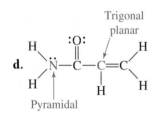

3.29 a. polar

H—Cl
⊢→

Polar

Because this molecule contains one bond, the bond and molecular dipoles are the same.

b. nonpolar

H—C—H with H above and H below

Nonpolar

C–H bonds are considered nonpolar.

c. polar

:C̈l—N̈—C̈l: ↑
     |
    :C̈l:

Polar

N–Cl bonds are nonpolar (same electronegativity), but the pair of electrons on the N gives the molecule polarity. The molecular dipole is shown in red.

d. polar

S̈=C=Ö
⊢→

Polar

C–O bonds are polar and C–S bonds are not. The Molecular dipole is shown in red.

e. polar

 ↑

Polar

S–H bonds are polar. This molecule has a bent shape. The lone pairs of electrons are on one side of the molecule in three dimensions. The Molecular dipole is shown in red.

f. polar

H—C—Br:
    |
   :Br:

Polar

C–Br bonds are polar. Considering the tetrahedral shape, the bromine side of the molecule is the negative side. The Molecular dipole is shown in red.

Additional Problems

3.31 a. 4   **b.** 7   **c.** 1   **d.** 3

3.33 a. 4   **b.** 1   **c.** 1   **d.** 3

3.35 a. fewer   **b.** valence   **c.** positive

3.37 a. 17 p, 18 e$^-$ **b.** 26 p, 23 e$^-$
    c. 24 p, 18 e$^-$ **d.** 7 p, 10 e$^-$
    e. 11 p, 10 e$^-$ **f.** 1 p, 0 e$^-$

3.39 a. iron(III)   **b.** chromium(VI)
    c. nitride   **d.** hydrogen ion, also proton

3.41 a. lithium ion, Li$^+$ **b.** bromide, Br$^-$

3.43 a. Kr   **b.** Ne   **c.** Ar   **d.** Ar

3.45 a. $C_2H_3O_2^-$ **b.** HCO_3^-
    c. nitrate   **d.** cyanide

3.47 a. $(NH_4)_2S$ **b.** NH_4Cl
    c. $(NH_4)_2SO_4$ **d.** NH_4OH

3.49 a. NaOH   **b.** Al_2O_3 **c.** KNO_3

3.51 a. sodium oxide   **b.** barium sulfate
    c. copper(II) chloride   **d.** magnesium nitrate
    e. iron(III) oxide   **f.** potassium fluoride

3.53 a. 2   **b.** 3, 2   **c.** 2   **d.** 2

3.55 1−

3.57 a. 4   **b.** 1   **c.** 3   **d.** 2

3.59 a. covalent, carbon tetrabromide
    b. covalent, silicon dioxide
    c. ionic, magnesium bromide
    d. covalent, nitrogen trichloride
    e. ionic, chromium(III) chloride

3.61 Ionic formulas are written for ionic compounds and use the smallest ratio of ions; molecular formulas are written for covalent compounds and indicate the exact number of atoms in the molecule.

3.63 a.   **b.**

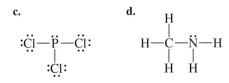

  c.   **d.**

3.65 a. selenium dioxide
    b. silicon tetrafluoride
    c. tetraphosphorous trisulfide
    d. oxygen difluoride

3.67

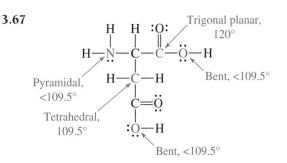

3.69 The two lone pairs on the oxygen in water repel each other more than a lone pair repels a bonded pair. This pushes the two O—H bonds in oxygen closer to each other than the N—H bonds in ammonia.

3.71

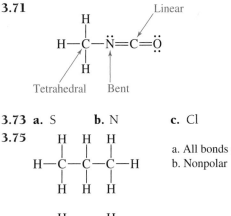

3.73 a. S   **b.** N   **c.** Cl   **d.** N

3.75

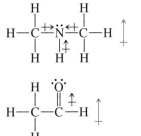

a. All bonds are nonpolar
b. Nonpolar

a. C–N bonds and N–H bond are polar
b. Polar, molecular dipole in red

a. C=O bond is polar
b. Polar, molecular dipole in red

3.77 $Ca_{10}(PO_4)_6(OH)_2$

3.79

a. Lewis structure shown at left.
b. Both carbons are trigonal planar and the bond angle is 120 degrees.
c. C—Cl bonds are polar, dipoles shown at left.
d. The polar bonds are pulling outward equally and oppositely so the molecule has no molecular dipole and is nonpolar.

Guided Inquiry Activities **FOR CHAPTER 4**

SECTION 4.2

EXERCISE 1 Representing Molecules on Paper

Information

The following structures are representations of two covalent compounds that are *structural isomers*: isopropyl alcohol (rubbing alcohol) and *n*-propyl alcohol.

Isopropyl alcohol

C_3H_8O
Molecular formula

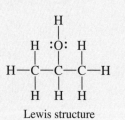 $CH_3CH(OH)CH_3$

Lewis structure   Condensed structure   Skeletal (or line) structure

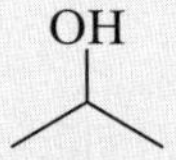

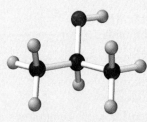

Ball-and-Stick Model

n-Propyl alcohol

C_3H_8O
Molecular formula

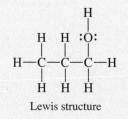 $CH_3CH_2CH_2OH$

Lewis structure   Condensed structure   Skeletal (or line) structure

Ball-and-stick model

Questions

1. How are *n*-propyl alcohol and isopropyl alcohol the same? How are they different? *Not connected in continuous chain.*

2. How do a molecular formula and a condensed structure differ?

3. How do a Lewis structure and a condensed structure differ?

4. a. What atoms do the black spheres in the ball-and-stick models represent?
   b. What two atoms do the red sphere and the small grey sphere to the right of it represent?

5. What atom is always at the intersection of two line segments in the skeletal structure?

6. Provide a definition for structural isomer. (**HINT:** *What is the same about the n-propyl alcohol and isopropyl alcohol molecules? What is different?*)

7. For the isobutane and *n*-butane molecules shown in their Lewis structure, provide the condensed structure and the skeletal structure.

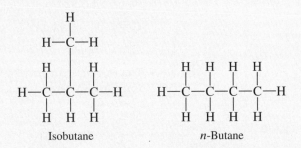

Isobutane   *n*-Butane

8. Provide the corresponding skeletal structure for the following hydrocarbons.

9. Explain how the number of hydrogens written after each carbon in the condensed structures in Problem 8 was determined.

10. Provide the condensed structure for the following skeletal structures.

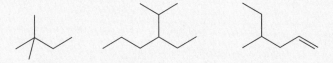

EXERCISE 2 Meet the Hydrocarbons

Information

Organic Compounds

- contain carbon
- form covalent bonds
- are grouped into families
  - The families, with the exception of the alkanes, are classified by the *functional group(s)* present.
  - Functional groups are the reactive part of the molecule. Polar molecules are more reactive than nonpolar molecules.

THE HYDROCARBON FAMILY OF ORGANIC COMPOUNDS	
Family Member	Functional Group
Alkane	None, all C and H are single bonded
Alkene	C=C double bond
Alkyne	C≡C triple bond
Aromatic	Planar, ring structures based on benzene. Can contain heteroatoms. Benzene

Hydrocarbon Molecule Set

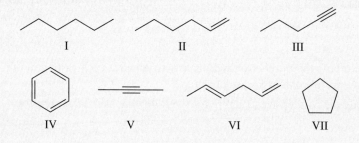

I   II   III

IV   V   VI   VII

Questions

1. How are all the molecules in the molecule set the same? What do all hydrocarbons have in common?
2. Which molecules are polar? Which are nonpolar?
3. Based on the information given, are hydrocarbons very reactive molecules?
4. Identify the molecules in the molecule set that belong to the following families.

Family	Molecule Number(s)
Alkane	
Alkene	
Aromatic	
Alkyne	

5. What is the molecular shape (geometry) of molecule IV?
6. Distinguish between an organic family of compounds and its functional group.

When you think of the word *organic* you may think of pesticide-free fruits and vegetables at the grocery store, but in chemistry the term has a broader meaning—*all things carbon*. In Chapter 4, we introduce the specific chemistry of carbon, the basis for life on Earth.

Introduction to Organic Compounds

The word *organic* can describe a set of environmentally friendly farming practices or imply that a product is in some way more natural, but chemists use *organic* to identify covalent compounds containing carbon. **Organic compounds** are composed primarily of carbon and hydrogen but may also include oxygen, nitrogen, sulfur, phosphorus, and a few other elements. The molecules of life, or **biomolecules**, such as proteins, carbohydrates, lipids, and DNA, belong to a class of compounds known as organic compounds. Compounds that do not contain carbon and hydrogen are classified as **inorganic compounds**. Using what we have learned about atoms (Chapter 2) and how they combine to form molecules (Chapter 3), we now bring these concepts together to begin the study of the molecules that make up living things and the chemistry of life.

Why are carbon-containing compounds singled out in this special category? Organic compounds, whether naturally occurring or synthesized in the laboratory, are found in familiar substances such as gasoline, cotton, plastics, vitamins, medicines, and cosmetics. Besides the importance of these compounds to life itself, carbon is unique in its ability to form bonds with other atoms of carbon to create chains and rings of various sizes and shapes. In fact, because of this unique characteristic of carbon, there are more than 25 million known organic compounds, compared to less than half a million total compounds for all of the other elements! Carbon's ability to form such a tremendous variety of compounds and our ability to manipulate these compounds in the laboratory contribute to our ever-increasing interest in **organic chemistry**, the field of chemistry dedicated entirely to studying the structure, characteristics, and reactivity of carbon-containing compounds.

The thought of learning the chemistry of more than 25 million individual organic compounds is more than just a little overwhelming. Luckily, to study and understand organic chemistry, we do not need to acquire such a huge volume of information. Over many years of studying organic chemistry, chemists have been able to group organic compounds into distinct families based on their molecular structure and composition. Furthermore, members of the same family behave similarly in chemical reactions. So, if we learn to recognize the structural similarities within the families of organic compounds, we can begin to understand a great deal of organic chemistry. This chapter focuses on the structures and reactions of those families that are biologically active and/or are relevant to the chemistry of life. To begin to understand organic chemistry, we start with the simplest family of organic compounds that, while not biologically active, form the structural foundation for all other organic compounds.

4.1 Alkanes: The Simplest Organic Compounds

We use fossil fuels for heating, transportation, and generating electricity. Most of these fossil fuels belong to the first family of organic compounds that we will consider. This family of compounds is known as the **alkanes**. These structurally simple organic compounds are made up solely of carbon and hydrogen. In fact, the alkanes are typically referred to as **saturated hydrocarbons**. The term *hydrocarbon* indicates that alkanes are made up entirely of hydrogen and carbon, and *saturated* indicates that these compounds contain only single bonds, with each carbon atom (in addition to being bonded to other carbon atoms) being bonded to (or saturated with) the maximum number of hydrogen atoms.

Straight-Chain Alkanes

A portable form of heating fuel often used in remote locations, in gas grills, and recreational vehicles is the alkane called propane. Compounds like propane belong to the simplest group of alkanes, known as **straight-chain alkanes**, and are made up of carbon atoms joined to one another to form continuous, unbranched chains of varying length. In order to distinguish individual alkanes from one another, each compound is given a name that is based on the number of carbon atoms in its chain. For example, propane is the name given to the alkane containing three carbon atoms connected in a straight chain and saturated with hydrogen.

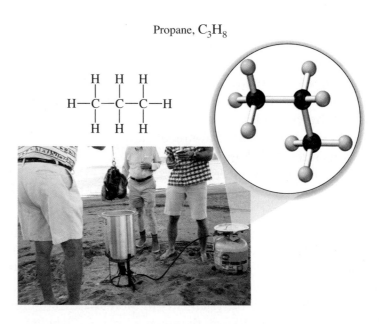

Propane, C_3H_8

The names and Lewis structures of the first 10 alkanes are shown in Table 4.1. The names of the first 4 alkanes are seemingly random, but they are in fact historical names still used today. The names of the straight-chain alkanes with five or more carbons are derived from a Greek numerical prefix (for example, *pent* = 5) followed by the ending *ane* (indicating the alkane family). It is important to be able to recognize the names of these first 10 alkanes because the names of the majority of organic compounds are derived from these names.

TABLE 4.1		NAMES OF THE FIRST 10 STRAIGHT-CHAIN ALKANES		
Number of Carbon Atoms	Prefix	Name of Alkane	Molecular Formula	Lewis Structure
1	meth	Methane	CH_4	
2	eth	Ethane	C_2H_6	
3	prop	Propane	C_3H_8	
4	but	Butane	C_4H_{10}	
5	pent	Pentane	C_5H_{12}	
6	hex	Hexane	C_6H_{14}	
7	hept	Heptane	C_7H_{16}	
8	oct	Octane	C_8H_{18}	
9	non	Nonane	C_9H_{20}	
10	dec	Decane	$C_{10}H_{22}$	

Cycloalkanes

In addition to forming long chains, carbon can also form rings. These compounds are called **cycloalkanes** and are also saturated hydrocarbons. As the family name *cyclo*-alkane indicates, the name of these compounds is derived by adding the prefix *cyclo* to the alkane name for the compound containing the same number of the number of carbon atoms. For example, the simplest cycloalkane is cyclopropane. Notice that the name indicates that this compound is a ring compound (cyclo) of three carbon atoms (propane).

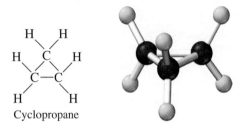

Cyclopropane

In theory, cycloalkanes (and similar ring-containing organic compounds) can exist in many different ring sizes. However, rings of five and six carbon atoms are the most common in nature. In our study of organic chemistry, we will focus on the more common rings of five and six carbon atoms. Table 4.2 gives the names and Lewis structures of cycloalkanes with three to six carbons.

Alkanes Are Nonpolar Compounds

In Chapter 3, a molecule was identified as polar or nonpolar by considering both the electronegativity of its atoms and the overall shape of the molecule. Nonpolar molecules are those that have an even distribution of electrons over their entire surface, while polar molecules have a partially negative part and a partially positive part (poles) due to an uneven distribution of electrons in the molecule.

The electronegativities of carbon and hydrogen are so similar that when these two elements form covalent bonds, they do so by sharing electrons equally. Because bonds

TABLE 4.2	CYCLOALKANES: NAMES AND COMMON STRUCTURES

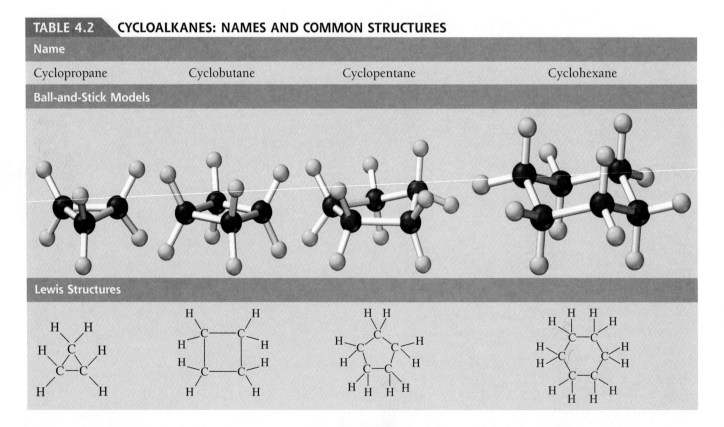

Name			
Cyclopropane	Cyclobutane	Cyclopentane	Cyclohexane

4.2 • Representing the Structures of Organic Compounds

formed between hydrogen and carbon are nonpolar and because alkanes are composed solely of carbon and hydrogen, alkane molecules, regardless of their shape, are also nonpolar. In Chapter 6, we'll discover that this property of alkanes dictates how these compounds behave.

4.2 Representing the Structures of Organic Compounds

In Tables 4.1 and 4.2 the Lewis structures of some common alkanes and cycloalkanes are shown. In Chapter 3 we saw that Lewis structures show the connectivity of the atoms in a molecule showing all atoms, bonds, and lone pairs in the molecule. All the information provided by a Lewis structure is important, but as molecules get larger, chemists often use simplified representations for the structures of organic compounds that still show the whole molecule but omit some of the details. For example, since we have established that an uncharged oxygen will always have two bonds and two lone pairs of electrons in a covalent compound, we often omit the lone pairs of electrons when drawing oxygen in an organic compound.

Condensed Structural Formulas

The first type of representation that we will consider is the **condensed structural formula** (also called condensed structure). Similar to Lewis structures, these structures show *all the atoms* in a molecule, but they show *as few bonds as possible*. Condensed structures may or may not show lone pairs and are not very useful for drawing cycloalkanes.

When drawing or interpreting condensed structures, it is helpful to remember the preferred covalent bonding patterns discussed in Chapter 3. For example, carbon always makes four bonds to other atoms. Because the bonds between atoms are not included in these structures, knowing these patterns will help you quickly interpret condensed structural formulas.

The condensed structural formula for propane is shown in the following figure between its Lewis structure and its molecular formula. Note the differences. The molecular formula shows only the type and number of atoms in the molecule while the Lewis structure shows the complete connectivity within the molecule—all atoms and all bonds. In terms of information provided, the condensed structural formula is somewhere between the two. While each of the formulas shows all of the atoms in the molecule, the information regarding connectivity increases as we move from molecular formula to condensed structural formula to Lewis structure.

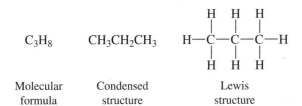

C_3H_8	$CH_3CH_2CH_3$	Lewis structure
Molecular formula	Condensed structure	Lewis structure

In the condensed structural formula, the atoms are listed from left to right in the order in which they are connected in the molecule, and it is understood that the octet of each atom is completed according to its preferred covalent bonding pattern. The condensed structure is based on the preferred bonding patterns, so it is reasonable that the structure of propane is not simply a string of atoms bonded together, for example, C—H—H—H—C—H—H—C—H—H—H. Rather, because hydrogen can make only one bond and carbon makes four bonds, the "string of atoms" structure has many problems and is not what is intended by a condensed structural formula.

From this example, notice that a carbon bonded to three hydrogen atoms in the Lewis structure is written as CH_3 in the condensed structure, and a carbon bonded to two hydrogen atoms is written as CH_2. In each case, the carbon atom's preferred bonding pattern of four bonds is completed by bonding to other carbon atoms.

SAMPLE PROBLEM 4.1

Drawing Condensed Structural Formulas

Draw a condensed structural formula for each of the Lewis structures in the following figures:

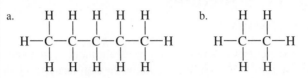

SOLUTION

The condensed structural formula shows all atoms but omits the bonds between atoms. To write a condensed formula, begin with the carbon atom farthest to the left in the structure. List that carbon and then the hydrogens bonded to it, using a subscript to indicate the number of hydrogens. Proceed in the same manner for each carbon atom in the Lewis structure.

a. The correct condensed structure is $CH_3CH_2CH_2CH_2CH_3$. Alternatively, this structure can be written as $CH_3(CH_2)_3CH_3$.

b. The correct condensed structure is CH_3CH_3.

SAMPLE PROBLEM 4.2

Converting Condensed Structures to Lewis Structures

Convert each of the following condensed structural formulas to a Lewis structure:
a. $CH_3CH_2CH_2CH_3$
b. $CH_3CH_2CH_2CH_2CH_2CH_2CH_3$

SOLUTION

Reviewing the preferred bonding patterns from Table 3.3, hydrogen can make only one bond, while carbon makes four. With these facts in mind, reverse the procedure that we used in previous Sample Problem 4.1. First, write the leftmost carbon atom and draw bonds for each of the hydrogen atoms listed after that carbon and add the hydrogens to those bonds. Then, draw a bond for the next carbon atom, add that carbon, and continue for the entire structure.

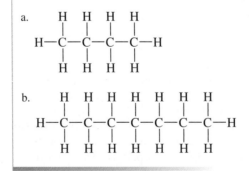

Skeletal Structures

The second type of structure that we will consider is commonly used on prescription drug inserts and on the labels of pesticides and herbicides. It is the main structure type used to represent cycloalkanes. This type of structure, called a **skeletal structure**, is truly a "bare bones" structure because it shows only the bonds between carbon atoms; carbon and hydrogen atoms are not shown explicitly. Propane is shown in Figure 4.1 as

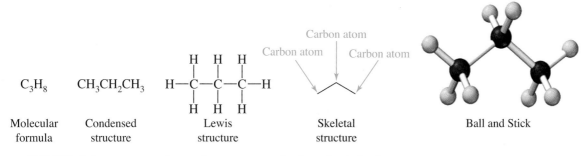

C_3H_8 $CH_3CH_2CH_3$ H—C—C—C—H

Molecular   Condensed   Lewis   Skeletal   Ball and Stick
formula   structure   structure   structure

▲ **FIGURE 4.1 Representations of propane.** Notice how the skeletal structure compares to the others. In the skeletal structure, the carbon and hydrogen atoms are not shown.

an example. Carbon atoms in skeletal structures are understood to exist at a corner/ angle created by the intersection of two bonds or at the termination of a bond. Propane contains three carbons, so is drawn with just two lines. All bonds to carbon are drawn in a skeletal structure, *except* those to hydrogen. In propane, the middle carbon atom shows two bonds to other carbons, implying that two hydrogens are also bonded to it. The end carbons show one bond; therefore, they would have three hydrogens bonded to them. Unless otherwise specified, each carbon atom is assumed to be saturated (see Section 4.1) with hydrogen, but hydrogens bonded to carbon are not shown, only implied. Because only bonds between carbon atoms are shown, it is not possible to draw a skeletal structure for CH_4, methane—it has only one carbon! In fact, we typically do not draw skeletal structures for alkanes with fewer than three carbon atoms.

Skeletal structures are particularly useful for representing the cycloalkanes. When these molecules are represented using skeletal structures, they appear as familiar geometric figures. So, the three-carbon ring of cyclopropane looks like a triangle, the four-carbon ring of cyclobutane looks like a square, the five-carbon ring of cyclopentane looks like a pentagon, and the six-carbon ring looks like a hexagon. Figure 4.2 shows the skeletal structures of the common cycloalkanes.

Cyclopropane   Cyclobutane   Cyclopentane   Cyclohexane   Cycloheptane

◀ **FIGURE 4.2 Skeletal structures of some common cycloalkanes.** The cycloalkane structures look like geometric figures. Each corner represents a carbon atom.

SAMPLE PROBLEM 4.3

Drawing Skeletal Structures

Draw a skeletal structure for each of the following compounds:

a. Pentane

b. Butane
$CH_3CH_2CH_2CH_3$

c. Cyclobutane

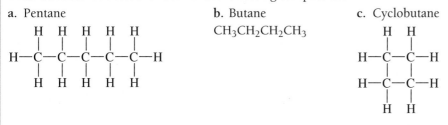

SOLUTION

To draw the skeletal structure for a straight-chain alkane, first count the number of carbon atoms in the chain. As you put your pen or pencil to the paper, count "one" for the first carbon atom. Then, move your pen or pencil to draw the bond to the next carbon and count "two." Continue the structure by drawing a zigzag line with each corner representing a carbon atom. The last line of the zigzag ends as you count the final carbon.

To draw the skeletal structure for a cycloalkane, count the number of carbon atoms in the ring and draw the corresponding geometric figure.

a. 〳〵〳   In the Lewis structure described, there are five carbon atoms. In its skeletal structure, notice that each of the three corners represents a carbon atom as does the beginning point on the left and the ending point on the right, giving a total of five carbon atoms.

b. 〳〵〵   A beginning, an end, and two angles = four carbon atoms

c. ☐   The four carbon atoms in the Lewis structure of this cycloalkane give a skeletal structure that looks like a square.

PRACTICE PROBLEMS

4.1 Describe the difference between a Lewis structure and a condensed structure in terms of atoms and bonds shown in the structures.

4.2 Describe the difference between a Lewis structure and a skeletal structure in terms of atoms and bonds shown in the structures.

4.3 Draw condensed structural formulas for the 10 straight-chain alkanes in Table 4.1.

4.4 Draw skeletal structures for the 10 straight-chain alkanes in Table 4.1.

4.5 Explain why it is not possible to draw a skeletal structure for methane.

4.6 Octane, C_8H_{18}, is a component of gasoline. Draw the Lewis structure, the condensed structure, and the skeletal structure of octane without referring back to Table 4.1.

4.7 Cyclohexane, C_6H_{12}, is used to clean greasy machines and parts in many industrial settings. Draw the Lewis structure and the skeletal structure of cyclohexane without referring back to Table 4.2.

4.3 Fatty Acids—Biological "Hydrocarbons"

In the popular press, we often read about the health issues related to a diet high in *saturated* fats. Earlier, alkanes were described as *saturated* hydrocarbons. In both cases, saturated has the same meaning: hydrocarbons with only carbon–carbon single bonds. **Saturated fatty acids** are the main components of saturated fats.

Fatty acids have structures and properties similar to the alkanes, and for this reason they are sometime called biological "hydrocarbons". They belong to a class of biomolecules called **lipids**. Lipids are structurally diverse but share the common property of being nonpolar biomolecules.

Structure and Polarity of Fatty Acid Molecules

Fatty acids are long-chain alkanelike compounds. The most common and biologically most important fatty acids are compounds containing from 12 to 22 carbon atoms. Most naturally occurring fatty acids have even numbers of carbon atoms. The names and structures of even-numbered saturated fatty acids are given in Table 4.3. Compare the structures of these compounds with those of the straight-chain alkanes in Table 4.1. Note that in skeletal structure noncarbon atoms like oxygen are shown, as are hydrogens, not bonded to carbon (OH, for example).

Notice that the structures of alkanes and saturated fatty acids both contain straight chains of carbon atoms, but the fatty acids have two oxygens bonded at one end. A group of atoms like this bonded in a particular way is called a **functional group**. Functional groups are used to classify organic compounds into various families based on reactivity. Fatty acids contain a functional group (a carbon, two oxygens, and a hydrogen bonded as shown in Figure 4.3) called a **carboxylic acid**; it is responsible for the "acid" part of the name of fatty acids. (More functional groups are introduced in section 4.4.)

▲ **FIGURE 4.3 A carboxylic acid functional group.** This functional group is common to all fatty acids.

TABLE 4.3 SATURATED FATTY ACIDS: NAMES AND STRUCTURES

Name	Carbon Atoms	Source	Structures
Lauric acid	12	Coconut	$CH_3CH_2CH_2CH_2CH_2CH_2CH_2CH_2CH_2CH_2CH_2COH$
Myristic acid	14	Nutmeg	$CH_3CH_2CH_2CH_2CH_2CH_2CH_2CH_2CH_2CH_2CH_2CH_2CH_2COH$
Palmitic acid	16	Palm	$CH_3CH_2CH_2CH_2CH_2CH_2CH_2CH_2CH_2CH_2CH_2CH_2CH_2CH_2CH_2COH$
Stearic acid	18	Animal fat	$CH_3CH_2CH_2CH_2CH_2CH_2CH_2CH_2CH_2CH_2CH_2CH_2CH_2CH_2CH_2CH_2CH_2COH$
Arachidic acid	20	Peanut	$CH_3CH_2CH_2CH_2CH_2CH_2CH_2CH_2CH_2CH_2CH_2CH_2CH_2CH_2CH_2CH_2CH_2CH_2CH_2COH$
Behenic acid	22	Canola	$CH_3CH_2CH_2CH_2CH_2CH_2CH_2CH_2CH_2CH_2CH_2CH_2CH_2CH_2CH_2CH_2CH_2CH_2CH_2CH_2CH_2COH$

Despite the presence of two electronegative oxygen atoms in the structure of a fatty acid, these compounds behave similarly to nonpolar alkanes. In Chapter 3 we saw that polar molecules have electronegative atoms like oxygen that draw electrons toward themselves, creating an uneven distribution of electrons in a molecule. The carboxylic acid part of a fatty acid *is* polar, but relative to the large nonpolar part it does not contribute as much to the overall polarity of the molecule. The nonpolar hydrocarbon chain of the fatty acid is so much larger than the polar carboxylic acid group that it dominates the character of the compound, making fatty acids mainly nonpolar. Molecules like fatty acids that have both polar and nonpolar parts are called **amphipathic** (from the Greek *amphi* meaning "both" and *pathic* meaning "ways") compounds.

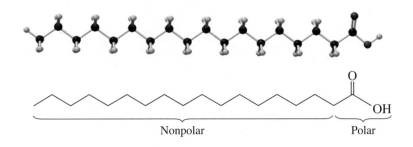

Nonpolar   Polar

Unsaturated Fatty Acids

Notice the difference in the structure of oleic acid (shown) and the structure of the saturated fatty acids from Table 4.3. Oleic acid and similar fatty acids are called **unsaturated fatty acids** because they contain less than the maximum number of hydrogen atoms per carbon atom. In these compounds, a carbon–carbon double bond is formed for each missing pair of hydrogen atoms so that each carbon atom maintains an octet. The carbon–carbon double bond is another functional group present in many organic compounds called an **alkene**. In the skeletal structure, the double bond is represented with a second line between the two carbons and in the fatty acids, the skeletal structure at the double bond is drawn in a U-shape. Fatty acids with one double bond are called **monounsaturated** and those with two or more double bonds are called **polyunsaturated**.

Oleic acid

$CH_3CH_2CH_2CH_2CH_2CH_2CH_2CH_2CH=CHCH_2CH_2CH_2CH_2CH_2CH_2CH_2COOH$

U-shape at double bond

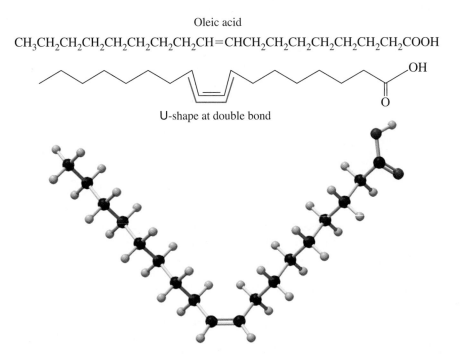

The presence of the carbon–carbon double bond does not affect the nonpolar nature of the unsaturated fatty acids. Table 4.4 shows the structures and names of six common unsaturated fatty acids. Take particular note of the ω (omega) designations of the unsaturated fatty acids. These designations are commonly used in nutrition literature and in the popular press. In this system, the carbon farthest from the carboxylic acid is numbered 1,

TABLE 4.4	UNSATURATED FATTY ACIDS: NAMES AND STRUCTURES		
Name	**Carbon Designation**	**Source**	**Structures**
Monounsaturated Fatty Acids			
Palmitoleic acid	[16:1], ω-7	Butter	
Oleic acid	[18:1], ω-9	Olives, corn	
Polyunsaturated Fatty Acids			
Linoleic acid* (commonly called LA)	[18:2], ω-6	Soybean, safflower, corn	
α-Linolenic acid* (commonly called ALA)	[18:3], ω-3	Flaxseed, canola	
γ-Linolenic acid (commonly called GLA)	[18:3], ω-6	Formed from ALA, evening primrose, spirulina	
Arachidonic acid (commonly called AA)	[20:4], ω-6	Formed from LA, lard, eggs	

*Indicates essential fatty acid.

the next carbon is 2, and so forth. The ω number indicates the carbon positioning of a double bond in the structure of the fatty acid. The unsaturated fatty acid α-linolenic acid commonly referred to as ALA (see Table 4.4) is given a designation of [18:3], ω-3. This indicates that the fatty acid has 18 carbons and 3 double bonds with the first double bond located between carbons 3 and 4. Two more double bonds are understood to occur at intervals of three carbons for a total of three. So, the other two double bonds of α-linolenic acid are located between carbons 6 and 7, and 9 and 10.

FATTY ACIDS IN OUR DIETS

Years of research by nutrition scientists have shown that a diet low in *overall* fat content is healthier. However, we do need some fats in our diets because they play important roles as insulators and protective coverings for our internal organs and nerve fibers. The Food and Drug Administration (FDA) recommends that a maximum of 30% of the calories in a normal diet come from fatty acid–containing compounds.

At the beginning of this section, we mentioned the health concern associated with a diet high in saturated fats, fats composed mainly of saturated fatty acids. The FDA also recommends that the majority of our fat intake come from foods containing a higher percentage of unsaturated fatty acids. The polyunsaturated fatty acids linoleic and α-linolenic are considered **essential fatty acids** because our bodies cannot produce these compounds so they must be obtained in the diet. Table 4.5 shows the fatty acid composition of some common fats.

TABLE 4.5 FATTY ACID COMPOSITION OF COMMON DIETARY FATS			
	Saturated	Monounsaturated	Polyunsaturated
	%*	%*	%*
Butter	68	28	4
Canola oil	7	61	32
Coconut oil	91	7	2
Corn oil	13	28	59
Lard	43	47	10
Olive oil	15	73	12
Palm oil	49	40	11
Soybean oil	15	25	60
Sunflower oil	13	21	66

*Approximate percentages by weight of total fatty acids

SAMPLE PROBLEM 4.4

Drawing Fatty Acids

Draw the skeletal structure of the following fatty acids:

a. Myristic acid [14:0]   b. Eicosenoic acid [20:1], ω-9

SOLUTION

The bracket notation indicates

[Total number of carbons:number of double bonds]

All fatty acids will have the carboxylic acid functional group at one end.

a. For myristic acid, this becomes

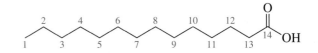

b. The double bonds in naturally occurring fatty acids will be represented in skeletal structure as a U-shape. The omega number tells us that the double bond will be between carbons 9 and 10 counting from the hydrocarbon end.

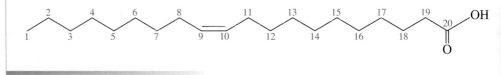

PRACTICE PROBLEMS

4.8 The most common fatty acid found in animals is the saturated fatty acid stearic acid, [18:0]. Draw stearic acid in skeletal structure.

4.9 Stearidonic acid [18:4] is an essential omega-3 fatty acid found in black currant and the cyanobacteria spirulina. Draw stearidonic acid in skeletal structure.

4.4 Families of Organic Compounds— Functional Groups

Organic compounds are classified into families based on common functional groups. (Remember that alkanes are defined as the family with no functional group.) As with the carboxylic acid and alkene groups introduced previously, a functional group is a common arrangement of atoms bonded in a particular way. Each functional group has specific properties and reactivity, and organic compounds that contain the same functional group behave similarly. To understand the properties and reactivity of each of the millions of different organic compounds, we really need to identify the properties and reactivity of only about a dozen common functional groups.

Some common functional groups are listed in Table 4.6. In this chapter, we will consider only the families of compounds that contain just carbon and hydrogen (the first four groups listed in Table 4.6). Note that most of the functional groups contain atoms other than carbon or hydrogen called **heteroatoms** in their structure. The most common heteroatoms are oxygen, nitrogen, sulfur, and the halogens. We will encounter the other functional groups listed in later chapters as they appear in the structures of biomolecules.

TABLE 4.6 FAMILIES OF ORGANIC COMPOUNDS: COMMON FUNCTIONAL GROUPS

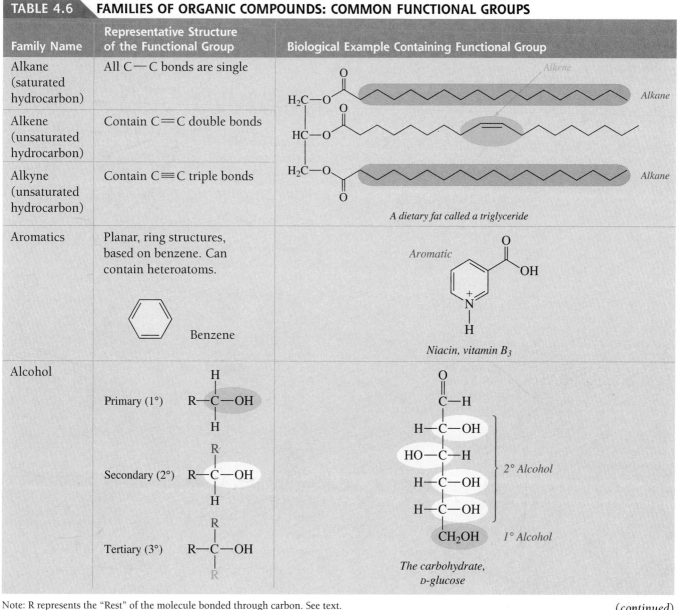

Family Name	Representative Structure of the Functional Group	Biological Example Containing Functional Group
Alkane (saturated hydrocarbon)	All C—C bonds are single	
Alkene (unsaturated hydrocarbon)	Contain C=C double bonds	
Alkyne (unsaturated hydrocarbon)	Contain C≡C triple bonds	*A dietary fat called a triglyceride*
Aromatics	Planar, ring structures, based on benzene. Can contain heteroatoms.	*Niacin, vitamin B₃*
Alcohol	Primary (1°), Secondary (2°), Tertiary (3°)	*The carbohydrate, D-glucose*

Note: R represents the "Rest" of the molecule bonded through carbon. See text.

(*continued*)

TABLE 4.6	FAMILIES OF ORGANIC COMPOUNDS: COMMON FUNCTIONAL GROUPS (Continued)	
Family Name	**Representative Structure of the Functional Group**	**Biological Example Containing Functional Group**

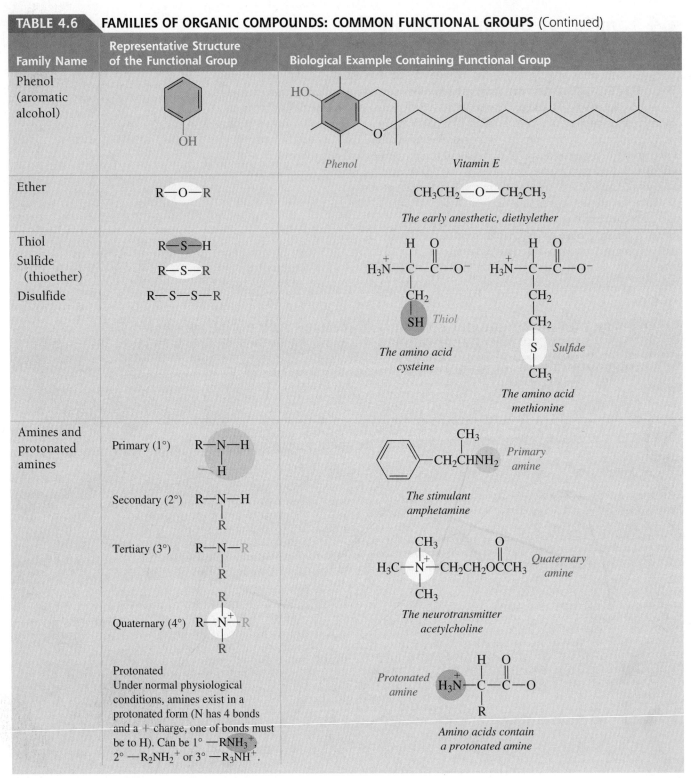

Note: R represents the "Rest" of the molecule bonded through carbon. See text.

(continued)

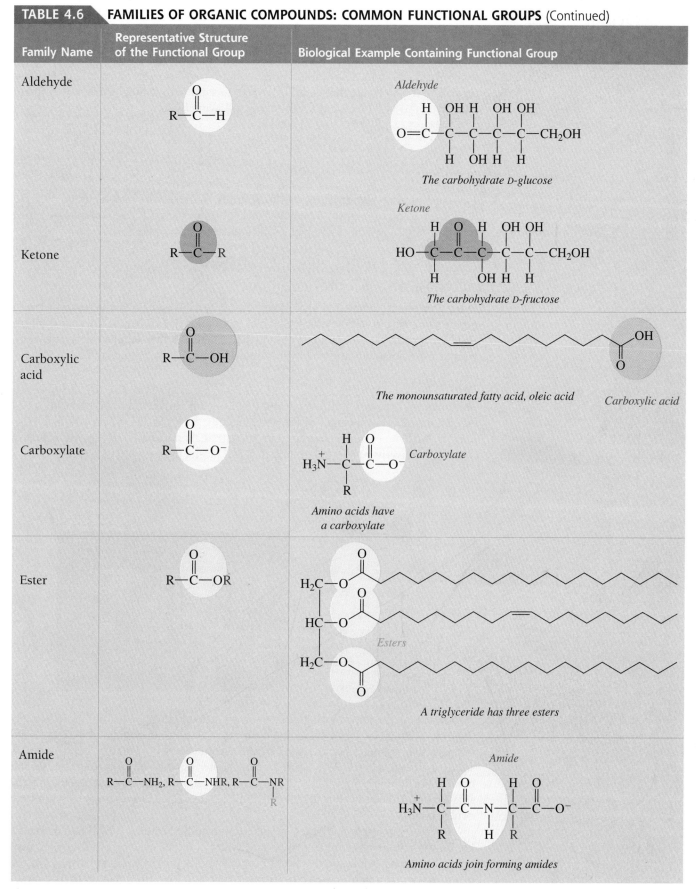

TABLE 4.6   FAMILIES OF ORGANIC COMPOUNDS: COMMON FUNCTIONAL GROUPS (Continued)

Note: R represents the "Rest" of the molecule bonded through carbon. See text.

The functional group is the reactive part of an organic molecule. To keep the focus on the functional group, an *R* is often used generically to represent the *Rest* of the molecule. The R part is usually bonded to the functional group through a carbon bond. For example, a carboxylic acid, like the fatty acids discussed in Section 4.3, may contain 20 or more carbons in its hydrocarbon portion, but the carboxylic acid functional group is the part of the molecule that identifies it with its family. To highlight this, we represent the compound as shown on the right of the following figure:

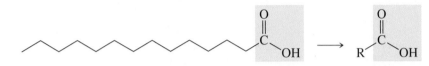

This helpful abbreviation allows us to simplify the structure and highlight just the functional group of interest. Remember that the abbreviation R can represent anything from one carbon to a more complex group containing many carbons.

Unsaturated Hydrocarbons—Alkenes, Alkynes, and Aromatics

In the previous section we saw that unsaturated fatty acids contain double bonds between one or more pairs of carbon atoms. Organic compounds that contain at least one multiple bond are broadly characterized as unsaturated. There are three families of organic compounds that contain unsaturated hydrocarbons: alkenes, alkynes, and aromatics.

Alkenes

As tomatoes, bananas, and other fruits ripen on the vine or tree, the plant hormone ethene, the simplest member of the alkene family, controls this process. Ethene (also known as ethylene) is also an important industrial chemical used in the production of the plastic polyethylene found in milk jugs and plastic bags.

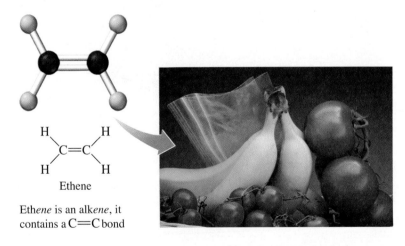

Ethene

Eth*ene* is an alk*ene*, it contains a C=C bond

Carbon atoms joined by a double bond are held closer together and more strongly than those joined by just a single bond. In other words, a double bond is shorter and stronger than a single bond. However, the second bond of a double bond is not as strong as the first bond. As we will see in Chapter 6, when alkenes react, the second bond of the double bond is broken, but the carbon atoms remain joined together by the single bond. This means that alkenes are more reactive than alkanes, which do not have double bonds.

Naturally occurring alkenes are also found in β-carotene, the alkene responsible for the orange pigment in carrots and sweet potatoes used by our bodies to produce vitamin A. Other alkenes like limonene and α-pinene are scents found in the oils of citrus fruits and pine trees, respectively, and give these plants their characteristic smells. Other biologically important molecules contain alkenes (fatty acids, cholesterol, and some metabolic intermediates) in addition to other functional groups.

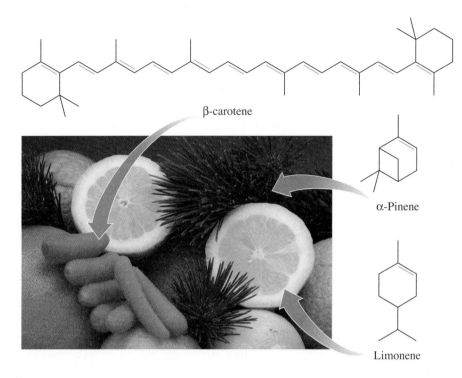

β-carotene

α-Pinene

Limonene

Alkynes

The fuel used in high-temperature welding torches to melt iron and steel is called ethyne or, more commonly, acetylene. Compounds, like acetylene, which contain one or more carbon–carbon triple bonds are members of the **alkyne** family.

$$H—C≡C—H$$

Ethyne

Eth*yne* is an alk*yne*, it contains a C≡C bond

As you might expect, the alkyne's triple bond is even shorter and stronger than the alkene's double bond. However, the two additional bonds of the triple bond mean that alkynes are even more reactive (and, therefore, less stable) than alkenes. Because they are so reactive, alkynes are rare in nature, occurring mainly in short-lived compounds. The poison dart frog secretes a poison that contains two alkyne functional groups as a defense mechanism against predators.

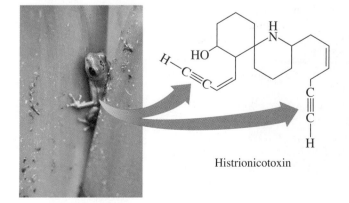

Histrionicotoxin

Aromatics

One of the more unique families of unsaturated hydrocarbons is the aromatic family. These compounds were originally dubbed "aromatic" because many of first ones discovered had pleasant aromas. Oil of spearmint and peppermint are compounds in the aromatic family. **Aromatic compounds** have a structure similar to benzene. With so many double bonds, these compounds may be presumed to be very reactive. However, the members of this family are unusually stable and not very reactive at all.

Many members of the aromatic family contain the unsaturated six-membered ring that is called a benzene ring. (When it is part of a much larger molecule, it is also called *phenyl*.) Benzene has the structure shown in the figure. Notice that benzene appears to have a series of alternating carbon–carbon single and double bonds. From the previous discussion of the lengths of double versus single bonds, we would expect benzene to have three longer single bonds and three shorter double bonds in an alternating pattern. We might also expect such a structure to have reactivity similar to that of a compound with three carbon–carbon double bonds. However, experimental evidence shows that all the carbon–carbon bonds of benzene are the same length and that benzene is an exceptionally stable compound. In fact, benzene is resistant to reactions that would break any (or all) double bonds. Unsaturated cyclic compounds, like benzene, which are unusually and unexpectedly stable (more stable than normal alkenes) are said to exhibit **aromaticity**.

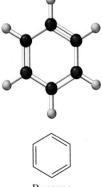

Benzene

The properties and behavior of benzene and other aromatic compounds arise from the fact that the double bonds between adjacent carbon atoms are not "static." In other words, the electrons shared to create those double bonds could be in either of the two arrangements shown in the following figure. (Note the difference as demonstrated by the type of bond between the starred carbon atoms in each structure.) In fact, the electrons in the double bonds are shared evenly by all six carbons. This is why you will often see benzene and benzene rings represented by the structure on the far right in the figure. This representation is called the **resonance hybrid** form and implies equal sharing of electrons in benzene instead of distinct double and single bonds.

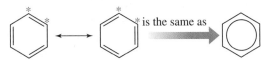

Because these electrons are free to "roam" about the molecule, they are much less apt to react with other molecules. This decreased reactivity translates to the enhanced stability exhibited by aromatic compounds.

There are many aromatic compounds containing two or more phenyl (benzene) rings present in tobacco smoke. These compounds are called polycyclic aromatic hydrocarbons or PAHs. Phenanthrene and benzo[a]pyrene are two such molecules. Polycyclic aromatic hydrocarbons are often formed when organic matter, such as tobacco leaves, is burned. Many of these PAHs have been shown to be carcinogenic. Aromatic compounds are also components of many plastics and pharmaceuticals.

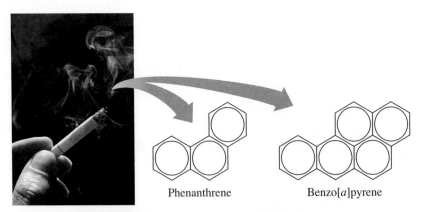

Phenanthrene   Benzo[a]pyrene

Tobacco smoke produces many polycyclic aromatic hydrocarbons

PRACTICE PROBLEMS

4.10 Identify the family of hydrocarbon present in the following:

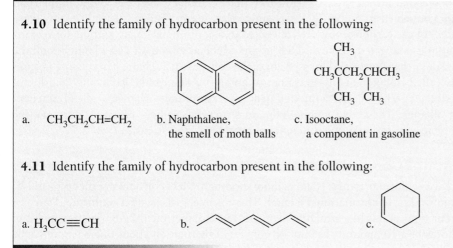

$$CH_3CCH_2CHCH_3$$

a. $CH_3CH_2CH=CH_2$ b. Naphthalene,
the smell of moth balls   c. Isooctane,
a component in gasoline

4.11 Identify the family of hydrocarbon present in the following:

a. $H_3CC\equiv CH$ b.   c.

4.5 Isomerism in Organic Compounds, Part 1

When you pull into a gas station to fill your car with fuel, you may choose between two or more types of gasoline with different octane ratings. What we call "regular" gasoline has an octane rating of 87, and the premium gasolines have ratings of 89, 92, or higher. Despite the name *octane rating*, these numbers have nothing to do with the straight-chain alkane octane introduced in Section 4.1. Octane ratings indicate how readily and completely a fuel burns (known as combustion efficiency) when compressed in the engine of your car. The ratings are based on the combustion efficiency of an eight-carbon alkane, but not the straight-chain octane.

Because of carbon's unique ability to form bonds with one, two, three, or even four other carbon atoms, compounds with the same number of carbon atoms can have the carbon atoms connected many different ways. Compounds with the same molecular formula but a different connectivity are called **structural isomers** (from the Greek *iso* meaning same, and *meros* meaning "parts").

Structural Isomers

The simplest alkane that has a structural isomer is butane. Butane is a straight chain alkane with four carbon atoms and has a molecular formula of C_4H_{10}. Now, take a look at the following two compounds:

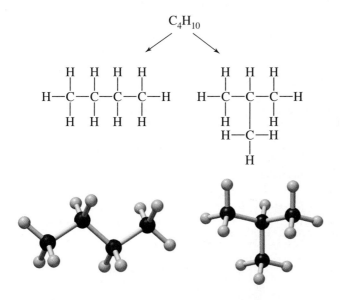

Both compounds contain 4 carbon atoms and 10 hydrogen atoms. They have the same molecular formula, C_4H_{10}. However, the 4 carbons in the compound on the right are not connected in a continuous chain as they are in butane. Instead, in this compound, 3 of the carbon atoms are connected in a continuous chain, and the fourth carbon atom is connected to the middle carbon. Much as a limb on a tree branches off the tree's trunk, that fourth atom branches off the main chain. This alkane is commonly called isobutane (it is an isomer of butane). Alkanes like isobutane that do not have all their carbon atoms connected in a single continuous chain are called **branched-chain alkanes**. These two compounds are structural isomers because they have the same molecular formula, but different connectivity of the atoms.

Conformational Isomers

Before we explore structural isomers more thoroughly, let's consider another type of isomer known as a **conformational isomer**. These isomers, also called **conformers**, are not different compounds like structural isomers; instead, they are different arrangements of the atoms of a compound that can be converted by rotation about one or more single bonds.

Let's consider butane again. The structure on the left in the following figure is the structure of butane that you've seen before. The structure on the right looks different. Are these two representations of the same molecule, or structural isomers?

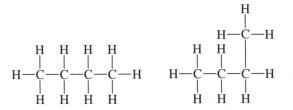

Because the connectivity of the carbons is the same, these two representations of the same molecules are conformational isomers (conformers). To convince yourself that these are the same molecule, put your finger on the first carbon of the structure on the left and trace the carbon chain to the fourth carbon. Now do the same thing with the structure on the right. You can trace the chain from the first to the fourth carbon without picking up your finger for both structures. This means that these two compounds both have four carbon atoms connected in a continuous chain. Since both have a molecular formula of C_4H_{10}, these two Lewis structures, though they look different, actually represent the same compound—butane.

But why do the structures look so different, and how can we make them look the same? The key to understanding the differences between these two structures is remembering that molecules are not rigid. Take some time to study Figure 4.4, which shows three-dimensional models of each of these compounds and demonstrates that by rotating about a carbon–carbon single bond, the compounds can be converted into each other. If you have access to a molecular model kit, build models of these compounds and see how rotating about one more bond can convert one conformational isomer into the other.

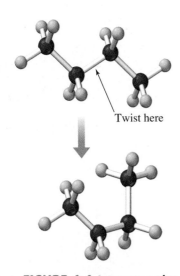

▲ **FIGURE 4.4 Interconverting two conformational isomers of butane.**

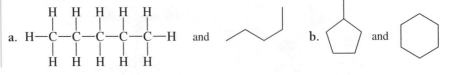

S A M P L E   P R O B L E M   4 . 5

Determining Structural and Conformational Isomers

For each of the following pairs, indicate whether they are structural isomers, conformational isomers, or not related.

a. (structure) and (structure)   **b.** (structure) and (structure)

SOLUTION

To determine the isomeric relationship between any two compounds, first determine the molecular formula of each. If the molecular formulas are not identical, the two compounds are not isomers and are unrelated.

a. Do not be misled by the different representations of these two compounds (Lewis versus skeletal structures). The molecular formula of each of these compounds is C_5H_{12}, so they are isomers of some type. Notice that both structures have a continuous chain of five carbon atoms. The only difference is that the atoms are oriented differently on the page. These are conformational isomers.

b. The molecular formula of each of these cycloalkanes is C_6H_{12} so they are isomers. The compound on the right has a ring containing all six of the carbon atoms. The compound on the left has a ring containing only five of the carbon atoms with the sixth carbon atom branched off of the ring. These are structural isomers.

Nomenclature of Simple Organic Compounds

One easy way to determine whether two compounds are structural or conformational isomers is to systematically name each compound. Conformational isomers will have the same name because they are different representations of the same compound. Structural isomers will have different names because they are different compounds.

The system used to name organic compounds was developed by the IUPAC (International Union of Pure and Applied Chemistry) and is used by chemists around the world. This system provides a unique name for any organic compound based on a straightforward series of rules and ensures that every compound has just one correct systematic name.

Branched-Chain Alkanes

Let's use the structural isomer of butane, isobutane, (shown in the following figure) that we encountered earlier in this section to illustrate the steps used in the IUPAC system to determine the name of a branched-chain alkane.

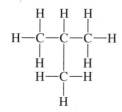

STEP 1. Find the longest continuous chain of carbon atoms. This is called the parent chain. Name the parent chain according to the alkane names given in Table 4.1. (*Hint:* Count from each end to every other end to make sure you find the longest chain.)

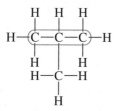

In this example, the name of the parent chain is *propane* since the longest chain has three carbons.

STEP 2. Identify the groups bonded to the main chain but not included in the main chain (ignore hydrogen, of course). These groups are known as **substituents** and, in the case of branched-chain alkanes they are called **alkyl groups**. The name of each alkyl group is derived from the alkane with the same number of carbon atoms by changing the *ane* ending of the name of the corresponding alkane to *yl*. For example, a one carbon alkyl group is named as a *methyl* group. [Methane (CH_4) minus *ane* = meth   and   meth plus *yl* = methyl ($-CH_3$)].

Similarly, a two-carbon alkyl group is an ethyl group, a three-carbon alkyl group is propyl, and so forth (see Table 4.7). So in our example, the substituent is a methyl group.

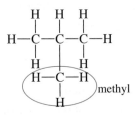

STEP 3. Number the parent chain starting at the end nearer to a substituent.
In this example, the substituent is on carbon number 2 no matter which end you start numbering.

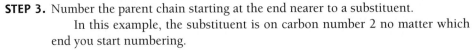

STEP 4. Assign a number to each substituent based on location, listing the substituents in alphabetical order at the beginning of the name. Separate numbers and words in the name by a dash.
The IUPAC name of the example compound is 2-methylpropane.
(Note: If more than one of the same type of substituent is present in the compound, you indicate this using the Greek prefixes *di*, *tri*, and *tetra* but ignore these prefixes when alphabetizing. For example, two methyl groups branching from a parent chain would be named *dimethyl*.)

TABLE 4.7   FOUR SIMPLEST ALKYL SUBSTITUENTS

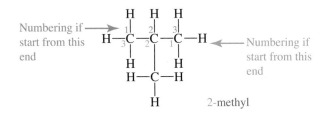

	Methyl	Ethyl	Propyl	Isopropyl
Lewis structure				
Condensed structure	–ξ–CH₃	–ξ–CH₂CH₃	–ξ–CH₂CH₂CH₃	CH₃CHCH₃
Line structure				
Ball-and-stick structure				

ξ represents the point where the group is bonded to the parent chain.

SAMPLE PROBLEM 4.6

Naming Branched-Chain Alkanes

Determine the correct IUPAC name for each of the following branched-chain alkanes:

a. $CH_3CH_2CH_2CHCH_3$
          |
          CH_3

b.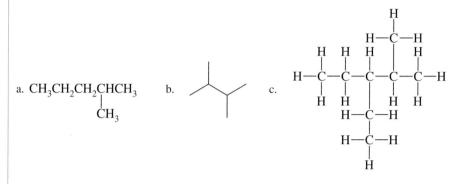

c.

SOLUTION

Carefully follow steps 1 to 4 to determine the correct name for each compound.

a. **STEP 1**

$\boxed{CH_3CH_2CH_2CHCH_3}$ pentane
              |
              CH_3

STEP 2

$CH_3CH_2CH_2CHCH_3$
              |
          $\boxed{CH_3}$ methyl

STEP 3

5   4   3   2   1
$CH_3CH_2CH_2CHCH_3$ 2-methyl
              |
              CH_3

STEP 4

2-methylpentane

b. **STEP 1**

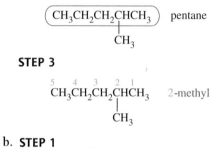   butane

STEP 2

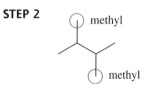 methyl

methyl

STEP 3

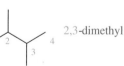   2,3-dimethyl

STEP 4

2,3-dimethylbutane

Note that the two numbers are separated by a comma.

c. **STEP 1**

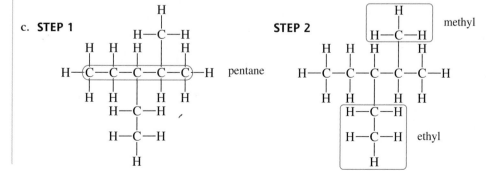   pentane

STEP 2

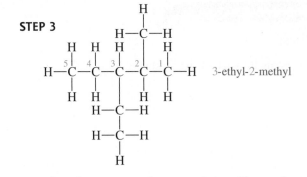

STEP 3

STEP 4
3-ethyl-2-methylpentane

3-ethyl-2-methyl

Note that there is more than one chain of five carbons in length. All five-carbon chains will produce the same IUPAC name.

Haloalkanes

The members of Group 7A on the periodic table, collectively known as the halogens, can also serve as substituents on alkane chains. Alkanes with a halogen substituent in place of a hydrogen atom are known as **haloalkanes** or alkyl halides. When a halogen is a substituent, the name is changed by replacing the *ine* ending of the name of the element with an *o*. The substitutent names for the halogens become *fluoro, chloro, bromo,* and *iodo.* The rules for naming haloalkanes are the same as those for naming branched-chain alkanes with the halogen being the substituent.

Cycloalkanes

The rules for determining the IUPAC names for cycloalkanes are much the same as those for the branched-chain alkanes with a couple of modifications to the steps given previously as follows:

STEP 1. The ring serves as the parent name as long as it has more carbons than any substituent.

STEP 2. As before, identify the substituents.

STEP 3. Number the carbons in the ring. Carbon 1 will always have a substituent.

STEP 4. Assign numbers to the substituents. On a ring bearing a single substituent, that substituent is assumed to be on position 1 of the ring. If only one substituent is present, a 1 is implied and need not be listed (See Sample Problem 4.7c). When more than one substituent is present, the ring should be numbered to give the substituents the lowest possible combination of numbers.

SAMPLE PROBLEM 4.7

Naming Haloalkanes and Cycloalkanes

Determine the correct IUPAC name for each of the following compounds:

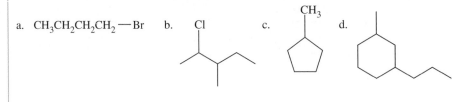

a. $CH_3CH_2CH_2CH_2$ —Br   b.   c.   d.

SOLUTION

For the haloalkanes, follow steps 1 to 4 given for the branched-chain alkanes to determine the correct name for each compound.

a. **STEP 1**   **STEP 2**

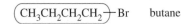

$\overline{(CH_3CH_2CH_2CH_2}$—Br   butane $CH_3CH_2CH_2CH_2$ —(Br) bromo

STEP 3

CH₃CH₂CH₂CH₂—Br   1-bromo
 4   3   2   1

STEP 4
1-bromobutane

b. **STEP 1**

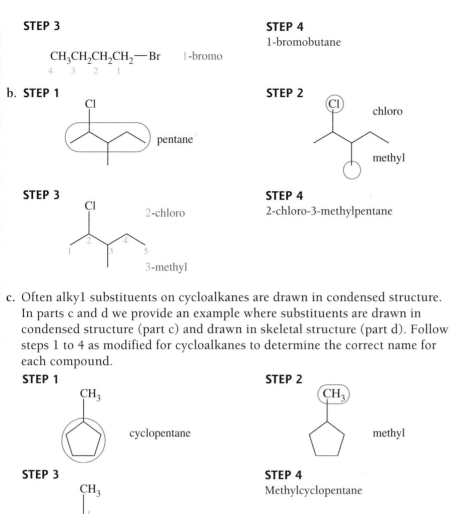

Cl

pentane

STEP 2

Cl   chloro

methyl

STEP 3

Cl   2-chloro

 1   2   3   4   5

3-methyl

STEP 4
2-chloro-3-methylpentane

c. Often alkyl substituents on cycloalkanes are drawn in condensed structure. In parts c and d we provide an example where substituents are drawn in condensed structure (part c) and drawn in skeletal structure (part d). Follow steps 1 to 4 as modified for cycloalkanes to determine the correct name for each compound.

STEP 1

CH₃

cyclopentane

STEP 2

CH₃

methyl

STEP 3

CH₃

 1
5   2
4   3

Single substituent on a cycloalkane will always be on carbon #1.
methyl

STEP 4
Methylcyclopentane

d. **STEP 1**

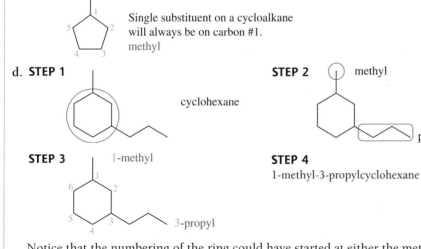

cyclohexane

STEP 2

methyl

propyl

STEP 3

1-methyl

 1
6   2
5   3
 4

3-propyl

STEP 4
1-methyl-3-propylcyclohexane

Notice that the numbering of the ring could have started at either the methyl or the propyl substituent to get the combination of 1 and 3. In such a case, give the lower number to the substituent that comes first alphabetically.

SAMPLE PROBLEM 4.8

Determining Isomeric Relationships by Naming

Each of the pairs of the following compounds has the same molecular formula. Determine the IUPAC name of each compound and then decide whether they are structural or conformational isomers.

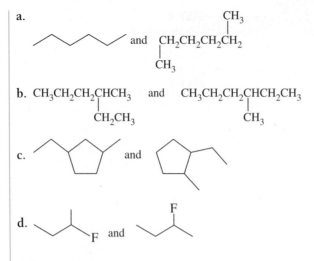

a.

b. CH₃CH₂CH₂CHCH₃   and   CH₃CH₂CH₂CHCH₂CH₃
              | |
           CH₂CH₃   CH₃

c. [cyclopentane structure]   and   [cyclopentane structure]

d. [structure with F]   and   [structure with F]

SOLUTION

a. The compound on the left is a straight-chain alkane with six carbon atoms and is, therefore, hexane. Because of the way that the structure on the right is drawn, you may be tempted to name the compound as a butane with two methyl substituents (1,4-dimethylbutane); however, consider that answer carefully. A single methyl group on either end of a compound is not a substituent but is actually a part of the parent chain. Like the compound on the left, this compound is also a continuous chain of six carbon atoms. It is also hexane. Since these two compounds have the same name, they are the same compound, and the structures shown are simply conformational isomers.

b. Do not assume that the longest continuous chain is always horizontal on the page. In this problem, both compounds have a parent chain of six carbon atoms—hexane—as shown.

Following the steps outlined previously, both of these compounds have an IUPAC name of 3-methylhexane and are, therefore, conformational isomers.

c. The parent of both of these compounds is cyclopentane, and each also has an ethyl substituent and a methyl substituent. So, these two compounds are very similar. But, when we number the rings to get the IUPAC name of each, we notice that the compound on the left has a lowest combination of numbers of 1 and 3 for the substituents, while the one on the right has a lowest combination of numbers of 1 and 2. The correct names, 1-ethyl-3-methylcyclopentane (for the compound on the left) and 1-ethyl-2-methylcyclopentane (for the compound on the right), confirm that these two are not the same compound but are structural isomers.

d. The parent chain of both of these compounds is butane. Recall that the chain must be numbered to give the substituent the lowest possible number. The compound on the left would be numbered from right to left to give the carbon bearing the fluorine the number 2.

[structure numbered 1, 2, 3, 4 with F]

The compound on the right would be numbered from right to left and would also give the carbon bearing the fluorine the number 2.

The name of both compounds is 2-fluorobutane, so they are the same compound *and therefore conformational isomers.*

PRACTICE PROBLEMS

4.12 a. Describe what structural isomers have in common.
   b. Describe how structural isomers are different.

4.13 a. Describe what conformational isomers have in common.
   b. Describe how conformational isomers are different.

4.14 Determine the relationship between each of the pairs of the following compounds. Are they structural isomers, conformational isomers, or not related?

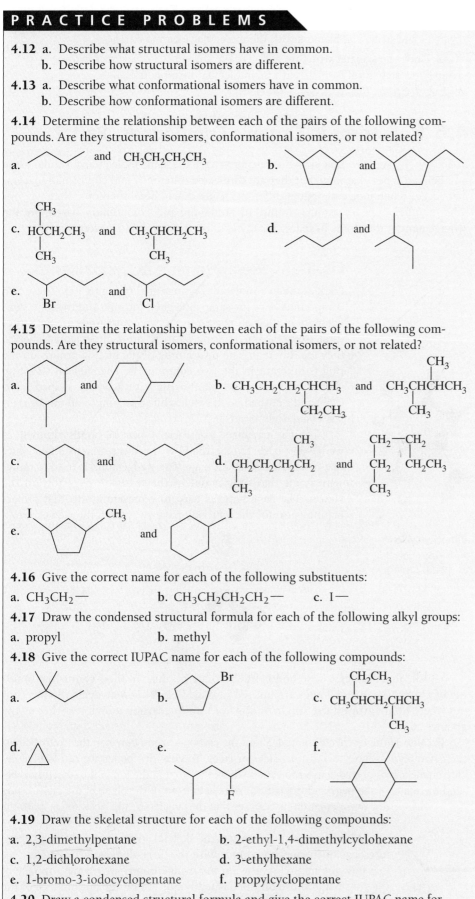

a.   and $CH_3CH_2CH_2CH_3$

b.   and

c. $HCCH_2CH_3$ and $CH_3CHCH_2CH_3$

d.   and

e.   and

4.15 Determine the relationship between each of the pairs of the following compounds. Are they structural isomers, conformational isomers, or not related?

a.   and

b. $CH_3CH_2CH_2CHCH_3$ and $CH_3CHCHCH_3$
                CH_2CH_3 CH_3

c.   and

d. $CH_2CH_2CH_2CH_2$ and CH_2—CH_2
            CH_3 CH_2 CH_2CH_3
                                         CH_3

e.   and

4.16 Give the correct name for each of the following substituents:

a. CH_3CH_2—   **b.** $CH_3CH_2CH_2CH_2$—   **c.** I—

4.17 Draw the condensed structural formula for each of the following alkyl groups:

a. propyl   **b.** methyl

4.18 Give the correct IUPAC name for each of the following compounds:

a.   **b.**   **c.** $CH_3CHCH_2CHCH_3$
                                        CH_2CH_3
                                        CH_3

d.   **e.**   **f.**

4.19 Draw the skeletal structure for each of the following compounds:

a. 2,3-dimethylpentane   **b.** 2-ethyl-1,4-dimethylcyclohexane
c. 1,2-dichlorohexane   **d.** 3-ethylhexane
e. 1-bromo-3-iodocyclopentane   **f.** propylcyclopentane

4.20 Draw a condensed structural formula and give the correct IUPAC name for the three alkane structural isomers with the molecular formula C_5H_{12}.

4.21 Draw the skeletal structure and give the correct IUPAC name for three cycloalkane structural isomers with the molecular formula C_5H_{10}.

4.22 Draw the skeletal structure and give the correct IUPAC name for three haloalkane structural isomers with the molecular formula C_4H_9Br.

4.6 Isomerism in Organic Compounds, Part 2

In Section 4.5, we saw that structural isomers have the same molecular formula (same number and types of atoms), but they are connected differently and conformational isomers occur when molecules adopt different shapes due to bond rotations. Here we consider another type of isomer found in some organic compounds based on the arrangement of the atoms in space. This type of isomer is called a **stereoisomer**.

Cis–Trans Stereoisomers in Cycloalkanes

Look at the two molecules represented in Figure 4.5. These two compounds have the same molecular formula and the *same* connectivity of their atoms. So these two compounds are *not* structural isomers. But looking at their three-dimensional shapes, they are different. The positioning of the methyl groups on the ring is different in the two compounds. The skeletal structure shows the compound on the left with a wedge-shaped bond (◄) and a dashed bond (⸽⸽⸽), while the compound on the right has two dashed bonds.

Organic chemists use these types of bonds, known as wedge-and-dash bonds, to show the three-dimensional nature of a molecule on the flat page. The wedge bond is used to represent atoms that project out of the textbook page toward the viewer. The dash bond is used to represent atoms that project behind (or into) the textbook page away from the viewer.

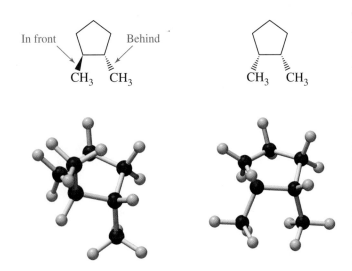

▲ **FIGURE 4.5 Stereoisomers of 1,2-dimethylcyclopentane.** The ball-and-stick structure shows two spatial arrangements. In the skeletal structures this spatial arrangement is represented by wedges (in front of the paper page) and dashes (behind the paper page)

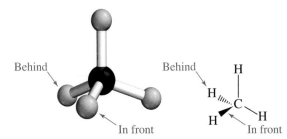

Unlike the alkanes whose bonds freely rotate because of their chain-like bonding of atoms, the cycloalkanes are limited in their flexibility. The cycloalkanes cannot freely rotate around the single bonds holding the carbon atoms together as the alkanes can.

Because of the restricted rotation about the carbon–carbon bonds in the cycloalkanes, these compounds have two distinct sides or faces. To view this better, we can look at cyclopentane across one edge in a projection where the point at the top of the structure is behind the plane of the page and the bottom edge is in front of the plane of the page. Using this representation, we can see that the ring has a "top" face and a "bottom" face.

With this in mind, compare the skeletal structures of the two substituted cyclopentane compounds with the three-dimensional models. Notice how the methyl groups on the wedge bonds are on the top face of the ring while those on the dashed bonds are on the bottom face of the ring. The wedges and dashes allow us to show the three-dimensional nature of a molecule in the two-dimensional space of a page.

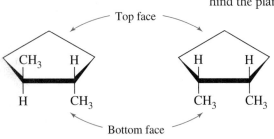

If we were to name either of the two molecules using the IUPAC rules developed previously, the name of both molecules would be 1,2-dimethylcyclopentane. In the representations shown, the two molecules are clearly different. How can we distinguish the two by name?

The two cyclopentane compounds shown in Figure 4.5 belong to a class of isomers called **cis–trans stereoisomers**. In this example, the stereoisomer with the methyls on the same face of the cyclopentane ring (both on top in this case) is called the cis stereoisomer. (Use this memory trick to remember cis: same side is cis. Notice all of the "s" sounds). The other isomer with one methyl on the top face and the second on the bottom face is called the trans stereoisomer (trans comes from the Latin meaning "across" as in the word "transatlantic.") In the trans stereoisomer, the methyls are across the ring from each other. The prefixes *cis* and *trans* in italics are included at the beginning of the compound names to denote the arrangement of the substituents in the compound and to give each compound a unique name. So in Figure 4.5, the complete name of the isomer on the left is *trans*-1,2-dimethylcyclopentane and the isomer on the right is *cis*-1,2-dimethylcyclopentane.

S A M P L E   P R O B L E M   4 . 9

Distinguishing Structural Isomers and Cis–Trans Stereoisomers

For each of the following pairs, indicate whether they are structural isomers or cis–trans stereoisomers or the same compound.

a.

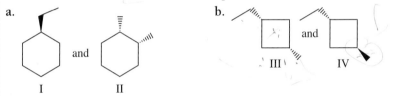

b.

(handwritten annotation: not Structural Isomer connected the same.)

I   II   III   and   IV

S O L U T I O N

First determine the molecular formula of each compound. If the molecular formulas are not identical, the two compounds are not isomers.

a. These compounds both have the molecular formula of C_8H_{16}, so they are either isomers or the same molecule. Notice that compound I has an ethyl group bonded to the cyclohexane ring while compound II has two methyl groups bonded to the cyclohexane ring. Two methyls are not the same as one ethyl. These two compounds do not have the same connectivity, so they are structural isomers.

b. These compounds both have the molecular formula of C_7H_{14}, so they are either isomers or the same molecule. Now check the connectivity. Both have a cyclobutane ring with an ethyl group on one carbon and a methyl group two carbons away. So, these two compounds have the same connectivity. Both compounds also have the same name, 1-ethyl-3-methylcyclobutane, so they are not structural isomers. Finally, check the arrangement of the atoms in space. In compound III, the ethyl and methyl are cis to each other—they are both on the bottom face of the ring. In compound IV, the methyl is on the top face of the ring and the ethyl is on the bottom face of the ring, so the substituents are trans to each other. These two compounds are cis–trans stereoisomers.

Cis–Trans Stereoisomers in Alkenes

Unlike the freely rotating single bonds in a straight-chain alkane, the double bond of an alkene cannot freely rotate. To help you understand, consider the carnival trinket called the Chinese handcuff shown in Figure 4.6. If you join the index fingers of each hand with one of these toys (and try not to pull your hands apart), your two hands can still rotate independently, despite the fact that they are "bonded" together. The same is true for two carbon atoms joined by a single bond.

▶ **FIGURE 4.6 Rotation in single versus double bonds.** Like a Chinese handcuff, the rotation about a single bond (one handcuff) is allowed (a), but the rotation about a double bond (two handcuffs) is restricted (b).

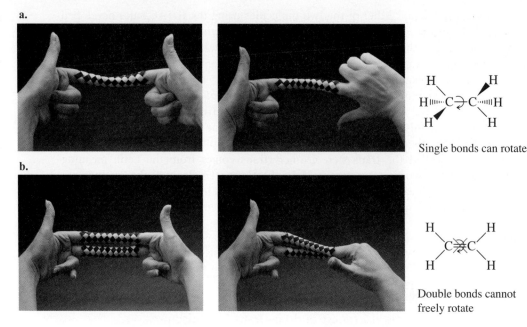

Single bonds can rotate

Double bonds cannot freely rotate

Now consider what would happen if you used two Chinese handcuffs and joined both of your index fingers *and* both of your middle fingers. With two "bonds" now joining your hands, they can no longer rotate independently of each other. The same is true for two carbon atoms joined by a double bond. The double bond between the two atoms will not allow them to rotate independently of each other. Organic chemists say that the carbon–carbon double bond is rigid or that it has **restricted rotation**.

Previously we saw that the rings of cycloalkanes have both a top face and a bottom face because of their inability to rotate freely. Because of the rigidity of the carbon–carbon double bond, alkenes also have two "faces" and some can have cis–trans stereoisomers.

Consider the two alkenes shown in Figure 4.7. The dashed line separates the two faces of the double bond into a top face (above the line) and a bottom face (below the line). Notice that the compound on the left has both similar groups (usually carbon-containing groups) on the same side (face) of the double bond, while the one on the right has them on opposite sides of the double bond. Since the double bond is rigid, these two compounds are not identical, nor are they conformers (they cannot be

▶ **FIGURE 4.7 Cis–trans stereoisomers of alkenes.**

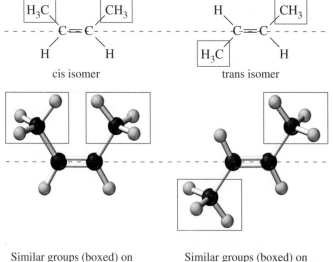

cis isomer

trans isomer

Similar groups (boxed) on the same side of the line through the double bond.

Similar groups (boxed) on opposite sides of the line through the double bond.

changed into the other by rotating around a bond). These two compounds are cis–trans stereoisomers. The compound with the similar groups on the same side of the double bond is the cis stereoisomer. The compound with the similar groups on the opposite sides of the double bond is the trans stereoisomer.

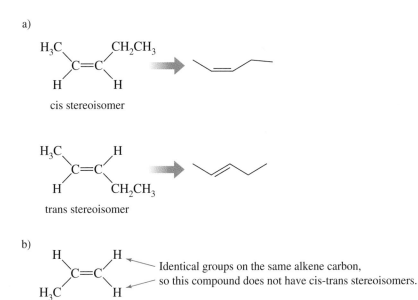

a)

cis stereoisomer

trans stereoisomer

b)

Identical groups on the same alkene carbon, so this compound does not have cis-trans stereoisomers.

◀ **FIGURE 4.8 Structures of cis–trans stereoisomers.** (a) Structures of cis–trans stereoisomers and (b) structure showing why some alkenes do not have cis–trans stereoisomers.

Figure 4.8 shows some other alkenes and their isomers, as well as the skeletal structures of each. Notice that not all alkenes exhibit cis–trans stereoisomerism. If one of the alkene carbons has two identical groups bonded to it (hydrogen in the case shown), this compound cannot exist as cis–trans stereoisomers.

Now think back to the unsaturated fatty acids mentioned in Section 4.3 that contained carbon–carbon double bonds. Do you see that these U-shaped skeletal structures found in unsaturated fatty acids are cis double bonds (see Figure 4.9)?

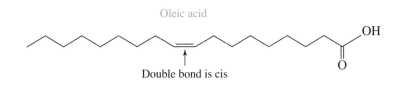

Oleic acid

Double bond is cis

◀ **FIGURE 4.9 Natural fatty acids are cis stereoisomers.** The naturally occurring fatty acids discussed in Section 4.3 contain only cis double bonds.

SAMPLE PROBLEM 4.10

Identifying Cis–Trans Stereoisomers

Identify each of the following alkenes as either a cis or trans stereoisomer:

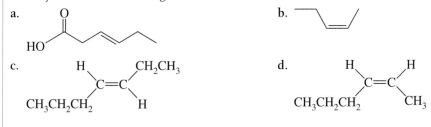

a.

b.

c.

d.

SOLUTION

To solve these problems, start by drawing a line through the double bond as shown in the following table. Then, determine whether the carbon groups are on the same side (cis) or opposite sides (trans) of the line.

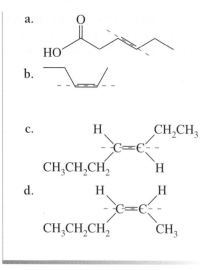

a. The carbon groups bonded to the alkene carbons are on opposite sides of the line. This is a trans stereoisomer.

b. The carbon groups bonded to the alkene carbons are on the same side of the line. This is a cis stereoisomer.

c. The carbon groups bonded to the alkene carbons are on opposite sides of the line. This is a trans stereoisomer.

d. The carbon groups bonded to the alkene carbons are on the same side of the line. This is a cis stereoisomer.

Stereoisomers—Chiral Molecules and Enantiomers

When you enjoy the refreshing taste of spearmint chewing gum or the pleasantly spicy, licorice-like flavor of caraway seeds on fresh rye bread, your taste buds are actually distinguishing between two stereoisomers. The flavors of spearmint and caraway come from a pair of stereoisomers called the carvones. Figure 4.10 shows that the carvone in spearmint is slightly different than the carvone in caraway seeds. Even though stereoisomers are identical except for the arrangement of groups in space, because they have different shapes, our bodies can actually distinguish between them!

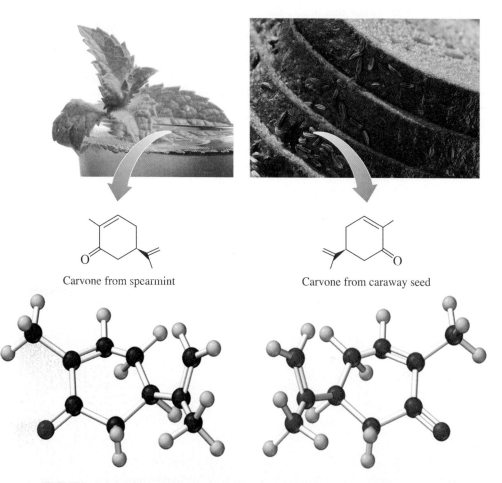

Carvone from spearmint

Carvone from caraway seed

▲ **FIGURE 4.10 The carvone molecules.** The two stereoisomers from caraway seeds and spearmint are identical except for their spatial arrangement.

From Figure 4.10 we can determine that both carvones have a molecular formula of $C_{10}H_{16}O$, so they are isomers. The connectivity of the atoms in the two structures is also the same, meaning that they are not structural isomers, so they are stereoisomers. However, with only one wedge-and dash-type bond present in each compound, the two carvones cannot be cis–trans isomers because cis–trans isomers of cycloalkanes must have two such bonds. With all those similarities, we might begin to think that the two carvones are identical. If that is the case, these two molecules would be superimpos-able. In other words, if we placed one directly on top of the other, every atom on each compound would align or stack on top of one another like two baseball caps. Let's try this with the car-vones. First, we'll take the carvone from spearmint and flip it over so that its ketone car-bonyl group is oriented in the same direction as the carvone from caraway seeds.

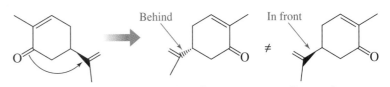

Carvone from spearmint   "Flipped" carvone   Carvone from caraway
                                from spearmint

Mentally slide one so that the molecules are stacked on top of each other and no-tice that the molecules are the same in all but one detail. Before we flipped the carvone from the spearmint, the group at the right of the structure was in front of the page, but when we flipped the carvone over, that group was then behind the page. The identical group on the carvone from caraway seeds was still in front of the page. Therefore, the two compounds are not superimposable because of the difference in the location in space of this group on each molecule.

In fact, if you look carefully, you can see that the molecules are mirror images of each other. Compounds like the two carvones, which have the same molecular formula and the same connectivity but are nonsuperimposable mirror images of each other, belong to a class of stereoisomers called **enantiomers**.

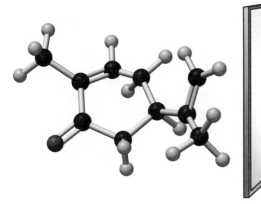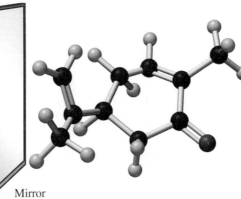

can not flip because not the same.

Mirror

In our everyday life, we encounter numerous common objects that have nonsuper-imposable mirror images. Your shoes are a good example. Notice that your two shoes look very much the same, but they are not identical. Your left shoe is the mirror image of your right shoe. The same holds true for a right-handed and left-handed baseball glove. Objects such as these that have nonsuperimposable mirror images are termed **chiral** (from the Greek *chiros* meaning "of the hand").

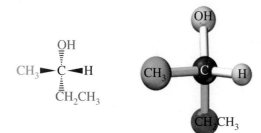

▲ **FIGURE 4.11 A chiral carbon.**
Four different atoms or groups of atoms are bonded to the chiral carbon (shown in black).

Molecules can also be chiral. The two carvones, which exist as a pair of enantiomers, can be thought of as a right-handed and left-handed form of the same molecule. Most molecules that exist as enantiomers contain a **chiral center**; a tetrahedral carbon atom bonded to four different atoms or groups of atoms. In Figure 4.11, the colored spheres around the atoms or groups of atoms highlight the four different groups on the chiral carbon in the molecule. Notice that the chiral carbon atom (shown in black) is bonded to the four different groups: hydroxyl (—OH), hydrogen (—H), methyl (—CH₃), and ethyl (—CH₂CH₃).

There are two forms of this molecule and they are mirror images that are not superimposable on each other (see Figure 4.12). Like the carvones, this molecule can exist in two forms that are enantiomers of each other. On paper we can represent the attachments to the chiral center with wedges and dashes and indicate the location of a chiral center with an asterisk.

▶ **FIGURE 4.12 The mirror image forms.** These forms are not superimposable on each other, so this is a chiral compound.

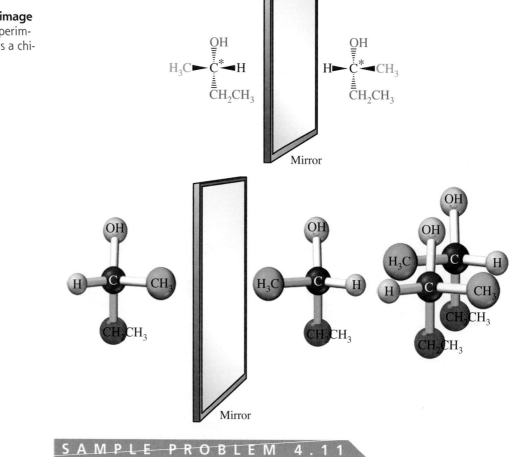

Identifying Chiral Carbon Atoms

Place an asterisk (*) next to the chiral carbon atoms (if any) in each of the following compounds:

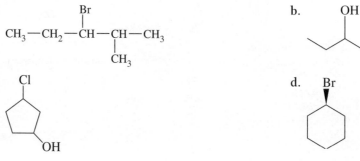

S O L U T I O N

Locate the tetrahedral carbons in the molecule. (Carbons with double or triple bonds typically are not chiral centers since they do not have four groups attached.) Determine if four different groups are attached to the tetrahedral carbons.

a. All carbon atoms bonded to two or more hydrogen atoms cannot be a chiral center since two hydrogens are the same group, so cross those out to remove them from consideration. In this case, that leaves only two carbon atoms (highlighted in blue) to look at more carefully.

$$CH_3-CH_2-CH-CH-CH_3$$
with Br above the third carbon and CH_3 below the fourth carbon (the CH_3 groups at ends and the CH_2 are crossed out)

The blue-highlighted carbon atom on the right is bonded to two methyl groups. Since these two groups are the same, this carbon cannot be chiral.

$$CH_3-CH_2-CH-CH-CH_3$$
with Br above, CH_3 below, and "same" arrow indicating the two CH_3 groups are the same

Expanding the structure to show the bond between the carbon and the hydrogen and then comparing the groups, we see that the carbon atom bonded to Br is bonded to four distinct groups and, therefore, is a chiral carbon.

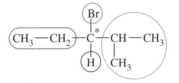

Answer:

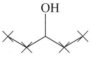

$$CH_3-CH_2-\underset{*}{CH}-CH-CH_3$$
with Br above the starred carbon and CH_3 below the adjacent carbon

b. Identifying chiral carbons on skeletal structures may seem more difficult at first, but with practice, these types of structures can make the process quicker. Remember, in a skeletal structure, a bend or corner is a $-CH_2-$ if no other bonds exist at the corner. You can immediately rule out these carbons as well as those at the end of a line ($-CH_3$).

This leaves only one carbon atom to consider further. At this point, you may find it useful to draw in the hydrogen atom that we typically omit in a skeletal structure and perhaps even add the symbol for the carbon itself.

HO and H with a C at center in a skeletal structure

Now we can see that the OH group and the H are not the same, but notice that the other two groups bonded to this carbon atom are identical: Both are ethyl groups.

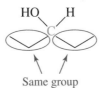

Same group

This compound does not contain a chiral center.
Answer: no chiral carbons

c. As before, eliminate those carbon atoms that have two or more hydrogen atoms bonded to them.

Following the steps from the previous example, consider the remaining carbon atoms one at a time, adding the hydrogen and the C.

On this carbon atom, the H and Cl are different, but what about the other groups? In order to determine if these groups are the same or different, list the atoms and what is bonded to them moving clockwise around the ring starting at the carbon you are considering. Then do the same in the counterclockwise direction. For this carbon, in the clockwise direction the atoms are CH_2, CH—OH, CH_2, CH_2. In the counterclockwise direction the atoms are CH_2, CH_2, CH—OH, CH_2. Notice that the lists are *not* identical. The CH—OH is second in the clockwise direction, but third in the counterclockwise direction. So, the two groups do not have the same connectivity of atoms and, therefore, are not the same. This means that this is a chiral carbon.

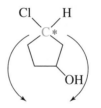

Groups are not identical
in the clockwise and
counterclockwise directions
around the ring.

Following the same procedure for the carbon atom bearing the OH group verifies that it is a chiral carbon atom as well.

Answer:

d. Crossing out all the "corners" in this molecule, we quickly see that only one carbon remains for further consideration.

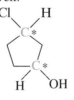

Don't let the presence of the wedge-and-dash bonds in this structure lead you to the quick conclusion that this carbon must be chiral. Instead, follow the procedure we used in Problem 4.11c. Doing this we discover that in the clockwise direction, the atoms bonded are five consecutive CH_2's. The same is true for the counterclockwise direction—five consecutive CH_2's. Since the atoms are the same and in the same order in both the clockwise and counterclockwise directions around the ring, the groups are identical and this is not a chiral carbon atom.

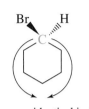

Groups are identical in the
clockwise and counterclockwise
directions around the ring.

Answer: no chiral carbons

The Consequences of Chirality

Why does one carvone give the taste of spearmint and the other the taste of caraway? The taste receptors on your tongue cells are also "handed." Much as you cannot put your right hand into a left-handed baseball glove, a chiral molecule can "fit" only into a complementary receptor—one that fits its shape. This is also the case with many pharmaceuticals where only a single enantiomer has biological activity. One of the enantiomers of a drug called L-Dopa is effective in the treatment of Parkinson's disease, while its mirror image has no biological effect. The anti-inflammatory, ibuprofen, found in Advil®, has one active enantiomer. The other enantiomer is readily converted to the active form by the body.

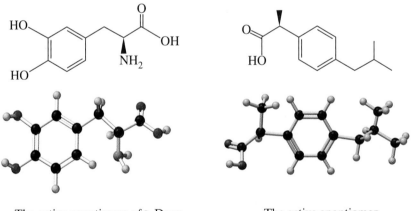

The active enantiomer of L-Dopa used to treat Parkinson's

The active enantiomer of Ibuprofen, an anti-inflammatory

In some cases, one enantiomer of a drug can be beneficial and the other harmful. This was the case with the anti-morning-sickness drug thalidomide, which was marketed outside the United States in the late 1950s and early 1960s. One enantiomer of the drug was effective in alleviating the symptoms of morning sickness. The mirror image, as was tragically discovered later, was teratogenic—it caused birth defects. Because the drug was initially sold as a 50:50 mixture of the enantiomers, many mothers who took it later gave birth to babies with severe birth defects including shortened arms or legs and sometimes no limbs. Thalidomide was subsequently removed from markets worldwide. Recently, thalidomide has reemerged in the pharmaceutical market as a treatment for HIV, leprosy, and some forms of cancer, but it is prescribed under very tight restrictions.

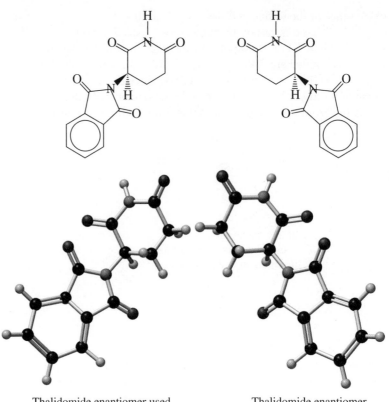

Thalidomide enantiomer used
to treat morning sickness

Thalidomide enantiomer
that causes birth defects

PRACTICE PROBLEMS

4.23 Determine if each of the following cycloalkanes can exist as cis–trans stereoisomers. For those that can, use wedge-and-dash structures to draw the two isomers. Label each of the isomers you drew as the cis stereoisomer or the trans stereoisomer.

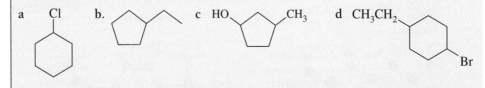

a   Cl   b.   c HO   CH₃   d CH₃CH₂   Br

4.24 Determine if each of the following cycloalkanes can exist as cis–trans stereoisomers. For those that can, use wedge-and-dash structures to draw the two isomers. Label each of the isomers you drew as the cis stereoisomer or the trans stereoisomer.

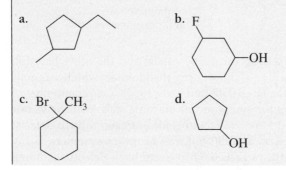

a.

b. F   —OH

c. Br   CH₃

d.   OH

4.25 Mark the chiral carbons in the following molecules, if any, with an asterisk (*):

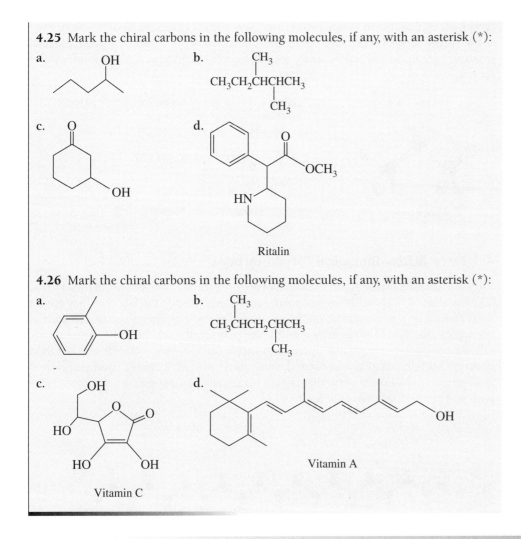

Ritalin

4.26 Mark the chiral carbons in the following molecules, if any, with an asterisk (*):

a.

b.
CH₃
|
CH₃CHCH₂CHCH₃
|
CH₃

c.
Vitamin C

d.
Vitamin A

SUMMARY

4.1 Alkanes: The Simplest Organic Compounds

An organic compound is any compound that is composed primarily of carbon and hydrogen and the simplest of these are the alkanes, containing *only* single-bonded carbon and hydrogen. Alkanes can exist as straight-chain alkanes or as cycloalkanes.

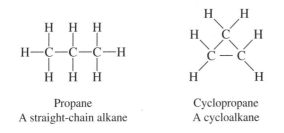

Propane
A straight-chain alkane

Cyclopropane
A cycloalkane

4.2 Representing the Structures of Organic Compounds

Organic compounds can be represented with Lewis structures showing all atoms, bonds, and lone pairs of electrons in a molecule. Condensed structural formulas show all atoms in an organic molecule and their relative positioning, but they show bonds only when necessary for conveying the correct structure. Lone pairs may or may not be a part of a condensed structure. Skeletal structures show the bonding "skeleton" of an organic molecule by showing all carbon-to-carbon bonds. Hydrogen atoms bonded to carbon are not shown in skeletal structures, and carbon atoms are understood to exist at the corner formed when two bonds meet or at the termination of a bond. Skeletal structures are useful for representing cycloalkanes, but condensed structural formulas are not.

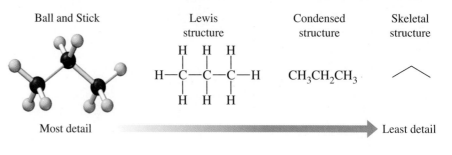

Ball and Stick	Lewis structure	Condensed structure	Skeletal structure

Most detail ➞ Least detail

4.3 Fatty Acids—Biological "Hydrocarbons"

Fatty acids are alkanelike biomolecules that are the primary components of dietary fats. Like the alkanes, fatty acids are nonpolar compounds, despite the fact that they have a polar carboxylic acid functional group on one end of the molecule. Compounds such as fatty acids that have a polar region and a nonpolar region on the same molecule are called amphipathic compounds. Fatty acids with a carbon–carbon double bond in their structure are referred to as unsaturated, while those without a double bond are referred to as saturated. Naturally occurring fatty acids usually contain an even number of carbons and cis double bonds (see Section 4.6).

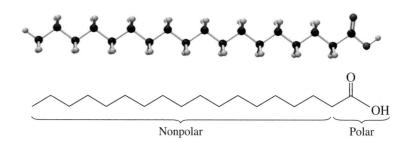

Nonpolar   Polar

4.4 Families of Organic Compounds: Functional Groups

Organic compounds are grouped into families based on the identity of the functional group(s) present. A functional group is a common grouping of atoms bonded in a particular way. Functional groups have specific properties and reactivity. Compounds with the same functional group behave similarly. Organic compounds may contain a hydrocarbon portion in addition to their functional group. Since the functional group is the part of the molecule that is of interest, we typically represent the hydrocarbon portion as R (the Rest of the molecule). The following table shows the functional groups highlighted in this chapter:

Family Name	Representative Structure of the Functional Group
Alkane (Saturated hydrocarbon)	All C—C bonds are single
Alkene (Unsaturated hydrocarbon)	Contain C=C double bonds
Alkyne (Unsaturated hydrocarbon)	Contain C≡C triple bonds
Aromatics	Planar, ring structures, based on benzene. Can contain heteroatoms.

4.5 Isomerism in Organic Compounds, Part 1

Organic compounds with the same molecular formula, but a different connectivity, are known as structural isomers. Straight-chain alkanes and branched-chain alkanes with the same molecular formulas are examples of structural isomers. These should not be confused with conformational isomers, which are actually different representations of the same compound. Conformational isomers can be interconverted simply by rotating about single bonds. Structural isomers cannot be interconverted in this way. They are different molecules. One of the quickest ways to distinguish between structural and conformational isomers is to determine the name of each compound because conformational isomers will have the same name whereas structural isomers will not.

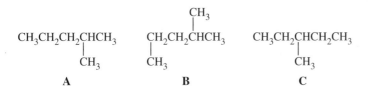

Compounds **A** and **B** are conformational isomers of 2-methylpentane, but compound **C**, 3-methylpentane, is a structural isomer of **A** and **B**.

4.6 Isomerism in Organic Compounds, Part 2

Stereoisomers are two molecules with the same molecular formula and same connectivity, but a different arrangement of the atoms in space. Cycloalkanes with two substituents on different ring carbons can exist as cis–trans stereoisomers. Since cycloalkane rings have a top face and a bottom face, they can have substituents on the same ring arranged differently on these faces. The cis stereoisomer has both substituents on the same face of the ring, while the trans stereoisomer has the substituents on opposite faces. Because rotation is also restricted around the double bond in alkenes, these compounds can exist as cis–trans stereoisomers. In the case of alkenes, we look at the similar groups (typically carbon-containing groups) bonded to the alkene carbons to determine the cis stereoisomer (same face) or the trans stereoisomer (opposite faces). Organic compounds can also exist as stereoisomers if they contain a chiral center, which is a carbon atom bonded to four different atoms or groups of atoms. Compounds with a single chiral center exist as a pair of stereoisomers known as enantiomers. Enantiomers are related to each other as nonsuperimposable mirror images.

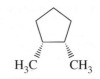

trans-1,2-dimethylcyclopentane   *cis*-1,2-dimethylcyclopentane

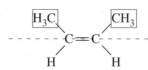

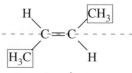

cis stereoisomer
Similar groups (in red
boxes) on the same side
of the line through the
double bond.

trans stereoisomer
Similar groups (in red
boxes) on opposite sides
of the line through the
double bond.

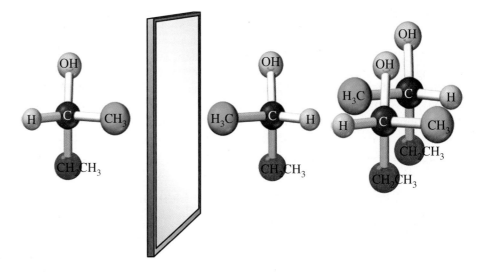

KEY TERMS

alkanes—A family of organic compounds whose members are composed of only singly bonded carbons and hydrogens; this family has no functional group.

alkenes—A family of organic compounds whose functional group is a carbon–carbon double bond (C=C).

alkyl group—A hydrocarbon group minus one hydrogen atom.

alkynes—A family of organic compounds whose functional group is a carbon–carbon triple bond (C≡C).

amphipathic—A compound that has both polar and non-polar regions.

aromatic compounds—A family of cyclic organic compounds whose functional group is a benzene ring.

aromaticity—A term used to describe the unexpected stability of aromatic compounds.

biomolecules—The molecules of life, including carbohydrates, lipids, nucleic acids, and proteins.

branched chain alkanes—Alkanes that have more than one chain of carbon atoms; the shorter chains are considered branches from the main, longer chain.

carboxylic acids—A family of organic compounds whose functional group is a carbonyl (C=O) bonded to an OH; the functional group is often called a carboxyl group (—C) and is abbreviated as COOH.

chiral—A term derived from the Greek word meaning "hand" or "handed."

chiral center—Also chiral carbon; a carbon atom with four different atoms or groups of atoms bonded to it.

cis–trans stereoisomers—A type of isomerism in which compounds differ only in the position of two substituents on a ring or a double bond; the cis isomer has the two substituents on the same face of the ring or double bond and the trans isomer has them on opposite faces.

condensed structural formula—A representation of organic compounds that shows all atoms and their arrangement, but as few bonds as necessary to convey the correct arrangement of atoms.

conformational isomer—An isomer that differs only by rotation about one or more bonds.

conformers—See *conformational isomer*.

cycloalkanes—Hydrocarbon compounds containing a ring of carbon atoms.

enantiomers—Compounds containing at least one chiral carbon atom that are nonsuperimposable mirror images.

essential fatty acids—Simple lipid compounds that must be obtained from our diet.

fatty acids—Simple lipid compounds with a long alkanelike hydrocarbon chain bonded to a carboxyl group.

functional group—A group of atoms bonded in a particular way that is used to classify organic compounds into the various families; each functional group has specific properties and reactivity.

haloalkanes—Hydrocarbon compounds in which a hydrogen atom has been replaced by a halogen atom (F, Cl, Br, or I).

heteroatoms—Atoms in an organic compound other than carbon and hydrogen.

inorganic compounds—Compounds composed of elements other than carbon and hydrogen.

lipid—A class of biomolecules whose structures are nonpolar.

monounsaturated—A term used to describe organic compounds that have one double bond in their structure.

organic chemistry—The field of chemistry dedicated to studying the structure, characteristics, and reactivity of carbon-containing compounds.

organic compounds—Compounds composed primarily of carbon and hydrogen but that may also include oxygen, nitrogen, sulfur, phosphorus, and a few other elements in their structures.

polyunsaturated—A term used to describe organic compounds that have more than one double or triple bond.

resonance hybrid—A representation used to indicate electron sharing among several atoms.

restricted rotation—The inability of carbon atoms in a double bond to rotate relative to each other.

saturated fatty acids—Fatty acids that have no carbon–carbon double bonds; each carbon of the alkyl portion of a fatty acid is bonded to the maximum number of hydrogen atoms.

saturated hydrocarbons—Organic compounds containing only carbon and hydrogen in which each carbon is bonded to the maximum number of hydrogen atoms.

skeletal structure—A representation of organic compounds that shows only the bonding of the carbon framework.

stereoisomers—Compounds differing only in the arrangement of their atoms in space.

straight chain alkanes—Alkanes that have all their carbon atoms connected in a single continuous chain.

structural isomers—Compounds with the same molecular formula (number and type of atoms), but a different connectivity of the atoms.

substituent—Atom or group of atoms that are not part of the main chain or ring in an organic compound but are bonded to it.

unsaturated fatty acids—Fatty acids that contain one or more carbon–carbon double bonds.

ADDITIONAL PROBLEMS

4.27 Identify the following formulas as organic or inorganic compounds:

    a. KCl   **b.** C_4H_{10} **c.** CH_3CH_2OH
    d. H_2SO_4 **e.** $CaCl_2$ **f.** CH_3CH_2Cl

4.28 Identify the following formulas as organic or inorganic compounds:

    a. $C_6H_{12}O_6$ **b.** Na_2SO_4 **c.** I_2
    d. C_2H_5Cl **e.** $C_{10}H_{22}$ **f.** CH_4

4.29 Alkanes are also referred to as *saturated hydrocarbons*. Explain the meaning of the term *hydrocarbon*. Why are alkanes called saturated hydrocarbons?

4.30 Give the molecular formula and draw the Lewis structure of pentane.

4.31 Give the molecular formula and draw the Lewis structure of cyclopentane.

4.32 Compare the molecular formulas of pentane and cyclopentane from your answers to problems 4.30 and 4.31. What is the difference in their molecular formulas? Is cyclopentane a saturated hydrocarbon? Why or why not?

4.33 Give the name of the alkane with 10 carbons.

4.34 Give the name of the cycloalkane with seven carbons.

4.35 Convert each of the Lewis structures shown into a condensed structural formula:

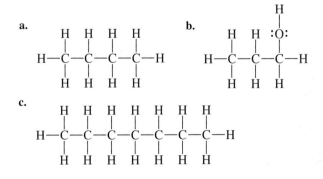

4.36 Convert each of the Lewis structures in problem 4.35 into a skeletal structure.

4.37 Lewis structures, condensed structural formulas, and skeletal structures are used to represent the structure of an organic compound. Each of the following compounds is shown in one of these representations. Convert each compound into the other two structural representations not shown.

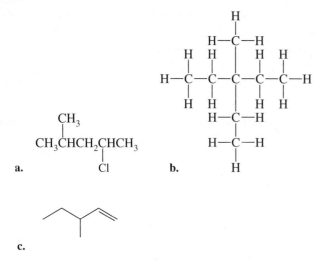

a.

b.

c.

4.38 The molecular formula of an organic compound gives the least detail about the compound's structure. Give the molecular formula of each of the compounds in problem 4.37.

4.39 Explain the structural difference between a saturated and an unsaturated fatty acid.

4.40 Draw the condensed structural formula and skeletal structure of the saturated fatty acid with 16 carbon atoms. What is the name of this fatty acid?

4.41 Give the name and omega number of the following fatty acid:

$CH_3CH_2CH_2CH_2CH_2HC=CHCH_2HC=$
$CHCH_2CH_2CH_2CH_2CH_2CH_2CH_2COOH$

4.42 Draw the structure of γ-linolenic acid [18:3] and give its omega number.

4.43 Despite the fact that fatty acids have a polar −COOH group as part of their structure, they are mainly nonpolar compounds. Explain why these compounds are mostly nonpolar.

4.44 Identify all of the functional groups in each of the following molecules:

a.
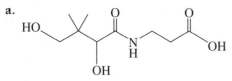

Pantothenic acid (vitamin B5)

b.
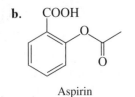

Aspirin

4.45 Identify all of the functional groups in each of the following molecules:

a.
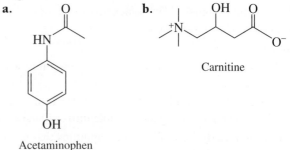

b.

Carnitine

Acetaminophen

4.46 Name the four functional groups circled in the following molecule:

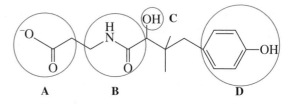

A   B   D

4.47 Zantac™ is used to treat the overproduction of stomach acid for patients with chronic heartburn or gastric ulcers. Identify the four functional groups circled in the structure of Zantac.

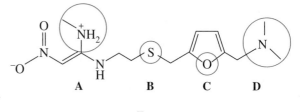

A   B   C   D

Zantac

4.48 In the introduction to Section 4.5, we pointed out that octane ratings are based on an eight-carbon alkane that is a structural isomer of octane called 2,2,4-trimethylpentane. Draw the skeletal structure for this compound.

4.49 Write the condensed formula for each of the following molecules:
  a. 3-ethylhexane
  b. 1,3-dichloro-3-methylheptane
  c. 4-isopropyl-2,3-dimethylnonane

4.50 Draw skeletal structures for each of the following molecules:
  a. ethylcyclopropane
  b. cis-1-chloro-3-methylcyclohexane
  c. isopropylcyclohexane

4.51 Using condensed structural formulas, draw three conformational isomers of hexane.

4.52 Using skeletal structures, draw two conformational isomers of butane.

4.53 Draw the two alkane structural isomers with molecular formula C_4H_{10}. Give the IUPAC name of each compound.

4.54 Draw the two cycloalkane structural isomers with the molecular formula C_4H_8. Give the IUPAC name of each compound.

4.55 How many structural isomers are possible for the molecular formula $C_5H_{11}F$? Draw the skeletal structure and give the IUPAC name of each compound.

4.56 Explain the difference between a structural isomer and a stereoisomer.

4.57 Using wedge-and-dash bonds, draw both the cis and trans stereoisomers for each of the following compounds:
a. 1,3-dimethylcyclohexane
b. 1-bromo-2-ethylcyclopentane

4.58 Using wedge-and-dash bonds, draw both the cis and trans stereoisomers for each of the following compounds:
a. 1-chloro-4-fluorocyclohexane
b. 1,3-diethylcyclobutane

4.59 For each of the following compounds, indicate whether or not it can exist as cis–trans stereoisomers. If it can exist as the two isomers, draw both as a condensed structure.
a. $H_2C=CHCH_2CH_3$
b.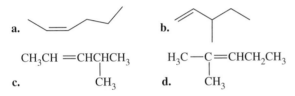
c. $CH_3CH_2CH=CHCH_2CH_3$
d.

4.60 For each of the following compounds, indicate whether or not it can exist as cis–trans stereoisomers. If it can exist as the two isomers, draw both as a condensed structure.

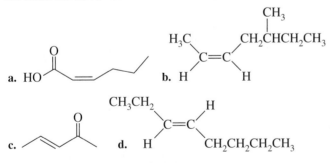

a.
b.
$CH_3CH=CHCHCH_3$
        CH_3
c.
$H_3C-C=CHCH_2CH_3$
        CH_3
d.

4.61 Determine whether each of the following is the cis or the trans stereoisomer:

a. HO (with O double bond)
b. H_3C and $CH_2CHCH_2CH_3$ with CH_3
c.
d. CH_3CH_2 and $CH_2CH_2CH_2CH_3$

4.62 Determine whether each of the following is the cis or the trans stereoisomer:

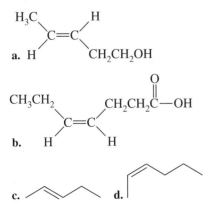

a. H_3C, H / CH_2CH_2OH
b. CH_3CH_2 / CH_2CH_2C-OH (with O)
c.
d.

4.63 Mark the chiral carbons in the following molecules, if any, with an asterisk (*).

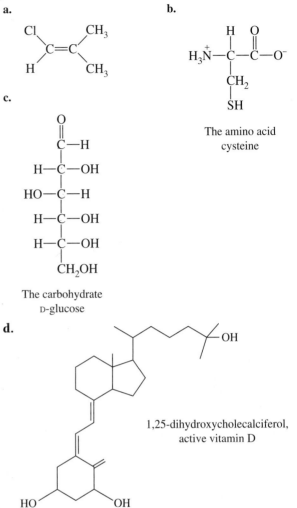

a. (Cl, CH₃, H, CH₃ alkene)

b. The amino acid cysteine

c. The carbohydrate D-glucose

d. 1,25-dihydroxycholecalciferol, active vitamin D

4.64 Mark the chiral carbons in the following molecules, if any, with an asterisk (*).

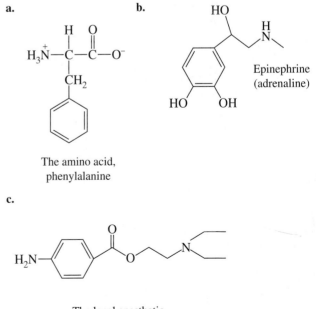

a. The amino acid, phenylalanine

b. Epinephrine (adrenaline)

c. The local anesthetic, novocaine

d.

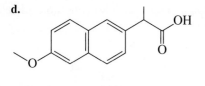

The anti-inflammatory,
naproxen (Aleve, Naprosyn)

4.65 Are the following compounds structural isomers, cis–trans isomers, or enantiomers?

a.

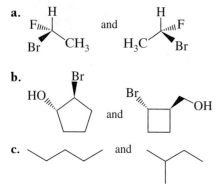

b.

c.

4.66 Draw the enantiomer of each of the following compounds. If the compound is not chiral, state that fact.

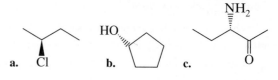

a. **b.** **c.**

CHALLENGE PROBLEMS

4.67 Certain omega–3 fatty acids can be found only in animal sources, such as fatty fish. Two of these are eicosapentaenoic acid (EPA) [20:5] and docosahexaenoic acid (DHA) [22:6], both of which are ω-3 fatty acids. DHA has been shown to be important in healthy brain development, so it has recently been added to infant formulas. Breastmilk is rich in DHA as long as the mother maintains a healthy diet that includes fish. Draw skeletal structures of the fatty acids EPA and DHA.

4.68 For each pair of molecules, in the following figures, identify the pair as below:

A. structural isomers.   **B.** the same molecule.
C. cis–trans stereoisomers.   **D.** different molecules.

a.

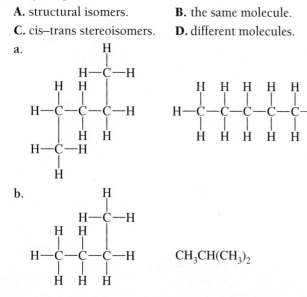

b.

c.

$$H-\overset{\overset{\displaystyle H}{|}}{\underset{\underset{\displaystyle H}{|}}{C}}-\overset{\overset{\displaystyle H}{|}}{\underset{\underset{\displaystyle H}{|}}{C}}-\overset{\overset{\displaystyle H}{|}}{\underset{\underset{\displaystyle H}{|}}{C}}-\overset{\overset{\displaystyle H}{|}}{\underset{\underset{\displaystyle H}{|}}{C}}-H$$

d.

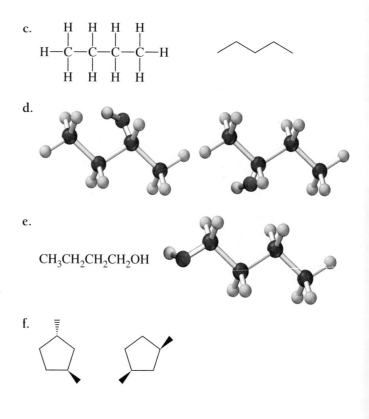

e.

$CH_3CH_2CH_2CH_2OH$

f.

ANSWERS TO ODD-NUMBERED PROBLEMS

Practice Problems

4.1 A Lewis structure shows all atoms, bonds, and all non-bonding electrons. A condensed structure shows all atoms, but as few bonds as possible.

4.3 CH_4
CH_3CH_3
$CH_3CH_2CH_3$
$CH_3CH_2CH_2CH_3$
$CH_3CH_2CH_2CH_2CH_3$
$CH_3CH_2CH_2CH_2CH_2CH_3$
$CH_3CH_2CH_2CH_2CH_2CH_2CH_3$
$CH_3CH_2CH_2CH_2CH_2CH_2CH_2CH_3$
$CH_3CH_2CH_2CH_2CH_2CH_2CH_2CH_2CH_3$
$CH_3CH_2CH_2CH_2CH_2CH_2CH_2CH_2CH_2CH_3$

4.5 Skeletal structures show mainly bonds between carbon atoms. Since methane has only one carbon, it is not possible to draw its skeletal structure.

4.7

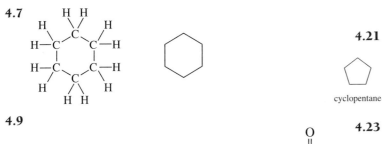

4.9

4.11 a. alkyne   **b.** alkene   **c.** alkene (a cycloalkene)

4.13 a. Conformational isomers have the same molecular formula and connectivity.
   b. Conformational isomers differ by rotation about one or more single bonds.

4.15 a. structural isomers   **b.** not related
   c. not related   **d.** conformational isomers
   e. structural isomers

4.17 a. $CH_3CH_2CH_2-$ **b.** CH_3-

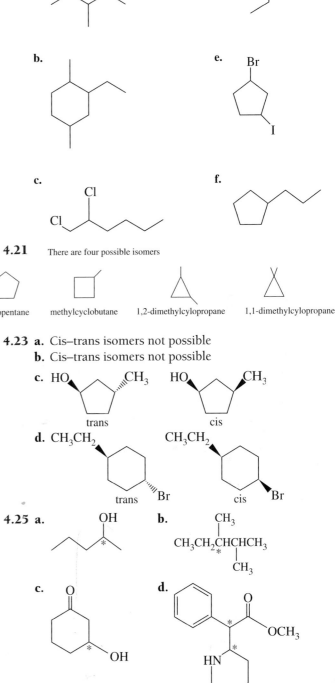

4.19 a.   **d.**
b.   **e.** Br ... I
c. Cl ... Cl   **f.**

4.21 There are four possible isomers
cyclopentane   methylcyclobutane   1,2-dimethylcylopropane   1,1-dimethylcylopropane

4.23 a. Cis–trans isomers not possible
   b. Cis–trans isomers not possible
   c. HO...CH₃ trans   HO...CH₃ cis
   d. CH_3CH_2...Br trans CH_3CH_2...Br cis

4.25 a. OH   **b.** CH_3
   $CH_3CH_2\overset{*}{C}HCHCH_3$
   CH_3
   c. O ... OH   **d.** OCH₃ ... HN
   Ritalin

Additional Problems

4.27 a. inorganic   **b.** organic   **c.** organic
   d. inorganic   **e.** inorganic   **f.** organic

4.29 Hydrocarbons are organic compounds composed only of carbon and hydrogen. Saturated refers to the fact that each carbon is bonded to the maximum number of hydrogens.

4.31 C_5H_{10}

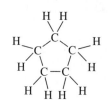

4.33 Decane

4.35 a. $CH_3CH_2CH_2CH_3$ **b.** $CH_3CH_2CH_2OH$
   c. $CH_3CH_2CH_2CH_2CH_2CH_2CH_3$

4.37 a.

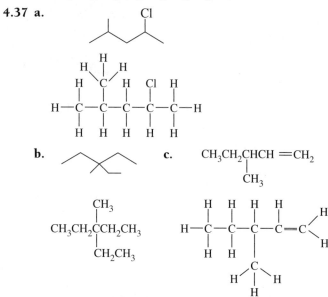

b.   **c.** $CH_3CH_2CHCH=CH_2$
                                CH_3

4.39 An unsaturated fatty acid contains one or more carbon–carbon double bonds, but a saturated fatty acid contains no double bonds.

4.41 linoleic acid, [18:2], ω-6

4.43 The nonpolar hydrocarbon portion of the molecule is so much larger than the small polar COOH group that it dominates the polarity of the compound, making it nonpolar.

4.45 a.

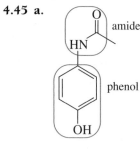

Acetaminophen

b.

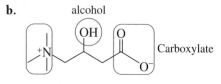

Amine   Carnitine

4.47 A = protonated amine, B = sulfide (thioether), C = ether, D = amine

4.49 $CH_3CH_2CHCH_2CH_2CH_3$
     a. CH_2CH_3

b.
                    CH_3
                 $CH_2CH_2CCH_2CH_2CH_2CH_3$
            Cl   Cl

c.
                   CH_3
         $CH_3CHCHCHCH_2CH_2CH_2CH_2CH_3$
           CH_3 $CHCH_3$
                    CH_3

4.51

$CH_3CH_2CH_2CH_2CH_2CH_3$ CH_3CH_2 CH_3CH_2
                       $CH_2CH_2CH_2CH_3$ $CH_2CH_2CH_2$
                                            CH_3

4.53

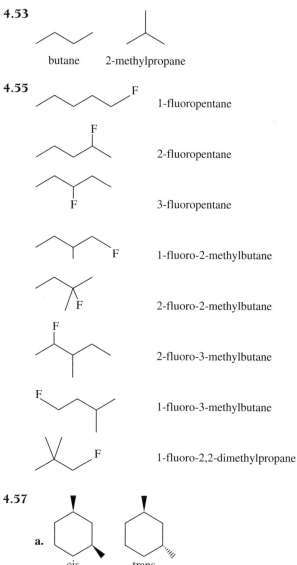

butane   2-methylpropane

4.55

1-fluoropentane

2-fluoropentane

3-fluoropentane

1-fluoro-2-methylbutane

2-fluoro-2-methylbutane

2-fluoro-3-methylbutane

1-fluoro-3-methylbutane

1-fluoro-2,2-dimethylpropane

4.57

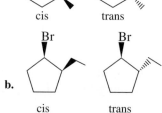

a.   cis   trans

b.   cis   trans

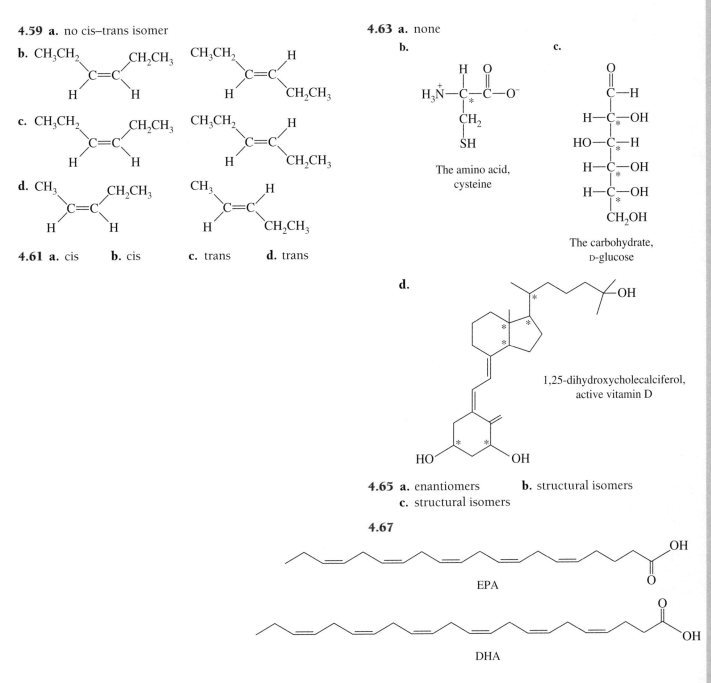

4.59 a. no cis–trans isomer

b.

c.

d.

4.61 a. cis   **b.** cis   **c.** trans   **d.** trans

4.63 a. none

b.

The amino acid, cysteine

c.

The carbohydrate, D-glucose

d.

1,25-dihydroxycholecalciferol, active vitamin D

4.65 a. enantiomers   **b.** structural isomers
c. structural isomers

4.67

EPA

DHA

Guided Inquiry Activities FOR CHAPTER 5

SECTION 5.2

EXERCISE 1 Fischer Projections

Information

In Chapter 4 we saw that a chiral carbon is a carbon in a molecule with four different groups bonded to it. When a chiral carbon appears in a molecule, there are two different ways that the four different groups can arrange themselves around the carbon. The two resulting molecules are related to each other as nonsuperimposable mirror images, a special type of stereoisomer called an enantiomer. The simplest carbohydrate is glyceraldehyde, $C_3H_6O_3$. Glyceraldehyde contains an aldehyde functional group, and so it is referred to as an *aldose*. In carbohydrates, the D-isomer has the chiral center farthest from the carbonyl arranged like D-glyceraldehyde, and the L-isomer is a mirror image of the D-isomer.

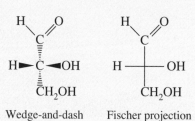

Wedge-and-dash   Fischer projection   Ball-and-stick

▲ FIGURE 1 D-Glyceraldehyde shown in three different representations.

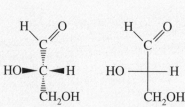

Wedge-and-dash   Fischer projection   Ball-and-stick

▲ FIGURE 2 L-Glyceraldehyde shown in three different representations.

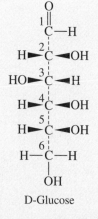

D-Glucose

▲ FIGURE 3 D-Glucose shown in wedge-and-dash.

Questions

1. In D-glyceraldehyde (Figure 1), list the groups attached to the chiral carbon that are in front of the plane of the paper. Which groups attached to the chiral carbon are behind the plane of the paper?

2. Where is the chiral carbon in the Fischer projection? Identify it on the glyceraldehyde molecules (Figures 1 and 2) with an asterisk.

3. What does the horizontal line in the center of the Fischer projections shown represent?

4. What does the vertical line in the center of the Fischer projections shown represent?

5. How are D- and L-glyceraldehyde different?

6. Identify the chiral centers on D-glucose (Figure 3) with an asterisk.

7. Draw the Fischer projection of D-glucose.

8. Draw L-glucose (the mirror image of D-glucose) in the Fischer projection next to D-glucose that you drew in question 7.

9. D-Galactose is an *epimer* of D-glucose: All the chiral centers are identical, except for C4, which is reversed. Draw the Fischer projection of D-galactose.

SECTION 5.4

EXERCISE 2 Ring Formation

Part 1. Information

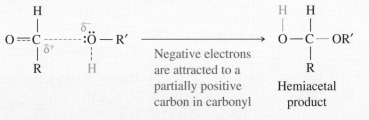

▲ **FIGURE 4** Hemiacetal formation reaction.

Questions

1. In Figure 4, name the functional groups of the compounds on the reactant side of the chemical equation.

2. Which oxygen in the hemiacetal product in Figure 4 (right or left) was the carbonyl oxygen from the aldehyde?

3. Which oxygen in the hemiacetal product in Figure 4 (right or left) was from the alcohol, R′OH?

Part 2. Information

The carbonyl carbon (C1) in a monosaccharide is referred to as the *anomeric* carbon.

Conventions for Drawing the Ring Form of a Six Carbon D-Aldose

- Carbon 6 will always be on the top side of the ring.
- Hydroxyls that do not react and are on the right of a Fischer projection will be on the bottom side of the ring.
- Hydroxyls that do not react and are on the left of a Fischer projection will be on the top side the ring.

Questions

4. Monosaccharides chemically react within themselves (intramolecularly) due to polar opposites strongly attracting each other within the molecule.
   a. Referring to Figure 3, if the O on carbon 5 of D-glucose (O5) is attracted to the carbonyl carbon 1 (C1), reacting to form a bond between them (making a hemiacetal functional group), how many atoms would be enclosed in a ring?
   b. How many of those atoms are carbon?
   c. How many atoms are oxygen? Recall from Section 4.6 that a carbon ring can be represented with a top and bottom face. Considering this, draw the ring form of D-glucose.

5. Where did you place the OH for C1 (top or bottom)?

6. If you placed it on the bottom, you drew α-D-glucose. If you placed the OH on C1 on the top, you drew β-D-glucose. Which one did you draw?

7. Can you devise a rule for identifying these two ring forms (α and β anomers) of D-glucose relative to the position of C6?

Most people know that carbohydrates are found in sweet and starchy foods, but did you know that there are also carbohydrates in milk? Did you know that the ABO blood markers are also carbohydrates? Chapter 5 explores the unique structures and functions of carbohydrates.

Carbohydrates
LIFE'S SWEET MOLECULES

If you were to ask a friend, "What do you know about carbohydrates?" you might get such varied answers as "carbohydrates are sugar," "they provide energy," or "there are *simple* carbohydrates and *complex* carbohydrates." Let's take a closer look at these common statements.

Carbohydrates *are* sugar, and they *do* provide energy. Most people associate carbohydrates with sugary sweet foods that give the body quick energy. Our bodies break down carbohydrates to extract energy. Carbon dioxide and water are released in the process. The carbohydrate glucose is the primary nutrient that our body uses to produce energy. Many carbohydrates can be recognized by their name, which often ends in *ose*. Carbohydrates are organic molecules essential to life, and so they are classified as biomolecules.

Simple carbohydrates are referred to as simple sugars and are often sweet to the taste. We will start our exploration of carbohydrate structure with these smaller, simple sugars. If we eat more simple sugars than we need for energy, our body chemically converts them to fat. More complex carbohydrates include starches and the plant and wood fiber known as cellulose. As we will see, these molecules are much larger than simple sugars and break down more slowly, thus allowing more of the molecules to be used for energy before excesses might be stored as fat.

What many people do not realize about carbohydrates is that they have functions in the body other than simply serving as an energy source. Carbohydrates are found on the surfaces of cells where they can act as "road signs," allowing molecules to distinguish one cell from another. For example, the ABO blood markers are carbohydrates found on the surface of red blood cells that allow us to distinguish our body's blood type from a foreign blood type. Our bodies also contain a carbohydrate that prevents blood clots (heparin). Carbohydrates are also found in larger molecules, such as our genetic material, DNA and RNA.

5.1 Classes of Carbohydrates

The simplest carbohydrates are **monosaccharides** (*mono* is Greek for "one," *saccharum* is Latin for "sugar"). These often sweet-tasting sugars cannot be broken down into smaller carbohydrates. The common carbohydrate glucose, $C_6H_{12}O_6$, is a monosaccharide. **Disaccharides** consist of two monosaccharide units joined together. A disaccharide can be split into two monosaccharide units. Ordinary table sugar, sucrose, $C_{12}H_{22}O_{11}$, is a disaccharide that can be split, or hydrolyzed, into the two monosaccharides glucose and fructose.

Oligosaccharides are carbohydrates containing anywhere from three to nine monosaccharide units. The blood-typing groups known as ABO are oligosaccharides.

When 10 or more monosaccharide units are joined together, the large molecules that result are termed **polysaccharides** (*poly* is Greek for "many"). In polysaccharides the sugar units can be connected in one continuous chain or the chain can be branched. Starch, a polysaccharide in plants, contains large chains of glucose that can be broken down to produce energy.

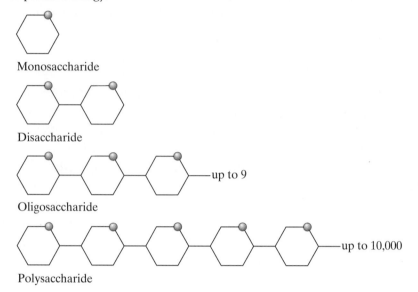

Monosaccharide

Disaccharide

Oligosaccharide — up to 9

Polysaccharide — up to 10,000

5.2 Monosaccharides

The simple sugars known as monosaccharides contain the elements carbon, hydrogen, and oxygen and have the general formula $C_n(H_2O)_n$ where n is a whole number 3 or higher. Monosaccharides contain several functional groups. The functional groups of glucose are shown in Figure 5.1. Carbohydrates contain the functional group alcohol (or hydroxyl) represented as —OH. They also contain a functional group with an oxygen double-bonded to a carbon that is commonly called **carbonyl** (pronounced "carbonEEL"). The carbonyl is found as an aldehyde or ketone in the monosaccharides. Before discussing monosaccharides further, let's look at these functional groups more closely.

Carbonyl

Functional Groups in Monosaccharides—Alcohol, Aldehyde, and Ketone

Alcohol

Ethanol, a commercially important compound produced from the fermentation of the simple sugars in grains and fruits, is one of the simplest members of the family of organic compounds known as **alcohols**. Ethanol is the alcohol present in liquor, beer, and wine. It is also a primary additive in the alternative fuel blends such as gasohol and E85, which is 85% ethanol and only 15% gasoline.

Alcohols are classified by the number of alkyl groups attached to the carbon atom that is bonded to the hydroxyl group. (This carbon atom is sometimes referred to as the alcoholic carbon.) The number of alkyl groups attached to the alcoholic carbon directly impacts the reactivity of the alcohol. A **primary (1°) alcohol** has *one* alkyl group attached to the alcoholic carbon, a **secondary (2°) alcohol** has *two* such alkyl groups, and

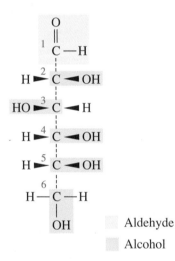

▲ **FIGURE 5.1 Functional groups in the monosaccharide glucose.** Glucose contains the carbonyl-containing functional group aldehyde and several alcohol groups.

Aldehyde

Alcohol

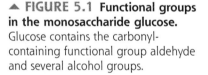

CH_3CH_2OH ←— Alcohol

Ethanol, an alcohol

a **tertiary (3°) alcohol** has *three*. Monosaccharides contain both primary and secondary alcohols.

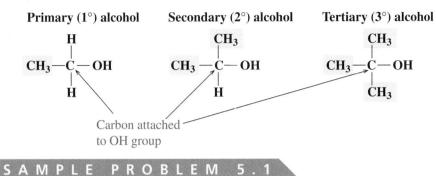

Primary (1°) alcohol **Secondary (2°) alcohol** **Tertiary (3°) alcohol**

Carbon attached to OH group

S A M P L E   P R O B L E M   5 . 1

Classifying Alcohols

Classify each of the following alcohols as a primary (1°), secondary (2°), or tertiary (3°) alcohol:

a. $CH_3CH_2CH_2OH$ b.

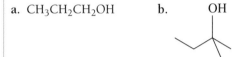

S O L U T I O N

To determine whether an alcohol is primary, secondary, or tertiary, answer the following question: How many carbons are directly bonded to the C—OH carbon?

a. One carbon is bonded to the C with the OH; therefore this is a primary (1°) alcohol.

b. Three carbons are bonded to the C with the OH; therefore this is a tertiary (3°) alcohol.

Aldehydes

Benzaldehyde, the compound responsible for the aroma of almonds and cherries, is a member of the simplest family of carbonyl-containing organic compounds known as the **aldehydes**. The presence of the benzene ring in its structure further classifies benzaldehyde as an aromatic aldehyde.

Members of the aldehyde family always have a carbonyl group with a hydrogen atom bonded to one side of the carbonyl and an alkyl or aromatic group bonded to the other. The lone exception to this is formaldehyde, which has a hydrogen bonded to each side of the carbonyl. (Formaldehyde was once used as a preservative for biological specimens, but it is no longer used because it was found to cause cancer in some animals.) Monosaccharides can contain an aldehyde functional group at one end of the molecule (in addition to multiple hydroxyl groups).

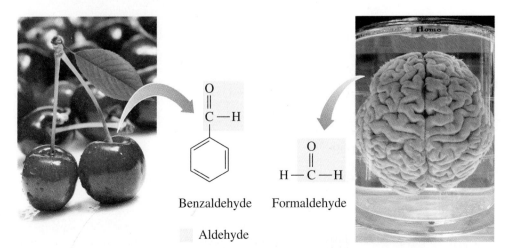

Benzaldehyde Formaldehyde

Aldehyde

Ketones

The **ketone** family of organic compounds is structurally similar to the aldehydes. The difference is that ketones have an alkyl or aromatic group on *both* sides of the carbonyl. The simplest ketone is acetone, which was once the main component of fingernail polish remover. Because of its tendency to cause dry skin, acetone has now been largely replaced in many formulations of nail polish remover.

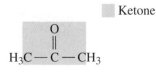

Ketone

Acetone

In addition to monosaccharides, ketones occur in a wide variety of biologically relevant compounds. For example, pyruvate is a ketone-containing compound formed during the metabolic breakdown of carbohydrates. Butanedione, the flavor of butter, contains two ketone groups. Note that in the structures of pyruvate and butanedione, the carbonyl has a carbon as the *first* atom on each side—this distinguishes the functional group as a ketone regardless of the other atoms that may be present in the structure.

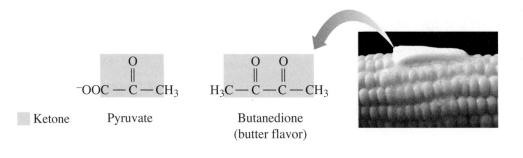

Ketone   Pyruvate   Butanedione
(butter flavor)

A monosaccharide that contains an aldehyde functional group is referred to as an **aldose**, and one that contains a ketone functional group is called **ketose**. The most common monosaccharides contain three to six carbons. A monosaccharide with three carbons is a *triose*, one with four carbons is a *tetrose*, one with five carbons is a *pentose*, and one with six carbons is a *hexose*. As examples, **glucose**, the most abundant monosaccharide found in nature, is an aldose containing six carbons and therefore classified as an aldohexose, while **fructose**, a common monosaccharide found in fruits, is a ketose containing six carbons and is classified as a ketohexose. Look at the monosaccharides shown and convince yourself that these are examples of an aldohexose and ketopentose by counting the carbons and identifying the functional groups.

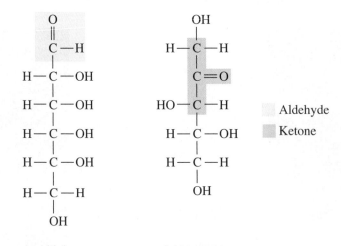

An aldohexose   A ketopentose

SAMPLE PROBLEM 5.2

Identifying Aldehydes and Ketones

Identify each of the following compounds as an aldehyde or a ketone:

a.

b.

c.

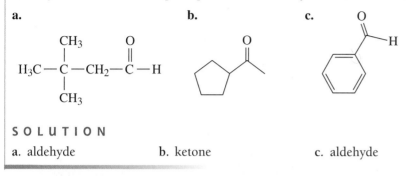

SOLUTION

a. aldehyde   b. ketone   c. aldehyde

PRACTICE PROBLEMS

5.1 Classify each of the following alcohols as a primary (1°), secondary (2°), or tertiary (3°) alcohol:

a.

b.

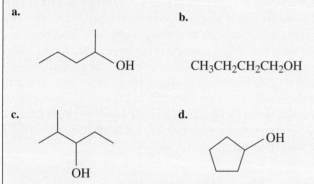

$CH_3CH_2CH_2CH_2OH$

c.

d.

5.2 Classify each of the following alcohols as a primary (1°), secondary (2°), or tertiary (3°) alcohol:

a.

b.

c.

d.

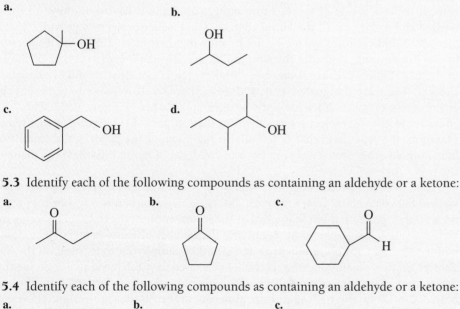

5.3 Identify each of the following compounds as containing an aldehyde or a ketone:

a.

b.

c.

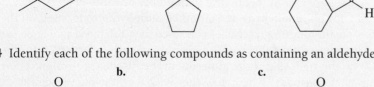

5.4 Identify each of the following compounds as containing an aldehyde or a ketone:

a.

b.

c.

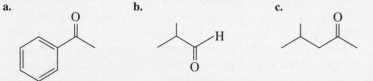

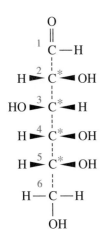

* Chiral centers in glucose

Stereochemistry in Monosaccharides

Multiple Chiral Centers

Let's examine the structures of carbohydrates more closely. First, look at the carbons in the molecule of glucose shown. The carbons bonded to the alcohol (—OH) groups all have a tetrahedral geometry. Recall from Chapter 4 that a carbon atom with tetrahedral geometry that has four different atoms or groups of atoms attached to it is known as a chiral carbon. As we also saw in Chapter 4, a compound with a single chiral carbon atom can exist in two different forms called enantiomers.

How many chiral carbons does a glucose molecule contain? If we examine the carbons from 1 to 6 (carbon 1 at the top), carbon 1 (the carbon of the aldehyde group) is not tetrahedral, so it cannot be chiral. Carbons 2 to 5 are tetrahedral and have four different atoms or groups of atoms attached, so they are chiral carbons. Carbon 6 is tetrahedral but does not have four *different* groups of atoms attached to it (two of the atoms attached are hydrogen), so it is not chiral. Therefore, glucose has a total of four chiral centers. The groups bonded to each chiral center of a glucose molecule could have two different arrangements (a pair of mirror images), so how many stereoisomers of glucose (possible arrangements around the chiral centers) are possible?

The number of stereoisomers possible increases with the number of chiral centers present in a molecule. For molecules with one chiral center, there are only two different ways the attached atoms or groups can be arranged spatially (and these are mirror images). For molecules with two chiral centers, there are a total of four different ways to arrange the attached atoms or groups differently, and for molecules with three chiral centers, there are eight different ways to attach atoms or groups. The general formula for determining the number of stereoisomers is 2^n, where n is the number of chiral centers present in the molecule. Because glucose has four chiral centers, ($2^4 = 2 \times 2 \times 2 \times 2 = 16$), *16* stereoisomers are possible! It is amazing then that *1* of these stereoisomers is our preferred energy source.

Representing Stereoisomers—The Fischer Projection

Considering all the stereoisomers possible for a molecule like glucose and the need for designating the position of the attachments on chiral carbons, it would be convenient to have a way of representing molecules without having to draw all those wedges and dashes on all the chiral centers. The Fischer projection provides a simpler way of indicating chiral molecules by showing their three-dimensional structure in two dimensions. Consider the simplest aldose, glyceraldehyde ($C_3H_6O_3$), shown in the next figure. Because it has one chiral center, it can exist as one of two enantiomers. These are designated as D- and L-enantiomers. If we were to represent the enantiomers of glyceraldehyde with wedges and dashes to show their shape, we could represent the molecules as shown in the top part of the figure. In this representation, the only atom that is actually in the plane of the paper is the chiral carbon, with all other atoms either in front of (wedges) or behind (dashes) the plane of the page.

In the **Fischer projection**, *horizontal lines on a chiral center represent wedges, and vertical lines on a chiral center represent dashes.* A chiral carbon is not shown as a "C" on a Fischer projection but is implied at the intersection of the lines. (This gives the viewer a quick and easy way of identifying the number of chiral centers.) The designation of D and L for glyceraldehyde and all other carbohydrates is based on the Fischer projection positioning in glyceraldehyde, used as a reference molecule for this designation. All **D-sugars** have the —OH on the chiral carbon farthest from the carbonyl (C=O) on the *right side* of the molecule in the Fischer projection. The enantiomer of this is the **L-sugar**, which has the —OH group on the chiral carbon farthest from the C=O on the *left side* of the Fischer projection. Most of the carbohydrates commonly found in nature and the ones we use for energy are D-sugars. D- and L-glyceraldehyde are represented by the Fischer projections shown in the bottom part of the figure.

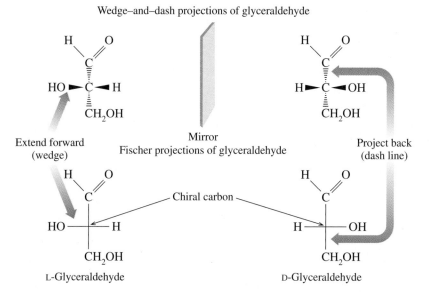

Wedge–and–dash projections of glyceraldehyde

Mirror
Fischer projections of glyceraldehyde

Extend forward (wedge)

Project back (dash line)

Chiral carbon

L-Glyceraldehyde

D-Glyceraldehyde

Similarly, Figure 5.2 shows the molecule D-glucose transformed into a Fischer projection.

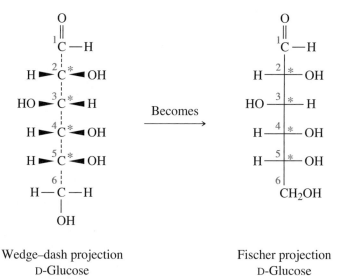

Wedge–dash projection
D-Glucose

Becomes

Fischer projection
D-Glucose

◀ FIGURE 5.2 D-Glucose represented in a Fischer projection. Attachments to chiral carbons on the horizontal point toward the viewer, and attachments to chiral carbons on the vertical point away from the viewer. Note: The dashes in the wedge-and-dash projection at left are drawn as vertical lines for clarity.

When drawing enantiomers in Fischer projection, they are written as if there is a mirror placed between the two molecules. The reflection of one molecule is identical to the second molecule, its enantiomer. All chiral centers have their horizontal groups switched in their enantiomer when viewed in Fischer projection. Attached atoms or groups on the right in one enantiomer appear on the left side of the other. The enantiomers D- and L-glucose are shown in the following figure:

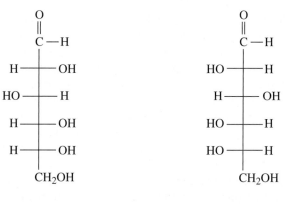

D-Glucose

L-Glucose

SAMPLE PROBLEM 5.3

Writing Enantiomers of a Fischer Projection

Draw the Fischer projection for the enantiomer (mirror image) of each of the following:

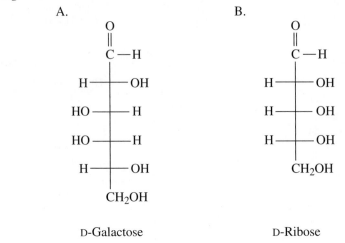

A.

D-Galactose

B.

D-Ribose

SOLUTION

To write an enantiomer of a Fischer projection, imagine that a mirror exists between the pair of molecules so that one enantiomer's reflection will be the other enantiomer. What appears on the right of one will appear on the left of the other. The L-enantiomer is shown to the right of the original.

A.

B.

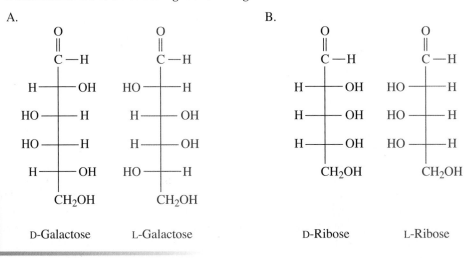

D-Galactose   L-Galactose

D-Ribose   L-Ribose

Stereoisomers That Are Not Enantiomers

So far we have seen one stereoisomer of D-glucose: its enantiomer L-glucose. Because there can be only one mirror image for any stereoisomer, how are the other 14 of 16 possible stereoisomers related to D-glucose? Stereoisomers that are not enantiomers are called **diastereomers**. Diastereomers are stereoisomers that are not exact mirror images. Figure 5.3 shows the monosaccharides D-galactose and D-talose in Fischer projection, both of which are diastereomers of D-glucose.

Some Important Monosaccharides

Several of the more common monosaccharides are hexoses (containing six carbons) produced and used by nature only as the D-isomers. The D-form is discussed here.

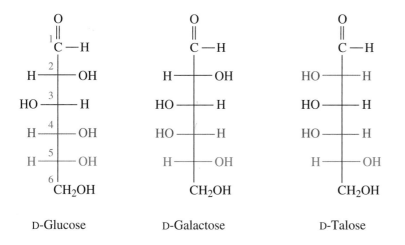

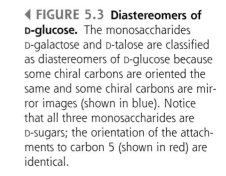

D-Glucose   D-Galactose   D-Talose

▸ **FIGURE 5.3 Diastereomers of D-glucose.** The monosaccharides D-galactose and D-talose are classified as diastereomers of D-glucose because some chiral carbons are oriented the same and some chiral carbons are mirror images (shown in blue). Notice that all three monosaccharides are D-sugars; the orientation of the attachments to carbon 5 (shown in red) are identical.

The most abundant monosaccharide found in nature is glucose. We commonly refer to D-glucose as dextrose, blood sugar, and grape sugar. It is found in fruits, vegetables, and corn syrup. This monosaccharide can be broken down inside cells to produce energy in the body through a series of chemical reactions referred to as **glycolysis**. Diabetics have difficulty getting glucose into their cells to undergo glycolysis, which is why they must regularly monitor their blood glucose levels. Glucose is also a sugar unit in the disaccharides sucrose (table sugar) and lactose (milk sugar) as well as the polysaccharides starch, glycogen, and cellulose.

Galactose (Figure 5.3) is found in nature combined with glucose in the disaccharide lactose, which is present in milk and other dairy products. Galactose has a single chiral center (carbon 4) arranged opposite that of glucose. Diastereomers that differ in just one chiral center (as compared with more than one chiral center) are called **epimers**. The body can chemically convert galactose into glucose for use in glycolysis with the help of an enzyme called an epimerase, which assists in switching the orientation of the groups attached to carbon 4 in galactose to that of glucose.

Mannose is a monosaccharide found in some fruits and vegetables that is not easily absorbed by the body. The most notable fruit that contains high amounts of mannose is the cranberry. Mannose has been shown to be effective against urinary tract infections (UTIs). When the level of mannose builds up in the bladder, the bacteria causing the UTI will attach themselves to the mannose in the urine instead of the other carbohydrates found on the outside of the cells lining the urinary tract and will be eliminated. Mannose is also an epimer of glucose.

The ketose fructose is also commonly referred to as fruit sugar or levulose and is found in fruits, vegetables, and honey. In combination with glucose it gives us the disaccharide sucrose (table sugar). Fructose is the sweetest monosaccharide (70% sweeter than table sugar), making it popular with dieters who can get the same sweet taste with fewer calories. Even though it is not an epimer of glucose, fructose can be broken down for energy production in the body by the chemical reactions of glycolysis as we will see in Chapter 12.

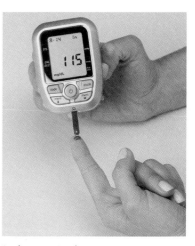

A glucose monitor.

CH_2OH
|
$C = O$
HO —— H
H —— OH
H —— OH
CH_2OH

D-Fructose

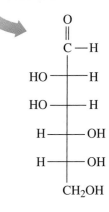

D-Mannose

The pentoses (five-carbon sugars) ribose and 2-deoxyribose are a part of larger biomolecules called nucleic acids that make up our genetic material, a topic we will address in Chapter 11. The nucleic acids are distinguished in their name by the monosaccharide they contain. *Ribo*nucleic acid (RNA) contains the sugar ribose, and *deoxyribo*nucleic acid (DNA) contains the sugar deoxyribose. Structurally, the only difference between the two pentoses is the absence of an oxygen atom on carbon 2 of deoxyribose. Ribose is also found in the vitamin *ribo*flavin and other biologically important molecules as we will see in Chapter 12.

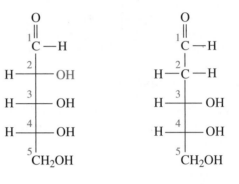

D-Ribose   D-2-Deoxyribose

SAMPLE PROBLEM 5.4

Distinguishing Stereoisomers

Classify structures A, B, and C in the figure as being either an enantiomer or a diastereomer of D-mannose.

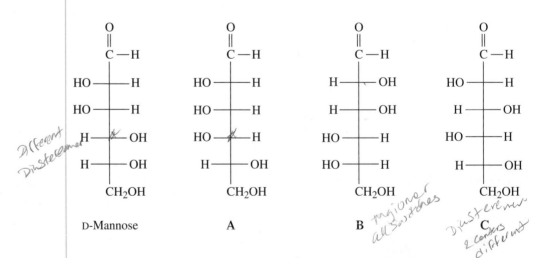

SOLUTION

Structure A has only one chiral center oriented differently from D-mannose, so it is therefore a diastereomer. Structure B is the mirror image of D-mannose, its enantiomer is L-mannose. Structure C has two chiral centers oriented differently from D-mannose, so it is a diastereomer.

PRACTICE PROBLEMS

5.5. How are disaccharides related to monosaccharides?

5.6. How are polysaccharides related to monosaccharides?

5.7. Name the functional groups present in aldoses.

5.8. Name the functional groups present in ketoses.

5.9. Draw the Fischer projection for the enantiomer (mirror image) of each of the following:

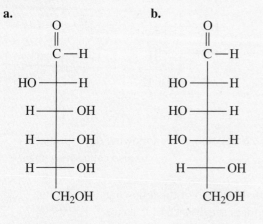

a.

D-Altrose

b.

D-Talose

5.10. Draw the Fischer projection for the enantiomer (mirror image) of each of the following:

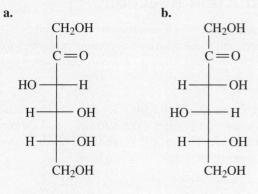

a.

D-Fructose

b.

D-Sorbose

5.11. Classify structures A, B, and C in the figure as being either an enantiomer or a diastereomer of D-galactose.

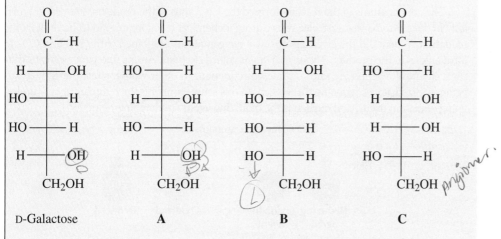

D-Galactose   **A**   **B**   **C**

5.12. Classify structures A, B, and C in the figure as being either an enantiomer or a diastereomer of D-glucose.

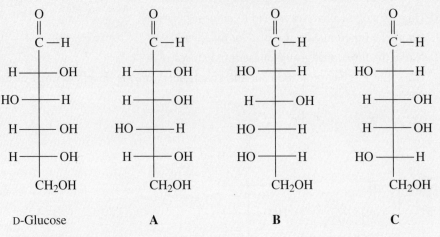

D-Glucose   A   B   C

5.13 Identify the monosaccharide that fits each of the following descriptions:
a. also referred to as dextrose   b. also called fruit sugar
c. used to treat urinary tract infections

5.14 Identify the monosaccharide that fits each of the following descriptions:
a. in combination with glucose produces the disaccharide lactose
b. also called blood sugar   c. also called levulose

5.3 Oxidation and Reduction Reactions

Chemical reactions called oxidation and reduction are essential to energy production and transfer in living systems. To understand these reactions in organic molecules like monosaccharides, let's start by examining some inorganic (nonliving) examples.

Oxidation and Reduction

Why is the Statue of Liberty green in color as shown in Figure 5.4 if it is made of copper metal that is not green? The copper reacts with water, carbon dioxide, and the oxygen in the air forming the green patina finish. Two copper compounds result, copper carbonate and copper hydroxide. The balanced equation is shown here:

▲ **FIGURE 5.4 The Statue of Liberty.** The Statue of Liberty appears green in color due to the oxidation of copper (Cu) metal.

$$2Cu(s) + H_2O(g) + CO_2(g) + O_2(g) \longrightarrow CuCO_3(s) + Cu(OH)_2(s)$$

Shiny orange   Sea green
metal

Let's examine the reactants and product a little more closely to see what happened to the copper and the oxygen during this reaction.

The copper atoms in the reactant formed the Cu^{2+} ions of the products. *The copper atoms lost electrons.* Atoms that lose electrons during a chemical reaction are said to be undergoing **oxidation**. Where did the electrons go? The oxygen atoms combined with H_2O and CO_2 to form anions in the product. *The oxygen atoms gained electrons.* Atoms that gain electrons during a chemical reaction are said to undergo **reduction**. In other words, while the copper was being oxidized, it was reducing the oxygen. In this sense, copper is the **reducing agent** (undergoing oxidation itself), and oxygen is the **oxidizing agent** (undergoing reduction).

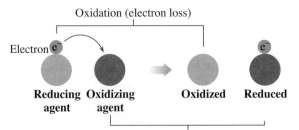

Oxidation and reduction reactions, commonly called *redox* reactions, are always coupled together because if a reactant is oxidized (losing one or more electrons), another reactant is simultaneously reduced (gaining one or more electrons). A mnemonic device that is helpful for remembering what is happening to the electrons in a redox reaction is to remember the letters in the word "OIL RIG", which in this case stands for Oxidation Is Loss (of electrons), Reduction Is Gain (of electrons).

An organic molecule is oxidized if it gains oxygen *or* loses hydrogen and is reduced if it either gains hydrogen *or* loses oxygen. Many biological reactions involve the transfer of oxygen or the transfer of hydrogen atoms and are classified as redox reactions. See Table 5.1 for a summary of these characteristics.

Monosaccharides and Redox

Monosaccharides add oxygen atoms (or lose hydrogens) during oxidation and add hydrogen (or lose oxygen) during reductions. These reactions can occur at the aldehyde carbonyl (carbon 1) in aldoses. The $C=O$ bond is highly polar, making it reactive. Aldoses can undergo both oxidation *and* reduction. The aldehyde functional group can oxidize to a carboxylic acid or can reduce to an alcohol. In monosaccharides an oxidation produces a sugar acid and a reduction produces a sugar alcohol.

TABLE 5.1

CHARACTERISTICS OF OXIDATION AND REDUCTION REACTIONS

Oxidation	
Always Involves	**May Involve**
Loss of electrons	Addition of oxygen
	Loss of hydrogen

Reduction	
Always Involves	**May Involve**
Gain of electrons	Loss of oxygen
	Gain of hydrogen

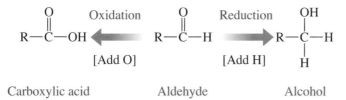

Carboxylic acid   Aldehyde   Alcohol

When molecules are oxidized, they act as a reducing agent and reduce a second reactant. One useful oxidation reaction for sugars occurs in the **Benedict's test**. This tests for the presence of an aldose in a solution. During the Benedict's test an aldehyde group reduces copper ions from Cu^{2+} to Cu^+ (copper gained one electron). The reacting Cu^{2+} ions are soluble, coloring the initial reaction solution blue. The aldehyde group is oxidized by Cu^{2+}, forming a sugar acid, while Cu^{2+} becomes reduced by the aldehyde, forming Cu^+ ions. The resulting copper(I) oxide (Cu_2O), is not soluble and forms a brick-red precipitate in solution (see Figure 5.5). Because aldoses are easily oxidized and can therefore serve as reducing agents, sugars capable of reducing substances like the Cu^{2+} are referred to as **reducing sugars**. Fructose and other ketoses are also reducing sugars, even though they do not contain an aldehyde group, because in the presence of oxidizing agents they can rearrange to aldoses.

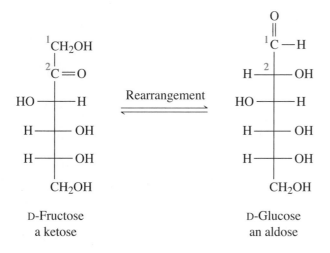

D-Fructose   D-Glucose
a ketose   an aldose

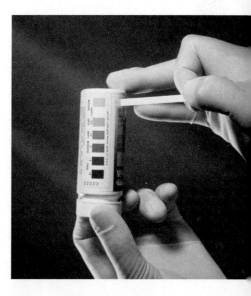

The Benedict's test can be used to monitor glucose levels in urine. Glucose test strips, like the one shown at right, produce a range of color changes if glucose is present. Excess glucose in urine suggests high levels of glucose in the bloodstream, an indicator for diabetes.

▶ **FIGURE 5.5 The oxidation of D-glucose.** In the presence of the Benedict's reagent (containing Cu^{2+} ions), D-glucose reacts to produce the sugar acid, D-gluconic acid, and a brick-red precipitate, Cu_2O.

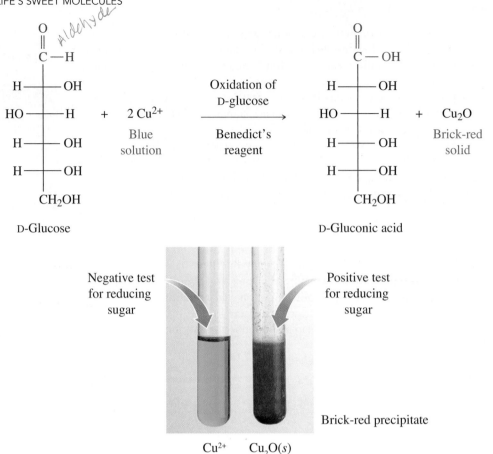

$$\begin{array}{c} O \\ \parallel \\ C-H \\ \end{array}$$ *Aldehyde*

	C—H	
H—	—OH	
HO—	—H	
H—	—OH	
H—	—OH	
	CH₂OH	

D-Glucose   + 2 Cu²⁺ Blue solution

Oxidation of D-glucose
Benedict's reagent

	C—OH	
H—	—OH	
HO—	—H	
H—	—OH	
H—	—OH	
	CH₂OH	

D-Gluconic acid   + Cu₂O Brick-red solid

Negative test for reducing sugar

Positive test for reducing sugar

Brick-red precipitate

Cu^{2+} $Cu_2O(s)$

An aldose or ketose can also be reduced to an alcohol when the carbonyl reacts with hydrogen under the right conditions (see Figure 5.6). Sugar alcohols are produced commercially as artificial sweeteners and are found in sugar-free foods (see Figure 5.7).

Sugar alcohols can be produced in the body when glucose levels remain high in the bloodstream. An enzyme called aldose reductase acts to reduce excessive glucose to the sugar alcohol sorbitol, which at high concentration can contribute to cataracts (clouding of the lens of the eye). These so-called "sugar cataracts" are commonly seen in diabetics.

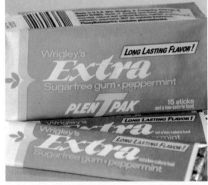

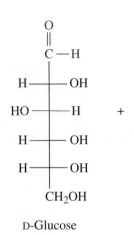

D-Glucose   + H₂

Reduction

D-Glucitol also called D-sorbitol

▲ **FIGURE 5.6 The reduction of D-glucose.** In the presence of a hydrogen source, D-glucose reacts to produce the sugar alcohol D-glucoitol, which is better known as D-sorbitol.

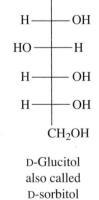

▲ **FIGURE 5.7 Sugar alcohols as sweeteners.** Sugar alcohols like sorbitol, xylitol, and mannitol are used in many sugar-free products.

SAMPLE PROBLEM 5.5

Distinguishing Oxidation and Reduction

Determine whether the following substances are undergoing oxidation or reduction:

a. Iron metal forms rust, Fe_2O_3.

b. Copper(II) ions in solution form solid copper metal on a surface.

c. Wine (containing ethanol, CH_3CH_2OH) sours to vinegar (CH_3COOH).

d. The biomolecule NAD^+ becomes NADH.

SOLUTION

a. Iron metal (Fe) is losing electrons when it ionizes to form rust, Fe^{3+}. Losing electrons represents an *oxidation*.

b. Copper(II) ions (Cu^{2+}) forming copper metal (Cu) require that copper gain two electrons. Gaining electrons represents a *reduction*.

c. When an alcohol forms a carboxylic acid, the alcohol first loses a hydrogen, then an oxygen is added. Both represent *oxidation*.

d. In the biological reaction, $NAD^+ \rightarrow NADH$, a hydrogen is added. This represents a *reduction*.

PRACTICE PROBLEMS

5.15 Are the substances shown in italics undergoing oxidation or reduction?

a. *D-Glucose* forms a brick-red solution when subjected to the Benedict's test.

b. *Aluminum* metal forms aluminum oxide (Al_2O_3).

c. A *carboxylic acid* ($RCOOH$) reacts to form an alcohol (RCH_2OH).

5.16 Are the substances shown in italics undergoing oxidation or reduction?

a. *Silver ions* (Ag^+) are electroplated into silver atoms on flatware.

b. *D-Glucose* reacts with hydrogen to form the sugar alcohol D-sorbitol.

c. The biomolecule *$FADH_2$* loses hydrogens becoming FAD.

5.17 Write the products if (a) carbon 1 is oxidized and (b) if carbon 1 is reduced in D-ribose.

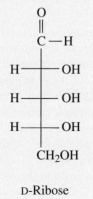

D-Ribose

5.18 Write the products if (a) carbon 1 is oxidized and (b) if carbon 1 is reduced in D-xylose.

$$
\begin{array}{c}
\text{O} \\
\parallel \\
\text{C}-\text{H} \\
\text{H}-\!\!\!-\!\!\!-\text{OH} \\
\text{HO}-\!\!\!-\!\!\!-\text{H} \\
\text{H}-\!\!\!-\!\!\!-\text{OH} \\
\text{CH}_2\text{OH}
\end{array}
$$

D-Xylose

5.4 Ring Formation—The Truth About Monosaccharide Structure

In addition to being reactive with oxidizing agents and reducing agents, a carbonyl group can react with a hydroxyl (—OH) functional group. The lone pairs of electrons on the oxygen of a hydroxyl make the oxygen partially negative (δ^-). In turn, the oxygen of the carbonyl makes the carbonyl carbon partially positive (δ^+). The partially negative (δ^-) oxygen of the hydroxyl is attracted to the partially positive (δ^+) carbon of the carbonyl group. When this happens, the functional group **hemiacetal** (pronounced hem-ee-ass-i-TAL) is formed as shown in the general reaction in Figure 5.8.

▶ **FIGURE 5.8 Formation of the hemiacetal group.** The electrons on the oxygen of the alcohol group (δ^-) are attracted to the partial positive charge (δ^+) of the carbon in the carbonyl (blue dashed line). The joining of these two functional groups results in a hemiacetal.

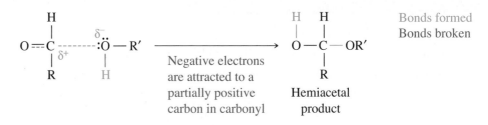

Monosaccharides contain *both* a carbonyl and several hydroxyl (alcohol) functional groups, and these two functional groups can react *within the same molecule*. Using D-glucose as an example, the hydroxyl on carbon 5 (C5) can curl around to react with the carbonyl at carbon 1 (C1), producing two possible ring structures as shown in Figure 5.9. When this occurs, five carbons and an oxygen form a ring, and one of the carbons (carbon 6) remains outside the ring. The truth is monosaccharides exist in a ring form most of the time because the carbonyl group reacts readily with a hydroxyl to form a hemiacetal ring.

Two structures are possible during ring formation. Recall that a carbonyl group is trigonal planar (flat), so the oxygen in the hydroxyl group on carbon 5 can form its bond on either the top or the bottom side of the carbonyl. Different stereoisomers (3-D arrangements) are then formed at C1. In any monosaccharide, the carbonyl carbon that reacts to form the hemiacetal in the reaction is referred to as the **anomeric carbon**. (Note that in the ring form, the anomeric carbon is the only carbon bonded directly to two oxygen atoms.)

Two different ring arrangements can be produced from a single linear monosaccharide chain. These two interconvertible forms are termed **anomers**. The two anomers are referred to as the alpha (α) and the beta (β) anomers. Anomers are distinguished by the positioning of the —OH group on the anomeric carbon relative to the position of the carbon outside the ring (for D-glucose carbon 6). In the six-member ring form of D-isomers called a **pyranose**, carbon 6 (C6) is always drawn on the top side of the ring.

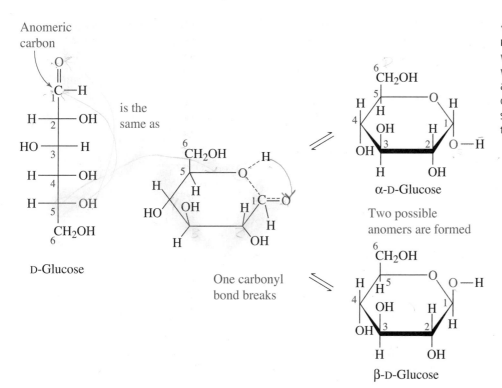

◀ FIGURE 5.9 Ring formation in D-glucose. D-Glucose forms a ring when the alcohol group on C5 reacts with the carbonyl (C1). α and β anomers are formed. C6 remains outside the ring. Dashed lines show bonds breaking and forming to create a ring.

- In the α anomer, the —OH on the anomeric carbon is trans to the carbon outside the ring (they are on opposite sides of the ring).
- In the β anomer, the —OH on the anomeric carbon is cis to the carbon outside the ring (they are on the same side of the ring).

D-Fructose contains *both* a ketone group and several hydroxyl groups. The —OH on carbon 5 (C5) can curl around and react with the carbonyl positioned at carbon 2 (C2), allowing two possible ring structures, as in D-glucose (see Figure 5.10). Four carbons and an oxygen form the five-member ring called a **furanose** and two of the carbons (C1 and C6) remain outside the ring.

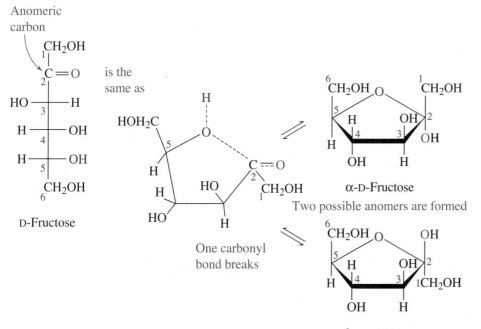

◀ FIGURE 5.10 Ring formation in D-fructose. D-Fructose forms a ring when the alcohol group on C5 reacts with the carbonyl C2. α and β anomers are formed. C6 and C1 remain outside the ring. Dashed lines show bonds breaking and forming to create ring.

The anomers are similarly distinguished for a five-member and six-member ring. Alpha and beta anomers are determined by the positioning of the —OH group on the anomeric carbon (in D-fructose this is C2) relative to the carbon outside the ring (for D-fructose C6).

- In the α anomer, the —OH on the anomeric carbon is trans to the carbon outside the ring (on opposite sides).
- In the β anomer, the —OH on the anomeric carbon is cis to the carbon outside the ring (on the same side).

Let's practice drawing monosaccharides in the pyranose ring form (six-member ring) from the linear Fischer projection. The following figure demonstrates how to draw the ring form of D-galactose as the β anomer.

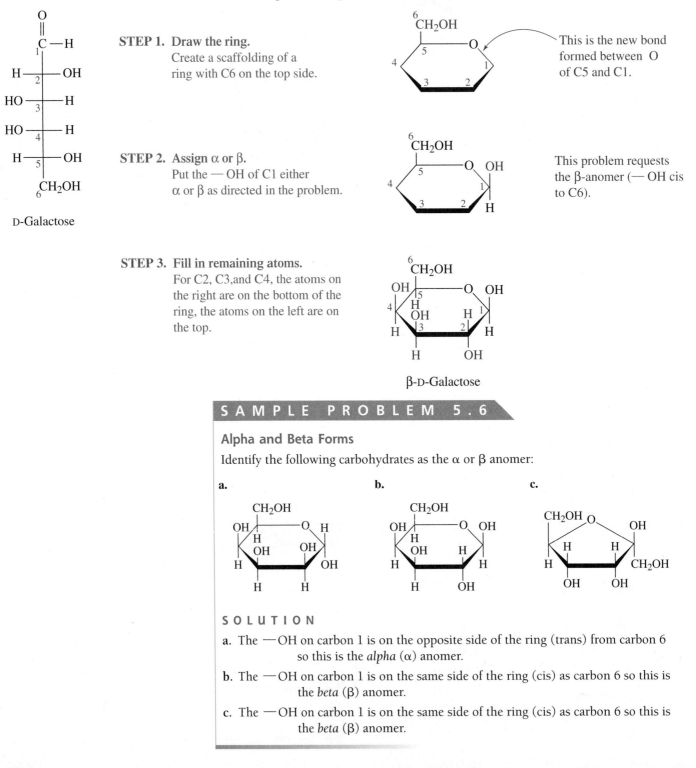

STEP 1. Draw the ring.
Create a scaffolding of a ring with C6 on the top side.

This is the new bond formed between O of C5 and C1.

STEP 2. Assign α or β.
Put the —OH of C1 either α or β as directed in the problem.

This problem requests the β-anomer (—OH cis to C6).

STEP 3. Fill in remaining atoms.
For C2, C3, and C4, the atoms on the right are on the bottom of the ring, the atoms on the left are on the top.

β-D-Galactose

D-Galactose

SAMPLE PROBLEM 5.6

Alpha and Beta Forms

Identify the following carbohydrates as the α or β anomer:

a. **b.** **c.**

SOLUTION

a. The —OH on carbon 1 is on the opposite side of the ring (trans) from carbon 6 so this is the *alpha* (α) anomer.

b. The —OH on carbon 1 is on the same side of the ring (cis) as carbon 6 so this is the *beta* (β) anomer.

c. The —OH on carbon 1 is on the same side of the ring (cis) as carbon 6 so this is the *beta* (β) anomer.

S A M P L E   P R O B L E M   5 . 7

Drawing Pyranose Rings from Linear Monosaccharides

Draw the alpha (α) anomer of D-mannose in pyranose ring form.

SOLUTION

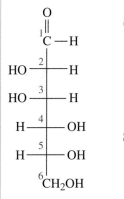

D-Mannose

STEP 1. Draw the ring.
Create a scaffolding of a
ring with C6 on the top side.

This is the new bond
formed between
O of C5 and C1.

STEP 2. Assign α or β.
Put the — OH of C1 either
α or β as directed in the problem.

This problem requests
the α-anomer
(— OH trans to C6).

STEP 3. Fill in remaining atoms.
For C2, C3,and C4, the atoms on
the right are below the ring, the
atoms on the left are above.

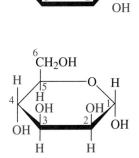

α-D-Mannose

P R A C T I C E   P R O B L E M S

5.19 Identify the following carbohydrates as the α or β anomer:

a.

alpha

b.

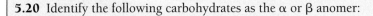

bottom =Inside
on R side of
Fischer

5.20 Identify the following carbohydrates as the α or β anomer:

a.   **b.**

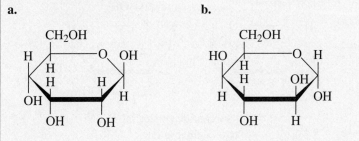

5.21 Draw the α and β anomer of D-talose in pyranose ring form:

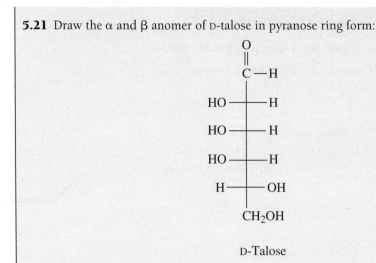

D-Talose

5.22 Draw the α and β anomer of D-altrose in pyranose ring form:

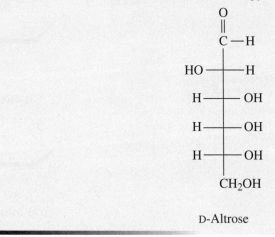

D-Altrose

5.5 Disaccharides

Condensation and Hydrolysis—Forming and Breaking Glycosidic Bonds

Disaccharides are formed when two monosaccharides are joined together through a chemical reaction. In a monosaccharide in ring form, the most reactive —OH in the molecule is on the anomeric carbon (C1 in an aldose). When this hydroxyl reacts with a hydroxyl on another monosaccharide a **glycosidic bond** forms. These glycosidic bonds connect monosaccharides to each other and in a more general sense connect monosaccharides to any alcohol. In the following reaction, two glucose units are joined to form the disaccharide maltose:

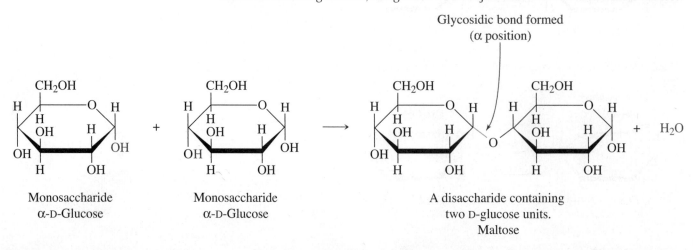

The formation of **glycosides** is an example of another common chemical reaction of organic molecules. During this reaction, a molecule of water is eliminated as two molecules are joined. When a small molecule like water is produced, this reaction is referred to as a **condensation** reaction (*condense*—"to bring together"). Often when H_2O is produced, this reaction is more specifically referred to as a **dehydration**. A condensation reaction operating in reverse is a **hydrolysis** reaction (*hydro* means "water," *lysis* means "break"; literally "to break with water"), where water is consumed as a reactant. Condensation reactions can occur between a number of functional groups that contain an —H in a polar bond (like O—H or N—H) and an —OH group that can be removed to form water.

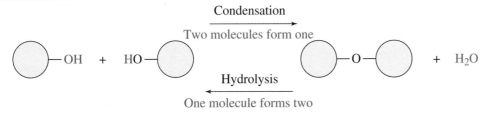

SAMPLE PROBLEM 5.8

Locating a Glycosidic Bond

Locate the glycosidic bond in the following carbohydrates:

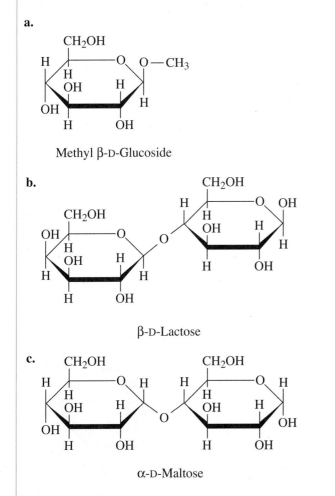

a.

Methyl β-D-Glucoside

b.

β-D-Lactose

c.

α-D-Maltose

SOLUTION

The glycosidic bond joins a monosaccharide through a condensation reaction to another alcohol group. One of the carbons bonded to the oxygen must be the anomeric carbon. The glycosidic bonds are blue in the figure.

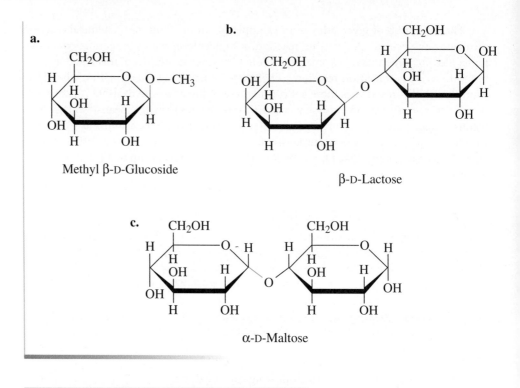

Methyl β-D-Glucoside

β-D-Lactose

α-D-Maltose

S A M P L E   P R O B L E M   5 . 9

Identifying Condensation and Hydrolysis

Identify the following reactions as representations of a condensation or a hydrolysis:

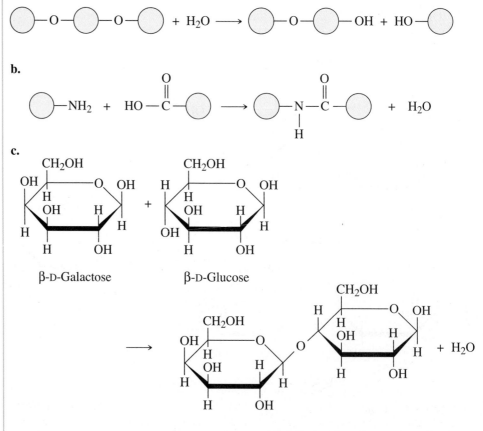

β-D-Galactose

β-D-Glucose

β-D-Lactose

SOLUTION

a. This representation indicates water as a reactant, breaking a larger molecule. This is a *hydrolysis*.

b. This representation produces water in the reaction when the functional groups amine and carboxylic acid are combined to form an amide. This is a *condensation*.

c. This reaction connects two monosaccharides into a disaccharide through a glycosidic bond. This is a *condensation*.

Naming Glycosidic Bonds

If we look back at the disaccharide maltose that was formed from two glucose units, we see that the glycosidic bond was formed in the α position. If the second monosaccharide had bonded to the β side (in this case top) of the first glucose, this would have created a different disaccharide molecule with a different shape (see Figure 5.11). Because of this, it becomes necessary to specify how the monosaccharides are bonded, that is, α or β. In addition to the specific configuration bonded (α or β), the carbons that were joined must also be specified. Any —OH group on the second monosaccharide can, in principle, react with the first monosaccharide's anomeric carbon. Because of this, it is necessary to specify the carbon atoms joined by the glycosidic bond. The convention for naming glycosidic bonds is to specify the anomer bonded and the carbons bonded. For maltose, the glycosidic bond is specified as $\alpha(1 \rightarrow 4)$ (stated "alpha-one-four").

Maltose

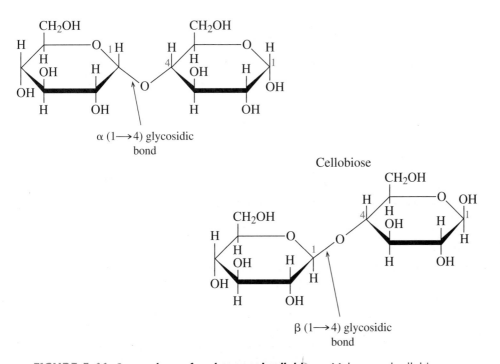

▲ **FIGURE 5.11 Comparison of maltose and cellobiose.** Maltose and cellobiose are disaccharides of D-glucose bonded $\alpha(1 \rightarrow 4)$ and $\beta(1 \rightarrow 4)$, respectively. The difference in the glycosidic bond positioning (alpha versus beta) makes the shape of these two molecules different.

Naming Glycosidic Bonds

Name the glycosidic bond in the following disaccharides:

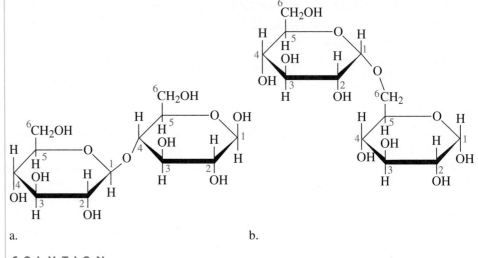

a.   b.

SOLUTION

a. The anomeric carbon (C1) in this glycosidic bond is in the β anomer form. The carbons that are connected are C1 on the left and C4 on the right. The glycosidic bond is designated β(1 → 4).

b. The anomeric carbon (C1) in this glycosidic bond is in the α anomer form. The carbons that are connected are C1 on the left and C6 on the right. The glycosidic bond is designated α(1 → 6).

Three Important Disaccharides—Maltose, Lactose, and Sucrose

Three common disaccharides formed through the condensation of two monosaccharides are maltose, lactose, and sucrose. Their formation is outlined as follows:

α-D-glucose + D-glucose → maltose + H_2O [glycosidic bond α(1 → 4)]

β-D-galactose + D-glucose → lactose + H_2O [glycosidic bond β(1 → 4)]

α-D-glucose + β-D-fructose → sucrose + H_2O [glycosidic bond α,β(1 → 2)]

In discussing saccharides larger than the monosaccharides, the "D" designation is often dropped from the name and assumed unless noted. For clarity, it is also common to see the ring form with the hydrogens not shown but implied. These shortcuts will be used from this point forward.

Maltose

Maltose, also known as malt sugar, is a disaccharide formed in the breakdown of starch. Malted barley, a key ingredient in beer, contains high levels of maltose. The malting process involves monitoring the germination of barley grains, during which the starch in the grain is converted to maltose through hydrolysis. This process is halted before the grains sprout by drying and roasting the grains. The glucose in the maltose of malted barley can then be converted to alcohol by yeast in the fermentation process.

The glycosidic bond between the glucose units in maltose is α(1 → 4). Because one of the anomeric carbons (C1 on the second glucose unit) is free, maltose is a reducing sugar (see Figure 5.12).

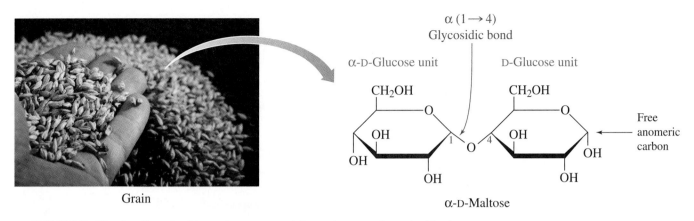

▲ FIGURE 5.12 **The dissacharide maltose.** One of the main sugars in malted barley (grains in photo) is maltose formed from hydrolyzed starch.

Lactose

Lactose, or milk sugar, is found in mammalian milk and commercial milk products. An intolerance to lactose can occur in people who inherit or lose the ability to produce the enzyme that hydrolyzes this disaccharide into its two monosaccharides. This hydrolytic enzyme, commonly known as lactase, is sometimes added to commercial products to assist in the digestion of milk products for lactose intolerant people. When lactose remains undigested, intestinal bacteria break down undigested lactose and in doing so produce excessive abdominal gas and cramping.

The glycosidic bond in lactose is $\beta(1 \rightarrow 4)$ because it occurs between C1 of a β-galactose and C4 of a glucose unit. Because one of the anomeric carbons (C1 on the glucose unit) is free, lactose is a reducing sugar (see Figure 5.13).

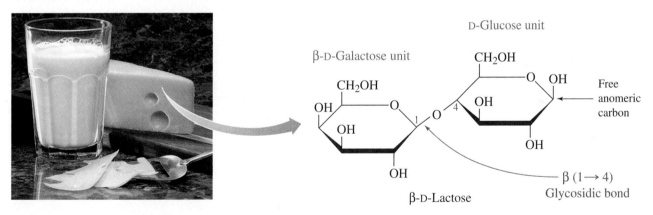
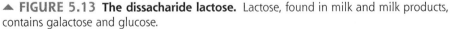

▲ FIGURE 5.13 **The dissacharide lactose.** Lactose, found in milk and milk products, contains galactose and glucose.

Sucrose

Sucrose, ordinary table sugar, is the most abundant disaccharide in nature. High quantities of sucrose are found in sugar cane and sugar beets, which are the main commercial sources.

When glucose and fructose join in an $\alpha,\beta(1 \rightarrow 2)$ glycosidic bond, sucrose is formed. In sucrose, both anomeric carbons are bonded (carbon 1 of glucose and carbon 2 of fructose), and, therefore, both are noted in the name of the glycosidic bond. Because there is no free anomeric carbon, sucrose does not react with the Benedict's reagent and is *not* a reducing sugar (see Figure 5.14).

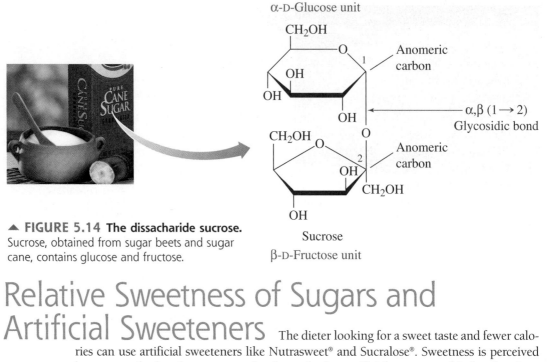

▲ **FIGURE 5.14 The dissacharide sucrose.** Sucrose, obtained from sugar beets and sugar cane, contains glucose and fructose.

Relative Sweetness of Sugars and Artificial Sweeteners

The dieter looking for a sweet taste and fewer calories can use artificial sweeteners like Nutrasweet® and Sucralose®. Sweetness is perceived by our taste buds, where sugar molecules and other artificial sweeteners bind and register this taste in the brain. Sweetness is referenced relative to sucrose (table sugar), which is given a value of 100. Fructose has a sweetness value nearly twice that of table sugar. For this reason, it is often found in dietetic foods because the same sweet taste requires half as much fructose as sucrose. Artificial sweeteners are hundreds of times sweeter than sucrose, so much less needs to be used. In addition, many artificial sweeteners lack calories because they cannot produce energy as many monosaccharides do through glycolysis. The sweetness of some sugars and other sweeteners compared to sucrose is shown in Table 5.2.

TABLE 5.2	RELATIVE SWEETNESS OF SUGARS AND ARTIFICIAL SWEETENERS	
Sweetener	**Sweetness Relative to Sucrose (=100)**	**Description**
Simple Sugars		
Fructose	140–175	Fruit sugar, a monosaccharide that is a component of sucrose
Invert sugar	120	Found in honey, hydrolyzed sucrose
Sucrose	100	Table sugar, a disaccharide containing glucose and fructose
Xylitol	100	A sugar alcohol, used in sugar-free products
Glucose	75	Dextrose, a monosaccharide that is a component of sucrose and lactose
Erythritol	70	A sugar alcohol used in sugar free products
Sorbitol	36–55	A sugar alcohol used in sugar-free products
Maltose	32	A disaccharide of glucose
Galactose	30	A monosaccharide that is a component of the disaccharide lactose
Lactose	15	Milk sugar, a disaccharide containing galactose and glucose
Other Sweeteners		
Sucralose (Splenda®)	60,000	Chlorinated disaccharide
Saccharin	45,000	An organic substance initially discovered as a by-product of research on dyes
Stevia (found in Truvia™)	25,000	Natural sweetener containing nonhydrolyzable glucosides found in the leaves of the plant *Stevia rebaudiana*
Aspartame (Nutrasweet®)	18,000	Contains two amino acids, aspartic acid and phenylalanine

PRACTICE PROBLEMS

5.23 Identify the following reactions as a condensation or hydrolysis:

a. two monosaccharides reacting to form a disaccharide

b. the formation of a glycosidic bond

c. a reaction in which one molecule breaks into two and an —H and —OH are added

5.24 Identify the following reactions as condensation or hydrolysis:

a. a disaccharide breaking into two monosaccharides

b. a reaction in which two molecules combine forming one and a molecule of water is produced

c. the breaking of a glycosidic bond

5.25 Name the glycosidic bond present in mannobiose shown in the following figure:

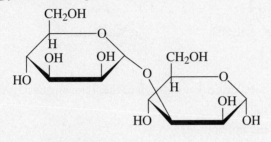

Mannobiose

5.26 Name the glycosidic bond present in melibiose, a disaccharide that has the sweetness of about 30 compared with sucrose.

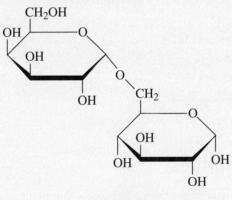

Melibiose

5.27 For each of the following disaccharides, name the glycosidic bond and draw the monosaccharide units produced by hydrolysis:

a.

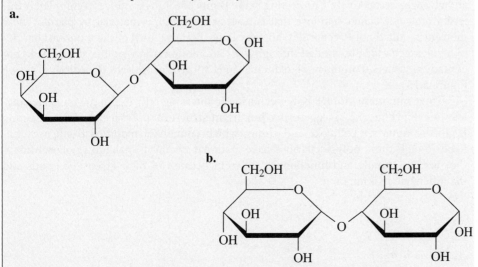

b.

5.28 For each of the following disaccharides, name the glycosidic bond and draw the monosaccharide units produced by hydrolysis:

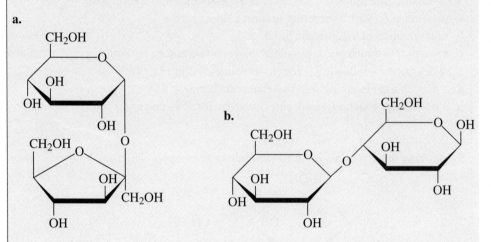

a.

b.

5.29 Identify a disaccharide that fits each of the following descriptions:
a. ordinary table sugar
b. found in milk and milk products
c. also called malt sugar
d. hydrolysis gives galactose and glucose

5.30 Identify a disaccharide that fits each of the following descriptions:
a. not a reducing sugar
b. composed of two glucose units
c. also called milk sugar
d. hydrolysis gives glucose and fructose

5.6 Polysaccharides

Some days there is more sunlight than other days—can plants store glucose for a rainy day? What happens if we eat more simple sugars (monosaccharides and disaccharides) than we need for energy—can we store that energy? Glucose can be stored in both plants and animals by connecting α-glucose units through glycosidic bonds as starch and glycogen respectively. Connecting many β-glucose units produces a completely different molecule called cellulose that is used as a structural material by plants. These molecules, called polysaccharides, are very large. Because most of the monosaccharides in a polysaccharide are bonded through their anomeric carbon, polysaccharides do not contain a sufficient number of reducing ends to give a positive Benedict's test (see Figure 5.15).

Three important **storage polysaccharides** containing only α-glucose units are amylose, amylopectin, and glycogen. Two important **structural polysaccharides** containing β-glucose units are cellulose and chitin, which contains a modified β-glucose unit. Even though these polysaccharides have a similar chemical makeup (glucose units), they are structurally and functionally different because of their glycosidic bonds and the difference in branching.

a.

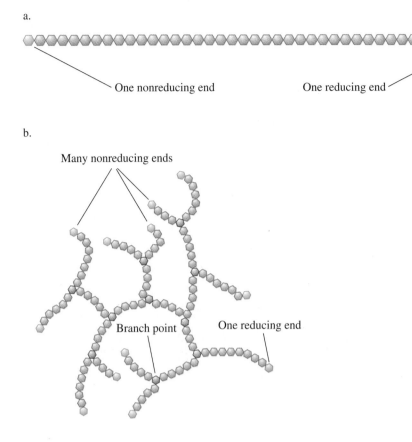

One nonreducing end   One reducing end

b.

Many nonreducing ends

Branch point   One reducing end

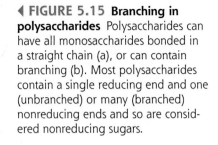

◀ **FIGURE 5.15 Branching in polysaccharides** Polysaccharides can have all monosaccharides bonded in a straight chain (a), or can contain branching (b). Most polysaccharides contain a single reducing end and one (unbranched) or many (branched) nonreducing ends and so are considered nonreducing sugars.

Storage Polysaccharides

Amylose and Amylopectin—Starch

Have you ever eaten an unripe banana? Its taste is less sweet than a ripe banana, and the texture is rather mealy. The mealy or grainy texture is due to the stores of starch in the banana's fruit that have yet to be hydrolyzed to glucose as the fruit ripens. Starch is a glucose storage polysaccharide stored in small granules in plant cells. Plant foods that we think of as starchy—like potatoes, grains, and beans—contain an abundance of these polysaccharides. Starch is a mixture of two polysaccharides, amylose and amylopectin. **Amylose,** which makes up about 20% of starch, is made up of anywhere from 250–4000 D-glucose units bonded $\alpha(1 \rightarrow 4)$ in a continuous chain. Long chains of D-glucose bonded $\alpha(1 \rightarrow 4)$ will tend to coil much like a telephone cord.

    Amylopectin makes up about 80% of plant starch. It also contains D-glucose units connected by $\alpha(1 \rightarrow 4)$ glycosidic bonds. However, unlike amylose, about every 25 glucose units along a glucose chain branches off through an $\alpha(1 \rightarrow 6)$ glycosidic bond (see Figure 5.16). During fruit ripening, starch granules undergo hydrolysis by the action of an enzyme called amylase that hydrolyzes the $\alpha(1 \rightarrow 4)$ glycosidic bonds in starch, producing glucose and maltose, which are sweet. When we eat starch, our digestive system breaks it down into glucose units for use by our bodies.

Glycogen

Animals store glucose as a polysaccharide too. **Glycogen** is the storage polysaccharide found in animals. Most glycogen stores are located in the liver and in muscles. Glycogen is identical in structure to amylopectin except that $\alpha(1 \rightarrow 6)$ branching occurs about every 12 glucose units. Glycogen is hydrolyzed to glucose in the liver and sent into the bloodstream to maintain constant glucose levels in the blood during fasting periods when sugars are not being consumed. The large amount of branching in this molecule allows for quick hydrolysis when glucose is needed.

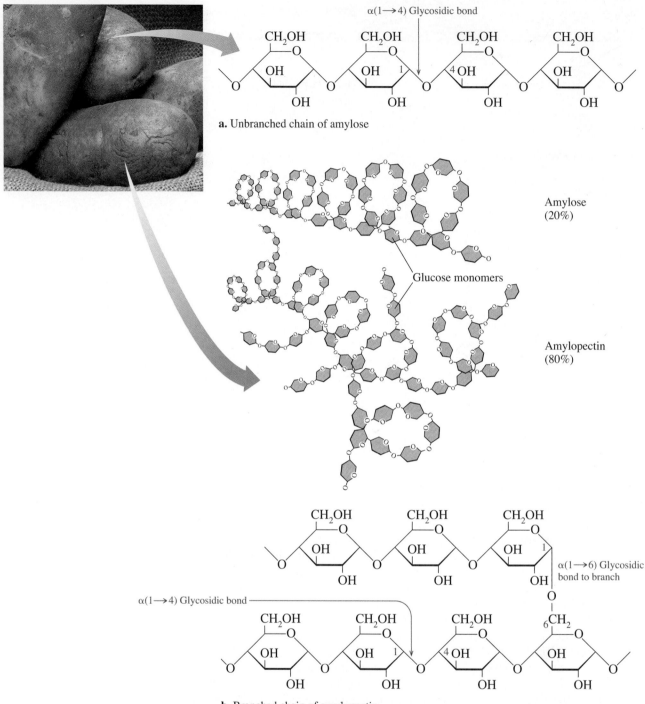

a. Unbranched chain of amylose

b. Branched chain of amylopectin

▲ **FIGURE 5.16 The storage polysaccharide starch.** The structure of (a) amylose is a straight-chain polysaccharide of glucose units and (b) amylopectin is a branched chain of glucose.

Structural Polysaccharides

Cellulose

Trees stand tall because of the structural polysaccharide cellulose. **Cellulose** also contains glucose units, but they are D-glucose units bonded $\beta(1 \rightarrow 4)$. This single change in glycosidic bond configuration completely changes the overall structure of cellulose compared with that of amylose. Where amylose coils in an α-bonded glucose chain, the β-bonded chain of cellulose is straight. Many of these straight chains of cellulose align next to each other, forming a strong, rigid structure (see Figure 5.17).

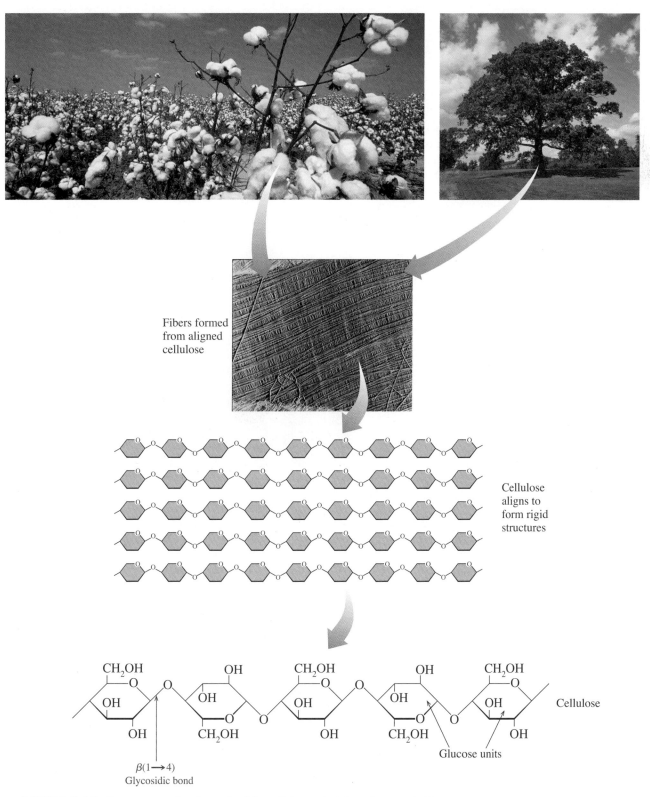

Fibers formed from aligned cellulose

Cellulose aligns to form rigid structures

$\beta(1 \longrightarrow 4)$
Glycosidic bond

Cellulose

Glucose units

▲ **FIGURE 5.17 The structural polysaccharide cellulose.** Cellulose forms a rigid structure when the straight chain of the single polysaccharide formed from β-glucose units linking $\beta(1 \rightarrow 4)$ aligns with neighboring cellulose molecules.

Nutritionally cellulose is considered an insoluble fiber because we lack the enzyme called cellulase that hydrolyzes the $\beta(1 \rightarrow 4)$ glycosidic bonds in cellulose. Even though we do not eat wood, we get cellulose in our diet from whole grain foods. We cannot digest cellulose, but it is still an important part of our diet because it assists with digestive movement in the small and large intestine. Some animals and insects—like sheep, cows, and termites—can digest cellulose because their digestive system contains bacteria and other microorganisms that produce cellulase.

Chitin

Chitin is a polysaccharide that makes up the exoskeleton of insects and crustaceans and the cell walls of some fungi. This polysaccharide is made up of a modified β-D-glucose called *N*-acetylglucosamine containing β(1 → 4) glycosidic bonds.

Like cellulose, chitin is a structurally strong material that has many uses, one of which is a surgical thread that biodegrades as a wound heals. Chitin is also present in many insects' exoskeletons and serves to protect them from water. Because of this property, chitin can be used to waterproof paper. When ground up, chitin becomes a powder that holds in moisture and it can be added to cosmetics and lotions. The structure of chitin is shown in Figure 5.18.

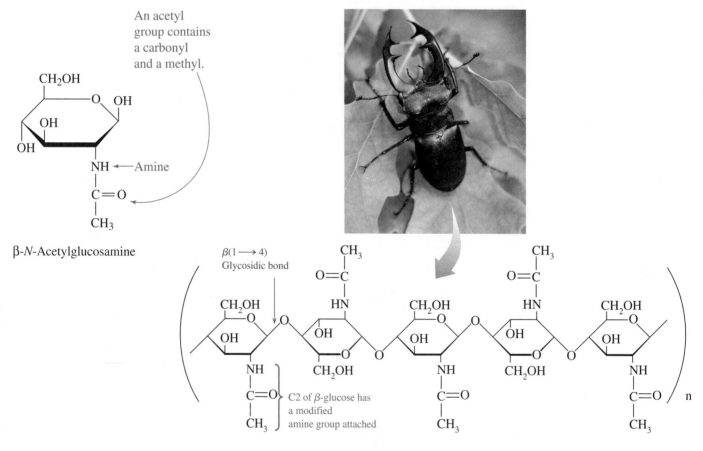

▲ **FIGURE 5.18 The structural polysaccharide chitin.** Chitin is formed from β-*N*-acetylglucosamine units bonded β(1 → 4).

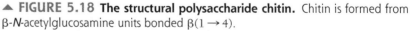

S A M P L E   P R O B L E M   5 . 1 1

Structures of Polysaccharides

Identify the polysaccharide described by each of the following:

a. a polysaccharide that is stored in the liver and muscle tissues

b. a component of starch containing only α(1 → 4) glycosidic bonds

c. a polysaccharide found in the exoskeleton of insects

S O L U T I O N

a. glycogen   **b.** amylose   **c.** chitin

5.31 Describe the similarities and differences in the following polysaccharides:

a. amylose and amylopectin

b. amylopectin and glycogen

5.32 Describe the similarities and differences in the following polysaccharides:

a. amylose and cellulose

b. cellulose and chitin

5.33 Give the name of one or more polysaccharides that matches each of the following descriptions:

a. not digestible by humans

b. the storage form of carbohydrates in plants

c. contains only $\alpha(1 \rightarrow 4)$ glycosidic bonds

d. glucose polysaccharide with the most branching

5.34 Give the name of one or more polysaccharides that matches each of the following descriptions:

a. the storage form of carbohydrates in animals

b. contains only $\beta(1 \rightarrow 4)$ glycosidic bonds

c. contains both $\alpha(1 \rightarrow 4)$ and $\alpha(1 \rightarrow 6)$ glycosidic bonds

d. produces maltose during digestion

5.7 Carbohydrates and Blood

ABO Blood Types

What is your blood type? Many people know their blood type, but most do not know that the blood types of A, B, AB, and O refer to carbohydrates. Your red blood cells have a number of chemical markers bonded to the cell surface identifying the cells as yours as they circulate through the bloodstream. One set of those chemical markers are the oligosaccharides known as the ABO blood markers, which contain either three or four sugar units.

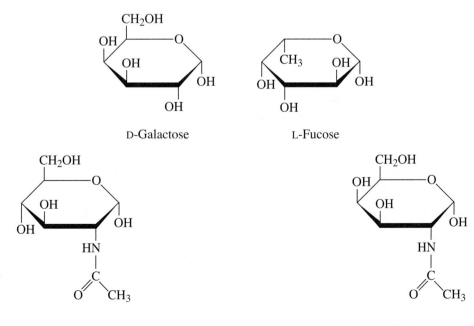

D-Galactose   L-Fucose

N-Acetylglucosamine   N-Acetylgalactosamine

All the blood types include the carbohydrates N-acetylglucosamine, galactose, and fucose, a carbohydrate that often appears in nature as its L-enantiomer. Type O blood contains only these three carbohydrates attached to the red blood cell surface through a glycosidic bond. Type A and type B blood contain these three plus a fourth carbohydrate. In type A, the fourth carbohydrate is an N-acetylgalactosamine bonded to the galactose unit while in type B blood the fourth carbohydrate is a second galactose bonded to the galactose. People with type AB blood have both the A carbohydrate set and the B carbohydrate set on their red blood cells.

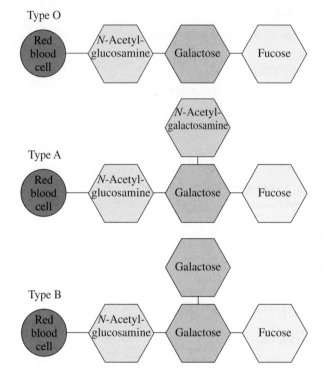

Universal Donors and Acceptors

The O blood type is considered the universal donor blood type, but why? The body can only recognize its own carbohydrate set (A, B, or O) and will try to destroy what it considers a foreign blood type by producing protein molecules called antibodies that act to destroy the foreign blood type. Because the trisaccharide on the cells of O-type blood (N–acetylglucosamine—galactose—L–fucose) are present on cells of *all* blood types (A, B, and AB), O-type blood can be donated to any blood type. No blood type recognizes the O carbohydrate set as foreign.

The AB blood type is considered the universal acceptor blood type for an analogous reason. AB blood contains all possible ABO combination types, so any blood type transfused will be accepted by the body. The compatibility of blood types is summarized in Table 5.3.

TABLE 5.3	COMPATIBILITY OF BLOOD GROUPS	
Blood Group	Can Receive Blood Types	Cannot Receive Blood Types
A	A,O	B, AB
B	B, O	A, AB
AB[a]	A, B, AB, O	Can receive all blood types
O[b]	O	A, B, AB

[a]AB universal acceptor.
[b]O universal donor.

Heparin is a medically important polysaccharide that prevents clotting (acts

Heparin

as an anticoagulant) in the bloodstream. Test tubes, tubing, and needles used for drawing blood are routinely coated with heparin to prevent the blood from clotting. Heparin is a highly ionic polysaccharide of many repeating disaccharide units; the disaccharide includes an oxidized monosaccharide and a D-glucosamine. This polysaccharide belongs to a group of polysaccharides called **glycosaminoglycans**, all of which have highly charged repeating disaccharide units. Negative charges are due to the presence of sulfate groups.

Oxidized monosaccharide   D-Glucosamine monosaccharide

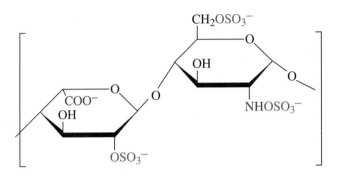

Heparin's repeating disaccharide

SAMPLE PROBLEM 5.12

ABO Blood Types

Explain whether the following blood types could be donated to a person with type A blood:

a. A   b. B   c. AB   d. O

SOLUTION

a. Two matched blood types can always be donated to each other; the transfused blood will not be recognized as foreign. Type A can be donated to type A.

b. Type B cannot be donated to type A because the person with type A blood would still recognize type B blood as foreign and try to eliminate it. Type B cannot be donated to type A.

c. Even though type AB contains the A carbohydrates, it also contains the B carbohydrates, which a person with type A blood would recognize as foreign. Type AB cannot be donated to type A.

d. Because the O carbohydrate set is present within the type A carbohydrates, type O blood will not be recognized as foreign and will be accepted. Type O can be donated to type A.

PRACTICE PROBLEMS

5.35 Explain whether the following blood types could be donated to a person with type B blood:

a. A   b. AB

5.36 Explain whether the following blood types could be donated to a person with type O blood:

a. B   b. AB

chapter five

SUMMARY

5.1 Classes of Carbohydrates

Carbohydrates are classified as monosaccharides (simple sugars), disaccharides (two monosaccharide units), oligosaccharides (three to nine monosaccharide units), and polysaccharides (many monosaccharide units).

5.2 Monosaccharide Structure

The simplest carbohydrates are the monosaccharides with a molecular formula of $C_n(H_2O)_n$ where $n = 3$–6. Monosaccharides can be drawn linearly in a representation called the Fischer projection that highlights their chiral centers. Most carbohydrates found in nature are the D-isomer. Stereoisomers that have multiple chiral centers can either be related to each other as enantiomers (mirror images of each other) or diastereomers (not enantiomers). Some important monosaccharides are glucose, galactose, mannose, fructose, deoxyribose, and ribose.

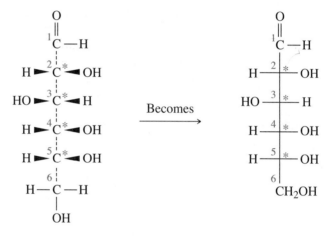

Wedge–dash projection
D-Glucose

Becomes

Fischer projection
D-Glucose

5.3 Oxidation and Reduction Reactions

Oxidation and reduction are common chemical reactions that occur in both inorganic compounds and organic molecules. Oxidation can be described as a loss of electrons and reduction as a gain of electrons. For organic molecules and biomolecules, oxidation is more easily described as the addition of oxygen to a molecule (or loss of hydrogen) and reduction as the addition of hydrogen to a molecule (or loss of oxygen). The anomeric carbon of carbohydrates (C1 of an aldose) is highly reactive and can undergo both oxidation to a carboxylic acid or reduction to an alcohol. Monosaccharides are considered reducing sugars because their anomeric carbon can react to reduce another compound in a redox reaction.

5.4 Ring Formation

A hydroxyl group and the carbonyl functional group of a linear monosaccharide can react to enclose the hydroxyl's oxygen in a ring. Because the carbonyl group is a planar structure, this reaction produces two possible ring arrangements about the anomeric carbon termed the α and the β anomer when a ring is formed.

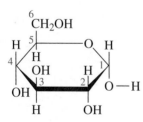

α-D-Glucose
in ring form

5.5 Disaccharides

Another pair of common reactions that occur in many biomolecules is condensation and hydrolysis. Condensation reactions produce a molecule of water while bonding two organic molecules together, whereas hydrolysis reactions consume a molecule of water while breaking up a larger organic molecule into two smaller ones. Carbohydrates form glycosides when an anomeric carbon reacts with a hydroxyl on a second organic molecule. This condensation results in the formation of a glycosidic bond. Glycosidic bonds are named by designating the anomer of the reacting monosaccharide and the carbons that are bonded, for example, $\alpha(1 \rightarrow 4)$. Some important disaccharides formed through condensation reactions are maltose, lactose, and sucrose.

Monosaccharides and disaccharides make up simple sugars, many of which are sweet to the taste. The sweetness of carbohydrates and other carbohydrate substitutes is indexed relative to the sweetness of sucrose (see Table 5.2).

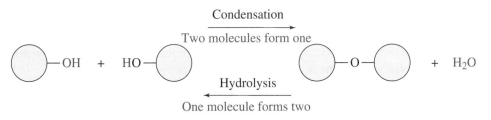

5.6 Polysaccharides

Polysaccharides consist of many monosaccharide units bonded together through glycosidic bonds. Glucose can be stored as a polysaccharide called starch in plants and glycogen in animals. Starch consists of two polysaccharides: amylose, a linear chain of glucose, and amylopectin, a branched chain of glucose. Glycogen is also a branched polysaccharide of glucose, but it contains more branching than does amylopectin. Two polysaccharides that are structurally important in nature include cellulose in plants (wood) and chitin in arthropods (exoskeleton) and fungi (cell wall). Cellulose is a linear chain of glucose, but in cellulose the glycosidic bonds are β, whereas in starch they are bonded α. Chitin is also a linear chain of a modified glucose, N-acetylglucosamine bonded β Both these structural polysaccharides form strong, water-resistant materials when the linear chains are aligned with each other.

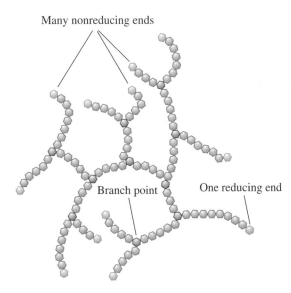

5.7 Carbohydrates and Blood

Carbohydrates are also used as recognition markers on the surfaces of cells and in other bodily fluids. The ABO blood groups are oligosaccharides of which one of the carbohydrate units is L-fucose, one of the few L-sugars found in nature. These ABO oligosaccharides are found on the surface of red blood cells. The A and B blood groups look like the O blood group except that they contain an additional monosaccharide. For this reason the O blood type is considered the universal donor. Heparin is a polysaccharide consisting of a repeating disaccharide containing an oxidized monosaccharide and a glucosamine. Heparin functions in the blood as an anticoagulant and is commonly found as a coating on medical tubing and syringes used during blood transfusions.

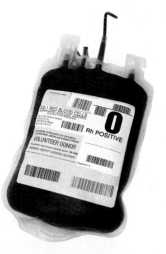

KEY TERMS

alcohols—A family of organic compounds whose functional group is an OH (hydroxyl).

aldehydes—A family of organic compounds whose functional group is a carbonyl (C=O) bonded to at least one hydrogen atom.

aldose—A monosaccharide that contains the aldehyde functional group.

amylopectin—An α-D-glucose polysaccharide with $\alpha(1 \rightarrow 4)$ glycosidic bonds and $\alpha(1 \rightarrow 6)$ branch points approximately every 25 glucose units; a component of starch.

amylose—An unbranched α-D-glucose polysaccharide with $\alpha(1 \rightarrow 4)$ glycosidic bonds; a component of starch.

anomeric carbon—In a monosaccharide, the carbon that was the carbonyl in the linear structure; for example, carbon 1 of D-glucose.

anomers—Sugars that are diasteromers, differing only in the position of the hydroxyl at the anomeric carbon.

Benedict's test—A common test for the presence of a reducing sugar. If reducing sugar is present, a brick-red precipitate results.

carbohydrate—A simple or complex sugar composed of carbon, hydrogen, and oxygen; a primary source of energy in a normal diet.

carbonyl—A functional group that contains a carbon double bonded to an oxygen (C=O).

cellulose—An unbranched β-D-glucose polysaccharide with $\beta(1 \rightarrow 4)$ glycosidic bonds; the main component of wood and plants.

chitin—An unbranched β-D-acetylglucosamine polysaccharide with $\beta(1 \rightarrow 4)$ glycosidic bonds; the main component in the exoskeletons of arthropods and the cell walls of fungi.

condensation—A reaction involving the combination of two organic molecules. A small molecule such as H_2O is also produced. The reverse reaction is hydrolysis.

dehydration—A condensation reaction in which H_2O is produced.

diastereomer—A stereoisomer that is not a mirror image. Diastereomers exist for molecules with more than one chiral center.

disaccharide—A carbohydrate composed of two monosaccharides joined through a glycosidic bond.

epimer—Stereoisomers with multiple chiral centers that differ only at one chiral center; a type of diastereomer.

Fischer projection—A projection that shows the three-dimensional shape of a molecule on a two-dimensional space. The horizontal lines from a chiral center indicate wedges (in front of the plane) and vertical lines from a chiral center indicate dashes (behind the plane). The chiral center is at the intersection of the lines.

fructose—A monosaccharide found in honey and fruit juices; it is combined with glucose in sucrose; also called levulose and fruit sugar.

furanose—The five-membered ring form of a monosaccharide containing four carbons and one oxygen atom in the ring.

galactose—A monosaccharide that occurs combined with glucose in lactose.

glucose—The most prevalent monosaccharide in the diet. An aldohexose found in fruits, vegetables, corn syrup, and honey; also known as blood sugar and dextrose.

glycogen—An α-D-glucose polysaccharide with $\alpha(1 \rightarrow 4)$ glycosidic bonds and $\alpha(1 \rightarrow 6)$ branch points approximately every 12 glucose units; storage polysaccharide found in animals.

glycolysis—A series of chemical reactions in the body that break down monosaccharides producing energy.

glycosaminoglycan—A highly ionic polysaccharide that contains a repeating disaccharide unit consisting of an oxidized monosaccharide and a glucosamine.

glycoside—See glycosidic bond.

glycosidic bond—Formed when two hydroxyl groups join, one of which is on the anomeric carbon of a monosaccharide in ring form. When one of the hydroxyls is part of a second monosaccharide, a disaccharide is formed.

hemiacetal—The functional group formed when an alcohol and aldehyde functional group combine. It contains a tetrahedral carbon bonded to —OH and —OR.

heparin—A medically important polysaccharide that prevents clotting in the bloodstream.

hydrolysis—A reaction involving the breaking of one large organic molecule into two smaller ones. One of the reactants is H_2O. The reverse reaction is condensation.

ketones—A family of organic compounds whose functional group is a carbonyl (C=O) bonded to two alkyl groups.

ketose—A monosaccharide that contains the ketone functional group.

lactose—A disaccharide consisting of glucose and galactose found in milk and milk products.

maltose—A disaccharide consisting of two glucose units; obtained from the hydrolysis of starch in germinating grains.

mannose—A monosaccharide that is not easily absorbed by the body, but prevalent in some fruits like cranberries.

monosaccharide—A simple sugar unit containing three or more carbons with the general formula $C_n(H_2O)_n$.

oligosaccharide—A carbohydrate composed of three to nine monosaccharide units joined through glycosidic bonds.

oxidation—The loss of electrons during a chemical reaction; in organic reactions, often appears as a gain of oxygen or a loss of hydrogen.

oxidizing agent—Responsible for oxidizing another reactant, the reactant undergoing reduction.

polysaccharide—A carbohydrate composed of many monosaccharide units joined through glycosidic bonds. The type of monosaccharide, glycosidic bond, and branching varies.

primary (1°) **alcohol**—An alcohol that has one alkyl group bonded to the carbon atom with the —OH.

pyranose—The six-membered ring form of a monosaccharide that contains five carbons and one oxygen atom in the ring.

reducing agent—Responsible for reducing another reactant: the reactant undergoing oxidation.

reducing sugar—A carbohydrate with a free aldehyde group capable of reducing another substance.

reduction—The gain of electrons during a chemical reaction; in organic reactions often appears as a gain of hydrogen or a loss of oxygen.

secondary (2°) alcohol—An alcohol that has two alkyl groups bonded to the carbon atom with the —OH.

storage polysaccharide—Polysaccharides that are stored in cells as a glucose energy reserve; examples include starch in plants and glycogen in animals.

structural polysaccharide—Polysaccharides whose function is to provide structure for an organism; examples include cellulose, which supports plant structure, and chitin in the exoskeleton of arthopods.

sucrose—A disaccharide composed of glucose and fructose; a nonreducing sugar, commonly called table sugar or "sugar."

D-sugar—Based on the arrangement of the attachments on D-glyceraldehyde, (in the Fischer projection the hydroxyl group on the chiral carbon farthest from the carbonyl is on the right side of the molecule); these carbohydrate orientations are the more common in nature.

L-sugar—Based on the arrangement of the attachments on L-glyceraldehyde, (in the Fischer projection the hydroxyl group on the chiral carbon farthest from the carbonyl is on the left side of the molecule); these carbohydrate orientations are rarely found in nature.

tertiary (3°) alcohol—An alcohol that has three alkyl groups bonded to the carbon atom with the —OH.

SUMMARY OF REACTIONS

Oxidation—Reduction of Organic Compounds

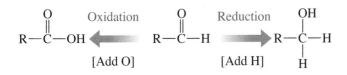

Carboxylic acid   Aldehyde   Alcohol

Oxidation of Monosaccharides (D-Glucose)

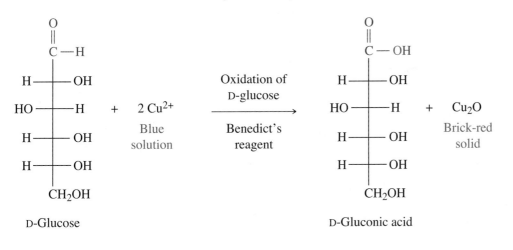

D-Glucose   D-Gluconic acid

Reduction of Monosaccharides (D-Glucose)

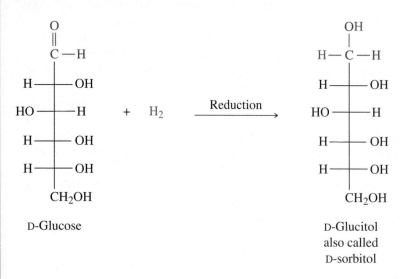

D-Glucose

D-Glucitol
also called
D-sorbitol

Ring Formation (D-Glucose)

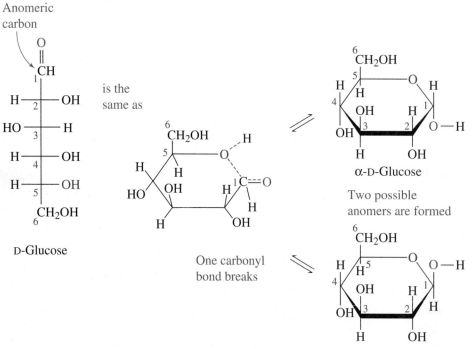

Anomeric carbon

D-Glucose

is the same as

One carbonyl bond breaks

α-D-Glucose

Two possible anomers are formed

β-D-Glucose

Glycosidic Bond Formation (D-Glucose)

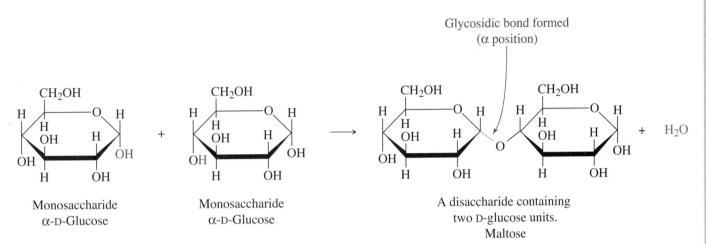

Disaccharide Condensation

glucose + glucose ⟶ maltose + H₂O
glucose + galactose ⟶ lactose + H₂O
glucose + fructose ⟶ sucrose + H₂O

ADDITIONAL PROBLEMS

5.37 Write the molecular formula for a carbohydrate containing four carbons.

5.38 What would be the molecular formula of a monosaccharide characterized as an aldopentose?

5.39 Explain the difference between an oligosaccharide and a polysaccharide.

5.40 Classify each of the following as primary, secondary or tertiary alcohols:

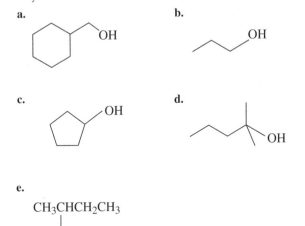

e.

CH₃CHCH₂CH₃
   |
   OH

5.41 Classify each of the following as primary, secondary or tertiary alcohols:

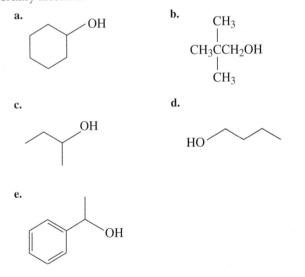

5.42 Explain the difference between an enantiomer and an epimer.

5.43 How are the following pairs of carbohydrates, shown in Fisher projection, related to each other? Are they structural isomers, enantiomers, diastereomers, or epimers?

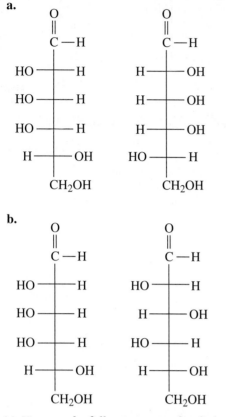

5.44 How are the following pairs of carbohydrates, shown in Fisher projection, related to each other? Are they structural isomers, enantiomers, diastereomers, or epimers?

5.45 Draw the Fischer projection of the product of the oxidation of D-galactose at C1.

5.46 Draw the Fischer projection of the product of the oxidation of the monosaccharide D-talose at C1.

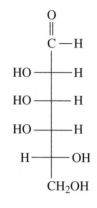

D-Talose

5.47 Draw the Fischer projection of the product of reduction reaction of D-galactose at C1.

5.48 Draw the Fischer projection of the product of the reduction reaction of D-talose at C1.

5.49 Will the following carbohydrates produce a positive Benedict's test?
 a. D-glucose b. lactose
 c. sucrose d. starch

5.50 Will the following carbohydrates produce a positive Benedict's test?
 a. D-mannose b. L-fucose
 c. maltose d. glycogen

5.51 Identify the following carbohydrates as the α or β anomer:

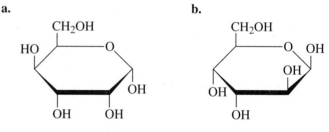

5.52 Identify the following carbohydrates as the α or β anomer:

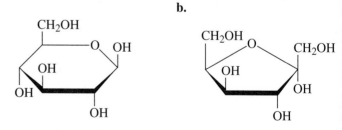

5.53 Draw the α and β anomer of D-mannose.

5.54 Draw the α and β anomer of D-fructose.

5.55 Draw the product of the following 1 → 4 condensation and name the glycosidic bond:

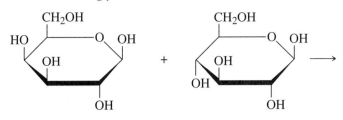

5.56 Draw the product of the following 1 → 4 condensation:

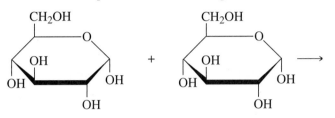

5.57 Give the name of one or more carbohydrates that match the following descriptions:
   a. ordinary table sugar
   b. dextrose
   c. an L-sugar found in the ABO blood types
   d. found in starch
   e. found in the cell walls of fungi

5.58 Give the name of one or more carbohydrates that match the following descriptions:
   a. a disaccharide of D-glucose bonded α(1 → 4)
   b. fruit sugar
   c. contains a sugar acid and sugar amine disaccharide repeat unit
   d. storage polysaccharide in animals
   e. insoluble fiber found in plants and trees

CHALLENGE PROBLEMS

5.59 Carbohydrates are also easily oxidized at C6. Provide the product of the oxidation of (a) D-galactose at C6 and (b) D-talose at C6.

5.60 Carbohydrates are abbreviated using a three-letter abbreviation followed by their glycosidic bond type. For example, maltose and sucrose can be written respectively as

$$Glc\alpha(1 \rightarrow 4)Glc \qquad Glc\alpha(1 \rightarrow 2)\beta Fru$$
$$\text{Maltose} \qquad\qquad \text{Sucrose}$$

Provide the structure for the O type blood carbohydrate set given the following abbreviation:

$$L\text{-}Fuc\alpha(1 \rightarrow 2)Gal\beta(1 \rightarrow 4)GlcNAc$$

5.61 The structure of sucralose, found in the artificial sweetener Splenda®, is shown in the figure. Its advertising slogan is "made from sugar, so it tastes like sugar" (www.splenda.com/page.jhtml?id=splenda/faqs/nocalorie.inc#q1). It consists of a chlorinated disaccharide made up of galactose and fructose. In the molecule shown,
   a. identify the galactose unit and the fructose unit.
   b. identify the type of glycosidic bond present.

c. determine if Splenda is a reducing sugar.

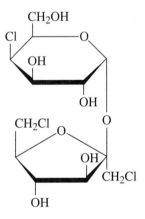

5.62 Which of the components in starch is more likely to be broken down more quickly in plants, amylose or amylopectin? Why?

ANSWERS TO ODD-NUMBERED PROBLEMS

Practice Problems

5.1 **a.** secondary **b.** primary
c. secondary **d.** secondary

5.3 **a.** ketone **b.** ketone **c.** aldehyde

5.5 Two monosaccharides make up one disaccharide.

5.7 aldehyde, hydroxyls (alcohol)

5.9 **a.** **b.**

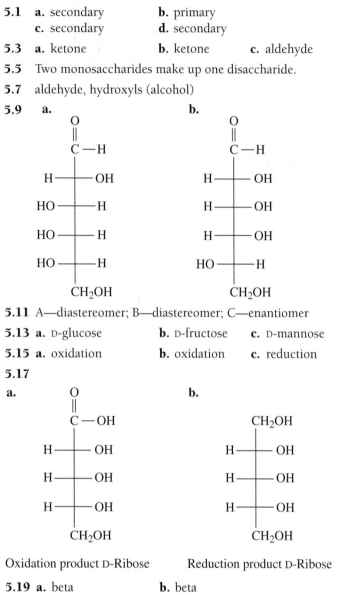

5.11 A—diastereomer; B—diastereomer; C—enantiomer

5.13 **a.** D-glucose **b.** D-fructose **c.** D-mannose

5.15 **a.** oxidation **b.** oxidation **c.** reduction

5.17

a. **b.**

Oxidation product D-Ribose   Reduction product D-Ribose

5.19 **a.** beta **b.** beta

5.21

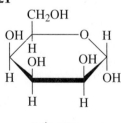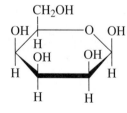

α-Anomer   β-Anomer

5.23 **a.** condensation **b.** condensation **c.** hydrolysis

5.25 α(1 → 3)

5.27
a. bond is β (1→ 4)

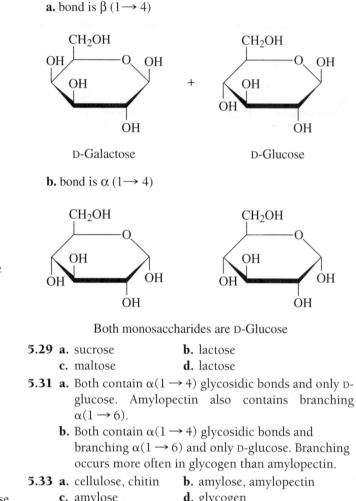

D-Galactose   D-Glucose

b. bond is α (1→ 4)

Both monosaccharides are D-Glucose

5.29 **a.** sucrose **b.** lactose
c. maltose **d.** lactose

5.31 **a.** Both contain α(1 → 4) glycosidic bonds and only D-glucose. Amylopectin also contains branching α(1 → 6).
b. Both contain α(1 → 4) glycosidic bonds and branching α(1 → 6) and only D-glucose. Branching occurs more often in glycogen than amylopectin.

5.33 **a.** cellulose, chitin **b.** amylose, amylopectin
c. amylose **d.** glycogen

5.35 **a.** No **b.** No

Additional Problems

5.37 $C_4H_8O_4$

5.39 An oligosaccharide is smaller, containing between 3 and 9 monosaccharide units, while a polysaccharide contains 10 or more monosaccharide units.

5.41 a. secondary **b.** primary **c.** secondary
d. primary **e.** secondary

5.43 a. enatiomer **b.** epimer

5.45

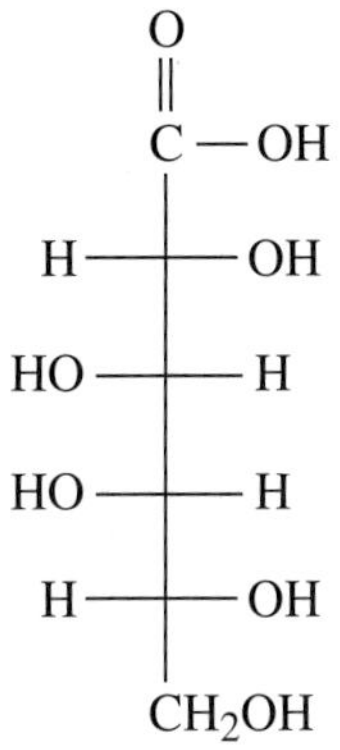

```
        O
        ||
        C—OH
   H ———— OH
  HO ———— H
  HO ———— H
   H ———— OH
       CH₂OH
```

Galactose oxidized at C1

5.47

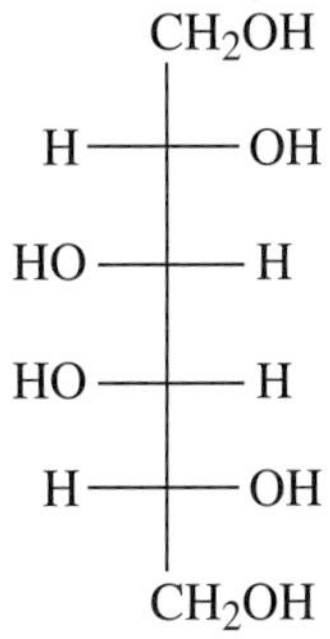

```
       CH₂OH
   H ———— OH
  HO ———— H
  HO ———— H
   H ———— OH
       CH₂OH
```

Galactose reduced at C1

5.49 a. yes **b.** yes **c.** no **d.** no

5.51 a. alpha **b.** beta

5.53

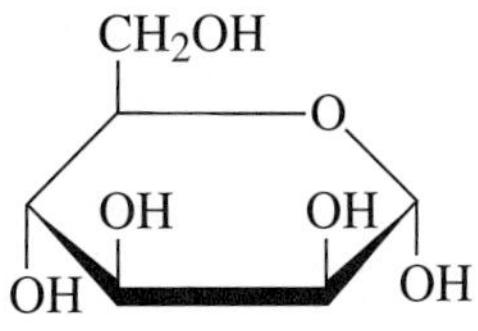

α-D-Mannose 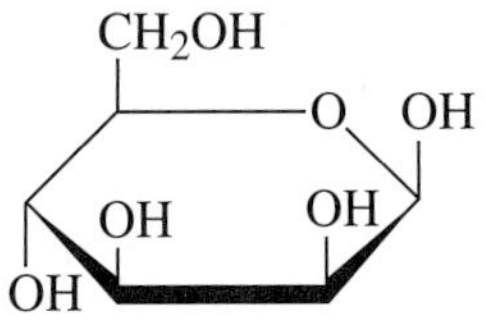   β-D-Mannose

5.55

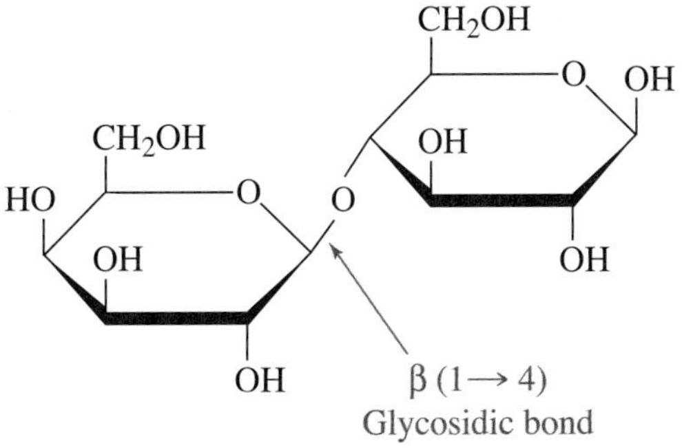

β (1 → 4) Glycosidic bond

5.57 a. sucrose **b.** glucose
c. L-fucose **d.** amylose, amylopectin
e. chitin

5.59 a.

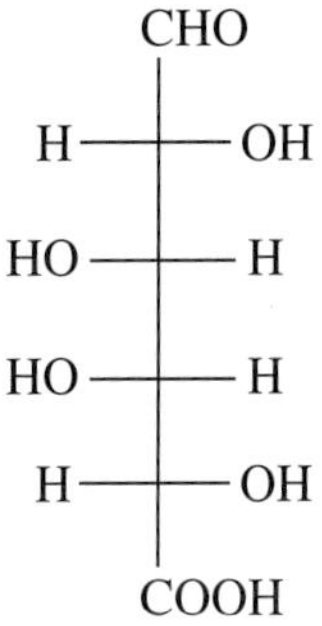

```
       CHO
   H ———— OH
  HO ———— H
  HO ———— H
   H ———— OH
       COOH
```

b.

```
       CHO
  HO ———— H
  HO ———— H
  HO ———— H
   H ———— OH
       COOH
```

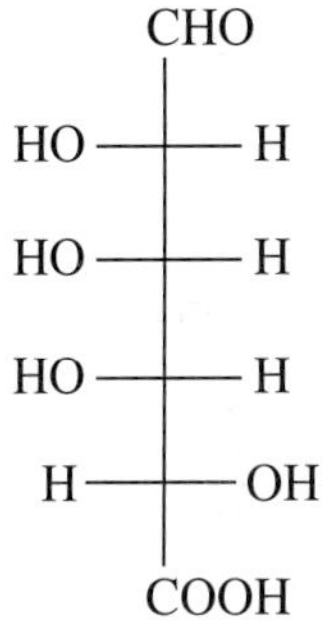

5.61

a. Galactose / Fructose

α,β (1 → 2) Glycosidic bond

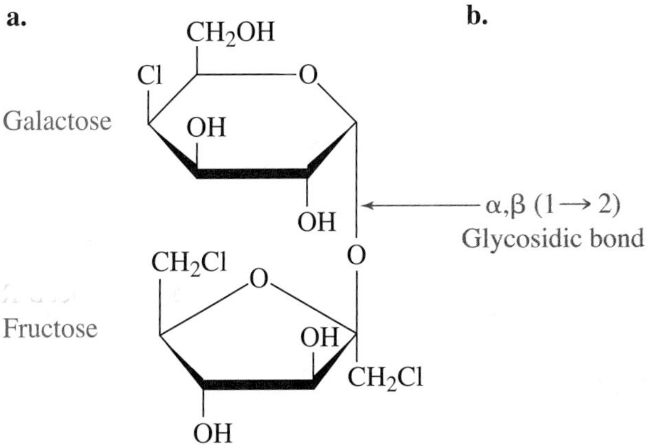

b.

c. Splenda is not a reducing sugar.

Guided Inquiry Activities **FOR CHAPTER 6**

EXERCISE 1 The Attractive Forces

Information

Molecules "stick" together in the liquid and solid states due to attractions between the molecules. This accounts for some of their physical properties like boiling point, melting point, viscosity, and surface tension. In solids, each molecule is fixed relative to the neighboring molecules and these attractions are optimized.

Questions

1. What is common about the attractions between all the forces shown circled on the next page in Table 1?

2. Which is stronger, a formal $+/-$ attraction, or a partial δ^+/δ^- attraction?

3. Which attraction can occur in nonpolar organic molecules?

4. Hydrogen bonds are a unique attractive force (not an actual covalent bond) present in some molecules. The attraction occurs between a δ^+ (also known as *polarized*) hydrogen in a polar bond with O, N, or F and a lone pair of electrons (:) on an O, N, or F. The hydrogen is called the *donor*, and the pair of electrons is referred to as the *acceptor*. This attraction is indicated with a dashed line between the polarized H and O, N, or F.

   a. How many polarized H's are in water?

   b. How many lone pairs of electrons are in a molecule of H_2O?

   c. How many H bonds can one water molecule make to neighboring water molecules? Draw them.

5. How many hydrogen bonds can one methanol molecule (CH_3OH) make to neighboring methanol molecules? Draw them.

6. The boiling point of water is $100\,°C$ and the boiling point of methanol is $65\,°C$. Provide an explanation for this difference based on the strength of their attractive forces.

7. Polar molecules can contain more than one type of attractive force. Fill in the following table with a yes or no answer to the question: "Does this attractive force operate between molecules of the given compound?"

Molecule	Ion Dipole	H Bond	Dipole–Dipole	London
CCl_4				
H_2O				
$CHCl_3$				
CH_3COO^-				

TABLE 1	EXAMPLES OF THE ATTRACTIVE FORCES BETWEEN MOLECULES (FROM STRONGEST TO WEAKEST)

Force Name	Example
Ionic (also known as a salt bridge)	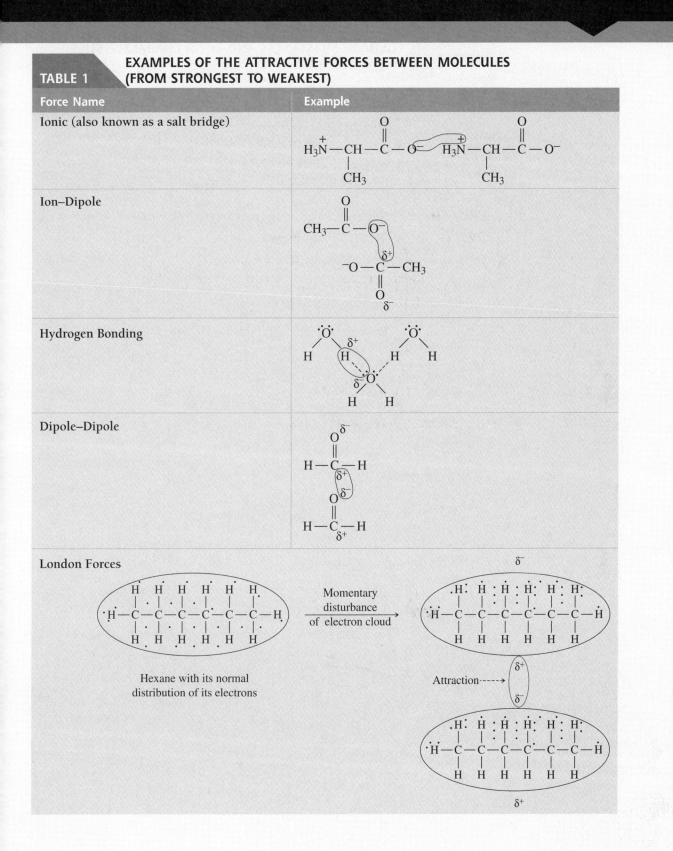
Ion–Dipole	
Hydrogen Bonding	
Dipole–Dipole	
London Forces	

Hexane with its normal distribution of its electrons

Momentary disturbance of electron cloud →

Attraction ----→

EXERCISE 2 Solubility in Water

TABLE 2	COMMON SUBSTANCES AND THEIR CHARACTERISTICS
Substance	Characteristics
NaCl (table salt)	Ionic compound
Sucrose (table sugar)	Polar covalent compound
Vegetable oil	Nonpolar covalent compound
Soap	Contains a highly polar part (often ionic) and a nonpolar part

1. Based on your experience, which of the compounds in Table 2 will dissolve in water?

2. Which of the characteristics listed applies to water?

3. The *golden rule of solubility* states that "like dissolves like." What is *alike* about the compounds you listed in question 1 and water?

4. Based on their characteristics, predict whether each of the following compounds might be soluble in water:

   a. acetone, $H_3C - \overset{\overset{\textstyle O}{\|}}{C} - CH_3$

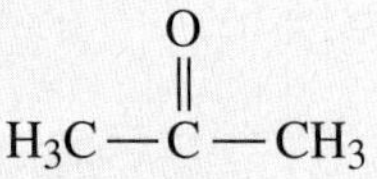

   b. ethanol, CH_3CH_2OH

   c. $NaHCO_3$

   d. cyclohexane,

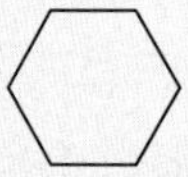

   e. octanol, ⌇⌇⌇OH

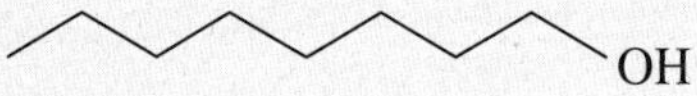

5. Draw and describe how an ionic compound like NaCl might interact with water through ion–dipole attractions.

6. Draw and describe how a polar covalent compound like formaldehyde $\left(\overset{\overset{\textstyle O}{\|}}{\underset{H \quad H}{C}} \right)$ might interact with water through dipole–dipole attractions.

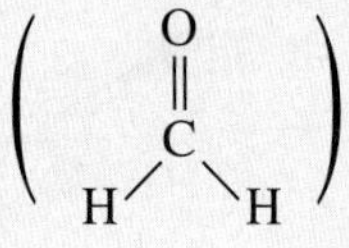

7. Many pharmaceuticals contain aromatic rings or large areas of hydrocarbon (nonpolar). Quite often, pharmaceuticals are synthesized in an ionic form to increase their solubility in aqueous solution. Ionic charges on organic molecules dramatically increase the compound's solubility. Predict whether the following pharmaceuticals would likely be soluble in water:

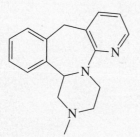

Pentolinium—an antihypertensive

Remeron—an antidepressant

Methamphetamine—mainly recreational
usage sometimes prescribed for ADHD

8. Devise a rule to predict the solubility of an organic compound in water.

O
U
T
L
I
N
E

The stickiness of plastic wrap exploits the attractive forces, called intermolecular forces, acting between molecules. Read on to see how intermolecular forces play a part in stain removal by soaps, the difference between a fat and oil and the unique organization of the cell membrane.

Intermolecular Forces
STATE CHANGES, SOLUBILITY, AND CELL MEMBRANES

Have you ever, after eating half of a sandwich, gone to the pantry to get a piece of plastic wrap to save the sandwich? You pull out the box of plastic wrap, unroll a few inches, and attempt to tear it on the built-into-the-box-but-never-seems-to-work-right plastic wrap cutter. You finally get the plastic wrap to cut, only to have it fold back onto itself. You take the now stuck-together sheet in hand and try to find an end so that you can separate the parts. When you are finally successful and begin to pull the parts away from each other, you notice the parts separate with a strange sound—much like the "static electricity sound" that is made when you remove a sock from the leg of your pants when both have just come out of the dryer. The sound you hear actually comes from your disruption of the forces attracting the molecules of the plastic wrap to each other!

In Chapter 3, we saw that the electrons in a molecule are not always evenly distributed over the entire molecule (section 3.6). This variance in electron distribution *within* a molecule gives rise to attractive forces *between* molecules known as **intermolecular forces**. In this chapter, we explore the different types of intermolecular forces and how they affect the behavior of molecules.

6.1 Types of Intermolecular Forces

Unlike covalent bonds, which involve the actual sharing of electrons between atoms *within* the same molecule, intermolecular forces are attractions *between* molecules. These intermolecular forces are weaker than even the weakest covalent bond. Intermolecular forces arise mainly from the attraction of an electron-rich area of one molecule (where electrons spend more of their time) to an electron-poor area of another molecule (where electrons spend less time.) In other words, the partially negative (δ^-) part of one molecule is attracted to the partially positive (δ^+) part of a second molecule. This attraction between two molecules is an intermolecular force.

We will consider three types of intermolecular forces in order of increasing strength as shown in the accompanying figure.

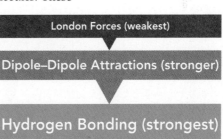

London Forces

The attractions that cause plastic wrap to stick to itself, which we discussed in the introduction to this chapter, are called **London forces**. This intermolecular force occurs momentarily in all molecules when electrons become unevenly distributed over a molecule's surface. If the distribution of electrons over a molecule's surface becomes temporarily unequal (for example, if all the electrons shift to one side), an unequal charge, or induced dipole, is momentarily created. The partially positive side of this temporary dipole attracts the electrons on a second molecule, creating an attraction between these two molecules and inducing a temporary dipole in the second molecule. The result is the attractive force that you feel as you pull apart two sheets of plastic wrap (see Figure 6.1). While *all molecules exhibit London forces*, these forces are significant only in the case of nonpolar molecules because these are the *only* intermolecular forces in which nonpolar molecules can participate.

Many plastic wraps are made of long, nonpolar hydrocarbon chains. When the plastic wrap folds back onto itself, these long chains interact with each other creating temporary dipoles throughout the material. The attraction between the temporary dipoles

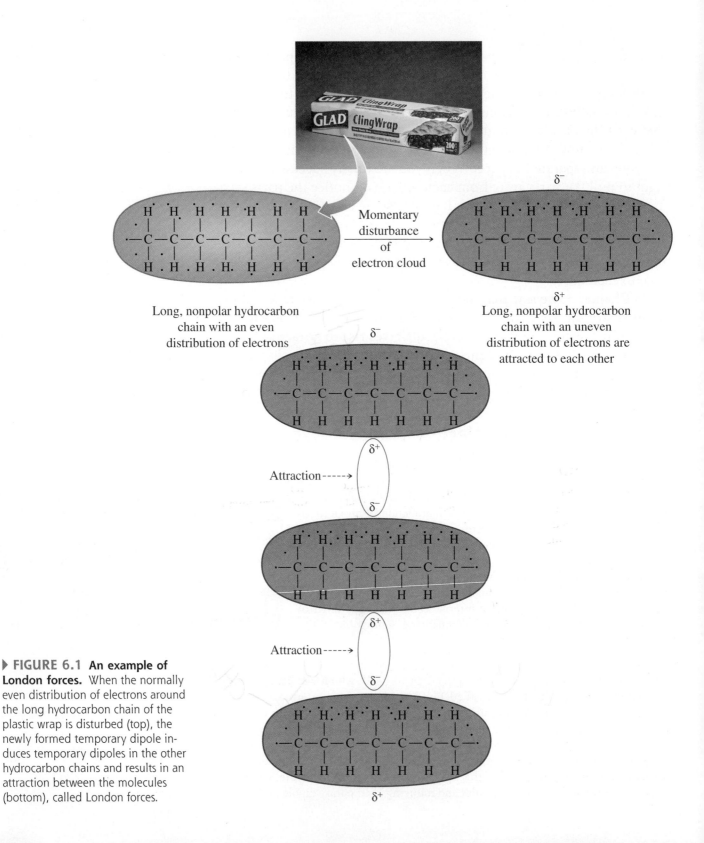

Long, nonpolar hydrocarbon chain with an even distribution of electrons

Momentary disturbance of electron cloud

Long, nonpolar hydrocarbon chain with an uneven distribution of electrons are attracted to each other

▶ FIGURE 6.1 **An example of London forces.** When the normally even distribution of electrons around the long hydrocarbon chain of the plastic wrap is disturbed (top), the newly formed temporary dipole induces temporary dipoles in the other hydrocarbon chains and results in an attraction between the molecules (bottom), called London forces.

causes the wrap to stick to itself. Although this is the weakest intermolecular force, many of these weak forces acting together can become quite strong. The terms *induced dipole* and *dispersion force* describe the same intermolecular force as London forces.

Dipole–Dipole Attractions

Some molecules—polar molecules—have a permanently uneven distribution of electrons caused by electronegativity differences in the atoms that make up the molecules. These molecules always have a partially positive part and a partially negative part. Such molecules have a *permanent dipole*. Because the dipole in these molecules does not come and go as it does in the case of London forces, the attraction of the partially positive end of one molecule for the partially negative end of another molecule is stronger and more pronounced than London forces. This type of attraction involves the interaction of two dipoles and is called a **dipole–dipole attraction** (see Figure 6.2).

Dipole–dipole attractions do not exist between nonpolar molecules because these molecules do not have permanent dipoles. Molecules with permanent dipoles will also have London forces, but the attraction of the dipoles is much stronger, making the London forces negligible.

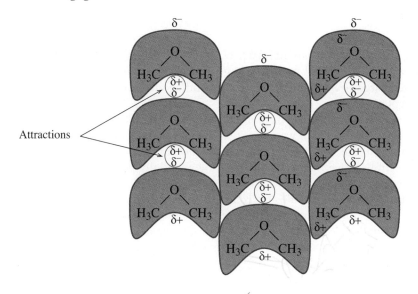

Attractions

◀ **FIGURE 6.2 An example of a dipole–dipole attraction.** Molecules of dimethyl ether are polar. The partially negative end of the dipole of one molecule of dimethyl ether is attracted to the partially positive end of the dipole of the second molecule.

Hydrogen Bonding

Certain polar molecules have an especially strong attraction for one another due to a large dipole that arises from partial charges on particular atoms in their structures. This force, called **hydrogen bonding**, is so prevalent in nature that even though it is just a very strong dipole–dipole attraction, it gets its own name. This attraction involves hydrogen and is much stronger than other dipole–dipole intermolecular forces; however, it is important to understand that this attraction is not a bond like the covalent bonds discussed in Chapter 3 because there is no sharing of electrons. Instead, the hydrogen bond is simply a very strong attraction between two molecules.

Hydrogen bonding requires the interaction of two players, a donor and acceptor, as described in the following table:

REQUIREMENTS FOR HYDROGEN BONDING	
Name	**Description**
Hydrogen-bond donor (δ^+)	A molecule with a hydrogen atom covalently bonded to an oxygen, nitrogen, or fluorine (O, N, or F)
Hydrogen-bond acceptor (δ^-)	A molecule with a nonbonding pair of electrons on an oxygen, nitrogen, or fluorine (O, N, or F)

▶ **FIGURE 6.3 Hydrogen bonding in water.** A single water molecule (center) can make up to four hydrogen-bonding interactions with neighboring water molecules. The hydrogen-bond donor hydrogen is designated d, and the hydrogen-bond acceptor electrons are designated a. Hydrogen-bonds are indicated with a red dashed line.

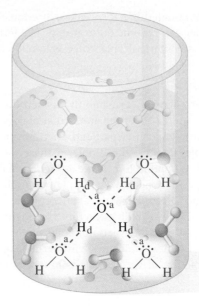

The high electronegativity of O, N, and F polarizes the hydrogen atom of the donor with a high partial positive charge (δ^+). This strongly attracts the high partial negative charge (δ^-) centered on the nonbonding electron pair of the O, N, or F on the acceptor.

Let's look at a glass of pure water as our first example of hydrogen bonding (see Figure 6.3). A molecule of water has two hydrogen atoms that are bonded to the oxygen. Both of these hydrogen atoms are attached to O by polar bonds (because O is more electronegative than H) with H having a partially positive charge (δ^+). Each of these hydrogen atoms can act as a hydrogen-bond donor (d) in a hydrogen-bonding interaction. A water molecule also contains two nonbonding pairs of electrons on the oxygen. These nonbonding pairs mean that each water molecule can also accept two hydrogen bonds (a). So, one water molecule can make up to four hydrogen bonds to other water molecules as shown in Figure 6.3. Hydrogen bonds are always illustrated using a straight dashed line.

Hydrogen bonds can occur between the same molecules as seen in pure water, between two different polar molecules as seen in the interactions of ethanol and acetone with water, or even between different parts of the same molecule (intramolecularly) as seen in the molecule ethylene glycol monomethyl ether.

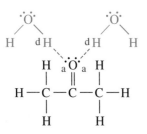

Acetone in water

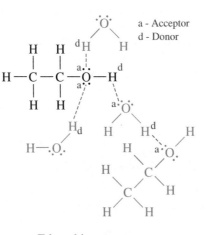

Ethanol in water

a - Acceptor
d - Donor

Intramolecular hydrogen bonding

S A M P L E   P R O B L E M   6 . 1

Drawing Hydrogen Bonds

Can the following molecules hydrogen-bond to water? If so, how many hydrogen-bonds can each make? Draw them.

$$\begin{array}{c} OH \\ | \end{array}$$
a. CH_3CHCH_3 b. CH_3CH_2Cl c. CH_3CN

S O L U T I O N

In order to hydrogen-bond to water, which contains both donors and acceptors for hydrogen bonding, the molecule must have either a hydrogen bonded to a N, O, or F (donor) or a nonbonding pair of electrons on an O, N, or F (acceptor).

a. This molecule has both a hydrogen-bond donor (O—H hydrogen) and two hydrogen-bond acceptor electron pairs (two on the O). This molecule can form three hydrogen bonds with water as shown:

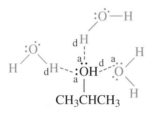

b. This molecule has no O, N, or F atoms so there are no hydrogen-bond acceptors present. The H atoms in this molecule are bonded to carbon, making them non-polar; therefore, there are no hydrogen-bonding donors present in this molecule. This molecule cannot form hydrogen-bonds to water.

c. This molecule has one nitrogen atom with one lone pair of electrons present that can act as a hydrogen-bond acceptor. The H atoms in this molecule are bonded to carbon, making them nonpolar, and therefore there are no hydrogen-bonding donors present in this molecule. This molecule can make one hydrogen bond to water as shown:

$$CH_3CN \underset{a}{:} --- \underset{d}{H} \overset{\overset{\cdot\cdot}{O}\cdot\cdot}{\diagup} \diagdown H$$

S A M P L E   P R O B L E M   6 . 2

Identifying Intermolecular Forces

Identify the strongest intermolecular force that could exist between each of the following pairs of molecules:

a. $CH_3CH_2OCH_2CH_3$ and $CH_3CH_2CH_2CH_2CH_2CH_3$

$$\begin{array}{c} O \\ \| \end{array}$$
b. H_3CCCH_3 and $CH_3CH_2OCH_2CH_3$

$$\begin{array}{c} O \\ \| \\ C \\ H \diagup \quad \diagdown H \end{array}$$
c.   and CH_3OH

S O L U T I O N

To solve this problem, first determine the types of intermolecular forces in which each of the molecules could possibly participate. (These are listed next to each of the molecules in the answers.) Then, comparing the lists, look for common interactions and choose the strongest, remembering that hydrogen bonding is the strongest and London forces are the weakest.

a. $CH_3CH_2OCH_2CH_3$	London forces and dipole attractions (this molecule is polar).	The strongest intermolecular force between these two molecules is *London forces*.
$CH_3CH_2CH_2\ CH_2CH_2CH_3$	London forces only (this molecule is nonpolar).	
b. $\underset{\displaystyle H_3CCCH_3}{\overset{\displaystyle O}{\overset{\displaystyle \|}{}}}$	London forces and dipole–dipole attractions; could act as a hydrogen-bond acceptor, but not a donor.	The strongest intermolecular force between these two molecules is *dipole–dipole attraction*.
$CH_3CH_2OCH_2CH_3$	London forces and dipole–dipole attractions; could act as a hydrogen-bond acceptor, but not a donor.	
c. $\underset{\displaystyle H \overset{\displaystyle C}{\diagup}\ {}^{\diagdown} H}{\overset{\displaystyle O}{\overset{\displaystyle \|}{}}}$	London forces and dipole–dipole attractions. This molecule can also be a hydrogen-bond acceptor, but not a donor.	The strongest intermolecular force these two molecules share is *hydrogen bonding*. As shown, the first molecule is the acceptor and other is the donor.
$CH_3-\overset{\displaystyle ..}{\underset{\displaystyle ..}{O}}-H$	London forces, dipole–dipole attractions, and hydrogen bonding, both donor and acceptor.	$\underset{\substack{\text{Acceptor} \qquad\quad \text{Donor}}}{\overset{\displaystyle H \diagdown}{\underset{\displaystyle H \diagup}{C}}=\overset{\displaystyle ..}{\underset{\displaystyle ..}{O}}:--H-\overset{\displaystyle ..}{\underset{\displaystyle ..}{O}}-CH_3}$

Ion–Dipole Attraction: A Similar Attractive Force

When you add a teaspoon of salt to a glass of water, what do you observe? The salt dissolves, right? What you are actually witnessing is another very strong attractive force in action. Because salt is made up of ions and not molecules, this force is technically not an inter*molecular* force, but it is a strong interaction found in many biomolecules like proteins. In fact, this attractive force is stronger than hydrogen bonding, the strongest intermolecular force. The interaction, which is called an **ion–dipole attraction**, results from the attraction of an ion to the opposite partial charge on a polar molecule. As shown in Figure 6.4, a cation, such as the sodium ion, is attracted to the partially negative end of the dipole of a polar molecule like water. Similarly, an anion, such as chloride, is attracted to the partially positive end of the dipole of water. As we will see later in the chapter, this interaction plays an important role in the solubility of ionic compounds.

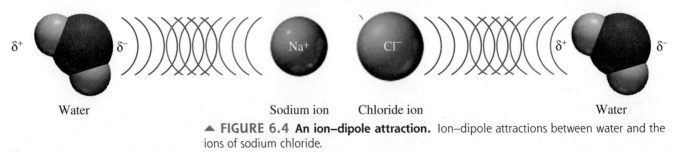

Water   Sodium ion   Chloride ion   Water

▲ **FIGURE 6.4 An ion–dipole attraction.** Ion–dipole attractions between water and the ions of sodium chloride.

PRACTICE PROBLEMS

6.1 Explain the difference between a covalent bond and an intermolecular force.

6.2 Polar molecules can contain more than one type of attractive force. Fill in the table with a yes or no answer to the following question: "Does this intermolecular force operate between molecules of the given compound?"

Molecule	H-Bond	Dipole-Dipole	London
CF_4			
H_2O			
CH_3NH_2			

6.3 Polar molecules can contain more than one type of attractive force. Fill in the table with a yes or no answer to the following question: "Does this intermolecular force operate between molecules of the given compound?"

Molecule	H-Bond	Dipole-Dipole	London
CH_3OH			
NH_3			
CH_3CH_3			

6.4 What type(s) of intermolecular forces exist between all molecules?

6.5 Explain the difference between the dipole in London forces and the dipole in dipole–dipole attractions.

6.6 Given that only polar molecules can participate as donors in hydrogen bonding, is it true that all polar molecules can be hydrogen-bond donors? Explain.

6.7 Explain the requirements for a molecule to be a hydrogen-bond acceptor.

6.8 Considering your answers to problems 6.6 and 6.7, can water and carbon dioxide form a hydrogen bond? Explain.

6.9 Why is the ion–dipole attraction not classified as an intermolecular force?

6.10 For each of the following pairs of compounds, identify the strongest attractive force that would exist between the two: London forces, dipole–dipole attractions, hydrogen bonding, or ion-dipole attractions.

a. NaF and CH_3OH b. CH_3OCH_3 and $CH_3CH_2CH_3$

c. CH_3OH and CH_3NH_2 d. CH_3OCH_3 and sodium acetate, CH_3COONa

6.11 For each of the following pairs of compounds, identify the strongest attractive force that would exist between the two: London forces, dipole–dipole attractions, hydrogen bonding, or ion–dipole attractions.

a. CH_3CH_3 and $CH_3CH_2CH_3$ b. $CH_3CH_2NH_2$ and KCl

c. $CH_3CH_2OCH_3$ and CH_3OH d.
$$\underset{H}{\overset{O}{\underset{\diagup \ \diagdown}{\overset{\parallel}{C}}}} \ H \quad \text{and} \quad NH_4Cl$$

6.12 For the pairs in problem 6.10 for which you listed ion–dipole attractions as a possibility, show how each of the ions would interact with the polar molecule. (Refer to Figure 6.4.)

6.13 For the pairs in problem 6.11 for which you listed ion–dipole attractions as a possibility, show how each of the ions would interact with the polar molecule. (Refer to Figure 6.4.)

6.2 Intermolecular Forces and Solubility

Are you a fan of oil and vinegar salad dressing? Even if you are not, you have seen that before adding it to a salad, you must vigorously shake the mixture and then quickly pour it. Why? Because oil and water (vinegar is mostly water) don't mix. But, have you ever wondered why oil and water don't mix? The answer lies in intermolecular forces!

The Golden Rule of Solubility

Oil and water don't mix, but sugar and water do. In both cases, we are witnessing the solubility (or lack thereof) of one substance in another. As we discussed in the previous section, the polarity of molecules (polar versus nonpolar) dictates the types of intermolecular forces displayed between molecules. The **golden rule of solubility**—*like dissolves like*—means that molecules that are similar will dissolve in each other. The "similarity" mentioned here relates to the character of the molecules—are they polar or nonpolar? In short, the golden rule of solubility says that *molecules that have similar polarity and participate in the same types of intermolecular forces will dissolve each other*.

Applying the Golden Rule to Nonpolar Compounds

Now, let's look more closely at oil and water. Dietary oils, known as **triglycerides**, are nonpolar organic compounds formed through the condensation reaction of three fatty acids with a compound called glycerol (see Figure 6.5). Organic chemists refer to this reaction as an **esterification** because an ester functional group is formed during the reaction (actually the triglyceride product contains three ester groups). As nonpolar

▶ **FIGURE 6.5 Condensation reaction of three fatty acids with glycerol to form a triglyceride.** Although the triglyceride contains some polar oxygen bonds in the ester functional groups, the nonpolar areas of the molecule dominate the polarity, making the molecule overall nonpolar.

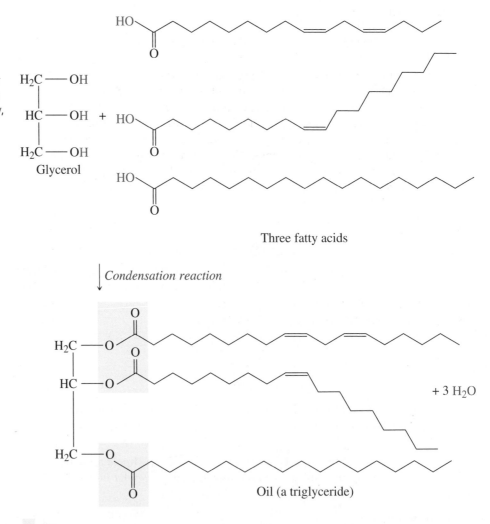

compounds, oils are attracted to neighboring molecules through London forces. Water, on the other hand, is a polar molecule and interacts with other molecules through dipole–dipole attractions and hydrogen bonding. This means that, in terms of intermolecular forces, oil and water are very *unlike* each other. In order for one molecule to dissolve another, the molecules must interact with each other. Oil and water do not interact with each other so they do not dissolve in each other. The attractions among the water molecules are much greater than the attraction between a water molecule and an oil molecule. Even after a bottle of oil and vinegar salad dressing is shaken vigorously, the contents of the bottle will separate back into two layers—a watery layer (vinegar) and an oil layer.

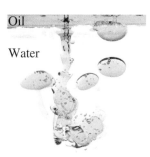

Applying the Golden Rule to Polar Compounds

Why does table sugar, also an organic compound, dissolve in water, but oil does not? In Chapter 5 we learned that the substance that we call table sugar is the disaccharide sucrose. Like all other carbohydrates, sucrose has multiple hydroxyl (—OH) groups. The key to the interaction of sucrose and water lies in these functional groups.

The hydroxyl groups of sucrose make it a polar compound and give it the ability to interact with water not only through dipole–dipole attractions, but also through hydrogen-bonding interactions as shown in Figure 6.6. Because table sugar and water are both polar and share these intermolecular forces in common, table sugar is an organic compound that *is* soluble in water.

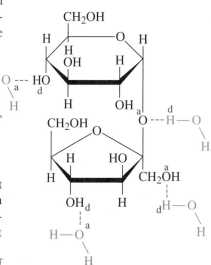

Applying the Golden Rule to Ionic Compounds

What about the solubility of ionic compounds in water? In section 6.1 we saw that ion–dipole attractions can exist between water and ions. These attractions are much stronger than the intermolecular attractions that exist between water and polar covalent compounds. So ionic compounds are often more soluble in water than covalent compounds.

In fact, for most ionic compounds, the ion–dipole attractions between the ions of an ionic compound and water are so strong that when enough water molecules interact with each ion, the ionic bond between the two ions is disrupted.

Individual ion–dipole attractions are not stronger than an ionic bond, but when multiple water molecules interact with an ion, the sum of these attractive forces is greater than the strength of the ionic bonds. The result, as shown in Figure 6.7 for sodium chloride, is that in water many ionic compounds break apart into their component ions, and each ion is surrounded by water molecules. This process is called **hydration**.

▲ **FIGURE 6.6 Sucrose hydrogen bonding with water.** A few of the many possible hydrogen bonds (shown in red) possible between water and sucrose (acceptors and donors labeled). The lone pair electrons on O are implied, and not shown for clarity.

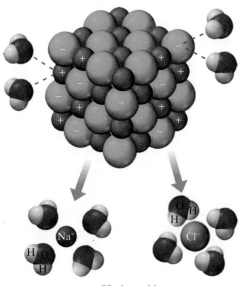

Hydrated ions

◀ **FIGURE 6.7 Hydration of NaCl.** When NaCl dissolves in water, the multiple ion–dipole attractions between water and the Na^+ and Cl^- ions cause the compound to break apart into its component ions.

The Unique Chemistry of Soap

One relevant application of these attractive forces to solubility occurs in soap. Soap is composed of fatty acid salts. **Salts** are ionic compounds. Unlike fatty acids, which we saw in Chapter 4 contain a carboxylic acid functional group $\left(R - \overset{\overset{\displaystyle O}{\|}}{C} - OH \right)$ and a long hydrocarbon chain making them largely nonpolar, fatty acid salts are ionic because they contain the carboxylate form $\left(R - \overset{\overset{\displaystyle O}{\|}}{C} - O^- \right)$ of the functional group (hydrogen removed) at one end. The charge on the carboxylate makes this end of the molecule ionic. This end is commonly called the polar *head*. The rest of the molecule behaves like a nonpolar compound. So, fatty acid salts have long nonpolar hydrocarbon *tails* and extremely polar (ionic) *heads*. Molecules like soap that have both polar and nonpolar parts are called **amphipathic** compounds.

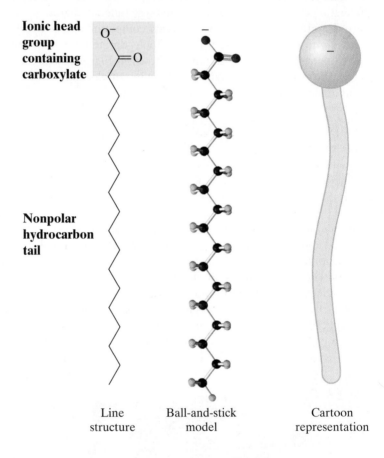

| Line structure | Ball-and-stick model | Cartoon representation |

Ionic head group containing carboxylate

Nonpolar hydrocarbon tail

Amphipathic compounds like soaps will not dissolve in water because of the large nonpolar tails present. The nonpolar tails are **hydrophobic** (water fearing) and will be excluded from the water, associating with each other and interacting through London forces. The ionic heads, which are **hydrophilic** (water loving), interact with the water mainly through ion–dipole attractions. Because the water interacts only with the ionic heads, the tails associate with each other, thereby creating the core of a spherical structure in the water called a **micelle** (see Figure 6.8). The polar heads are attracted to the water and form the shell of the micelle. Micelles form because the ion–dipole and hydrogen-bonding attractions between water molecules and the ionic heads are stronger (and preferred) to water interactions with the hydrocarbon tails.

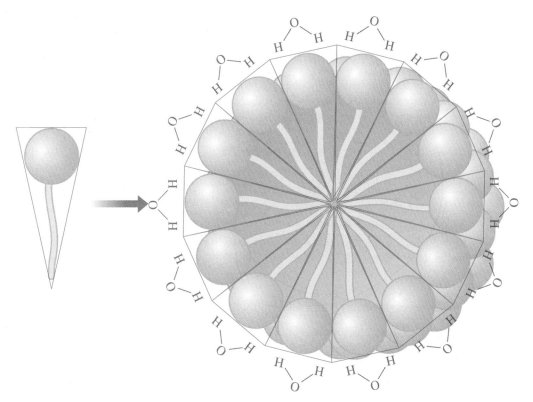

▲ **FIGURE 6.8 A spherical micelle.** A fatty acid salt has a conical shape so that when the nonpolar hydrocarbon tails are pushed to the center by the strong interactions that water can make to the ionic heads, a spherical-shaped micelle is formed.

How Soap Works

How does soap do its job of removing greasy dirt? Most stains, including grease and dirt, are nonpolar in nature, and so, based on the golden rule, greasy dirt is not soluble in water. When skin or clothing with a greasy dirt stain is washed with soapy water, the stain is attracted to the nonpolar hydrocarbon tails of the soap and is dissolved in the interior of the micelle formed by the soap molecules (see Figure 6.9). The ionic head groups surround the exterior surface of the micelle and form ion–dipole interactions with the water. Because the surface of the micelle is covered with the polar head groups, the entire micelle, with the greasy stain molecules now dissolved and "hidden" in its interior, is soluble in water and is washed down the drain.

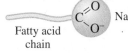
Soap molecule

C=O
O⁻ Na⁺

Fatty acid
chain

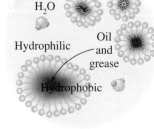
H_2O

Hydrophilic

Oil
and
grease

Hydrophobic

◄ **FIGURE 6.9 The cleaning action of soap.** Soap micelles dissolve oil and dirt in the nonpolar interior, which are then rinsed away with water.

SAMPLE PROBLEM 6.3

Predicting Solubility

For each of the following pairs of compounds, predict whether they are likely to be *soluble* or *not soluble* in each other. If soluble, identify the attractive forces involved in their interaction. If not soluble, explain why not.

a. H_2O and [hexagon]   b. CH_3OH and KBr   c. CH_3OCH_3 and CH_3OH

SOLUTION

To solve this problem, you will need to determine the types of intermolecular/attractive forces that the two compounds have in common. The more and stronger the attractive forces they share, the more likely they are to be soluble in each other.

a. H_2O and cyclohexane [hexagon] have only London forces in common. In polar molecules, the contribution of London forces compared to the polar attractive forces present is insignificant. Therefore, these two molecules are *not soluble* in each other.

b. CH_3OH is a polar compound, and KBr is ionic. These two compounds can interact via ion–dipole attractions and, therefore, are *soluble* in each other.

c. Both CH_3OCH_3 and CH_3OH are polar compounds and thus can interact via dipole–dipole attractions. In addition, CH_3OCH_3 can act as a hydrogen-bond acceptor, and CH_3OH can act as its hydrogen-bond donor, allowing them to interact through hydrogen bonding. Therefore, these two molecules are *soluble* in each other.

PRACTICE PROBLEMS

6.14 For each of the following pairs of compounds, predict whether they are likely to be *soluble* or *not soluble* in each other. If soluble, identify the attractive forces involved in their interaction. If not soluble, explain why not.

a. NH_3 and CH_4 b. CH_4 and
$$\overset{\displaystyle O}{\underset{H \quad\quad H}{\overset{\|}{C}}}$$

6.15 For each of the following pairs of compounds, predict whether they are likely to be *soluble* or *not soluble* in each other. If soluble, identify the attractive forces involved in their interaction. If not soluble, explain why not.

a. NaCl and [zigzag line]   b. CH_3OH and CH_3OCH_3

6.16 Predict which member of each of the following pairs of compounds will be more soluble in water:

a. fatty acid or fatty acid salt   b. KCl or $(CH_3)_3N$

6.17 Predict which member of each of the following pairs of compounds will be more soluble in water:

a. CH_3CH_2OH or CH_3CH_3 b. CH_3NH_2 or CS_2

6.18 Identify the ester functional groups in the triglyceride molecule shown in Figure 6.5.

6.3 Intermolecular Forces and Changes of State

You may remember the scene in the children's Christmas classic *Frosty the Snowman* when Frosty takes his little friend Karen into the greenhouse to help her warm up. Once they get inside, Professor Hinkle, the evil magician, locks them in, and Frosty changes from a snowman into a puddle of water. In scientific terms, he went from being a solid (snow or ice) to a liquid (water). But how did this happen? How did the greenhouse cause this major change in Frosty's life? Of course, you know that it was the heat in the greenhouse that changed Frosty from a solid to a liquid, but how did the heat cause that change?

In Chapter 1, we saw that matter exists as a solid, liquid, or gas—these are the three *states of matter*. We further saw that the state in which a given substance exists depends largely on the motion of its molecules: Molecules of a liquid move faster and with more types of motion than the molecules of a solid. So, did the heat in the greenhouse cause Frosty's water molecules to move faster? Yes, and by doing so it disrupted the intermolecular forces.

Heat and Intermolecular Forces

Just as holding hands with your sweetie is much easier when you are sitting together on a couch in a crowded room than when you are dancing on a crowded dance floor, intermolecular forces are stronger between two molecules moving slowly than between two molecules bouncing wildly about. When two molecules with low energy come near each other, there is a much greater possibility that an intermolecular attraction will develop (if possible) than when two molecules with high energy move quickly past each other. In other words, as a substance is heated, its molecules begin to move faster and faster; in turn, the intermolecular attractions between these two molecules decrease— they are no longer held together as tightly by intermolecular forces. When a solid substance is heated, it **melts** and becomes a liquid, and when a liquid is heated, it **evaporates** and becomes a gas. These transitions are called **changes of state** or **phase transitions**. Adding heat to a substance causes its molecules to move faster, which, in turn, disrupts the intermolecular forces holding the molecules together and causes the substance to change from one state to another.

Boiling Points and Alkanes

Pentane and octane are both alkanes with simple straight-chain structures and that exist normally as liquids. When heated, pentane boils at 36 °C, but octane does not boil until the temperature reaches 125 °C. Why do two compounds that are so similar boil at such different temperatures? In order to understand this, we first need to understand what happens at the molecular level in the process of boiling.

The temperature at which all molecules of a substance change from a liquid to a gas is called the **boiling point**. In order for a substance to boil, a couple of things must happen. The molecules of the substance must push back the molecules of the atmosphere at the liquid surface. In addition, each molecule must overcome the attractive forces of the other molecules in the liquid so that it can move into the gas phase *as a single molecule*. The heat supplied during boiling provides the energy necessary for the molecules of a substance to accomplish each of these tasks and move individually from the liquid into the gas phase.

Atmospheric pressure

Liquid pentane pressure

Vapor pentane pressure

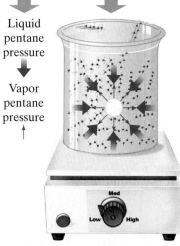

As heating begins, bubbles of vapor are formed in the liquid.

When boiling, vapor bubbles rise to the surface and escape. (36 °C is boiling point for pentane).

Because the molecules of the atmosphere stay the same no matter what the liquid surface is, the difference in the boiling points of two or more compounds must lie in their intermolecular forces. Let's see how this works for our alkane example.

The structures of pentane and octane are shown Figure 6.10. Like all other alkanes, both are nonpolar compounds and, therefore, exhibit only London forces. When interacting in the liquid phase, note that molecules of octane have more opportunities to interact than do molecules of pentane simply because a single octane has more carbon atoms and, therefore, a greater surface area. Recall that London forces between molecules are a result of a disturbance in the distribution of electrons over the surface of a molecule. A molecule with a larger surface area has more surface contact with other molecules as well as more electrons to disturb. In other words, the more carbon atoms in a straight-chain alkane, the stronger the attractions between molecules. And, because these attractions must be overcome in order for the compound to boil, more heat is necessary to disrupt these attractions, which means that the boiling point is higher.

Table 6.1 shows the boiling points for the first 10 straight-chain alkanes. Notice that as the number of carbon atoms in the chain (and therefore the surface area of the molecule) increases, the boiling point also increases.

Next consider two alkanes with the same number of carbon atoms, but whose structures are very different—hexane and 2,3-dimethylbutane. These two compounds

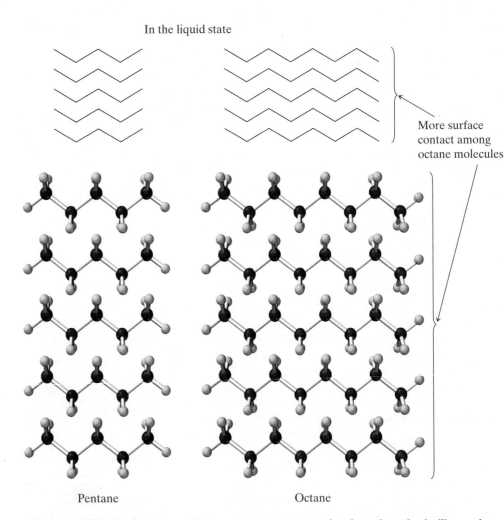

In the liquid state

More surface contact among octane molecules

Pentane   Octane

▲ **FIGURE 6.10 More surface contact between molecules raises the boiling point.**
Octane has a larger surface area and more potential surface for contact than pentane so its boiling point is higher.

TABLE 6.1	BOILING POINTS OF COMMON STRAIGHT-CHAIN ALKANES	
Name	**Structure**	**Boiling Point °C**
Methane	CH_4	−161
Ethane	CH_3CH_3	−89
Propane	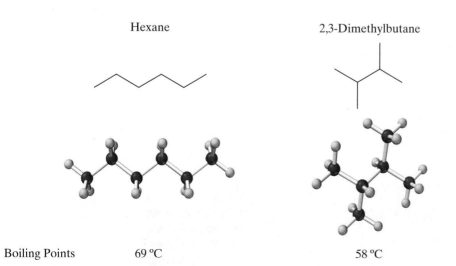	−42
Butane		−0.5
Pentane		36
Hexane		69
Heptane		98
Octane		125
Nonane		151
Decane		174

are structural isomers, meaning that they have the same molecular formula, but their connectivity is different. As shown in Figure 6.11, hexane is a straight-chain alkane, but 2,3-dimethylbutane is a branched-chain alkane.

In comparing their boiling points, hexane has a boiling point of 69 °C and 2,3-dimethylbutane has a boiling point of 58 °C. The boiling point of 2,3-dimethylbutane is 11 °C lower than that of hexane. Because these two molecules have the same number of carbon atoms, why would their boiling points be so different? The answer lies in intermolecular forces and in the surface area of the molecules.

Consider a plate of spaghetti and its accompanying bowl of meatballs. The long spaghetti noodles have more points of contact with each other and can line up with each other if you stretch them out on your plate. Similarly, the molecules of hexane, a straight-chain alkane, can "stack" close together like the spaghetti noodles. In contrast, molecules of 2,3-dimethylbutane, (see Figure 6.11) are, like meatballs, more spherical in their overall shape. When next to each other on your plate, two meatballs really touch at only one small point. Likewise, the molecules of 2,3-dimethylbutane have less surface contact between each other than do the hexane molecules. As mentioned earlier, the more contact between two molecules, the greater the attraction caused by the London forces between them. In turn, the greater London forces attraction means that

Hexane

2,3-Dimethylbutane

Boiling Points   69 °C   58 °C

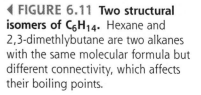

◀ **FIGURE 6.11 Two structural isomers of C_6H_{14}.** Hexane and 2,3-dimethlybutane are two alkanes with the same molecular formula but different connectivity, which affects their boiling points.

the boiling point of hexane is higher than that of 2,3-dimethylbutane. In fact, as a general rule, *for alkanes with the same number of carbon atoms, straight-chain alkanes have higher boiling points than do branched alkanes.*

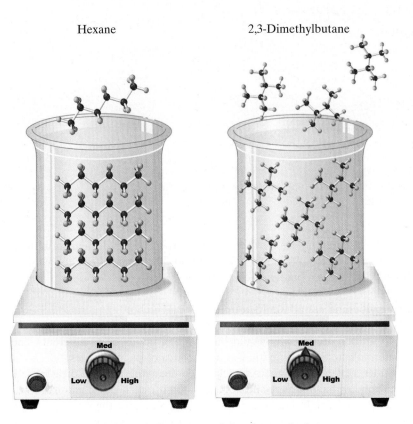

Hexane   2,3-Dimethylbutane

2,3-Dimethylbutane needs less heat to boil.

The Unusual Behavior of Water

Now, let's consider the case of water, H_2O. From its molecular formula, it is clear that water is a small molecule—it only has 3 atoms. Compare it to propane that has 11 atoms (C_3H_8). In terms of surface area, you might guess that water has a lower boiling point than propane. Yet, from Table 6.1 we can see that propane boils at $-42\,°C$, while water boils at $100\,°C$. There has to be something other than surface area involved to explain water's higher boiling point.

Earlier in the chapter, we discussed the fact that water is a polar molecule. Water's electrons are always unevenly distributed, unlike the alkanes. Water molecules strongly attract each other through hydrogen bonding, a much stronger intermolecular force than London forces. In order for water to boil, that is, to move from the liquid to the gas phase, there must be enough heat energy to disrupt the hydrogen-bonding interactions. This requires more heat energy than it does to disrupt the London forces in compounds like propane, leading to the unusually high boiling point of water.

Table 6.2 summarizes the trends in boiling points previously discussed.

TABLE 6.2	BOILING POINT TRENDS	
Nonpolar Molecules	Similar Size Molecules, Different Forces	Molecules with Same Forces
The greater the surface area, the higher the boiling point.	The stronger the forces, the higher the boiling point.	The stronger and more numerous the forces, the higher the boiling point.

SAMPLE PROBLEM 6.4

Predicting Boiling Points

For each of the following pairs of compounds, predict which substance would have the higher boiling point:

a. CH_3OH and CH_3CH_3

b. and

c. CH_3NHCH_3 and CH_3OH

SOLUTION

The first step in solving this problem is to determine the type of intermolecular forces possible for each molecule. Then, using the trends in Table 6.2, determine which molecule will have the higher boiling point.

a. CH_3OH is a polar molecule that can participate in hydrogen bonding. CH_3CH_3 is a nonpolar molecule and, therefore, can participate only in London forces. Boiling CH_3OH will require enough heat energy to disrupt its dipole–dipole attractions and the hydrogen bonding. Boiling CH_3CH_3 will require only enough heat energy to disrupt its London forces. Because London forces are weak intermolecular attractions, they can be disrupted with minimal input of heat energy. Therefore, we would predict that CH_3OH would have the higher boiling point. (In fact, our prediction is correct: CH_3OH boils at $65\,°C$ and CH_3CH_3 boils at $-89\,°C$.)

b. Both of these compounds are alkanes with a molecular formula of C_4H_{10}. Given that they have the same number of carbon atoms and participate in the same intermolecular forces (London forces), the boiling point will depend on surface area. Butane is a straight-chain alkane, but 2-methylpropane is a branched alkane, which means it is more compact, that is, it has a smaller surface area. Therefore, we would predict that because of its greater surface area, butane would have the higher boiling point. (Butane boils at $-0.5\,°C$ and 2-methylpropane boils at $-12\,°C$.)

c. Both CH_3NHCH_3 and CH_3OH are polar compounds and can participate in dipole–dipole and hydrogen bonding attractions. Applying the rule from the "Molecules with Same Forces" column in Table 6.2, we should determine how many hydrogen bonding contacts each molecule can make. CH_3NHCH_3 has one hydrogen-bond acceptor and one hydrogen-bond donor, so each molecule can form up to two hydrogen bonds. CH_3OH has two hydrogen-bond acceptors and one donor, so each molecule can form up to three hydrogen bonds. We would predict that CH_3OH would have a higher boiling point because it can form more hydrogen bonds than does CH_3NHCH_3. (CH_3NHCH_3 boils at $7\,°C$ and CH_3OH boils at $65\,°C$.)

PRACTICE PROBLEMS

6.19 In the following pairs of molecules, predict which one of the molecules in the pair would have the lower boiling point. Name the strongest intermolecular force present in each molecule.

a. CH_4 and H_2O

b. NH_3 and CO_2

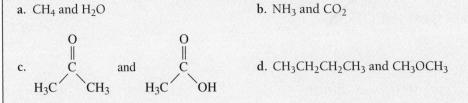

c.   and   **d.** $CH_3CH_2CH_2CH_3$ and CH_3OCH_3

6.20 In the following pairs of molecules, predict which one of the molecules in the pair would have the lower boiling point. Name the strongest intermolecular force present in each molecule.

a. CH_3OH and $CH_3CH_2OCH_2CH_3$

b.

$$\underset{H}{\overset{O}{\underset{\displaystyle |}{C}}}\!\!-\!CH_3$$ and $CH_3CH_2CH_3$

c. CH_3-N-CH_3 and $CH_3CH_2-NH_2$
          |
          H

d. CH_3CH_3 and CH_3CH_2OH

6.21 The formulas and boiling points of several similar compounds are listed in the following table. Determine the strongest intermolecular force present in each compound and explain the difference in their boiling points.

Formula	Boiling Point °C
H | H—C—H | H	−164
H | H—C—F | H	−78
H | H—C—OH | H	65

6.22 The formulas and boiling points of several similar compounds are listed in the following table. Determine the strongest intermolecular force present in each compound and explain the difference in their boiling points.

Formula	Boiling Point °C
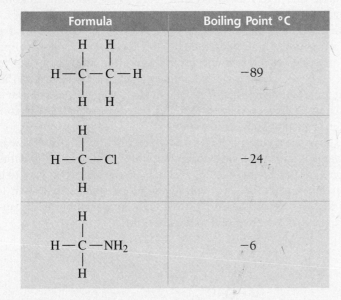	−89
	−24
	−6

6.23 Rank the following compounds in order of increasing boiling point (lowest to highest) and explain your ranking:

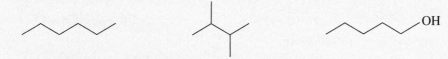

6.24 Rank the following compounds in order of decreasing boiling point (highest to lowest) and explain your ranking:

6.4 Fats, Oils, and Margarine—Solid to Liquid and Back Again: Melting

Much like boiling, melting (transition from solid to liquid) involves an increase in molecular motion and a disruption of intermolecular forces. In order to understand how this change of state occurs, let's look at the behavior of triglycerides—dietary fats and oils.

Fats

A juicy steak, a can of lard, or a tub of butter, all contain animal fats that you can find at your local grocery store. Animal **fat** is a solid or semisolid material at room temperature that, like the oils we saw earlier in the chapter, is a triglyceride made up of three fatty acids joined to a glycerol backbone. But if both fats and oils are triglycerides, why is one solid and the other liquid at room temperature? Intermolecular forces are key to the explanation.

In section 6.2, we saw the formation of a triglyceride from the condensation reaction of three fatty acids with glycerol. When the hydrocarbon chains (tails) of the fatty acids are mostly saturated, the triglyceride product is a fat. (In Chapter 4 we saw that a hydrocarbon containing all single C—C bonds was referred to as "saturated.")

In Figure 6.12a, you can see that in the structure of the fat molecule, the hydrocarbon tails of the three fatty acids are physically closer to each other than those of the oil in Figure 6.12b. This proximity creates disturbances in the electron distribution around each of the tails as they interact with the other tails. These disturbances result in London forces—the tails are attracted to one another. The saturated tails allow for many surface contacts, which increases the attractions between them and slows down (restricts) the molecular motions in the molecules, thereby allowing the molecules to form a solid.

So, fats are solids at room temperature because the motion of their hydrocarbon tails is restricted by the London forces holding them together. Instead of the unrestricted movement found in a single hydrocarbon tail like a soap, the tails of a fat are attracted to one another and their movement is more restricted. But why is the melting point of a fat low enough to melt at body temperature?

When a substance melts, it changes from a solid to a liquid. As we saw with boiling, when heat energy is added to a substance, intermolecular forces are disrupted, and molecular motion increases. So, in order for a fat to melt, the forces that have to be disrupted are weak London forces. Fats are typically solid substances, but with low melting points. Even the heat of your hand (37 °C) is enough to melt butter. Because the melting points are low, fats are often said to be semisolid.

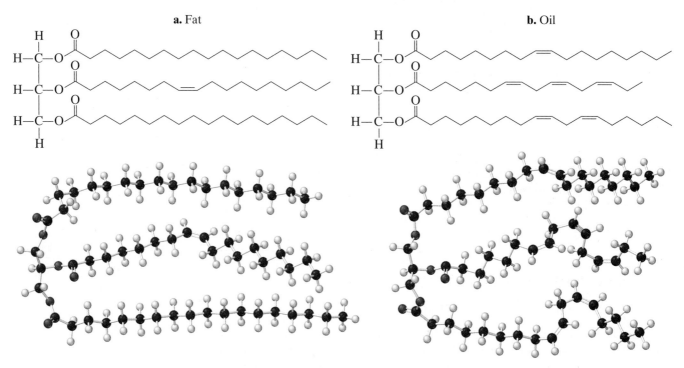

a. Fat

b. Oil

Tails are closer together in fat

Tails are farther apart in oil

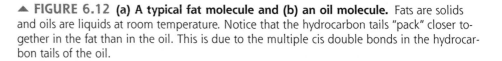

▲ **FIGURE 6.12 (a) A typical fat molecule and (b) an oil molecule.** Fats are solids and oils are liquids at room temperature. Notice that the hydrocarbon tails "pack" closer together in the fat than in the oil. This is due to the multiple cis double bonds in the hydrocarbon tails of the oil.

Oils

Dietary **oils** are derived from plants like corn, soybean, or canola oil and are also triglycerides—three fatty acids bonded to a glycerol. The difference in the name "fat" versus "oil" simply indicates the physical state at room temperature—oils are liquids and fats are solids. How can two molecules that are so similar in structure exist in different states?

The answer to this question lies in the structure of the hydrocarbon tails of the oil molecules. Compare the tails in the oil molecule to those in the fat molecule in Figure 6.12. The hydrocarbon tails in the oil molecules contain more cis double bonds than those in the fat. (Recall that naturally occurring fatty acids contain *only* cis double bonds—section 4.3.) Because they contain several carbon–carbon double bonds, the tails of the oil are called polyunsaturated. So how does this simple difference translate to a difference in the physical state (solid versus liquid) of these molecules?

Notice that the introduction of cis double bonds in the tails of the oil creates kinks in the normally straight hydrocarbon chain (see Figure 6.12b). The tails of the unsaturated oil cannot stack together as closely as those in the fat can. The kinked tails interact less with each other through London forces because they are not physically as close to one another. Less interaction means weaker forces of attraction, which in turn means that the hydrocarbon chains in the oil are less attracted to one another and, therefore, move more freely. The greater molecular freedom of motion among the hydrocarbon tails in the oil does not allow enough stacking of the tails for a solid to form, and oils thus remain liquid.

Margarine: Changing the Liquid to a Solid

Oils derived from plants have long been considered a healthier alternative for human dietary fat requirements. However, liquids are not always the most convenient form of triglycerides—imagine coating your morning toast with corn oil! Not only does it sound unappetizing, it would not be particularly easy to do. To overcome this problem, chemists have developed a method to convert the liquid, plant-derived triglycerides (oils) into solids to produce what we know today as margarine.

The production of margarine—a solid from a plant-derived oil such as corn oil—requires an understanding of the intermolecular forces involved. Based on our previous discussion, we can see that the only difference between fats and oils is the number of double bonds in the fatty acid chains. If double bonds could be removed and single bonds left in their place, the intermolecular attractions between the chains would increase, and the liquids would be solids.

The conversion of a double bond to a single bond can be accomplished by adding hydrogen atoms to the carbon–carbon double bond of an unsaturated compound through a reaction called **hydrogenation**.

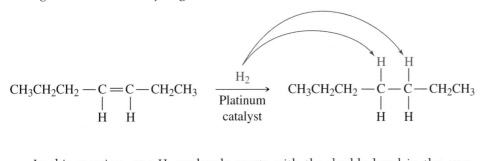

In this reaction, one H_2 molecule reacts with the double bond in the compound. The second bond between the two carbons is removed (leaving a single bond) and one hydrogen atom is added to each of the original carbon atoms of the double bond. This reaction, like many other organic reactions, progresses more quickly if a catalyst is included in the reaction mixture. A **catalyst** (in this case platinum metal) speeds up a reaction but remains unchanged at the completion of the reaction. In a hydrogenation reaction, the carbons of the unsaturated compound become *saturated* with hydrogen.

SAMPLE PROBLEM 6.5

Hydrogenation of Oils

How many hydrogen molecules (H_2) are required to convert the oil shown in the figure into a saturated fat?

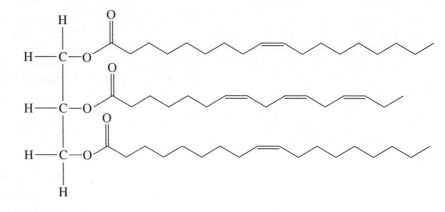

SOLUTION

In the previous discussion, we said that one molecule of H_2 is required to hydrogenate each carbon–carbon double bond in an unsaturated compound. Therefore, to solve this problem, count the number of carbon–carbon double bonds in the oil molecule (the carbon–oxygen double bonds will not react under these conditions). This number will be the same as the number of H_2 molecules needed to convert the oil molecule into a saturated fat molecule.

The oil molecule shown has five carbon–carbon double bonds, so five H_2 molecules will be needed for this reaction. The balanced equation for the complete reaction is:

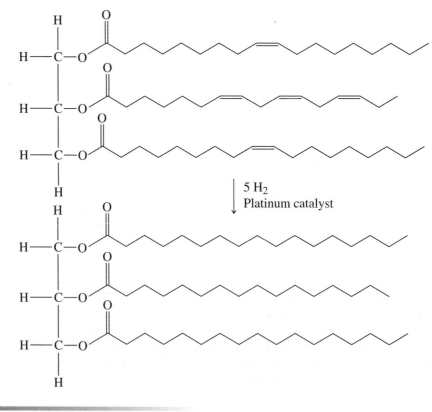

Trans Fats

If you like the flavor of margarine or butter on your morning toast or a warm biscuit, you may have noticed that cold margarine spreads much more easily than does cold butter. If not, try it. Because of its saturated fatty acid chains and increased London forces, butter, especially when it is refrigerated, is a firm solid that is difficult to spread. So why is the same not true for margarine?

In order to control the "spreadability" of margarine, the hydrogenation of plant-derived oils is controlled so that some double bonds become saturated while others remain intact. This process, known as *partial hydrogenation*, allows producers to create margarines that are solids that, although more saturated than oils, are less saturated than butter, making them easier to spread even when cold. One of the consequences of partial hydrogenation is that some of the double bonds are incompletely hydrogenated. Because hydrogenation is a difficult reaction to control and fully complete, some of the double bonds in these compounds reform, but when they do, they switch from the less favorable cis form to the more favorable trans form, resulting in the compounds known as *trans fats* (see Figure 6.13). While the trans double bonds do allow the fatty acid chains to stack more easily, enhancing the solidity of the fat, there is public concern because some studies show that trans fats have deleterious health effects. This has forced many manufacturers of partially hydrogenated oils to consider alternatives. Today all nutritional labeling must give the amounts of trans fat present in food.

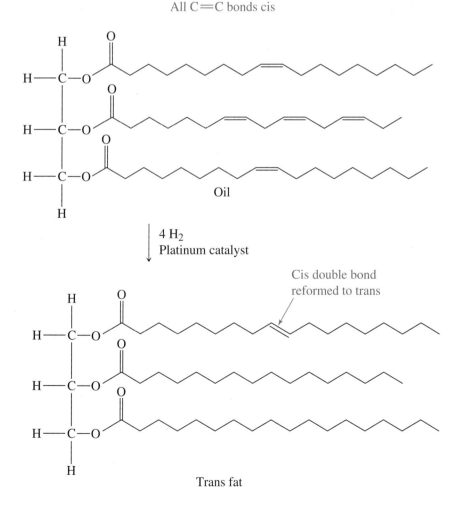

All C=C bonds cis

Oil

4 H₂
Platinum catalyst

Cis double bond
reformed to trans

Trans fat

◀ **FIGURE 6.13 Incomplete hydrogenation of an oil forming a trans fat.** The cis C=C bonds in the oil become hydrogenated to form alkanes or reform as trans C=C bonds.

PRACTICE PROBLEMS

6.25 What is a triglyceride? Draw the skeletal structure for a saturated triglyceride whose hydrocarbon tails contain six carbons.

6.26 Draw the structural formula for a simple triglyceride whose hydrocarbon tails contain eight saturated carbons, one of which is unsaturated between carbons 3 and 4 from the tail end.

6.27 Explain why fats are solids, but oils are liquids under normal conditions, despite the fact that both are triglycerides.

6.28 A sample of vegetable oil was reacted with enough hydrogen gas to completely saturate the oil. It was determined that each molecule of the oil required three molecules of H_2 for the reaction to be completed. Draw a possible structure of the oil that was hydrogenated.

6.29 Balance the following reaction equation by providing the missing number of hydrogen molecules:

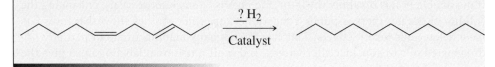

6.5 Intermolecular Forces and the Cell Membrane

The cells of our bodies are encased in a magnificently simple membrane. The structure of the cell membrane, which is similar in cells throughout the body, has the ability to be both flexible yet firm. As we will see in Chapter 7, the cell membrane serves to allow some molecules to move into and out of the cell while inhibiting passage of other molecules. What is it about the structure and composition of the cell membrane that enables it to be selective?

A Quick Look at Phospholipids

The primary structural components of cell membranes are molecules called **phospholipids**. Structurally similar to the fats and oils previously discussed, phospholipids have a glycerol backbone with fatty acids linked to it through an ester bond. However, unlike fats and oils, phospholipids have only *two* fatty acids on their glycerol backbone, arranged as shown in Figure 6.14(a). The third OH group of the glycerol backbone is linked to a phosphate-containing group instead of to a fatty acid. The phosphate-containing group is ionic (polar), which is in contrast to the fatty acid tails (nonpolar). There are many different phospholipids, but they all share this similar structure: a glycerol backbone with two nonpolar fatty acid tails and a polar phosphate-containing head. Because these molecules have both polar and nonpolar parts, they are amphipathic.

Phospholipids are often represented by a cartoonlike drawing (see Figure 6.14b) highlighting the dual character of these molecules (strongly polar head and two nonpolar tails). Earlier we saw that soap molecules have a similar structure with a polar head, but just one nonpolar tail. Having two nonpolar tails and a larger ionic head enhances both the nonpolar and polar characteristics, respectively, of phospholipids as compared to soap. The two tails of the phospholipid also affect the overall shape of the molecule. While soap molecules tend to be more conical in their overall shape due to their smaller polar head and form micelles, phospholipids tend to have a more cylindrical shape with a much larger polar head group. This cylindrical shape of the phospholipids hinders their ability to arrange themselves into micelles. So how do phospholipids behave in water?

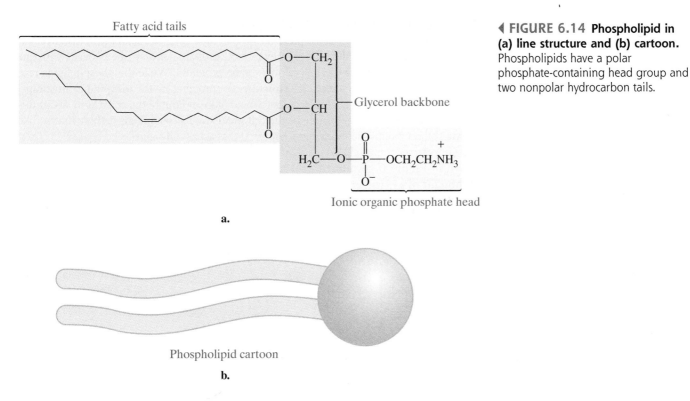

Fatty acid tails

Glycerol backbone

Ionic organic phosphate head

a.

◀ **FIGURE 6.14 Phospholipid in (a) line structure and (b) cartoon.** Phospholipids have a polar phosphate-containing head group and two nonpolar hydrocarbon tails.

Phospholipid cartoon

b.

The Cell Membrane Is a Bilayer

Consider the cell and its environment: Its surroundings are watery *and* its contents function in an aqueous environment. With water environments inside and outside the cell, how can phospholipids, molecules that have nonpolar hydrophobic tails, serve to surround and protect cells? The cylindrical shape of a phospholipid constrains them to lining up in a straight line instead of a sphere like a soap. No matter how the tails orient themselves—toward the inside or outside of the cell—the hydrocarbon portion would be forced into an aqueous environment.

A cell membrane composed of phospholipids cannot exist as a single layer. Instead, the phospholipids form a double layer called a *bi*layer as shown in Figure 6.15. The phospholipids of the "outer" layer of the membrane are oriented such that the polar heads are directed out into the surrounding aqueous environment. In turn, the phospholipids of the "inner" layer of the membrane are oriented with the polar heads directed into the aqueous environment of the interior of the cell. This arrangement leaves the nonpolar tails of both layers directed toward each other, creating a nonpolar interior region. The phospholipid bilayer, as it is known, is the foundation for a cell's membrane. Next let's examine the other molecules found in the membrane.

Aqueous environment outside the cell

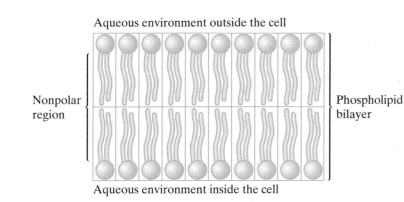

Nonpolar region

Phospholipid bilayer

Aqueous environment inside the cell

◀ **FIGURE 6.15 The cell membrane's bilayer of phospholipids.** The nonpolar hydrocarbon tails are excluded from the aqueous environment in the center of the bilayer, and the polar head groups interact with the aqueous environment.

Among the most important components in the phospholipid bilayer of the cell membrane are large biomolecules called proteins. Protein molecules can span the lipid bilayer extending outward on either side (integral membrane proteins) or they can associate with one polar surface of the bilayer (peripheral membrane proteins). While the phospholipids provide the basis for the cell membrane's structure, the proteins are the membrane's functional components, allowing selected molecules to move into and out of the cell. (More about this in Chapter 7.) The exterior surface of the cell membrane also contains carbohydrates (like the ABO blood groups mentioned in Chapter 5) that act as cell signals.

The current model for the cell membrane, called the **fluid mosaic model** shown in Figure 6.16, puts all these pieces together. This model creates what can be thought of as "icebergs" of protein floating in a "sea" of lipids, implying that the membrane itself is fluidlike because the phospholipids are able to move freely within the bilayer.

▶ **FIGURE 6.16 The fluid mosaic model of the cell membrane.** Proteins and cholesterol are embedded in a lipid bilayer of phospholipids. The bilayer forms a membrane-type barrier with polar heads at the membrane surface and the nonpolar tails in the center away from water.

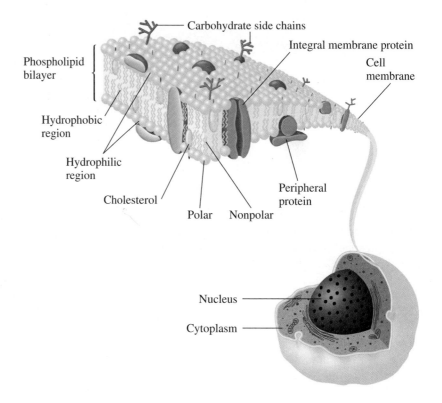

Cholesterol in Membranes

Take a close look at the membrane piece shown in Figure 6.17. Notice that there are cholesterol molecules within the phospholipid bilayer. Why is cholesterol included in a cell membrane?

Let's take a closer look at the structure of cholesterol shown. Cholesterol has a polar end (the OH group) and a nonpolar portion (the rest of the molecule), so like a phospholipid it is amphipathic. Cholesterol situates itself in the phospholipid bilayer so that the OH group protrudes out into the surrounding aqueous environment, while the rest of the rigid rings of the molecule are nestled in the nonpolar interior of the membrane.

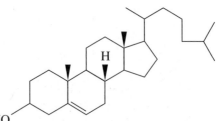

Cholesterol is an important component of the cell membrane. It modulates the fluidity or flexibility of a cell membrane depending on the amount present. Cholesterol can slip in between the hydrocarbon tails of the phospholipids, disrupting the London forces and increasing the membrane's fluidity. Cholesterol can also interact with the unsaturated hydrocarbon tails via London forces, decreasing their motion and increasing the rigidity of the membrane.

SUMMARY

6.1 Types of Intermolecular Forces

Intermolecular forces are weaker than covalent bonds. The three main types of intermolecular forces (from weakest to strongest) are London forces, dipole–dipole attractions, and hydrogen bonding. Each of these forces involves the attraction of an area of partially negative charge on one molecule to an area of partially positive charge on a second molecule. Similar to the intermolecular forces, the ion–dipole attraction is an attractive force between ions and molecules involving the attraction of the full charge on a cation or anion to the area of opposite partial charge on a polar molecule.

6.2 Intermolecular Forces and Solubility

The solubility of one substance in another is described by the golden rule of solubility, *like dissolves like*, which means that polar compounds dissolve polar compounds and nonpolar compounds dissolve other nonpolar compounds. In other words, the more intermolecular forces that two compounds have in common, the more soluble they will be in each other. Soap is able to dissolve nonpolar stains such as oil and grease and to be dissolved by water because it is an amphipathic compound. In water, soap molecules form spherical structures called micelles with the hydrophobic tails gathered together in the nonpolar interior of the micelle while the hydrophobic heads cover the surface and interact with the aqueous external environment.

Oil

Water

6.3 Intermolecular Forces and Changes of State

The boiling point is the temperature where the change of state from liquid to gas occurs. In order for a compound to boil, the molecules of the compound must overcome the intermolecular forces between molecules; therefore, molecules with stronger intermolecular forces require more heat energy to disrupt those attractions and therefore have higher boiling points. Water, which can form up to four hydrogen bonds for every water molecule, has a very high boiling point. In contrast, for alkanes that have only London forces between molecules, the boiling point depends largely on the surface area of the molecule—the greater the surface area, the higher the boiling point. This means that for alkanes with the same number of carbon atoms, the more branched the structure, the lower the boiling point.

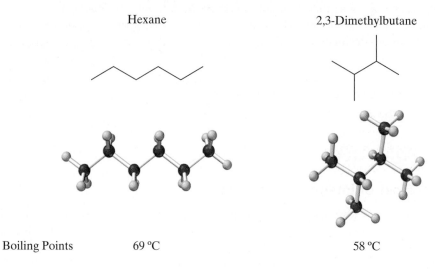

Hexane		2,3-Dimethylbutane
Boiling Points	69 °C	58 °C

6.4 Fats, Oils, and Margarine—Solid to Liquid and Back Again: Melting

As boiling moves molecules from the liquid into the gas state, melting is the change of state from solid to liquid. Both changes involve the disruption of intermolecular forces and an increase in the motion of the molecules in the substance. Fats are solid triglyceride compounds with low melting points because their molecules are held together only by London forces. Oils are also triglycerides, but they are liquids at room temperature. The fatty acid tails of oils are highly unsaturated, meaning they have multiple double bonds, while those in fats are mostly saturated. The double bonds so prevalent in the fatty acid components of oils exist solely in the cis form. The cis isomeric form of the double bonds, as opposed to the trans form, creates a kink in the fatty acid tails of the oils. This kink results in fewer and weaker London forces, which permits greater molecular motion and causes oils to be liquids at room temperature. Oils can be converted into fats by hydrogenation of the double bonds to increase the saturation of the fatty acid tails. This is the process used to create margarine from oils derived from plants. Unfortunately, hydrogenation is a difficult process to control and often results in the conversion of the naturally occurring cis double bonds to their trans isomers. The products of this process are the less healthy trans fats.

6.5 Intermolecular Forces and the Cell Membrane

Cell membranes are composed of phospholipids, which like soap molecules are amphipathic. However, phospholipids have a more strongly polar head and two nonpolar tails. Because of their shape, phospholipids of the cell membrane arrange themselves in a bilayer around the cellular contents with the polar heads of one layer oriented out into the surrounding aqueous environment and the polar heads of the second layer directed inward toward the aqueous interior of the cell. This results in a hydrophobic interior layer formed by the nonpolar tails. The fluid mosaic model describes the complete structure of the cell membrane, including not only the phospholipids, but also the protein components. Cell membranes also contain cholesterol molecules, which help modulate the fluidity of the membrane.

Aqueous environment outside the cell

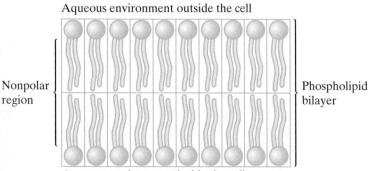

Nonpolar region

Phospholipid bilayer

Aqueous environment inside the cell

KEY TERMS

amphipathic—Term used to describe compounds with both polar and nonpolar parts.

boiling point—Temperature at which all molecules of a substance change from a liquid to a gas (boils).

catalyst—A substance that speeds up and participates in a chemical reaction but is unchanged at its completion.

change of state—The change that occurs when a substance that exists in one state of matter moves to another, for example, a solid becomes a liquid or a liquid becomes a gas.

dipole–dipole attraction—The attraction of the partially positive end of one polar molecule for the partially negative end of a second polar molecule.

esterification—A condensation reaction of a carboxylic acid and an alcohol to form an ester.

evaporate—The change in state from liquid to gas.

fat—A lipid molecule composed of three fatty acids joined to a glycerol backbone and which exists as a solid or semisolid at room temperature; also known as a triglyceride.

fluid mosaic model—A model used to describe the nature of the cell membrane. The model incorporates the structural components (phospholipids) and the functional components (proteins) of the membrane.

golden rule of solubility—Molecules that have similar polarities and participate in similar intermolecular forces will dissolve in each other.

hydration—For an ionic compound, the dissociation of the compound into its component ions by the action of many water molecules interacting with the ions and then surrounding each of the newly formed ions.

hydrogen bonding—The strong dipole–dipole attraction of a hydrogen, covalently bonded to an O, N, or F (the *hydrogen-*

bond donor), to a nonbonding pair of electrons on an O, N, or F (the *hydrogen-bond acceptor*).

hydrogenation—The addition of hydrogen (H_2) to a carbon–carbon double bond to convert it to a single bond.

hydrophilic—Literally, "water-loving"; polar compounds or parts of compounds that readily dissolve in water.

hydrophobic—Literally, "water-fearing"; nonpolar compounds or parts of compound that do not dissolve in water.

intermolecular forces—Attractive forces between two or more molecules that are caused by uneven distributions of electrons within a molecule.

ion–dipole attraction—An electrical attraction between an ion and a polar molecule.

London forces—The weakest intermolecular force; an attraction formed from temporary (or induced) dipoles on molecules.

melt—The change in state from solid to liquid.

micelle—A three-dimensional spherical arrangement of soap molecules in water with the hydrophobic tails pointing inside, away from the water, and the hydrophilic heads pointing out into the water.

oil—A lipid molecule composed of three fatty acids joined to a glycerol backbone and that exists as a liquid at room temperature; also known as a triglyceride.

phase transition—See change of state.

phospholipids—The primary structural component of cell membranes consisting of a glycerol backbone esterified with two fatty acids and a phosphate-containing group.

salt—An ionic compound.

triglyceride—The product of a condensation reaction of three fatty acids and glycerol; also known as a fat or an oil.

ADDITIONAL PROBLEMS

6.30 List the three types of intermolecular forces in order of increasing strength. Give a brief description of each interaction including the type(s) of molecules involved and how the attraction is formed.

6.31 Compare and contrast a dipole–dipole attraction and an ion–dipole attraction.

6.32 Show a possible hydrogen bond between methanol (CH_3OH) and dimethyl ether (CH_3OCH_3). Label the donor and the acceptor. Is there more than one possible hydrogen bond between these two molecules? Explain.

6.33 Show a possible hydrogen bond between methanol (CH_3OH) and water. Label the donor and the acceptor. Is there more than one possible hydrogen bond between these two molecules? Explain.

6.34 List all intermolecular forces that each of the following molecules can participate in with another identical molecule:

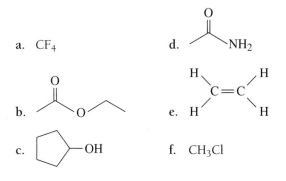

a. CF_4

b.

c.

d.

e.

f. CH_3Cl

6.35 List all intermolecular forces found in each of the pure substances:

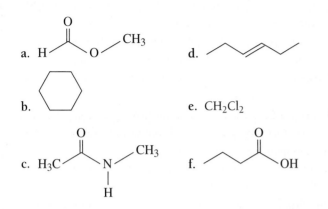

a. (formate ester structure, H—C(=O)—O—CH₃)

b. (cyclohexane)

c. H₃C—C(=O)—N(H)—CH₃

d. (alkene chain structure)

e. CH₂Cl₂

f. (carboxylic acid chain structure, ...—C(=O)—OH)

6.36 In terms of intermolecular forces, explain why oil and water do not mix.

6.37 In Chapter 4, we learned that fatty acids are not soluble in water because of their long hydrocarbon tails. However, fatty acid salts (soap molecules) are soluble and form micelles in water. Explain why fatty acid salts form micelles, but fatty acids do not.

6.38 Predict which member of each of the following pairs of compounds will be more soluble in vegetable oil:

  a. NH_3 or $CH_3CH_2CH_2CH_2CH_3$
  b. NaCl or $CH_3CH_2OCH_2CH_3$
  c. cyclopentane or $CH_3CH_2NH_2$
  d. CH_3OH or CH_3OCH_3

6.39 Predict which member of each of the following pairs of compounds will be more soluble in vegetable oil:

  a. fatty acid or fatty acid salt
  b. fat or water
  c. CH_3NH_2 or CS_2
  d. KCl or $CH_3CH_2CH_2CH_2CH_2CH_3$

6.40 In order to remove stains from delicate fabrics without using soap and water, dry cleaners once used perchloroethylene, which was banned in 1998 for environmental reasons.

Its structure is shown in the figure. Explain why perchloroethylene makes a good stain remover.

6.41 A stain on your shirt will not come out when you wipe it with a wet cloth. What clue does this give you to the nature of the stain?

6.42 Using your knowledge of intermolecular forces and solubility, explain why soap forms micelles in water.

6.43 Rank the four compounds in the figure in decreasing order of boiling point. Explain the reasoning behind your ranking.

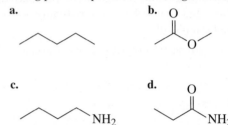

6.44 Rank the four compounds in the figure in decreasing order of boiling point. Explain the reasoning behind your ranking.

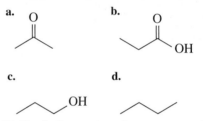

6.45 The boiling point of octane is much greater than that of hexane (see Table 6.1). Explain the difference in their boiling points.

6.46 The boiling point of heptane is over 60 °C higher than the boiling point of pentane. Draw the structure of both compounds and explain the difference in their boiling points.

6.47 Explain why a straight-chain alkane has a higher boiling point than a branched-chain alkane with the same molecular formula.

6.48 Draw the structure of the product of the following hydrogenation reaction:

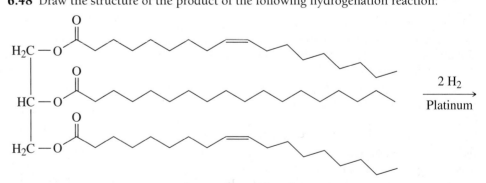

6.49 Draw the structure of the product of the following hydrogenation reaction:

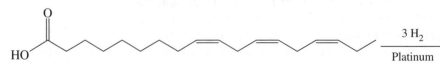

6.50 Sketch a cartoon model of the phospholipids bilayer of a cell membrane and label the polar and nonpolar parts.

6.51 Compare the structure of a soap molecule to a phospholipid and explain why a soap's polar head is smaller than that of a phospholipid.

6.52 Soap, phospholipids, and cholesterol are all amphipathic molecules. Provide a drawing of each and label the polar and nonpolar portion.

CHALLENGE PROBLEMS

6.53 If very small amounts of soap are dissolved in water, the soap molecules can form a structure known as a monolayer on the surface of the water with part of the soap molecule dissolved in the water and part sticking into the air. Using the cartoon figures for soap molecules, draw a small part of a monolayer showing the water, the air, and the soap molecules as they would be arranged.

6.54 Getem Clean, Inc. has hired you as chief chemist and put you in charge of a corporate project to build a better soap. Given your understanding of how soap works, what would you do to improve a fatty acid soap molecule to make it more effective in dissolving greasy dirt and in being dissolved by water?

6.55 Water and hydrogen sulfide are structurally similar and geometrically identical (both have a bent shape), yet H_2O is a liquid at room temperature and H_2S is a gas (rotten egg odor) at room temperature. In other words, water has a higher boiling point that hydrogen sulfide. Explain why two molecules that appear to be so similar have such different properties, especially in terms of their boiling points.

6.56 The structures and boiling points of cyclopentane and tetrahydrofuran are shown in the following figure. Explain why the boiling point of tetrahydrofuran is higher than that of cyclopentane.

Boiling point   49 °C   65 °C

6.57 Some other physical properties of liquids include viscosity, the resistance to flow, which can be thought of as the "thickness of a liquid," and surface tension, which measures the forces at a liquid surface. Consider the following liquids: *honey* (mainly glucose and fructose) versus *water*.
   a. In your experience, which liquid has a higher viscosity?
   b. What intermolecular forces are present? How many hydrogen-bonding interactions are possible in a molecule of glucose and in a molecule of water?
   c. Explain viscosity in terms of intermolecular forces.
   d. It is possible to float a small paper clip on the surface of a cup of water. Try it. What happens if you add soap to the water? Can you float the same paper clip?

ANSWERS TO ODD-NUMBERED PROBLEMS

Practice Problems

6.1 Covalent bonds involve the sharing of electrons within a molecule creating a chemical bond. Intermolecular forces involve attractions between positive and negative areas of different molecules. Actual electrons are not shared between atoms as in a chemical bond.

6.3

Molecule	H Bond	Dipole–Dipole	London
CH_3OH	Yes	Yes	Yes
NH_3	Yes	Yes	Yes
CH_3CH_3	No	No	Yes

6.5 The dipole in London forces is temporary; the dipole in dipole–dipole forces is permanent.

6.7 A nonbonded pair of electrons on an O, N, or F.

6.9 Because ions are not considered molecules (they are derived from dissociation of ionic compounds) technically this force is not between "molecules."

6.11 a. London forces   **b.** ion–dipole attraction
     c. hydrogen bonding   **d.** ion–dipole attraction

6.13

for 6.11b, (ion–dipole shown in red)   for 6.11d, (ion–dipole shown in red)

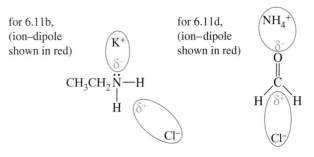

6.15 a. Likely not soluble. The only attractive force present in the alkane shown is the London force. The alkane can not form a permanent dipole to interact with the ionic compound, NaCl.

   b. Likely soluble. Both compounds shown have hydrogen-bond acceptors and CH_3OH has hydrogen-bond donors; since their forces are alike, they are likely soluble.

6.17 a. CH_3CH_2OH **b.** CH_3NH_2

6.19 a. CH_4, London force; H_2O, H-bonding; CH_4 has lower boiling point.

b. NH_3, H-bonding; CO_2, London force; CO_2 has lower boiling point

c.

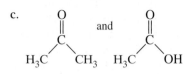

Dipole–dipole; H-bonding

has lower boiling point.

d. $CH_3CH_2CH_2CH_3$, London force; CH_3OCH_3, dipole–dipole; $CH_3CH_2CH_2CH_3$ has lower boiling point.

6.21 The highest boiling point is found in the compound with the strongest intermolecular forces (hydrogen bonding), the CH_3OH. The lowest boiling point is found in the compound with the weakest intermolecular forces (London force), CH_4.

Additional Problems

6.31 Both are attractive forces that occur between positive and negative areas. A dipole–dipole attraction occurs between two opposite partial charges. An ion–dipole attraction occurs between an ionic charge and a partial charge that are opposite.

6.33 H—C—O—H   The molecules can form more than

one hydrogen bond. Methanol can form three hydrogen bonds acting as an acceptor twice (lone pairs on oxygen) and a donor once.

6.35 a. London forces, dipole–dipole
b. London forces
c. London forces, dipole–dipole, hydrogen bonding
d. London forces
e. London forces, dipole–dipole
f. London forces, dipole–dipole, hydrogen bonding

6.37 Fatty acid salts can form ion–dipole interactions with water that are stronger than hydrogen bonding with water formed by fatty acids. This allows the micelles to form.

6.23 The molecule on the right has the highest boiling point since it has the strongest intermolecular forces. It is an alcohol and has hydrogen bonding and dipole–dipole attractions. Of the two alkanes, the one with the more branching has the lower boiling point because of its decreased surface area.

6.25

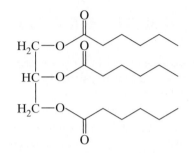

6.27 Even though fats and oils are both triglycerides, the large number of cis double bonds in the hydrocarbons tails of oils prevents them from interacting with each other as much as the saturated tails of the fats. More interactions mean that the tails move more slowly, becoming more like a solid, forming a fat.

6.29 2

6.39 a. fatty acid   **b.** fat
c. CS_2 **d.** $CH_3CH_2CH_2CH_2CH_2CH_3$

6.41 The stain must be hydrophobic because it is not soluble in water.

6.43 Highest to lowest boiling point:
d. 5 possible H-bonds per molecule (strongest)
c. 3 possible H-bonds per molecule
b. polar molecule
a. nonpolar molecule (weakest)

6.45 Because the hydrocarbon chain is longer in octane (meaning the molecule has a greater surface area), the London forces between molecules are stronger which makes the boiling point higher for octane.

6.47 A branched alkane has less surface contact with neighboring molecules than does a straight-chain alkane. The intermolecular forces are therefore stronger between the straight-chain alkane molecules, raising the boiling point.

6.49

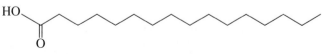

6.51 The phospholipid head group contains many more atoms than the three atoms in the carboxylate head group of a soap.

6.53

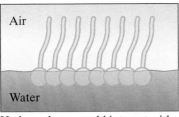

Hydrocarbons would interact with nonpolar air, polar heads with water at surface.

6.55 Sulfur cannot hydrogen bond, so H$_2$S molecules interact strongest through dipole–dipole interactions, which are weaker than hydrogen bonding.

6.57 a. Honey
   b. Hydrogen bonding, dipole–dipole attractions, and London forces. A glucose in ring form could make up to 17 possible hydrogen bonds; water can make 4 possible hydrogen bonds per molecule.
   c. Because there are so many more hydrogen bonds possible between sugar molecules in honey, this makes the honey more viscous than water.
   d. By adding soap to the water, the hydrogen bonding network between the water molecules is disrupted, and the paper clip can no longer float on the surface.

Guided Inquiry Activities **FOR CHAPTER 7**

EXERCISE 1 **Electrolytes**

Information

The warning in the figure appears on some electrical appliances.

nonelectrolyte—A substance that does not conduct electricity because it does not ionize (dissociate forming ions) in aqueous solution. A polar covalent compound like glucose is a nonelectrolyte.

strong electrolyte—A substance that conducts electricity because it completely dissociates into ions in aqueous solution. Ionic compounds that dissolve in water are strong electrolytes.

weak electrolyte—A substance that weakly conducts electricity because it partially ionizes in aqueous solution. A weak acid like a carboxylic acid is a weak electrolyte.

Questions

1. Electrical current is carried by a flow of electrons. This can only happen in an aqueous solution if whole (not partial) charges are present. Based on this information,
   a. does tap water conduct electricity? Explain in terms of charges in solution.
   b. does pure water conduct electricity? Explain in terms of charges in solution.

2. Are the following compounds likely to dissociate, partially dissociate, or not dissociate when dissolved in solution?
   a. $NaNO_3(s)$ b. Sucrose, $C_{12}H_{22}O_{11}(s)$ c. Formic Acid, $HCOOH(l)$

EXERCISE 2 **Osmosis**

Information

Cells and the solutions surrounding them are separated by the cell's membrane. The cell membrane is semipermeable. This means that some molecules are able to pass through while others are not. Water can spontaneously flow through the cell membrane in a direction that will equalize the concentrations inside and outside the cell. Three possibilities exist for a cell and its surrounding solution: The concentrations of both are the same, the exterior solution is of higher concentration than the solution inside the cell, or the interior solution is higher concentration than the solution outside the cell. The solution types are outlined in the following table and figure:

Solution Type	Amount of solute (amount of water) outside the cell	Amount of solute (amount of water) inside the cell	Movement of Water	Effect on the Cells
Isotonic	The same as inside	The same as outside	No net movement	No effect
Hypotonic	Lower than inside (higher than inside)	Higher than outside (lower than outside)	Water moves into the cells	Cells swell and can burst (lyse)
Hypertonic	Higher than inside (lower than inside)	Lower than outside (higher than outside)	Water moves out of the cells	Cells shrivel (crenate)

| Isotonic solution | Hypotonic solution | Hypertonic solution |

Red blood cells are isotonic with a NaCl solution that is 0.90% (m/v), also known as normal saline and 5% glucose, also known as D5W (dextrose 5% in water). Intravenous fluids must be isotonic to avoid cell lysis (by bursting) or the opposite effect, shriveling, also called crenation.

Demonstration—Common Osmosis

You will need two carrots and two lemon slices. Dissolve as much table salt as possible into some tap water to make a saturated salt solution. Place one carrot and one lemon slice into separate glasses of the salt solution and one carrot and one lemon into separate glasses of tap water. Allow them to sit overnight before answering the following questions.

Questions

1. Describe the appearance of the foods in (a) the tap water and (b) the salt water.
2. How did the foods change when they were placed in (a) the tap water and (b) the salt water?
3. In which glass did the water in the solution move into the food cells?
4. In which glass did the solution remove water from the food cells?
5. Which of the solutions (tap water or salt water) is (a) hypotonic, (b) hypertonic to the vegetables?
6. A cell is placed in a 10% (m/v) glucose solution.
   a. Which direction will water flow to equalize the concentrations: into the cell or out of the cell?
   b. Will the cell swell or crenate?
7. If a person pours a concentrated salt water solution on an earthworm, the worm will wriggle and die. What type of solution (isotonic, hypertonic, or hypotonic) was the salt solution? What are the earthworm's cells doing?
8. If a person drinks too much water too quickly, a condition known as hyponatremia will occur. In this condition, there is not enough sodium in the body fluids outside of the cell. What type of solution (isotonic, hypertonic, or hypotonic) would a person's cells be in? What are the cells likely to do?

Dissolving a teaspoon of sugar in a glass of iced tea may seem simple enough, but why do some substances dissolve in water while others don't? How do the properties of solutions affect the movement of molecules into a cell? Read Chapter 7 to find out.

Solution Chemistry
HOW SWEET IS YOUR TEA?

If you drink iced tea, you might sweeten it with sugar or a sugar substitute. Even before you add sweetener, a glass of iced tea has many different substances in it—water, caffeine, and other chemicals extracted from tea leaves—so iced tea represents a mixture. Once the sweetener has completely disappeared, all of the substances in a glass of iced tea are dissolved in water to make a refreshing drink. The liquid in the iced tea is now a homogeneous mixture called a solution. (See Section 1.3 for a review of homogeneous and heterogeneous mixtures.) If the tea is not sweet enough for you, you can change the taste by dissolving more sugar in it, or we could say you are increasing the sugar concentration in your tea. If you accidentally make your tea too sweet, you can always add more unsweetened tea and in so doing decrease the concentration of the sugar in the tea solution.

Understanding more about solutions and concentrations of solutions is important for many reasons. Body fluids, for example, are complex watery solutions containing dissolved substances like glucose and ions like K^+, Na^+, Cl^-, and HCO_3^-. Our bodies have some pretty sophisticated ways of keeping the concentration of dissolved molecules and ions constant.

7.1 Solutions Are Mixtures

A glass of iced tea represents a type of homogeneous mixture called a **solution**. A solution consists of at least one substance—the **solute**—evenly dispersed throughout a second substance—the **solvent**. The components in a solution do not react with each other, and so the sugar mixed into the tea is still sugar and still tastes sweet. The solute is the substance in a solution present in the smaller amount, and the solvent is the substance present in the larger amount. In the case of iced tea, the sugar is one of the solutes, and the water is the solvent. A glass of iced tea, although brownish in color, is transparent, and once the sugar is dissolved into the water, it will not "un"dissolve over time. These properties of solutions provide a quick way for you to determine whether a substance is a solution. See Table 7.1 for some examples of other properties of solutions.

TABLE 7.1	SOME PROPERTIES OF SOLUTIONS
Particles are evenly distributed.	
Components do not chemically react with each other.	
Aqueous solutions are transparent.	
Components do not separate upon standing.	
Concentration can be changed.	

Solute: The substance present in lesser amount

Salt

Water

Solvent: The substance present in greater amount

a.

b.

c.

▲ **FIGURE 7.1 Solutions can be solid, liquid, or gas.** (a) Brass is a solution of copper and zinc solids, (b) air is a solution of gases, and (c) carbonated beverages represent a solution of a gas in a liquid.

colloid

Phases of Solutes and Solvents

We tend to think of solutions as being solids, like sugar, dissolved in liquids like water; however, this does not have to be the case. Solutions can be homogeneous mixtures of gases. The air we breathe, for example, is a homogeneous mixture of gases, so it is also considered a solution in which nitrogen is the solvent and oxygen and other gases are the solutes. Brass metal is a solution of solids in solids because it is an alloy of the solute metal zinc in the solvent metal copper. The solute and solvent can either be a solid, liquid, or gas—the three phases of matter—in a solution (see Figure 7.1).

Our study of solutions will focus on dilute solutions where water is the solvent. These solutions are referred to as **aqueous solutions**.

SAMPLE PROBLEM 7.1

Identifying Solute and Solvent

Identify the solute and solvent in solutions composed of the following:
a. 20.0 mL of ethanol and 250 mL of water
b. a pinch of salt in a pot of boiling water
c. a 5-gallon tank filled with methane (natural gas) and a small amount of methyl mercaptan that gives a noticeable odor if any gas should leak

SOLUTION

Remember the solute is the substance in the smaller amount and the solvent is the substance in the larger amount.
a. Water is the solvent, and ethanol is the solute.
b. Water is the solvent, and salt is the solute.
c. Methane is the solvent, and methyl mercaptan is the solute.

Colloids and Suspensions

Is homogenized milk a solution? If we look back at the properties of solutions in Table 7.1, we would have to say that it is not, because milk is not a transparent liquid. Why is milk not a transparent liquid and therefore not a solution? Homogenized milk is uniform throughout and is therefore a homogeneous mixture, but it contains proteins and fat molecules that do not dissolve in water, the solvent for milk. Instead these molecules aggregate to form larger particles in solution much as the soap molecules, discussed in Chapter 6, formed micelles. The nonpolar areas collect in the center and the polar areas of the milk proteins face the exterior and interact with the solvent. Homogenized milk is a **colloid** (or colloidal mixture) because of the proteins and fat molecules that do not dissolve. By definition, the particles in a colloid must be between 1 nanometer (1×10^{-9} meter) and 1000 nanometers in diameter. Particles of this size remain suspended in solution, and so a colloid does not separate over time.

What kind of mixture is muddy water? The dirt in this mixture will separate from the water upon standing. If the diameter of the particles in a mixture is greater than 1000 nanometers (1 micrometer), the mixture is considered a **suspension**. Blood is another example of a suspension. Blood contains blood cells that are larger than 1 micrometer, and these cells will settle to the bottom of a test tube upon standing. Blood cells are routinely separated from plasma through centrifugation, a spinning process that accelerates settling (see Figure 7.2).

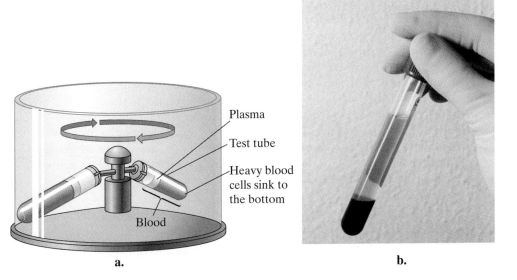

Blood after centrifugation

Plasma

Test tube

Heavy blood cells sink to the bottom

Blood

a.

b.

▲ **FIGURE 7.2 Blood is a suspension.** The cells in blood can be isolated using centrifugation (a) into the soluble components (plasma) and cells (b).

SAMPLE PROBLEM 7.2

Type of Mixture

Are the following solutions or colloids?

a. apple juice **b.** mayonnaise **c.** fog

SOLUTION

a. Apple juice is transparent and uniform throughout; therefore it is a solution.

b. Mayonnaise is opaque and uniform throughout; therefore, it is a colloid, not a solution.

c. Fog is translucent—that is, partially transparent in that it lets light pass through—and uniform throughout and contains small water droplets. It does not settle and, therefore, it is a colloid, not a solution.

PRACTICE PROBLEMS

7.1 Identify the solute and solvent in solutions composed of the following:

a. 80% nitrogen and about 20% oxygen in the air

b. 100.0 g of copper and 20.0 g of zinc

c. 500.0 mL of ethanol and a few drops of blue food coloring

7.2 Identify the solute and solvent in solutions composed of the following:

a. 0.80 L of N_2 and 0.01 L of CO_2

b. a spoonful of sugar and a pint of ice water

c. 3 tablespoons of vinegar (acetic acid) and 1 cup of water used in a salad dressing

7.3 Identify the following as solutions or colloids:

a. yogurt **b.** cola drink **c.** gasoline

7.4 Identify the following as solutions or colloids:

a. Kool-Aid® **b.** salt water **c.** cheddar cheese

7.2 Formation of Solutions

In Chapter 6, the golden rule of solubility, *like dissolves like*, was introduced to explain why some compounds dissolve in water and others do not. Polar covalent compounds, like glucose, and ionic compounds, like NaCl, will dissolve in water because strong attractive forces exist between them. Nonpolar covalent compounds, like hydrocarbons and oils, are not soluble in the polar molecules of water. Amphipathic molecules, like fatty acids that contain polar and nonpolar parts, can interact with water through their polar part, but these molecules typically do not form solutions. The dissolving process requires that individual solvent particles surround the solute molecules and interact through intermolecular forces or ion–dipole attractions. Because the particles are just redistributing themselves, dissolving is a physical change, not a chemical change. This process is referred to as **solvation** for all solutions, and more specifically as **hydration** in the case of an aqueous solution. For a review of this topic, see Section 6.2.

Factors Affecting Solubility and Saturated Solutions

How much sugar can you dissolve in your glass of iced tea before the sugar stops dissolving and only settles to the bottom of your glass? The maximum amount of a solute that can dissolve in a specified amount of solvent at a given temperature defines the **solubility**. You may have observed that it is easier to dissolve sugar in hot tea than in cold tea. Solubility depends on a number of factors including the polarity of the solute and the solvent (the attractive forces present), as well as the temperature of the solvent.

If a solution does *not* contain the maximum amount of the solute that the solvent can hold, it is referred to as an **unsaturated solution**. If a solution contains all the solute that can possibly dissolve, the solution is referred to as a **saturated solution**. If more solute is added to a saturated solution, the additional solute would remain undissolved. A solution that is saturated reaches an **equilibrium** state between the dissolved solute and undissolved solute. At equilibrium, the speed or rate of dissolving solute and the

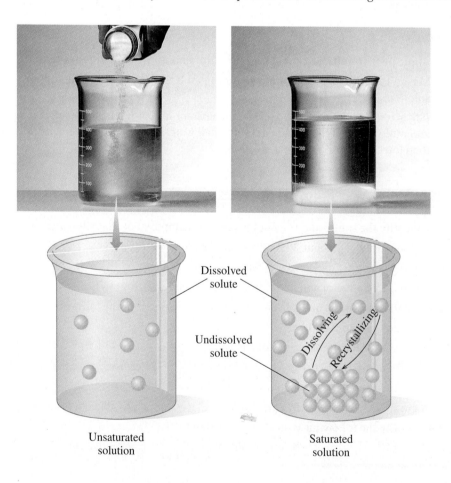

Dissolved solute

Undissolved solute

Dissolving

Recrystallizing

Unsaturated solution

Saturated solution

rate of dissolved solute reforming crystals is the same. This can be represented as a chemical equation where a double arrow or *equilibrium arrow* is used between the products and reactants in the chemical equation.

$$\text{Undissolved solute} \underset{\text{crystallization}}{\overset{\text{solvation}}{\rightleftharpoons}} \text{dissolved solute}$$

The conditions of gout and kidney stones occur when compounds build up in the body and exceed their solubility limits. Undissolved solids build up in bodily fluids. In the case of gout, the solid compound is uric acid. Insoluble needlelike crystals form usually in soft tissues or in cartilage and tendons in the lower extremities. Kidney stones typically contain the ionic compounds calcium phosphate or calcium oxalate and can form in the urinary tract, kidneys, ureter, or bladder when these calcium compounds do not remain dissolved in the urine. These conditions related to solubility can occur in individuals who are not forming urine normally. Both gout and kidney stones can be treated through changes in diet and drug therapy (see Figure 7.3).

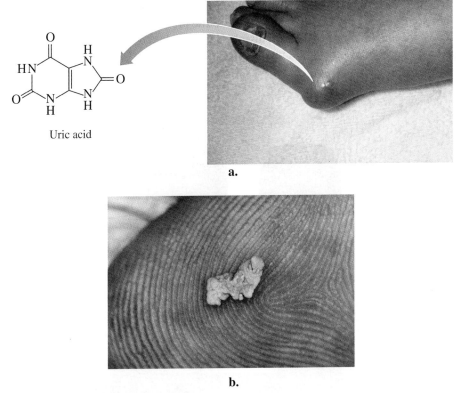

Uric acid

a.

b.

▲ **FIGURE 7.3 Exceeding solubility limits.** (a) When uric acid levels in the blood exceed the solubility limit, the excess forms the needlelike crystals responsible for gout. (b) Kidney stones, like the one shown here, are typically composed of the ionic compounds calcium phosphate or calcium oxalate. Even small crystals are painful when passing through the urethra.

Solubility and Temperature

It is easier to dissolve sugar in hot tea than in cold tea. In fact, more sugar will actually dissolve in hot tea than in cold tea. This demonstrates the effect of temperature on solubility. The solubility of most solids dissolved in water *increases* with temperature. This is an important property of solutions because solubility can be manipulated by changing the temperature of a solution.

Now, think about what might happen to a bottle of club soda left in the trunk of a car on a hot summer day. You may return to your car after several hours to find that the container exploded and the trunk of your car is soaking wet! Sodas are aqueous solutions containing dissolved CO_2 gas as one of the solutes. The reason the container burst

PRACTICE PROBLEMS

7.5 Do the following represent a saturated solution?

a. throwing a pinch of salt into a pot of boiling water

b. a layer of sugar forms at the bottom of a glass of tea as ice is added

7.6 Do the following represent a saturated solution?

a. a piece of rock salt does not change in size when added to a salt solution

b. a teaspoon of honey dissolves in hot tea

7.7 Fill in the blank in the statements below with the words "increase," "decrease," or "stay the same."

a. If the temperature of a solution increases, the solubility of most solid solutes will _____.

b. If the temperature of a solution increases, the solubility of a gaseous solute will _____.

c. If the pressure above a solution increases, the solubility of a gaseous solute will _____.

7.8 Fill in the blank in the statements below with the words "increase," "decrease," or "stay the same."

a. If the temperature of a solution decreases, the solubility of a gaseous solute will _____.

b. If the temperature of a solution decreases, the solubility of a solid solute will _____.

c. If the pressure above a solution decreases, the solubility of a gaseous solute will _____.

7.9 Explain what is happening in the following situations in terms of solubility:

a. An opened bottle of soda goes flat more slowly when recapped and stored in the refrigerator.

b. A bottle of club soda keeps its fizz longer if the lid is screwed on tightly.

c. Less sugar dissolves in cold iced tea than in hot tea.

7.10 Explain what is happening in the following situations in terms of solubility:

a. Fish die in a lake that gets heated with water from a power plant cooling tower.

b. Crystals form in a jar of honey that has been sitting for several months.

c. A person receives oxygen through a breathing device.

7.3 Chemical Equations for Solution Formation

A warning symbol on hair dryers and other small electrical appliances used in the bathroom warns us that we should not use them near the bathtub. Why? Because water conducts electricity, and if a plugged-in electrical appliance falls into a tub, a person in the tub could be electrocuted. Pure water does not conduct electricity, but tap water does. What is making the difference? Tap water contains dissolved ions, and those dissolved ions can conduct electricity. Solutes that produce ions in solution are called **electrolytes**.

We can place all substances that dissolve in water (hydrate) into one of three categories based on their ability to conduct electricity. Any aqueous solution can be tested to see if it will conduct electricity with a simple circuit consisting of a battery, wires to two electrodes, and a lightbulb (see Figure 7.5). If a solution can conduct electricity, the lightbulb glows when the electrodes are immersed in the solution. An ionic compound like NaCl completely dissociates into its charged ions when dissolved in water. The large number of charged ions present makes the lightbulb in the simple circuit burn brightly.

Ionic compounds dissolve in water are considered **strong electrolytes**. Covalent compounds like sugar do not **ionize** (dissociate to form ions) when they dissolve in solution; they exist as molecules, and no ions are formed upon dissolving. Soluble covalent compounds do not conduct electricity and are therefore termed **nonelectrolytes**. There are some covalent compounds that *can* partially ionize in water. These compounds contain a highly polar bond that can dissociate forming some ions in water. These covalent compounds weakly conduct electricity and are termed **weak electrolytes**. We can write balanced chemical equations to describe the hydration of each of these substances.

In Chapter 1, we introduced the basic format for a chemical equation: The reactants (represented by their chemical formula) always appear on the left side, and the products (also represented by their chemical formula) always appear on the right side. These two are separated by an arrow meaning *react to form* or *yield*.

$$\text{Reactants} \xrightarrow[\text{(react to form)}]{} \text{products}$$

Strong Electrolytes

Let's look at the balanced chemical equation for the hydration of the ionic compound magnesium chloride ($MgCl_2$), a strong electrolyte. The following equation can be verbally stated as "solid magnesium chloride is dissolved in water to yield magnesium ions and chloride ions."

$$MgCl_2(s) \xrightarrow{H_2O} Mg^{2+}(aq) + 2\,Cl^-(aq)$$

First, notice that the number of magnesium and chloride ions formed as products is the same as the number of each found in the reactant, regardless of whether the ions are together in solid form or dissolved in solution. This was first observed and stated by the French chemist Antoine Lavoisier (1743–1794) as the **law of conservation of mass**: Matter can neither be created nor destroyed. Matter merely changes forms. In other words, atoms cannot just appear or disappear. After writing a chemical equation, it is important to inspect it to make sure the same number of each element appears on each side of the equation; that is, the chemical equation must be *balanced*. Keep in mind that subscripts appear in formulas for compounds (like $MgCl_2$), and the full-size numbers in front of particles (atoms, ions, or compounds) indicate the total number of that particle present (like **2** Cl^-). These full-size numbers in front of particles are termed **coefficients**. In the preceding sample equation, the coefficient 2 in front of the chloride ion indicates that there are two chloride ions produced for every one $MgCl_2$ that dissociates. In addition, balancing requires that the charge be the same on both sides of the chemical equation. In this example, $MgCl_2$ has no net charge (reactant) and one Mg^{2+} and two Cl^- sum to a total charge of zero (products), and so the charges are also balanced.

$$MgCl_2(s) \xrightarrow{H_2O} Mg^{2+}(aq) + 2\,Cl^-(aq)$$

Charge in   Charge in products
reactants = 0   2 + 2(−1) = 0

Second, note the reaction arrow points in one direction, thereby implying that the process occurs in only one direction, that is, it is irreversible.

Third, note the phases of matter. For ionic compounds, the reactants will usually be a solid that dissolves. In the products, the phases will always be aqueous, because the substance is hydrated, that is, it is dissolved in water.

Finally, notice the water (H_2O) appearing below the reaction arrow. Substances that are not involved in the balanced equation, like the solvent in this case, are often placed at the arrow to give the reader more information regarding the conditions of the reaction. For the hydration equations in this chapter, water will appear here.

Nonelectrolytes

Nonelectrolytes are polar compounds that dissolve in water but do not ionize in water. As an example we can examine the following chemical equation for the carbohydrate glucose ($C_6H_{12}O_6$) dissolving in water:

$$C_6H_{12}O_6(s) \xrightarrow{H_2O} C_6H_{12}O_6(aq)$$

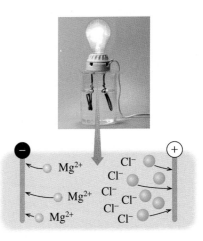

Strong electrolyte

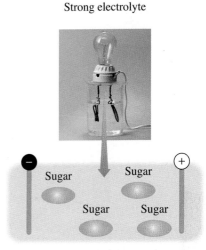

Nonelectrolyte

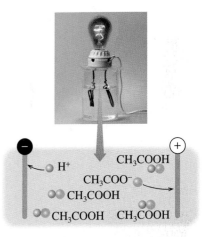

Weak electrolyte

▲ **FIGURE 7.5 Simple experiments for determining whether a solution contains strong, weak, or nonelectrolytes.** A pair of electrodes connected to a lightbulb and inserted in the solution will either conduct electricity strongly, weakly, or not at all depending on the number of ions present in the solution.

Covalent compounds like glucose do not dissociate. The only difference between the reactant and the product is the phase. Because glucose dissolves in water, we indicate this on the products side of the chemical equation by changing the phase of the glucose molecules to aqueous.

Weak Electrolytes

Which covalent compounds act as weak electrolytes? One of the organic functional groups that partially ionizes in solution is the carboxylic acid/carboxylate group.

$$R-\overset{\overset{\displaystyle O}{\|}}{C}-OH \quad / \quad R-\overset{\overset{\displaystyle O}{\|}}{C}-O^-$$

In Chapter 6 we saw that this functional group has two forms, the carboxylate form being ionic. A carboxylic acid group contains a very polar O—H bond that can dissociate to form carboxylate ions and H^+ ions when dissolved. We will use carboxylic acids as examples of weak electrolytes in this chapter. Consider the partial ionization of acetic acid, the compound in vinegar:

$$CH_3-\overset{\overset{\displaystyle O}{\|}}{C}-OH(l) \underset{H_2O}{\rightleftharpoons} CH_3-\overset{\overset{\displaystyle O}{\|}}{C}-O^-(aq) + H^+(aq)$$

As the number of H^+ and CH_3COO^- ions builds up in the solution, some of them will recombine to reform CH_3COOH. Eventually, the rates of the forward and reverse reactions equalize, and the amount of the acetic acid and the ions no longer changes—thus, an equilibrium exists. An equilibrium arrow, $\rightleftharpoons$ is used in this chemical equation to indicate this. Upon inspecting the equation, we see that the number of atoms and the total charge on each side is balanced, so the equation is balanced. The phase of the weak electrolyte before hydration may be solid, liquid, or gas.

Table 7.2 summarizes the steps for writing and balancing hydration equations for both ionic and covalent compounds.

TABLE 7.2	BASICS FOR WRITING AND BALANCING A CHEMICAL EQUATION FOR A HYDRATION REACTION

STEP 1. Determine whether the substance is a strong electrolyte, nonelectrolyte, or weak electrolyte.

Ionic compounds	Covalent compounds with no dissociating hydrogens, like sugars and alcohols.	Covalent compounds with highly polarized hydrogens, like a weak acid (for example, carboxylic acid).
⬇	⬇	⬇
Strong electrolyte	Nonelectrolyte	Weak electrolyte

STEP 2. Write the reactants and products separated by an arrow.

Strong electrolyte	Nonelectrolyte	Weak electrolyte
→	→	⇌

STEP 3. All products form an aqueous phase (*aq*).

STEP 4. For strong and weak electrolytes, make sure that the same number of each element or ion appears on both sides. Place coefficients *in front* of elements and compounds to balance the numbers.

STEP 5. Verify that the equation is balanced by confirming that charges on each side are equal.

SAMPLE PROBLEM 7.5

Strong, Weak, and Nonelectrolyte Solutions

Predict whether the following will fully dissociate, partially dissociate, or not dissociate when dissolved in solution:

a. $NaNO_3(s)$

b. Sucrose, $C_{12}H_{22}O_{12}(s)$

c. Formic acid, $HCOOH(l)$

SOLUTION

a. The ionic compound $NaNO_3$ will fully dissociate forming Na^+ and NO_3^- ions when dissolved in solution.

b. The covalent compound $C_{12}H_{22}O_{12}$ will not dissociate when dissolved in solution.

c. The carboxylic acid HCOOH will *partially* dissociate, forming the ions H^+, and $HCOO^-$, as well as HCOOH molecules.

SAMPLE PROBLEM 7.6

Balancing Hydration Equations

Write a balanced equation for the hydration of the following:

a. $KCl(s)$, a strong electrolyte

b. $CH_3OH(l)$, a nonelectrolyte

c. $HF(l)$, a weak electrolyte

SOLUTION

a. KCl is an ionic compound composed of K^+ and Cl^- ions. It will fully dissociate into its component ions when placed in water to produce the ions K^+ and Cl^-. There is one K^+ for every one Cl^- so the balanced equation is

$$KCl(s) \xrightarrow[H_2O]{} K^+(aq) + Cl^-(aq)$$

b. A nonelectrolyte will dissolve but not ionize, merely changing phase. The balanced equation is

$$CH_3OH(l) \xrightarrow[H_2O]{} CH_3OH(aq)$$

c. Weak electrolytes partially ionize. Some H becomes H^+ and some F becomes F^-. Some of the original HF compound is also present. An equilibrium is established.

$$HF(l) \underset{H_2O}{\rightleftharpoons} H^+(aq) + F^-(aq)$$

TABLE 7.3

THE RELATIONSHIP BETWEEN MOLES AND EQUIVALENTS EXPRESSED AS CONVERSION FACTORS

Ion	Conversion Factor
Na^+	$\dfrac{1 \text{ Eq } Na^+}{1 \text{ mole } Na^+}$
Ca^{2+}	$\dfrac{2 \text{ Eq } Ca^{2+}}{1 \text{ mole } Ca^{2+}}$
Cl^-	$\dfrac{1 \text{ Eq } Cl^-}{1 \text{ mole } Cl^-}$
SO_4^{2-}	$\dfrac{2 \text{ Eq } SO_4^{2-}}{1 \text{ mole } SO_4^{2-}}$

Ionic Solutions and Equivalents

Blood and other bodily fluids contain many electrolytes as dissolved ions like Na^+, K^+, Cl^-, and HCO_3^-. The amount of a dissolved ion found in fluids can be expressed by the unit **equivalent (Eq)**. An equivalent relates the charge in a solution to the number of ions or the moles of ions present (see Table 7.3). (Recall from Chapter 2 that moles can be related to the measurable quantity grams through the atomic mass.) For example, 1 mole of Na^+ has one equivalent of charge because the charge on a sodium ion is plus 1. One mole of Ca^{2+} has two equivalents of charge because one mole of calcium contains two charges (or equivalents) per mole. *The number of equivalents present per mole of an ion equals the charge on that ion.*

SAMPLE PROBLEM 7.7

Electrolytes in Solution

How many equivalents of Cl^- are present in a solution that contains 2.50 moles of Cl^-?

SOLUTION

Use the conversion factor that relates the number of moles of Cl^- to the number of equivalents.

$$2.50 \text{ moles } Cl^- \times \frac{1 \text{ Eq } Cl^-}{1 \text{ mole } Cl^-} = 2.50 \text{ Eq } Cl^-$$

SAMPLE PROBLEM 7.8

How many equivalents of Ca^{2+} are present in a solution that contains 2.50 moles of Ca^{2+}?

SOLUTION

Use the conversion factor that relates the number of moles of Ca^{2+} to the number of equivalents.

$$2.50 \text{ moles } Ca^{2+} \times \frac{2 \text{ Eq } Ca^{2+}}{1 \text{ mole } Ca^{2+}} = 5.00 \text{ Eq } Ca^{2+}$$

PRACTICE PROBLEMS

7.11 Predict whether the following will fully dissociate, partially dissociate, or not dissociate when dissolved in solution:

a. potassium fluoride, $KF(s)$, a strong electrolyte

b. hydrogen cyanide, $HCN(l)$, a weak electrolyte

c. fructose, $C_6H_{12}O_6(s)$, a nonelectrolyte

7.12 Predict whether the following will fully dissociate, partially dissociate, or not dissociate when dissolved in solution:

a. sodium sulfate, $Na_2SO_4(s)$, a strong electrolyte

b. ethyl alcohol, $CH_3CH_2OH(l)$, a nonelectrolyte

c. boric acid, $H_3BO_3(s)$, a weak electrolyte

7.13 Provide a balanced equation for the hydration of each compound in problem 7.11.

7.14 Provide a balanced equation for the hydration of each compound in problem 7.12.

7.15 For the following ionic compounds, write a balanced equation for their hydration in water:

a. $CaCl_2$ **b.** NaOH   **c.** KBr   **d.** $Fe(NO_3)_3$

7.16 For the following ionic compounds, write a balanced equation for their hydration in water:

a. $FeCl_3$ **b.** $NaHCO_3$ **c.** K_2SO_4 **d.** $Ca(NO_3)_2$

7.17 How many equivalents of K^+ are present in a solution that contains 4.25 moles of K^+?

7.18 How many equivalents of Mg^{2+} are present in a solution that contains 2.50 moles of Mg^{2+}?

7.4 Concentrations

Whether you drink a can of soda or a glass of the same soda from a 2-liter bottle, the soda tastes the same. Beverage manufacturers go to great lengths to maintain a consistent product; their beverage solutions must have exactly the same amount of ingredients (see Figure 7.6).

The amount of each ingredient dissolved in the soda determines the beverage's **concentration**. The concentration of a solution can be expressed in many different

◀ **FIGURE 7.6 Concentration and volume.** The concentration of the sugar and the taste of the beverage are the same in all containers, even though the volume is different.

TABLE 7.4

NORMAL CONCENTRATION RANGES FOR FIVE COMMON SUBSTANCES TESTED IN A BLOOD TEST**

Substance	Normal Range
Na^+	136–146 mEq/L
Cl^-	97–110 mEq/L
Glucose	65–110 mg/dL
Cholesterol	120–240 mg/dL
Hemoglobin (a protein)	12–16 g/dL

units, but the units we will encounter in this chapter relate the amount of each solute dissolved to the total amount of solution present.

$$\text{Concentration} = \frac{\text{amount of solute}}{\text{amount of solution}}$$

Note that the denominator is the *total* amount of solution (includes the solute), not just the amount of solvent.

Table 7.4 shows the normal concentration ranges for some substances typically tested in a blood profile. Notice that the units vary depending on the type of solute. Why all the different units? Different units are used so that all the numbers fall in ranges that are readable without using numbers after a decimal. For example, if the units for cholesterol were g/dL instead of mg/dL, the numbers in the normal range would be 0.120–0.240 g/dL, which is not as easy to read and discuss with a patient as a number ranging in the hundreds.

Milliequivalents per Liter (mEq/L)

The amount of electrolytes present in bodily fluids and typical intravenous fluid replacements are often represented by the concentration unit of milliequivalents per liter of solution (mEq/L).

$$1000 \text{ mEq} = 1 \text{ Eq}$$

Ionic solutions have a balance in the number of positive and negative charges present because they are formed by dissolving ionic compounds that have no net charge. Typical blood plasma has a total electrolyte concentration of 150 mEq/L, meaning that the total concentration of both positive and negative ions is 150 mEq/L (see Table 7.5). Examples of the concentrations of other ionic solutions used as intravenous fluids are shown in Table 7.6. These also show a balance of total charge.

TABLE 7.5

SOME TYPICAL CONCENTRATIONS OF ELECTROLYTES IN BLOOD PLASMA

Electrolyte	Concentration (mEq/L)
Cations	
Na^+	138
K^+	5
Mg^{2+}	3
Ca^{2+}	4
Total	150
Anions	
Cl^-	110
HCO_3^-	30
HPO_4^{2-}	4
Proteins	6
Total	150

TABLE 7.6 ELECTROLYTE CONCENTRATIONS IN INTRAVENOUS REPLACEMENT SOLUTIONS

Solution	Electrolytes (mEq/L)	Use
Sodium chloride (0.90%)	Na^+ 154, Cl^- 154	Replacement of fluid loss
Potassium chloride with 5.0% dextrose	K^+ 40, Cl^- 40	Treatment of malnutrition (low potassium levels)
Ringer's solution	Na^+ 147, K^+ 4, Ca^{2+} 4, Cl^- 155	Replacement of fluids and electrolytes lost through dehydration
Maintenance solution with 5.0% dextrose	Na^+ 40, K^+ 35, Cl^- 40, lactate$^-$ 20, HPO_4^{2-} 15	Maintenance of fluid and electrolyte levels
Replacement solution (extracellular)	Na^+ 140, K^+ 10, Ca^{2+} 5, Mg^{2+} 3, Cl^- 103, acetate$^-$ 47, citrate^{3-} 8	Replacement of electrolytes in extracellular fluids

SAMPLE PROBLEM 7.9

mEq/L

A Ringer's solution for intravenous fluid replacement has a concentration of 155 mEq Cl^- per liter of solution. If a patient receives 1250 mL of Ringer's solution, how many equivalents of Cl^- were given?

SOLUTION

This problem asks for the number of equivalents of Cl^- present, which can be determined from the number of mEq given through the conversion factor $\dfrac{1\ Eq}{1000\ mEq}$.

Using conversion factors, the information given (concentration and volume), and canceling appropriate units, the problem is solved as

$$\frac{155\ \cancel{mEq\ Cl^-}}{1\ \cancel{L\ solution}} \times \frac{1\ Eq}{1000\ \cancel{mEq}} \times \frac{1\ \cancel{L}}{1000\ \cancel{mL}} \times 1250\ \cancel{mL\ solution} = 0.194\ Eq\ Cl^-$$

Millimoles per Liter (mmol/L) and Molarity (M)

Sometimes the units for electrolytes like Na^+ are given in mmoles/L instead of mEq/L. The charge on an ion is the number of equivalents present in 1 mole (Section 7.3). So for an ion with a 1+ charge like sodium, the number of equivalents is equal to the number of moles, and the units mEq/L and mmole/L are the same. We can show this mathematically by converting mEq/L to mmole/L using the following conversion factor:

$$\frac{1\ mmole\ Na^+}{1\ mEq\ Na^+}$$

Using this conversion factor, we can determine the concentration of a 135 mEq/L Na^+ solution in mmole/L:

$$\frac{135\ \cancel{mEq\ Na^+}}{1\ L} \times \frac{1\ mmole\ Na^+}{1\ \cancel{mEq\ Na^+}} = \frac{135\ mmole\ Na^+}{1\ L}$$

Chemists use a unit related to mmole/L to describe the concentrations of solutions prepared in the laboratory. This unit is called **molarity, M**, and it is defined as

$$M = \frac{mole\ solute}{L\ solution}$$

Molarity (M) = moles of solute/liters of solution

The *mole* is directly related to the number of *molecules* present (through Avogadro's number). A 5.0 M (read "five molar") solution of ethanol (CH_3CH_2OH) has the same number of molecules present as a 5.0 M glucose ($C_6H_{12}O_6$) solution, even though the *mass* of ethanol and *mass* of glucose used to make the two solutions are different. The unit molarity is useful for chemists because they want to know how many particles of a solute are available to react in a chemical reaction.

$$M = \frac{moles\ of\ solute}{liter\ of\ solution} = \frac{5.0\ moles\ of\ glucose}{1\ L\ of\ solution} = 5.0\ M\ glucose$$

SAMPLE PROBLEM 7.10

Molarity

What is the molarity of a solution prepared by dissolving 0.50 moles of NaCl in water to give a total volume of 250 mL?

SOLUTION

To solve this problem, remember the units for molarity are mole/L. Here the solution is given in mL, so this must be converted to L. The moles of solute (in this case NaCl) must be divided by the volume of the solution:

$$\frac{0.50\ moles\ NaCl}{250\ \cancel{mL}\ solution} \times \frac{1000\ \cancel{mL}}{1\ L} = \frac{2.0\ moles}{1\ L} = 2.0\ M$$

SAMPLE PROBLEM 7.11

Molarity

a. How many moles of KNO_3 are present in 125 mL of a 2.0-M KNO_3 solution?

b. How many grams of KNO_3 are present?

SOLUTION

a. In this problem, the concentration is known, but the actual number of moles is not. The units on the final answer are moles. By definition, 2.0 M is 2.0 moles/L. Set up an equation to solve for moles with the given information:

$$\frac{2.0 \text{ moles } KNO_3}{1 \text{ L}} \times \frac{1 \text{ L}}{1000 \text{ mL}} \times 125 \text{ mL} = 0.25 \text{ mole } KNO_3$$

b. The number of grams can be determined from the number of moles found in part a by multiplying by the molar mass of KNO_3. The molar mass is the sum of a compound's atomic masses.

$$\text{Molar mass } KNO_3 = 39.10 + 14.01 + 3(16.00) = 101.11 \text{ g/mole}$$

$$0.25 \text{ mole } KNO_3 \times \frac{101.11 \text{ g } KNO_3}{1 \text{ mole } KNO_3} = 25 \text{ g } KNO_3$$

Percent (%) Concentration

100 cents (1 dollar)

10 cents

10%

A dime is 10% of a dollar.

There are three common concentration units that use percent: mass/volume percent, mass/mass percent, and volume/volume percent. The concentration unit percent is no different in definition from the everyday definition of percent, which indicates the number of *parts per one hundred parts*.

$$\% \text{ Concentration} = \frac{\text{parts of solute}}{100 \text{ parts of solution}}$$

As an example, a dime (10 pennies) is 10% of a dollar (100 pennies) or 10 parts out of a total 100 parts.

Percent (%)

If your chemistry class has 40 students enrolled and 20 students got an A on the first test, you might comment that one-half, or 50 percent, of the class got an A. Percent, represented by the symbol %, means part out of 100 total, or hundredths. So, 50% means 50 As out of a total 100 grades or 50 hundredths (0.50) which reduces to 1/2.

A fraction can be converted to a percent by dividing the numerator by the denominator, multiplying by 100, and adding a percent sign. A decimal number can be converted to a percent by multiplying by 100 and adding a percent sign.

Suppose a couple is trying to lose weight. The husband is told by his doctor to lose 40 pounds. The wife is told by her doctor to reduce her weight by 12%. The

MATH MATTERS

husband weighs 315 pounds and the wife weighs 285 pounds. Which one actually has to lose more weight to meet their goal? To determine the percent of the husband's total weight that he has to lose, a fraction is created to represent parts out of the whole. This will give a decimal answer that can then be converted to a percentage by multiplying by 100. In this example, the percent of weight the husband must lose is

$$\frac{\text{Weight to lose}}{\text{Total weight}} = \frac{40 \text{ pounds}}{315 \text{ pounds}} = 0.13$$
$$0.13 \times 100 = 13\%$$

To determine the number of pounds the wife must lose from 12%, you would first change the percent to a decimal and then multiply by her total weight. In this example the wife is told to lose 12% of her total weight. Twelve percent of 285 is

$$12\% = 0.12$$
$$0.12 \times 285 \text{ pounds total weight} = 34 \text{ pounds to lose}$$

Therefore, the husband has to lose more weight (40 pounds) than the wife does (34 pounds).

SAMPLE PROBLEM 7.12

Expressing Fractions and Decimals as Percent

a. Express 1/5 as a percent.   b. Express 0.35 as a percent.

SOLUTION

a. For a fraction, divide the numbers of the fraction, multiply by 100, and add a percent sign:

$$\frac{1}{5} = 0.20 \times 100 = 20\%$$

b. For a decimal, multiply by 100 and add the percent sign:
$$0.35 \times 100 = 35\%$$

SAMPLE PROBLEM 7.13

Expressing Percents as Fractions and Decimals

a. Express 25% as a decimal.   b. Express 25% as a fraction.

SOLUTION

a. To express 25% as a decimal, divide the value by 100:
$$\frac{25}{100} = 0.25$$

b. Percent means parts out of a total 100 parts, so the fraction is
$$\frac{25}{100} \text{ which reduces to } \frac{1}{4}$$

SAMPLE PROBLEM 7.14

Determining the Percent of a Number

a. What is 40.0% of 275?   b. What is 80.0% of 65?

SOLUTION

To determine the percent of a number, change the percent to a decimal and multiply by the number.

a. 40.0% = 0.400
   0.400 × 275 = 110

b. 80.0% = 0.800
   0.800 × 65 = 52

SAMPLE PROBLEM 7.15

Determining the Percentage One Number Is of Another Number

a. What percent of 1500 is 30?   b. 35 is what percent of 140?

SOLUTION

To determine what percent one number is of a second number, a fraction is created to represent parts out of the whole. This gives a decimal answer which can then be converted to a percentage by multiplying by 100.

a. $\frac{30}{1500} = 0.02$
   0.02 × 100 = 2%

b. $\frac{35}{140} = 0.25$
   0.25 × 100 = 25%

Percent Mass/Mass, % (m/m)

A solution can easily be prepared in a hospital pharmacy or laboratory with the unit % (m/m) by measuring both the solute and the solvent on a balance and mixing, realizing that mass of solute + mass of solvent = mass of solution. The concentration unit percent mass/mass, % (m/m), can be determined by

$$\% \text{ (m/m)} = \frac{\text{g solute}}{\text{g solution}} \times 100\%$$

This unit is also known as percent weight/weight (% wt/wt). Suppose we weighed out 8.0 g of glucose (the solute) on a scale and mixed 42.0 g of water (the solvent) with it. The total mass of the solution is the mass of the solute and the solvent (8.0 g + 42.0 g = 50.0 g). The concentration of the resulting solution would be solved as follows:

$$\frac{8.0 \text{ g glucose}}{50.0 \text{ g solution}} \times 100\% = 16\%$$

8.0 g glucose + 42.0 g water
(Solute)   (Solvent)

Percent Volume/Volume, % (v/v)

The unit % (v/v) is typically used when liquids or gases are the solute, for example, ethanol and water. A bottle of wine that is 14% (v/v) alcohol means that 14 mL of alcohol is present in 100 mL of the wine. The concentration unit percent volume/volume, % (v/v), can be determined by

$$\% \text{ (v/v)} = \frac{\text{mL solute}}{\text{mL solution}} \times 100\%$$

Percent Mass/Volume % (m/v)

The unit % (m/v) is often used in the preparation of intravenous fluids. Normal saline (NS) solution used to replace body fluids is 0.90% (m/v) NaCl, which means the solution contains 0.90 g of NaCl in 100 mL of solution. The concentration unit percent mass/volume, % (m/v), can be determined by

$$\% \ (m/v) = \frac{g \ solute}{mL \ solution} \times 100\%$$

This unit is also known as percent weight/volume (% w/v). In a hospital pharmacy, a solution could be prepared by weighing the solute on a balance and mixing it with a solvent to form a final total volume of solution.

Relationship to Other Common Units

The unit typically used for measuring the oxygen carrying protein hemoglobin in the blood is g/dL, which is the same unit as % (m/v). A deciliter is equal to 100 mL, so the unit g/dL is the same as g/100 mL. Hemoglobin levels between 13–18 g/dL (13–18 g per 100 mL of blood) for men and 12–16 g/dL for women are considered in the normal range.

A common unit used when measuring molecules like glucose and cholesterol levels in the blood is milligrams per deciliter (mg/dL). This unit is also referred to as mg% (milligram percent). The mg in front of the % symbol indicates that the definition is mg per 100 mL instead of the usual g per 100 mL definition of % (m/v). Glucose levels over 110 mg% (110 mg per 100 mL blood) are considered higher than normal. Total cholesterol levels below 200 mg/dL (200 mg per 100 mL blood) are considered desirable.

Glucose monitor measuring mg/dL

Parts per Million (ppm) and Parts per Billion (ppb)

Parts per million (ppm) and parts per billion (ppb) are convenient concentration units used to describe very dilute solutions. To get an idea of how small a part per million is, consider that there are one million pennies in $10,000, or a penny is a ppm of $10,000. In terms of volume, about 5 drops of food coloring in a bathtub of water represents a part per million, and about 1 drop of food coloring in an Olympic-sized swimming pool of water is about a part per billion—a very dilute solution.

Fluoride is often added to tap water at a level of less than 4 parts per million to promote strong teeth. The maximum contaminant level of lead in drinking water is 15 parts per billion. In the case of the fluoride and ppm, 4 ppm means 4 g of fluoride in every million mL of tap water. In the case of the lead and ppb, 15 ppb means 15 g of lead in every billion mL of tap water. It is easier to think in terms of liters of water versus millions or billions of mL of water, so the unit ppm is sometimes referred to as 1 mg/L and ppb as 1 μg/L.

$$1 \ ppm = \frac{1 \ g \ solute}{1,000,000 \ mL \ solution} \times \frac{1000 \ mg}{1 \ g} \times \frac{1000 \ mL}{1 \ L} = \frac{1 \ mg \ solute}{1 \ L \ solution}$$

A percent mass/volume (% m/v) is *parts per hundred*, so similarly parts per million and parts per billion can be determined by multiplying by a million or a billion, respectively:

$$ppm = \frac{g \ solute}{mL \ solution} \times 1,000,000 \leftarrow 1 \times 10^6$$

$$ppb = \frac{g \ solute}{mL \ solution} \times 1,000,000,000 \leftarrow 1 \times 10^9$$

SAMPLE PROBLEM 7.16

Calculating Percent Concentration

What is the % (m/m) concentration of a NaCl solution prepared with 60.0 g of NaCl and 500.0 g of water?

SOLUTION

The mass of the solute is 60.0 g of NaCl. The mass of the solution is the sum of the solute and the solvent:

$$60.0 \text{ g NaCl} + 500.0 \text{ g water} = 560.0 \text{ g solution}$$

$$\% \text{ m/m} = \frac{60.0 \text{ g NaCl}}{560.0 \text{ g solution}} \times 100\% = 10.7\%$$

SAMPLE PROBLEM 7.17

Solute in Percent Concentration

How many grams of glucose are present in exactly 500 mL of a 5.0% m/v solution?

SOLUTION

To solve this problem, remember the definition of this concentration unit.
A 5.0% m/v solution means that there are 5.0 g of glucose per 100 mL of solution, or

$$\frac{5.0 \text{ g glucose}}{100 \text{ mL solution}}$$

A simple ratio can be set up and the number of grams in 500 mL of solution determined:

$$\frac{5.0 \text{ g glucose}}{100 \text{ mL solution}} = \frac{? \text{ g glucose}}{500 \text{ mL solution}}$$

$$\frac{5.0 \text{ g glucose}}{100 \text{ mL solution}} \times 500 \text{ mL solution} = ? \text{ g glucose}$$

$$? = 25 \text{ g glucose}$$

Note: Because the volumes of 500 and 100 mL are exact, the number of significant figures in this problem is determined by 5.0%, which is two significant figures.

SAMPLE PROBLEM 7.18

Calculating ppm and ppb

What is the concentration in (a) parts per million (ppm) and (b) parts per billion (ppb) of a solution that contains 15 mg of lead in 1250 mL of solution?

SOLUTION

a. A part per million is equivalent to a mg per L so convert the units given to mg/L:

$$\frac{15 \text{ mg lead}}{1250 \text{ mL solution}} \times \frac{1000 \text{ mL}}{1 \text{ L}} = 12 \text{ mg/L} = 12 \text{ ppm}$$

b. A part per billion is equivalent to a μg per L (it is 1000 times smaller than a ppm) so convert the units given to μg/L There are 1000 μg per 1 mg.

$$\frac{15 \text{ mg lead}}{1250 \text{ mL solution}} \times \frac{1000 \text{ mL}}{1 \text{ L}} \times \frac{1000 \text{ μg}}{1 \text{ mg}} = 12{,}000 \text{ μg/L} = 12{,}000 \text{ ppb}$$

PRACTICE PROBLEMS

7.19 A sodium chloride intravenous (IV) fluid contains 154 mEq of Na^+ per liter of solution. If a patient receives exactly 500 mL of the IV solution, how many moles of Na^+ were given?

7.20 A 5% dextrose IV fluid contains 35 mEq of K^+ per liter of solution. If a patient receives 235 mL of the IV solution, how many moles of K^+ were delivered?

7.21 What is the concentration in mmole/L of a Ca^{2+} solution that is 2.50 mEq/L?

7.22 What is the concentration in mEq/L of a Mg^{2+} solution that is 100 mmole/L?

7.23 What is the molarity of 250 mL of solution containing 4.20 moles of $ZnCl_2$?

7.24 What is the molarity of 750 mL of solution containing 2.10 moles of Na_2SO_4?

7.25 What is the molarity of 1.0 L of solution containing 25 g of NaCl?

7.26 What is the molarity of 500 mL of solution containing 45 g of $NaHCO_3$?

7.27 How many grams of KBr are present in 300.0 mL of a 1.00 M solution?

7.28 How many grams of $CaCl_2$ are present in 25.0 mL of a 0.500 M solution?

7.29 What is the concentration in % (m/v) of a solution containing 0.30 g of glucose and 60 mL of solution?

7.30 Calculate the percent mass/volume (% m/v), for the solute in each of the following solutions:

a. 75 g of Na_2SO_4 in 250 mL of Na_2SO_4 solution

b. 39 g of sucrose in 355 mL of a carbonated drink

7.31 Calculate the percent mass/volume (% m/v), for the solute in each of the following solutions:

a. 2.50 g of KCl in 50.0 mL of solution   b. 7.5 g of casein in 120 mL of low-fat milk

7.32 What is the concentration in % (m/m) of a solution prepared by mixing 10.0 g of KCl with 100.0 g of distilled water?

7.33 How many grams of insulin are present in 26.0 g of a 0.450 % (m/m) solution?

7.34 How many grams of NaCl must be combined with water to prepare 500 mL of a 0.90% (m/v) physiological saline solution?

7.35 What is the concentration in ppm of a solution containing 0.30 mg of fluoride and 60 mL of tap water? What is the concentration in ppb?

7.36 What is the concentration in ppm of a solution containing 1.50 mg of copper and 100.0 mL of tap water?

7.5 Dilution

If you have ever prepared orange juice from a frozen can of concentrate by adding 3 cans of water, you have made a dilute solution from a more concentrated one. One way to prepare solutions of lower concentration is to dilute a solution of higher concentration by adding more solvent.

| 1 can of orange juice concentrate | + | 3 cans of water | = | 4 cans of orange juice |

More concentrated juice solution   Less concentrated juice solution

When you add water to the can of orange juice, the amount of orange juice present does not change even though you have a lot more solution present. The *amount of solute* stayed the same, but the *volume* of solution increased, so the *concentration* of the solution decreased (see Figure 7.7).

The following dilution equation represents this mathematically where $C_{initial}$ represents the initial concentration, C_{final} represents the final concentration, $V_{initial}$ represents the initial volume, and V_{final} represents the final volume. If three of the four variables are known, the fourth one can be determined:

$$C_{initial} \times V_{initial} = C_{final} \times V_{final}$$

For example, if you diluted 150 mL of a normal saline solution (0.90% m/v) to a final volume of 450 mL with water, what would the concentration of the final, diluted solution be?

$C_{initial}$ = 0.90% m/v

$V_{initial}$ = 150 mL

V_{final} = 450 mL dilute solution

C_{final} = ?

▶ **FIGURE 7.7 Dilution.** The amount of orange juice (solute) is the same both before and after adding water (solvent) to the concentrated can of orange juice.

Amount of juice in can = Amount of juice in pitcher

$$C_{initial} \times V_{initial} = C_{final} \times V_{final}$$

$$\frac{C_{initial} \times V_{initial}}{V_{final}} = C_{final}$$

$$\frac{(0.90\%) \times (150 \text{ mL})}{(450 \text{ mL})} = C_{final}$$

$$0.30\% = C_{final}$$

The dilution equation works with any concentration unit where the amount of solution (the denominator) is expressed in volume units. (The equation will not work with % m/m.) The dilution equation is useful in the health fields because many pharmaceuticals are often prepared as concentrated aqueous solutions and must be diluted before being administered to the patient.

SAMPLE PROBLEM 7.19

The Dilution Equation

Hydrocortisone is used as an anti-inflammatory for localized pain. You need to prepare a 50-mg/mL solution of hydrocortisone for an injection. You have 5 mL of a 200-mg/mL stock solution of hydrocortisone available. How many milliliters of the 50-mg/mL solution can you prepare?

SOLUTION

Establish the three quantities given and solve for the fourth unknown quantity.

$C_{initial}$ = 200 mg/mL

$V_{initial}$ = 5 mL

C_{final} = 50 mg/mL

V_{final} = ?

$$C_{initial} \times V_{initial} = C_{final} \times V_{final}$$

$$\frac{C_{initial} \times V_{initial}}{C_{final}} = V_{final}$$

$$\frac{(200 \text{ mg/mL}) \times (5 \text{ mL})}{(50 \text{ mg/mL})} = V_{final}$$

$$20 \text{ mL} = V_{final}$$

Does your answer make sense? Because you are diluting, the final volume determined should be more than the initial volume.

SAMPLE PROBLEM 7.20

Preparing a Solution Using the Dilution Equation

How would you prepare 500 mL of a 5% (m/v) glucose solution from a 25% (m/v) stock solution?

SOLUTION

As in Sample Problem 7.19, establish the three quantities given and solve for the fourth, unknown quantity.

$$C_{initial} = 25\%$$
$$V_{initial} = ?$$
$$C_{final} = 5\%$$
$$V_{final} = 500 \text{ mL}$$

$$C_{initial} \times V_{initial} = C_{final} \times V_{final}$$
$$V_{initial} = \frac{C_{final} \times V_{final}}{C_{initial}}$$
$$V_{initial} = \frac{(5\%) \times (500 \text{ mL})}{(25\%)}$$
$$V_{initial} = 100 \text{ mL}$$

The answer to the equation tells us that 100 mL of the stock solution should be used. The question asks how we would prepare the solution, so the final answer requires that we interpret our answer to the dilution equation by stating that to prepare the 5% solution we should combine 100 mL of the 25% solution and enough water for a total solution volume of 500 mL.

PRACTICE PROBLEMS

7.37 How many liters of a 0.90% (m/v) NaCl solution can be prepared from 600 mL of a 9.0% (m/v) stock solution?

7.38 How many mL of a 0.45% (m/v) KCl solution can be prepared from 500 mL of a 9.0% (m/v) stock solution?

7.39 What would the concentration of the resulting glucose solution be if 250 mL of a 12% (m/v) glucose solution were diluted to a final volume of 1 L with distilled water?

7.40 If 40 mL of a 6 M NaOH solution is diluted to a final volume of 200 mL, what is the resulting concentration of the solution?

7.41 How would you prepare 250 mL of a 0.225% m/v NaCl solution from a 0.90% (m/v) NaCl stock solution?

7.42 How would you prepare 2 L of 1 M $MgCl_2$ from a 5 M $MgCl_2$ stock solution?

7.6 Osmosis and Diffusion

Osmosis

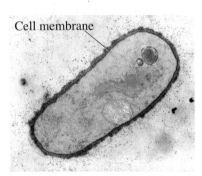
Cell membrane

Because our bodies are mostly water, we could consider ourselves to be composed of a set of specialized aqueous solutions. These solutions are found both inside and outside cells. The concentration of these solutions is highly controlled by the cells. The solutions are separated by a semipermeable barrier called the cell membrane whose structure was described in Chapter 6. A **semipermeable membrane** allows some molecules to pass through the barrier but not others. Under normal physiological conditions, these solutions are considered to be **isotonic solutions**, meaning that the concentration of dissolved solutes is the same on both sides of the membrane.

Consider this: Why is it that there is a limit to the amount of water that we can drink? And why is it that we cannot survive by drinking sea water? The short answer is that tap water is not as concentrated as the body's solutions, and sea water is more concentrated. Let's look more closely at these two questions.

A person can drink too much fresh water. It is recommended that we drink eight to twelve 8-ounce glasses of water a day. The kidneys can process up to 15 liters of water a day, so the average person would really have to overdo it to drink too much water. However, people engaging in vigorous exercise and infants whose formula is overly diluted can actually drink too much water. What is happening in the body? In this instance, the concentration of dissolved solutes in the body's internal solutions is higher than the concentration of dissolved solutes in tap water or dilute formula, so when a person drinks large quantities of tap water, it dilutes the blood. The concentration of solutes in the blood goes down, resulting in an imbalance between the concentration of bodily solution outside the cells (lower concentrated solution, less dissolved solutes) and bodily solution inside the cells (higher concentrated solution, more dissolved solutes). In this case, the solution outside of the cells is said to be a **hypotonic solution** (*hypo* is a prefix meaning "lower than").

When the solution concentrations inside and outside the cells are different, *water* will travel across the cell membrane in an attempt to equalize the concentrations. This passage of *water* through a semipermeable membrane such as a cell membrane is called **osmosis**. So, if we drink too much water, a condition known as hyponatremia (low sodium concentration here caused by excess water) can result from the bloodstream becoming diluted by all the ingested water. Through osmosis, too much water enters the cells in an attempt to equalize the concentrations, and as a result the cells swell up and could even burst (a phenomenon called lysing).

As water flows through a semipermeable membrane during osmosis, the water molecules in the more concentrated solution exert a certain amount of pressure (recall that pressure is force over an area) on the membrane as they attract the water molecules from the less concentrated solution to the more concentrated solution. This pressure is termed **osmotic pressure**. The greater the number of solute molecules present in a solution (that is, the more concentrated the solution), the greater the number of water molecules that will pass into the more concentrated solution to equalize the concentrations, and, as a result, the higher the osmotic pressure. Pure water has an osmotic pressure of zero. Applying pressure in opposition to the osmotic pressure will stop osmosis.

If we were marooned on a desert island without fresh water, why could we not drink sea water to quench our thirst? The concentration of dissolved ions in salt water is about three times that of the blood, so when sea water is consumed, it actually draws water out of the cells through osmosis in an effort to equalize the concentrations. This dehydrates the cells. If a person were to drink sea water, the concentration of solutes in the bloodstream would go up, resulting in an imbalance between the concentration outside the cells (higher concentrated solution, more dissolved solutes) and inside the cells (lower concentrated solution, less dissolved solutes). The solution outside the cells is said to be a **hypertonic solution** (*hyper* is a prefix meaning "higher than"). During dehydration, the cells shrivel in a process known as **crenation**. A summary of how cells behave in different types of solutions is shown in Figure 7.8.

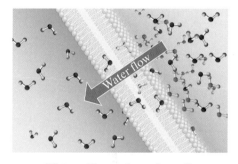

Water will pass through a cell membrane in an attempt to equalize the concentration on either side.

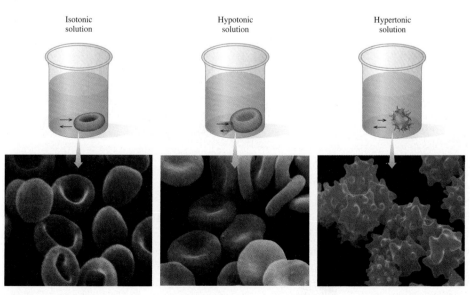

| Isotonic solution | Hypotonic solution | Hypertonic solution |

(a) Normal　　　　**(b)** Hemolysis　　　　**(c)** Crenation

◀ **FIGURE 7.8 How cells behave in different types of solutions.** (a) Healthy cells in isotonic solution. (b) Cells in hypotonic solution will swell. (c) Cells in hypertonic solution will shrink (crenate).

Let's look more closely at how osmosis occurs across a cell membrane. Given that cell membranes separate solutions of different concentration, consider the situation in the following illustration. In which direction would osmosis occur (water flow) to equalize the concentration?

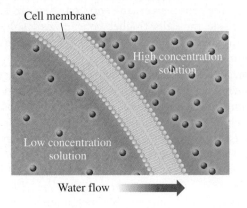

The net flow of water will be from the solution with the lower solute concentration into the solution of higher solute concentration to dilute the solute. Water moves from a solution containing more water molecules (dilute solution) into a solution containing fewer water molecules (more concentrated solution) to equalize the concentrations.

Because cell membranes are semipermeable, osmosis is an ongoing process and is used by the cells to maintain the concentrations inside and outside the cells at about the same level. Body fluids like blood, plasma, and lymph all exert some osmotic pressure. Intravenous (IV) solutions that are delivered into patients' bloodstreams are isotonic; they have solute concentrations equal to the solute concentrations inside of cells. Isotonic solutions minimize osmosis, which is desirable when introducing IV solutions into the blood and eventually to cells. Common isotonic IV solutions used in hospitals include 0.90% (m/v) NaCl (normal saline, NS) and a 5% (m/v) D-glucose (dextrose) solution commonly referred to as D5W ("Dextrose 5% in Water"). These solutions are called **physiological solutions** because they exert the same osmotic pressure as the cells and are isotonic with cells.

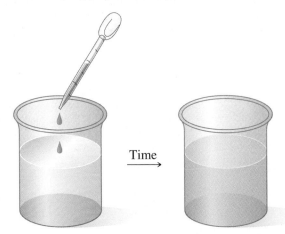

Diffusion and Dialysis

Suppose we put a drop of green food coloring into a large beaker of water. Over time, the green dye molecules (solute) will mix with the water (solvent) and the resulting solution will have a uniform light green tinge to it. The two solutions (one with a high concentration of green molecules and one with no green molecules) spontaneously mix, and the green solute molecules *diffuse* into the water to form one dilute solution with a final green color intermediate between green food coloring from the dropper bottle and water.

Kidney Dialysis

Small molecules and ions can also pass through the cell membrane. If a membrane is permeable to other molecules in addition to water, those molecules will also move across the membrane by a process similar to osmosis to equalize the concentrations of the two solutions. The movement of *solute* molecules across a semipermeable membrane is a diffusion process, not osmosis. **Diffusion** is the movement of molecules in a direction that lowers the concentration. One notable place where diffusion occurs in the body is in the kidneys. The kidneys act to remove small waste molecules out of the blood through diffusion across membranes in the kidneys. Cells and larger molecules, such as proteins found in the blood, are too large to pass through the membrane. These larger molecules are reabsorbed into the bloodstream after passing through the kidney. At the same time, small molecules like urea diffuse out of the blood (higher concentration) and move into urine (lower concentration) in a process called **dialysis**. If the kidneys cannot dialyze waste products out of the bloodstream, increased levels of urea and other wastes in the bloodstream can become life-threatening. A person whose kidneys are failing can undergo artificial dialysis—called hemodialysis—to cleanse the blood (see Figure 7.9). In this process, blood is removed from the patient and passes through one side of a semipermeable membrane in contact on the opposite side with a dialyzing solution that is isotonic with normal blood solute concentrations. Urea and small waste molecules are present in greater concentration in the patient's blood and diffuse out of the passing blood and into the dialyzing solution, and the dialyzed blood returns to the patient.

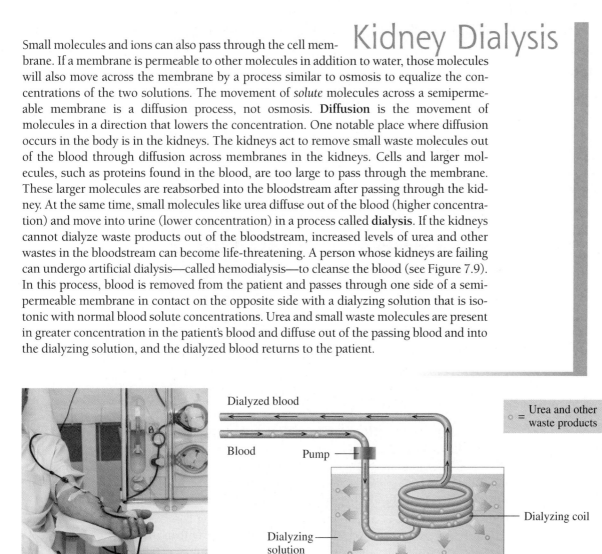

▲ **FIGURE 7.9 Hemodialysis.** Hemodialysis removes waste products from the blood while outside the body, much as a healthy kidney removes waste products from the blood inside the body.

SAMPLE PROBLEM 7.21

Osmosis and Osmotic Pressure

A cell with concentration equal to 5% (m/v) glucose is placed in a 10% (m/v) glucose solution.

a. In which direction will water flow to equalize the concentrations (into the cell or out of the cell)?

b. Which solution is exerting the greater osmotic pressure?

c. Will the cell swell or crenate?

SOLUTION

a. Water will flow from the area where there is more water—the 5% (m/v) solution—to the area where there is less water—the 10% (m/v) solution—so water will flow out of the cell.

b. The 10% (m/v) glucose solution has the higher solute concentration, so it has the greater osmotic pressure.

c. Because water is moving out of the cell, the cells will crenate.

Isotonic, Hypotonic, and Hypertonic Solutions

Are each of the following solutions considered isotonic, hypotonic, or hypertonic with respect to body fluids?

a. 3% (m/v) NaCl   b. 0.90% (m/v) NaCl   c. 0.09% (m/v) NaCl

S O L U T I O N

Compare each solution to physiological saline, which is 0.90% (m/v) NaCl.

a. hypertonic   b. isotonic   c. hypotonic

P R A C T I C E   P R O B L E M S

7.43 A cucumber placed in a briny salt water solution makes a pickle.

a. Does water leave or enter the cucumber's cells?

b. Does the cucumber swell or crenate?

c. Is the salt water solution hypertonic or hypotonic to the cucumber?

7.44 A lettuce leaf in the produce aisle at the grocery is accidentally submerged in water.

a. Will water exit or enter the lettuce cells?

b. Will the cells swell or crenate?

c. Is tap water hypertonic or hypotonic to the lettuce leaf cells?

7.45 A patient is undergoing hemodialysis. As the blood leaves the patient, its solute concentrations are _____ (hypertonic, hypotonic, isotonic) to the dialyzing solution. As the blood re-enters the patient, its solute concentrations are _____(hypertonic, hypotonic, isotonic) to the dialyzing solution.

7.46 A patient is undergoing hemodialysis. As the blood leaves the patient, it has _____ (higher, lower, the same) osmotic pressure than the dialyzing solution. As the blood reenters the patient, its osmotic pressure is _____(higher, lower, the same as) the dialyzing solution.

7.47 Are the following solutions isotonic, hypotonic, or hypertonic with physiological solutions?

a. 10% (m/v) NaCl   b. 0.90% (m/v) glucose

c. 5% (m/v) NaCl   d. 5% (m/v) glucose

7.48 Will a red blood cell—isotonic with 0.90% (m/v) NaCl and 5% (m/v) glucose—undergo crenation, lysis, or no change in each of the following solutions?

a. 10% (m/v) NaCl   b. distilled water

c. 5% (m/v) NaCl   d. 5% (m/v) glucose

7.7 Transport Across Cell Membranes

We saw in Section 7.6 that water can cross the cell membrane through a process called osmosis, but how do polar molecules like glucose or ions like Na^+ move across? As we discussed in Section 6.5, the main structural components of a cell membrane are the phospholipids, which have a polar (hydrophilic) "head" containing a phosphate and a nonpolar (hydrophobic) part containing long hydrocarbon "tails." Figure 7.10 shows the arrangement of the phospholipids in a cell membrane creating a nonpolar barrier enclosing its contents.

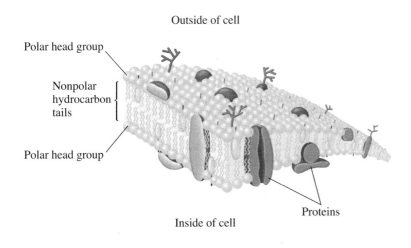

Outside of cell

Polar head group

Nonpolar hydrocarbon tails

Polar head group

Proteins

Inside of cell

◀ **FIGURE 7.10 The cell membrane.** Phospholipids create a nonpolar barrier to the passage of polar molecules.

Ions, nonpolar molecules, and polar molecules move across cell membranes in different ways. We will focus on three main forms of transport across cell membranes: passive diffusion, facilitated transport, and active transport.

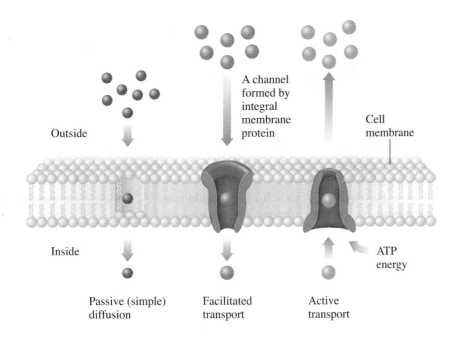

A channel formed by integral membrane protein

Cell membrane

Outside

Inside

ATP energy

Passive (simple) diffusion

Facilitated transport

Active transport

Small molecules that are in high concentration like water and the nonpolar molecules O_2, N_2, and CO_2 can diffuse directly through the cell membrane. Diffusion moves solutes in a direction that attempts to equalize the concentrations on either side of a membrane. This process does not require any additional energy, so this simple diffusion process is also referred to as **passive diffusion**. Other small nonpolar molecules like steroids can also passively diffuse through cell membranes.

Because of their polar character, ions and small polar molecules diffuse very slowly across the nonpolar barrier (refer to Figure 7.10) found in the center of the cell membrane, water being one exception. To enable small molecules and ions to pass through the cell membrane, some proteins found in the cell membrane have polar channels that open and close, allowing small polar molecules and ions to be transported across the cell membrane, to equalize concentrations. These proteins are often integral membrane proteins, spanning the phospholipid bilayer (Chapter 6). This type of transport is called **facilitated transport** and does not require energy. Glucose transporter proteins are found in virtually all cell membranes and facilitate the transport of glucose into the cell during times when glucose concentrations in the bloodstream are high (for example, after a meal).

7.4 Concentration

The concentration of a solution is the amount of solute dissolved in a certain amount of solution. Fluid replacement solutions are often expressed in units of mEq/L or in some cases mmole/L. Molarity is the moles of solute per liter of solution. Percent mass/volume expresses the ratio of the mass of solute (in g) to the volume of solution (in mL) multiplied by 100. This percent mass/volume is equivalent to the unit g/dL. Percent concentration is also expressed as mass/mass and volume/volume ratios. Parts per million and parts per billion are useful units for describing very dilute solutions.

7.5 Dilution

When a solution is diluted, the amount of solute stays the same while the volume of solution increases. The concentration of the solution decreases.

Amount of juice in can = Amount of juice in pitcher

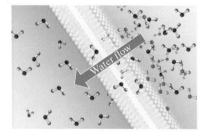

Water will pass through a cell membrane in an attempt to equalize the concentration on either side.

7.6 Osmosis and Diffusion

In osmosis, solvent (water) passes through a semipermeable membrane from a solution of lower solute concentration to a solution of higher solute concentration. The osmotic pressure exerted on the membrane is directly related to the number of water molecules pushing against that membrane. Isotonic solutions have osmotic pressures equal to those of bodily fluids. Cells maintain their volume in an isotonic solution, but they swell and may burst (lyse) in a hypotonic solution and shrink (crenate) in a hypertonic solution. In dialysis, water and small solute particles pass through a dialyzing membrane in a related process called diffusion while large particles like proteins are retained.

7.7 Transport Across Cell Membranes

The semipermeable membrane surrounding cells separates the cellular contents from the external fluids. The membrane is composed of a bilayer of phospholipids that arrange themselves so that the interior is nonpolar and the exterior is polar to interface with an aqueous environment. Molecules can be transported across the cell membrane by passive diffusion, facilitated transport, or active transport depending on their concentration inside and outside the cell and their polarity.

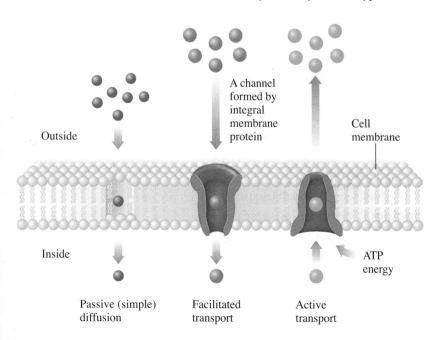

Outside

A channel formed by integral membrane protein

Cell membrane

Inside

ATP energy

Passive (simple) diffusion

Facilitated transport

Active transport

KEY TERMS

active transport—The transport of ions or small molecules across the cell membrane in a direction opposite to diffusion. This process involves an integral membrane protein and requires energy.

aqueous solution—A solution where water is the solvent.

colloid—A homogeneous mixture containing particles ranging from 1 to 1000 nm in diameter. Colloidal mixtures do not separate upon standing.

coefficient—In a chemical equation, the full-sized numbers in front of the chemical formulas that represent the number of each type of particle in the reaction.

concentration—The amount of solute dissolved in a certain amount of solution.

crenation—The shriveling of a cell due to water loss when the cell is placed in a hypertonic solution.

dialysis—A process where water and small molecules pass through a selectively permeable membrane.

diffusion—The movement of solutes that lowers the concentration of a solution. This movement does not require energy (it is a passive process).

electrolyte—A substance that conducts electricity due to the fact that it dissociates into ions in aqueous solution.

equilibrium—A state where two processes are occurring oppositely at the same rate. In a saturated solution the rate of a solid dissolving and the rate of a dissolved solid reforming crystals is the same.

equivalent—A unit of concentration that relates the charge in a solution to the number of ions or the moles of ions present in the solution.

facilitated transport—The diffusion of ions or small polar molecules across the cell membrane with the assistance of a protein.

hydration—The solvation process when water is the solvent.

Henry's law—Law stating that the solubility of a gas in a liquid is directly related to the pressure of that gas over the liquid.

hypertonic solution—A solution outside of a cell having a higher concentration of solutes than the solution inside the cell.

hypotonic solution—A solution outside of a cell having a lower concentration of solutes than the solution inside the cell.

ionize—To dissociate forming ions when dissolved.

isotonic solution—A solution outside of a cell having the same concentration of solutes as the solution on the inside of a cell.

law of conservation of mass—Law stating that matter can neither be created nor destroyed; it merely changes forms.

molarity—A unit of concentration defined as the number of moles of solute dissolved per liter of solution.

nonelectrolyte—A substance that does not conduct electricity because it does not ionize in aqueous solution.

osmosis—The passage of water across a semipermeable membrane in an effort to equalize the solution concentrations on either side. This passage does not require energy (it is a passive process).

osmotic pressure—The pressure that water exerts during osmosis. This amount of pressure applied to the more concentrated solution of the two separated solutions would stop osmosis.

passive diffusion—Diffusion of solutes across the cell membrane to equalize their concentration.

physiological solution—A solution that is isotonic with normal body fluids.

saturated solution—A solution containing the maximum amount of a solute capable of dissolving in the solution at a given temperature.

semipermeable membrane—A membrane that allows passage of some solutes and blocks the passage of others.

solubility—The maximum amount of a solute capable of dissolving in 100 g of solvent at a given temperature.

solute—The substance in a solution present in the smaller amount. Solutions can have more than one solute.

solution—A homogeneous mixture where particles (ions or molecules) are dispersed individually and evenly throughout.

solvation—The process involved when a solute dissolves in a solvent.

solvent—The substance in a solution present in the larger amount. There can be only one solvent in a solution.

strong electrolyte—A substance that conducts electricity because it completely ionizes in aqueous solution.

suspension—A mixture containing particles greater than 1000 nm in diameter. Suspensions separate upon standing.

unsaturated solution—A solution containing less than the maximum amount of solute capable of dissolving.

weak electrolyte—A substance that weakly conducts electricity because it partially ionizes in aqueous solution.

IMPORTANT EQUATIONS

$$\text{Concentration} = \frac{\text{amount of solute}}{\text{amount of solution}}$$

$$\text{Molarity, M} \quad M = \frac{\text{mole solute}}{\text{L solution}}$$

Percent concentration

$$\% \text{ concentration} = \frac{\text{parts of solute}}{100 \text{ parts of solution}}$$

$$\% \text{ (m/v)} = \frac{\text{g solute}}{\text{mL solution}} \times 100\%$$

$$\% \ (m/m) = \frac{g \ solute}{g \ solution} \times 100\%$$

$$\% \ (v/v) = \frac{mL \ solute}{mL \ solution} \times 100\%$$

$$ppm = \frac{g \ solute}{mL \ solution} \times 1,000,000$$

$$ppb = \frac{g \ solute}{mL \ solution} \times 1,000,000,000$$

$$C_{initial} \times V_{initial} = C_{final} \times V_{final}$$

ADDITIONAL PROBLEMS

7.51 Identify the following as a homogeneous or heterogeneous mixture:
  a. whipped cream   **b.** wine
  c. coffee with cream and sugar

7.52 Identify the following as a homogeneous or heterogeneous mixture:
  a. hair spray   **b.** gasoline   **c.** hand lotion

7.53 Identify the mixtures in problem 7.51 as solutions, colloids, or suspensions.

7.54 Identify the mixtures in problem 7.52 as solutions, colloids, or suspensions.

7.55 Does the solubility of the solute increase or decrease in each of the following situations?
  a. Sugar is dissolved in iced tea instead of hot tea.
  b. A bottle of soda (solute is CO_2 gas) is placed in the refrigerator instead of in a pantry at room temperature.
  c. An opened bottle of champagne (solute is CO_2 gas) is allowed to sit at the same temperature open to the air for hours.

7.56 Does the solubility of the solute increase or decrease in each of the following situations?
  a. Ice is placed in a glass of cola (solute is CO_2 gas).
  b. Honey is dissolved in hot water instead of cold water.
  c. The pressure of O_2 over a solution is increased (solute is O_2 gas).

7.57 Predict whether the following will fully dissociate, partially dissociate, or not dissociate when dissolved in solution:
  a. sodium iodide, $NaI(s)$, a strong electrolyte
  b. formic acid, $HCOOH(l)$, a weak electrolyte
  c. glucose, $C_6H_{12}O_6(s)$, a nonelectrolyte

7.58 Predict whether the following will fully dissociate, partially dissociate, or not dissociate when dissolved in solution:
  a. potassium nitrate, $KNO_3(s)$, a strong electrolyte
  b. isopropyl alcohol, $CH_3CH(OH)CH_3(l)$, a nonelectrolyte
  c. boric acid, $H_3BO_3(s)$, a weak electrolyte

7.59 Provide a balanced equation for the hydration of each compound in problem 7.57.

7.60 Provide a balanced equation for the hydration of each compound in problem 7.58.

7.61 How many equivalents of Ca^{2+} are present in a solution that contains 1.34 moles of Ca^{2+}?

7.62 How many equivalents of Cl^- are present in a solution that contains 6.00 moles of Cl^-?

7.63 A Ringer's solution for intravenous fluid replacement contains 4 mEq Ca^{2+} per liter of solution. If a patient receives 1750 mL of Ringer's solution, how many milliequivalents of Ca^{2+} were given?

7.64 An extracellular replacement fluid contains 3 mEq/L citrate. Citrate's conversion factor relating equivalents to moles is 3 Eq citrate/1 mole of citrate. If a patient receives 550 mL of this solution, how many millimoles of citrate were delivered?

7.65 What is the concentration in mmole/L of a SO_4^{2-} solution that is 25.0 mEq/L?

7.66 What is the concentration in mEq/L of a Na^+ solution that is 115 mmole/L?

7.67 What is the molarity of 1.50 L of solution containing 2.80 moles KCl?

7.68 What is the molarity of 750 mL of solution containing 6.2 moles $CaCl_2$?

7.69 What is the molarity of 0.50 L of solution containing 40.0 g of $CaCO_3$?

7.70 What is the molarity of 450 mL of solution containing 15 g NaCl?

7.71 A 750 mL bottle of wine contains 12% (v/v) ethanol. How many milliliters of ethanol are in the bottle of wine?

7.72 What is the concentration in % (v/v) of a methanol solution prepared by mixing 30.0 mL of methanol with 650 mL of distilled water?

7.73 How many grams of dextrose must be combined with water to prepare 500 mL of a 5.0% (m/v) dextrose maintenance solution?

7.74 The normal range for blood hemoglobin in females is 12–16 g/dL. What is this value in % (m/v)?

7.75 The normal level of urea nitrogen in adults is 7–18 mg/dL. What is this value in ppm?

7.76 The normal creatinine levels in adults are 0.6–1.2 mg/dL. What is this value in ppm?

7.77 Express the following numbers as percents:
  a. 1/4   **b.** 3/8   **c.** 66/100
  d. 0.58   **e.** 0.36   **f.** 0.125

7.78 Express the following numbers as decimals:
  a. 4.5%   **b.** 13.0%   **c.** 66%   **d.** 78%

7.79 Express the following numbers as fractions:
  a. 20%   **b.** 75%   **c.** 40%   **d.** 12%

7.80 Determine the numbers from the percentages given:
  a. What is 25% of 80?   **b.** What is 0.9% of 1000?
  c. What is 5.0% of 750?   **d.** What is 66.6% of 200?

7.81 Determine the percentage from the two numbers given:
  a. 50 is what percent of 125?
  b. 6 is what percent of 600?
  c. What percent of 300 is 15?
  d. What percent of 400 is 30?

7.82 An aspirin tablet contains 325 mg of an active ingredient, acetylsalicylic acid. If one tablet weighs 2.00 g, what percentage of the tablet is the active ingredient?

7.83 Calculate the final concentration of each of the following diluted solutions:
  a. 2.0 L of a 3.0 M HNO_3 solution is added to water so that the final volume is 6.0 L.
  b. Water is added to 0.50 L of a 6.0 M KOH solution to make 4.0 L of a diluted KOH solution.

7.84 Calculate the final concentration of each of the following diluted solutions:
  a. a 50.0-mL sample of 4.0% (m/v) NaOH is diluted with water so that the final volume is 100.0 mL.
  b. a 15.0-mL sample of 35% (m/v) acetic acid (CH_3COOH) solution is added to water to give final volume of 25 mL.

7.85 What is the final volume in liters that can be prepared from each of the following concentrated solutions?
  a. a 1.00 M HCl solution prepared from 150.0 mL of a 6.0 M HCl solution
  b. a 4.00% (m/v) $NaHCO_3$ solution prepared from 250 mL of a 20.0% (m/v) $NaHCO_3$

7.86 What is the initial volume in liters needed to prepare each of the following diluted solutions?
  a. 2.0 L of a 0.90% NaCl using a 18.0% (m/v) NaCl stock solution
  b. 1.5 L of a 5.0% glucose solution using a 15.0% glucose solution

7.87 Consider a cell placed in solution as shown in the following figure:

If the inside of the cell has a concentration equivalent to 0.90% (m/v) NaCl, is the cell in a hypertonic, hypotonic, or isotonic solution if the solution outside of the cell is the following:
  a. 5% NaCl   **b.** 0.090% NaCl   **c.** 0.90% NaCl

7.88 Under the conditions in problem 7.87 a–c, in which direction will water move by osmosis: into the cell, out of the cell, or will no net movement occur?

7.89 Edema, more commonly referred to as water retention, is characterized by a swelling of the tissues. Individuals with kidney disease will not excrete normal amounts of Na^+ leading to higher levels of Na^+ in the tissues. In terms of osmosis, explain how high levels of Na^+ in the tissues can lead to edema.

7.90 Many people gain relief from swollen feet at the end of the day by soaking their feet in Epsom salts which creates a hypertonic solution of $MgSO_4$. In terms of osmosis explain how soaking in Epsom salts can reduce swelling in the feet.

7.91 A process called reverse osmosis purifies tap water by pushing water through a set of semipermeable membrane filters. Explain why this process is called *reverse* osmosis.

7.92 Consider the cell in the following figure. In which direction (into the cell or out of the cell) would the Na^+ be transported if the transport protein in the cell membrane was

Na^+ | Cell membrane

  a. an ATPase pump?
  b. an ion transport protein operating under facilitated transport?

7.93 Do the following processes require energy when transporting molecules across the cell membrane?
  a. passive diffusion   **b.** facilitated transport
  c. active transport

7.94 Describe the concentration of the solution outside the cell as hypertonic or hypotonic if that solute is being transported across the cell membrane by
  a. passive diffusion.   **b.** facilitated transport.
  c. active transport.

CHALLENGE PROBLEMS

7.95 A scuba diver diving 100 m down in the ocean experiences 10 times as much pressure on her body than a person swimming at sea level experiences from the atmosphere. The high pressure of the air in a scuba diver's air tank affects the solubility of gases in the bloodstream. A condition called "the bends," which is related to nitrogen gas solubility, can occur in a scuba diver who ascends too quickly from a deep dive to the surface. Can you provide an explanation for this condition?

7.96 How would you prepare 500 mL of a 6.0 M NaOH solution using solid NaOH pellets? What is the % (m/v) concentration of a 6.0 M NaOH solution?

7.97 Two containers of equal volume are separated by a membrane that allows free passage of water but completely restricts passage of solute molecules. Solution A has 3 molecules of the protein albumin (molecular weight 66,000) and solution B contains 15 molecules of glucose (molecular weight 180). Into which compartment will water flow, or will there be no net movement of water? Which solution is at the higher concentration? Which chamber is exerting the higher osmotic pressure?

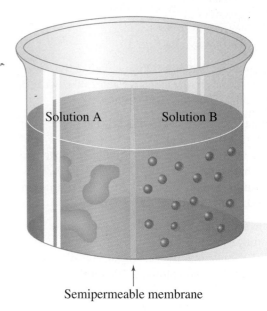

Semipermeable membrane

7.98 Proteinuria is a condition where excessive protein is found in the urine. What must be happening to the kidney's membrane filters (the glomeruli) if such large molecules are found in the urine?

ANSWERS TO ODD-NUMBERED PROBLEMS

Practice Problems

7.1 **a.** solute, oxygen, solvent, nitrogen
b. solute, zinc, solvent, copper
c. solute, blue food coloring, solvent, ethanol

7.3 **a.** colloid **b.** solution **c.** solution **d.** colloid

7.5 **a.** No, not saturated **b.** Yes, saturated

7.7 **a.** increase **b.** decrease **c.** increase

7.9 **a.** Decreasing the temperature increases the solubility of the gas in the soda, so more of the gas stays in the solution and capping the bottle increases the CO_2 pressure over the solution and increases the solubility of CO_2.
b. If the lid is on tightly, no gas can escape and gas that escapes from the solution will build up a pressure above the solution. At some point an equilibrium is reached where no more gas will escape from solution (until of course the bottle is opened).
c. The solubilities of most solid solutes decrease with temperature, so less sugar will dissolve in the iced tea.

7.11 **a.** fully dissociate **b.** partially dissociate
c. not dissociate

7.13 **a.** $KF(s) \xrightarrow[H_2O]{} K^+(aq) + F^-(aq)$

b. $HCN(l) \underset{H_2O}{\rightleftharpoons} H^+(aq) + CN^-(aq)$

c. $C_6H_{12}O_6(s) \xrightarrow[H_2O]{} C_6H_{12}O_6(aq)$

7.15 **a.** $CaCl_2(s) \xrightarrow[H_2O]{} Ca^{2+}(aq) + 2Cl^-(aq)$

b. $NaOH(s) \xrightarrow[H_2O]{} Na^+(aq) + OH^-(aq)$

c. $KBr(s) \xrightarrow[H_2O]{} K^+(aq) + Br^-(aq)$

d. $Fe(NO_3)_3(s) \xrightarrow[H_2O]{} Fe^{3+}(aq) + 3NO_3^-(aq)$

7.17 4.25 Eq

7.19 0.0770 mole Na^+

7.21 1.25 mmoles Ca^{2+}/L

7.23 17 M

7.25 0.43 M

7.27 35.7 g KBr

7.29 0.5% (m/v)

7.31 **a.** 5.00% **b.** 6.3%

7.33 0.117 g insulin

7.35 5 ppm, 5000 ppb

7.37 6 L

7.39 3% (m/v)

7.41 Add 62.5 mL of the 0.90% NaCl stock solution to enough distilled water for a total volume of 250 mL solution.

7.43 **a.** leave **b.** crenate **c.** hypertonic

7.45 hypertonic, isotonic

7.47 a. hypertonic   **b.** hypotonic
     c. hypertonic   **d.** isotonic

7.49 a. passive diffusion   **b.** facilitated transport
     c. active transport   **d.** facilitated transport

Additional Problems

7.51 a. homogeneous   **b.** homogeneous
     c. homogeneous

7.53 a. colloid   **b.** solution   **c.** colloid

7.55 a. decrease   **b.** increase   **c.** decrease

7.57 a. fully dissociate   **b.** partially dissociate
     c. not dissociate

7.59 a. $NaI(s) \xrightarrow{H_2O} Na^+(aq) + I^-(aq)$

     b. $HCOOH(l) \xrightleftharpoons[H_2O]{} HCOO^-(aq) + H^+(aq)$

     c. $C_6H_{12}O_6(s) \xrightarrow{H_2O} C_6H_{12}O_6(aq)$

7.61 2.68 Eq Ca^{2+}

7.63 7 mEq Ca^{2+}

7.65 12.5 mmole/L

7.67 1.87 M

7.69 0.80 M

7.71 90 mL ethanol

7.73 25 g dextrose

7.75 70–180 ppm urea nitrogen

7.77 a. 25%   **b.** 37.5%   **c.** 66%
     d. 58%   **e.** 36%   **f.** 12.5%

7.79 a. $\frac{1}{5}$ **b.** $\frac{3}{4}$ **c.** $\frac{2}{5}$ **d.** $\frac{3}{25}$

7.81 a. 40%   **b.** 1%   **c.** 5%   **d.** 7.5%

7.83 a. 1.0 M HNO_3 **b.** 0.75 M KOH

7.85 a. 0.90 L   **b.** 1.25 L

7.87 a. hypertonic   **b.** hypotonic   **c.** isotonic

7.89 The cells will attempt to dilute the higher than normal concentration of Na^+ in the tissues by moving more water into the tissues causing fluid retention.

7.91 If osmosis is water moving from a lower concentrated solution to a higher concentrated solution, balancing out the concentrations, the reverse would require water moving in the opposite direction. Less pure water (higher concentration of solutes) is forced (requires energy) through a filter (semi-permeable membrane) removing more impurities, making the water even purer for drinking.

7.93 a. No   **b.** No   **c.** Yes

7.95 According to Henry's law, more nitrogen would dissolve in the bloodstream at lower depths (higher pressure) than at the surface (lower pressure). When a diver ascends to the surface quickly, the pressure of the air in the tank (and therefore the air in the lungs) lessens more rapidly than a person can expel the nitrogen that is unable to stay dissolved in the bloodstream and nitrogen gas bubbles begin forming in the bloodstream causing the condition.

7.97 Water will move from solution A to solution B because solution B is a higher concentrated solution (more solute/solution). It does not matter that the albumin particles are huge; osmotic flow depends on the concentration, not the size of the particles.

SECTION 8.4

EXERCISE 1 Weak Acids

Information

Arrhenius definition

Arrhenius acid—A compound that dissociates in water to produce H^+.

Arrhenius base—A compound that dissociates in water to produce OH^-.

Brønsted–Lowry definition

acid—Any substance that can donate a proton (H^+) to another substance.

When an acid donates a proton, a conjugate base is produced.

base—Any substance that can accept a proton (H^+) from another substance.

When a base accepts a proton, a conjugate acid is produced.

strong acid—Dissociates most of its protons in water (~100% ionized).

Six Strong Acids: $HClO_4$, HNO_3, H_2SO_4, HCl, HBr, HI

Example: $HCl + H_2O \longrightarrow H_3O^+ + Cl^-$

strong base—Contains OH^- and will completely dissociate (ionize) in solution.

Example: $NaOH \longrightarrow Na^+ + OH^-$

weak acid—Dissociates some of its protons in water (~5% ionized). An equilibrium is established.

Example:

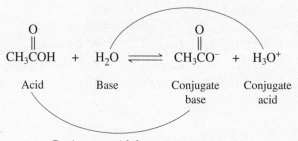

Conjugate acid–base pair

Conjugate acid–base pair

Questions

1. In the preceding examples of strong and weak acids (HCl and CH_3COOH), is the reacting water acting as an acid or a base?

2. Provide a definition for conjugate acid–base pair.

3. a. In the following reaction, is water acting as an acid or a base?

$$NH_3 + H_2O \rightleftharpoons NH_4^+ + OH^-$$

   b. Label the acid, base, conjugate acid, and conjugate base in the reaction.

   c. Provide the acid–base reaction for lactic acid ($CH_3CH(OH)COOH$), a weak acid, when it reacts with water. Label the acid, base, conjugate acid, and conjugate base.

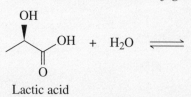

Lactic acid

d. Provide the
  1. conjugate base of H_2S _____
  2. conjugate acid of OH^- _____
  3. conjugate acid of HCO_3^- _____
  4. conjugate base of HCO_3^- _____

EXERCISE 2 The Bicarbonate Buffer System

Information

The bicarbonate buffer system is an effective physiological buffer in the blood, helping maintain a normal physiological pH of 7.4.

$$2H_2O + CO_2(g) \overset{\text{Blood}}{\rightleftharpoons} HCO_3^- + H_3O^+ \overset{\text{Blood}}{\rightleftharpoons} 2H_2O + CO_2(g)$$

Lungs expel CO_2 Tissues produce CO_2

The bicarbonate buffer system moves CO_2 produced at the tissues through the blood to be exhaled at the lungs.

The ventilation rate (rate of breathing) controls the amount of CO_2 present and therefore the amount of acid (H_3O^+) in the blood. When normal ventilation rates are altered, the normal pH balance becomes disrupted, causing distress. An increase in ventilation rate would expel more CO_2 than normal.

Consider the following bicarbonate equilibrium:

$$2H_2O + CO_2(g) \overset{\text{Blood}}{\rightleftharpoons} HCO_3^- + H_3O^+$$

Lungs expel CO_2

Questions

1. If the lungs fail to expel normal amounts of CO_2 due to shallow exhaling that can occur in lung disease, a condition called hypoventilation occurs. According to Le Châtelier's principle, if more CO_2 is added to the system illustrated in the equilibrium (not enough expelled), what happens to the acidity of the blood? Which of the four substances in the preceding equilibrium is a measure of acidity? How might a patient with this condition be treated?

2. If the lungs expel CO_2 faster than normal (ventilation rate becomes high—hyperventilation), which can occur under heavy exercise, and more CO_2 is removed from the system than normal, what will happen to the pH of the blood? How might a patient with this condition be treated?

3. Diabetics may produce excess acid in their tissues as a by-product of the fact that they metabolize more fats for energy versus sugars (metabolic acidosis). How would a diabetic's ventilation rate adjust to correct this condition?

4. Suppose a person drank too much of a baking soda solution (sodium bicarbonate) in order to combat heartburn (which is really excess stomach acid). What would happen if too much acid were removed from the bloodstream?

OUTLINE

The pool water that Olympic swimmer Michael Phelps swims in is maintained at a constant temperature and pH, a measure of acidity. Similarly his body regulates his temperature and blood pH through biochemical reactions and his breathing. Read Chapter 8 to understand more about pH and what happens if your blood pH deviates from normal.

Acids, Bases, and Buffers in the Body

When you eat a dill pickle, how does it taste? Sour, right? Citrus fruits like lemons and grapefruit also have a sour taste. Pickles, citrus fruits, and even sour gummy candies taste sour because they all contain acid. Our stomachs produce acid to help digest the food we eat and our muscles produce lactic acid when we exercise. An acid can be equalized—or, as chemists say, neutralized—by a substance called a base. Soaps are mild bases, and, like all other bases, they feel slippery to the touch.

People who maintain swimming pools use test kits to keep the concentration of dissolved ions and the pH of the water constant. The pH refers to the level of acidity of a solution. Life in general operates under very strict pH conditions. For example, amino acids, the building blocks of proteins, will change form if the acidity of a solution changes. Proteins change their shape and their ability to function if the pH of their surroundings changes. Our bodily fluids, including blood and urine, contain compounds that maintain pH. These compounds are called buffers. In this chapter we discuss the bicarbonate buffer system in the blood and the physiological conditions of acidosis and alkalosis, but first we consider acids, bases, and equilibrium.

8.1 Acids and Bases—Definitions

Acids

The first person to describe **acids** was the Swedish chemist Svante Arrhenius (pronounced *Ar-RAY-nee-us*). He described them as substances that dissociate producing hydrogen ions (H^+) when dissolved in water. The presence of hydrogen ions (H^+) gives acids their sour taste and allows acids to corrode some metals. In the early twentieth century Johannes Brønsted and Thomas Lowry independently expanded the definition of an acid to include the concept that an acid is a compound that *donates* a proton.

$$HCl(g) \xrightarrow[\text{H}_2\text{O}]{} H^+(aq) + Cl^-(aq)$$

Dissociates into ions

How is the hydrogen ion (H^+) referred to above by Arrhenius related to the proton used in the Brønsted–Lowry definition of an acid? We know a proton is a subatomic particle and that H^+ is an ion. Because most hydrogen atoms contain one proton and one electron, a hydrogen ion (H^+), which is a hydrogen atom that has lost its electron, and a proton are one and the same.

In an aqueous solution, a free proton (H^+) rarely exists. The partial negative charge on the oxygen atom in water is strongly attracted to the positive charge of a single proton. This attraction is so strong that the oxygen atom in water forms a third covalent bond, giving water a positive, ionic charge creating **hydronium ion, H_3O^+**.

The hydronium ion can be thought of as a hydrogen ion with a water molecule attached to it.

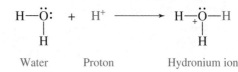

Water   Proton   Hydronium ion

Bases

According to Arrhenius, **bases** are ionic compounds that dissociate to form a metal ion and a hydroxide ion (OH^-) when dissolved in water. Most Arrhenius bases are formed from Group 1A and 2A metals, such as NaOH, KOH, LiOH, and $Ca(OH)_2$. These hydroxide bases are characterized by a bitter taste and a slippery feel. The Brønsted–Lowry definition of a base mirrors their acid definition in that a Brønsted–Lowry base is a compound that *accepts* a proton.

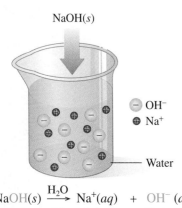

NaOH(s)

— OH^-
⊕ Na^+

— Water

$NaOH(s) \xrightarrow{H_2O} Na^+(aq) \;+\; OH^-(aq)$

Ionic   Dissociation   Hydroxide
compound   ion

Acids and Bases Are Both Present in Aqueous Solution

The Brønsted–Lowry definition states that acids *donate* protons and bases *accept* protons, implying that a proton is usually transferred in an acidic or basic solution. Often water can act as an acid or a base by donating or accepting a proton. We can write the formation of a hydrochloric acid solution as a transfer of a proton from hydrogen chloride to water. By accepting a proton in the reaction, water is acting as a base here according to the Brønsted–Lowry theory.

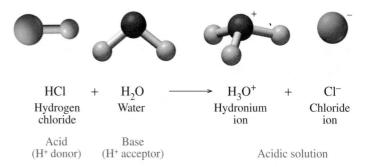

HCl	+	H_2O	⟶	H_3O^+	+	Cl^-
Hydrogen chloride		Water		Hydronium ion		Chloride ion

Acid   Base
(H^+ donor)   (H^+ acceptor)   Acidic solution

In another reaction, ammonia (NH_3) reacts with water. Because the nitrogen atom of NH_3 has a stronger attraction for a proton than the oxygen of water, water acts as an acid in this case by donating a proton.

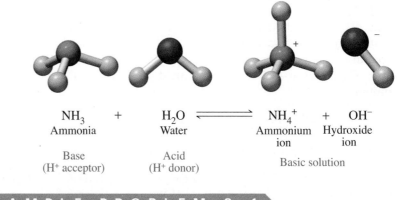

NH_3	+	H_2O	⇌	NH_4^+	+	OH^-
Ammonia		Water		Ammonium ion		Hydroxide ion

Base   Acid
(H^+ acceptor)   (H^+ donor)   Basic solution

SAMPLE PROBLEM 8.1

Acids and Bases

In each of the following equations, identify the reactant that is an acid (H^+ donor) and the reactant that is a base (H^+ acceptor):

a. $HBr(aq) + H_2O(l) \longrightarrow H_3O^+(aq) + Br^-(aq)$

b. $H_2O(l) + CN^-(aq) \longrightarrow HCN(aq) + OH^-(aq)$

SOLUTION

a. Examine what is happening to each of the reactants as it changes to product. When HBr forms Br^-, a hydrogen is donated without its electron (H^+) forming Br^-, so HBr is acting as the acid in this equation. When H_2O forms H_3O^+, it is accepting an H^+, which makes water a base in this equation.

b. Examine what is happening to each of the reactants as it changes to product. When H_2O forms OH^-, a hydrogen is donated without its electron (H^+) forming OH^-, so H_2O is acting as the acid in this equation. When CN^- forms HCN, it is accepting an H^+ (+ and − charges make HCN a neutral compound) making CN^- a base in this equation.

PRACTICE PROBLEMS

8.1 Indicate whether each of the following statements is characteristic of an acid or a base:

a. has a sour taste

b. accepts a proton

c. produces H^+ ions in water

d. is named potassium hydroxide

8.2 Indicate whether each of the following statements is characteristic of an acid or a base:

a. neutralized by a base

b. produces OH^- in water

c. has a slippery feel

d. donates a proton

8.3 In your own words, explain how a proton and the hydrogen ion represent the same thing.

8.4 In your own words, explain how in aqueous solution H^+ and H_3O^+ represent similar things.

8.5 In each of the following equations, identify the acid (proton donor) and base (proton acceptor) for the reactants:

a. $HI(aq) + H_2O(l) \longrightarrow H_3O^+(aq) + I^-(aq)$

b. $F^-(aq) + H_2O(l) \longrightarrow HF(aq) + OH^-(aq)$

8.6 In each of the following equations, identify the acid (proton donor) and base (proton acceptor) for the reactants:

a. $CO_3{}^{2-}(aq) + H_2O(l) \longrightarrow HCO_3{}^-(aq) + OH^-(aq)$

b. $H_2SO_4(aq) + H_2O(l) \longrightarrow H_3O^+(aq) + HSO_4{}^-(aq)$

8.2 Strong Acids and Bases

Acids and bases are classified by their ability to donate or accept protons, respectively. There are six common **strong acids** (see Table 8.1) that completely (~100%) dissociate, meaning they break up into ions, when placed in water, forming hydronium ions and anions.

$$HCl(g) + H_2O(l) \longrightarrow H_3O^+(aq) + Cl^-(aq)$$

All other acids that only partially dissociate (~5%) are considered **weak acids** and will be examined more closely in Section 8.5. One such example is acetic acid (CH_3COOH), which is the main component of vinegar (see Figure 8.1).

TABLE 8.1	
SIX COMMON STRONG ACIDS	
Acid Name	**Formula**
Perchloric acid	$HClO_4$
Sulfuric acid*	H_2SO_4
Hydroiodic acid	HI
Hydrobromic acid	HBr
Hydrochloric acid	HCl
Nitric acid	HNO_3

*Only the first proton is 100% dissociated. The product, $HSO_4{}^-$ is a weak acid.

▶ **FIGURE 8.1 Strong versus weak acid.** A strong acid such as HCl is completely dissociated (100%), whereas a weak acid such as CH_3COOH contains mostly molecules and few ions in solution.

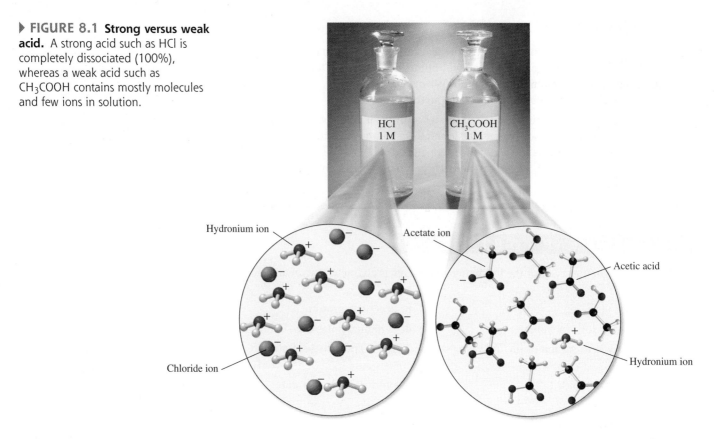

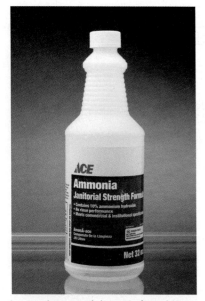

Ammonia, a weak base, is found in many cleaning supplies.

Strong bases, like NaOH (also known as lye), are used in household products such as oven cleaners or drain openers. The Arrhenius bases such as LiOH, KOH, NaOH, and $Ca(OH)_2$ are **strong bases** that dissociate completely (100%) in water. Because they are ionic compounds, they dissociate in water to give an aqueous solution of a metal ion and a hydroxide ion.

All other bases that only partially dissociate ($\sim$5%) are considered **weak bases**. Many common weak bases, such as those found in household cleaners, contain ammonia (NH_3).

Neutralization

What happens when a strong acid and strong base are mixed? Consider the HCl and NaOH used as examples previously. Because both completely dissociate to form ions in water, the water contains sodium ions and chloride ions as well as hydronium ions and hydroxide ions. The protons in the hydronium ions are strongly attracted to the hydroxide ions and combine to form water molecules. This chemical reaction produces a lot of energy as heat and is therefore an **exothermic** (*exo* means "to give off"; *thermo* means "heat") **reaction**.

$$H_3O^+(aq) + OH^-(aq) \longrightarrow 2H_2O(l) + heat$$

The sodium ions and the chloride ions remain in solution. If the water were removed, the ionic compound, sodium chloride (NaCl), would remain. The reaction of a strong acid and strong base therefore always produces water and an ionic compound called a **salt**. This reaction is called a **neutralization** reaction because the acid and base neutralize each other when they react to produce water. We can write the neutralization reaction for HCl and NaOH as follows:

$$HCl(aq) + NaOH(aq) \longrightarrow NaCl(aq) + H_2O(l)$$

Strong acid   Strong base   Salt   Water

Note that the chemical equation is balanced as written; that is, the number of atoms on the reactant side is equal to the number of atoms on the product side.

SAMPLE PROBLEM 8.2

Neutralization Reactions

Complete and balance the following neutralization reactions:

a. $HNO_3(aq) + KOH(s) \longrightarrow$

b. $HBr(aq) + Ca(OH)_2(s) \longrightarrow$

SOLUTION

When completing a neutralization reaction,

STEP 1. Form the products. The products will always be (a) a salt and (b) H_2O. The salt produced must be a neutral ionic compound (refer to Section 3.3 for a refresher on forming ionic compounds like salts).

STEP 2. Balance the chemical equation. The same number of atoms must appear in both the reactants and products. After forming the salt, inspect the reaction to see that it is balanced. Balance the reaction by adding coefficients in front of the product or reactant compounds where appropriate.

a. The products of this neutralization reaction will be H_2O and an ionic compound. The potassium ion has a 1+ charge (a Group 1A element) and the nitrate anion has a 1− charge (see Table 3.1), so the ionic compound has a 1:1 ratio of cations to anions with a formula of KNO_3.

STEP 1. Form the products.

$$HNO_3(aq) + KOH(s) \longrightarrow KNO_3(aq) + H_2O(l)$$

Strong acid ・ Strong base ・ Salt ・ Water

STEP 2. Balance the chemical equation. Inspecting the number of each atom on the reactant side versus the product side indicates that the same number of each atom is present (two H, four O, one N, and one K) so no further balancing is needed.

b. The products of this neutralization reaction will be H_2O and an ionic compound. The calcium ion has a 2+ charge (a Group 2A element), and the bromide anion has a 1− charge (a group 7A element), so the ionic compound combines in a 1:2 ratio of cations to anions with a formula of $CaBr_2$.

STEP 1. Form the products.

$$HBr(aq) + Ca(OH)_2(s) \longrightarrow CaBr_2(aq) + H_2O(l)$$

Strong acid ・ Strong base ・ Salt ・ Water

STEP 2. Balance the chemical equation. Two atoms of Br appear on the product side, and therefore a coefficient of 2 is needed on the reactant side in front of the HBr. This produces a total of 4 H atoms and 2 O atoms on the reactant side, so a coefficient of 2 is also need in front of the H_2O on the product side. Now the equation is balanced.

$$2HBr(aq) + Ca(OH)_2(s) \longrightarrow CaBr_2(aq) + 2H_2O(l)$$

Antacids

Antacids are substances that are used to neutralize excess stomach acid (HCl). Some antacids are mixtures of aluminum hydroxide and magnesium hydroxide. These

hydroxides are not very soluble in water, so the levels of available OH^- are not damaging to the intestinal tract. However, aluminum hydroxide has the side effects of producing constipation and binding phosphate in the intestinal tract, which may cause weakness and loss of appetite. Magnesium hydroxide has a laxative effect. These side effects are less likely when a combination is used.

$$Al(OH)_3(aq) + 3HCl(aq) \longrightarrow AlCl_3(aq) + 3H_2O(l)$$
$$Mg(OH)_2(s) + 2HCl(aq) \longrightarrow MgCl_2(aq) + 2H_2O(l)$$

Some antacids use carbonates to neutralize excess stomach acid. When carbonates are used to neutralize acid, the reaction produces a salt, water, *and* carbon dioxide gas. When calcium carbonate is used, about 10% of the calcium is absorbed into the bloodstream, where it elevates the levels of serum calcium. Calcium carbonate is not recommended for people who have peptic ulcers or a tendency to form kidney stones.

$$CaCO_3(s) + 2 HCl(aq) \longrightarrow CaCl_2(aq) + CO_2(g) + H_2O(l)$$

Sodium bicarbonate can affect the acidity level of the blood and elevate sodium levels in the body fluids. It is also not recommended in the treatment of peptic ulcers.

$$NaHCO_3(s) + HCl(aq) \longrightarrow NaCl(aq) + CO_2(g) + H_2O(l)$$

The neutralizing substances in some antacid preparations are shown in Table 8.2.

TABLE 8.2	BASIC COMPOUNDS IN SOME ANTACIDS
Antacid	**Base(s)**
Amphojel	$Al(OH)_3$
Milk of magnesia	$Mg(OH)_2$
Mylanta, Maalox, Di-Gel, Gelusil, Riopan	$Mg(OH)_2$, $Al(OH)_3$
Bisodol	$CaCO_3$, $Mg(OH)_2$
Titralac, Tums, Pepto-Bismol	$CaCO_3$
Alka-Seltzer	$NaHCO_3$, $KHCO_3$

PRACTICE PROBLEMS

8.7 Which of the following are strong acids?

a. H_2SO_4 b. HCl   c. HF   d. HNO_3 e. H_3PO_4

8.8 Which of the following compounds completely ionize in water?

a. HBr   b. CH_3COOH c. HCN   d. HI   e. NH_3

8.9 Complete and balance the following neutralization reactions:

a. $HNO_3(aq) + LiOH(s) \longrightarrow$ b. $H_2SO_4(aq) + Ca(OH)_2(s) \longrightarrow$

8.10 Complete and balance the following neutralization reactions:

a. $HNO_3(aq) + Mg(OH)_2(s) \longrightarrow$ b. $HCl(aq) + NaHCO_3(s) \longrightarrow$

8.11 Complete and balance the following neutralization reactions:

a. $HBr(aq) + Al(OH)_3(s) \longrightarrow$ b. $HI(aq) + CaCO_3(s) \longrightarrow$

8.12 Complete and balance the following neutralization reactions:

a. $H_2SO_4(aq) + NaOH(s) \longrightarrow$ b. $HBr(aq) + LiOH(s) \longrightarrow$

8.3 Chemical Equilibrium

Before we continue to weak acids and bases, we need to consider the general principles of chemical equilibrium.

Imagine a rock concert with general admission seating. Everyone tries to arrive in plenty of time to get a seat. When the doors open, people rush in and fill up all the seats in the concert hall. When the hall is full, event security stops people from entering the hall despite the fact that some people who waited in line did not get a seat. During the concert, the security personnel allow people to leave and others to enter to keep the hall full, but not overfilled. This steady flow of people in and out of the concert is a state of dynamic equilibrium where the rate of people entering the hall and the rate of people leaving is the same. Note that the number of people in the concert hall stays the same once the hall fills, even though the actual individuals in the concert hall are different.

People entering the concert   People leaving the concert

Just like the concert, some chemical reactions will, after forming product (people entering the concert), reverse and reform reactants (people leaving the concert). These are reversible reactions. A reversible chemical reaction can occur in either the forward or reverse direction depending on the conditions. To explore these reversible reactions more, let's start with a chemical reaction for the generation of ammonia,

$$N_2(g) + 3H_2(g) \rightleftharpoons 2NH_3(g)$$

Reactants   Equilibrium   Products
                    arrow

The generation of ammonia is a reversible reaction. Once ammonia is formed, the reaction will reverse, reforming nitrogen and hydrogen. Eventually, the rate of the formation of ammonia (the forward reaction) and the rate of reformation of nitrogen and hydrogen gases (the reverse reaction) become equal. This balance of the rates of the reactions is called **chemical equilibrium**. A special type of reaction arrow, called an *equilibrium arrow* (shown in the equation), is used in this chemical equation to indicate that both the forward and reverse reactions can take place simultaneously. This was introduced on page 253 in Chapter 7. This does not mean that the *amounts* of ammonia and of nitrogen and hydrogen gas are equal, but because the rates of the forward and reverse reactions are equal, there is no net change in amounts. In other words, the amounts of products versus reactants stay the same.

The Equilibrium Constant *K*

Going back to the rock concert, suppose the concert hall remains completely full for the duration of the concert and the number of people waiting to get in remains the same (although the actual individuals waiting may have changed). The concert attendance has reached equilibrium, and we could observe that the fraction of people inside the concert versus people outside the concert would be constant.

Inside

Outside

$= $ Constant

Similarly, the ammonia reaction shown reaches equilibrium. If we measure the concentrations of ammonia, nitrogen, and hydrogen present, the fraction of products to reactants would be a constant value. This value is called the **equilibrium constant**, K, and it is a characteristic of equilibrium reactions at a given temperature. K is defined as

$$K = \frac{[\text{Products}]}{[\text{Reactants}]}$$

The brackets, [ ], mean "molar concentration of." So, the equation states that the equilibrium constant, K, is equal to the molar concentration of the products divided by the concentration of the reactants. If there is more than one reactant or product, the concentrations are multiplied together. For the generation of ammonia, the expression for K is shown. The superscripts in the expression come from the coefficients (number of moles of each) found in the chemical equation.

$$K = \frac{[NH_3]^2}{[N_2][H_2]^3}$$

In general, for an equilibrium reaction of the form

$$aA + bB \rightleftharpoons cC + dD$$

the general equilibrium expression is given as

$$K = \frac{[C]^c[D]^d}{[A]^a[B]^b}$$

Only substances whose concentrations change appear in an equilibrium expression. Solids (s) and pure liquids (l) have constant concentrations and so do not appear in the equilibrium expression.

The value of K for the formation of ammonia is 9.6 at 300 °C. A value of 9.6 indicates a ratio of 9.6:1 for products:reactants. Actual values for K vary greatly depending on temperature and reaction. If K has a value equal to 1, the ratio of products:reactants is 1:1 or the [products] = [reactants]. A value of K greater than (>) 1 indicates that the amount of products (numerator) is larger than the amount of reactants (denominator) or [products] > [reactants]. A value of K less than (<) 1 indicates that the amount of reactants (denominator) is larger than the amount of products (numerator) or [products] < [reactants] (see Table 8.3).

TABLE 8.3

INTERPRETING VALUES OF K

Value of K	Predominant Species at Equilibrium
$K = 1$	Equal amounts of products and reactants
$K > 1$	Products
$K < 1$	Reactants

SAMPLE PROBLEM 8.3

Equilibrium Constant Expressions

Write an equilibrium constant expression for the following reactions:

a. $CH_4(g) + H_2O(g) \rightleftharpoons CO(g) + 3H_2(g)$

b. $CaCO_3(s) \rightleftharpoons CO_2(g) + CaO(s)$

SOLUTION

K is defined as the ratio of product concentrations to reactant concentrations. Any coefficients in front of molecules appear as superscripts in the expression.

a. $K_{eq} = \dfrac{[CO][H_2]^3}{[CH_4][H_2O]}$

H_2O appears in this expression because it appears as a gas in the equilibrium equation.

b. $K_{eq} = [CO_2]$

Because $CaCO_3$ and CaO are solids, they are considered to have constant concentrations and do not appear in the equilibrium expression.

SAMPLE PROBLEM 8.4

Interpreting Values of *K*

K for the following reaction at 25 °C is 1×10^2.

$$CO(g) + H_2O(g) \rightleftharpoons H_2(g) + CO_2(g)$$

Based on the value for *K*, at equilibrium are the products or reactants greater?

SOLUTION

Because the value of $K > 1$, the products are present in greater amounts.

Effect of Concentration on Equilibrium—Le Châtelier's Principle

Suppose we had our reaction for the generation of ammonia sitting in a reaction vessel, and it was at equilibrium. What would happen if we decided to inject more nitrogen into the vessel?

$$N_2(g) + 3H_2(g) \rightleftharpoons 2NH_3(g)$$

According to **Le Châtelier's** (pronounced *leh-shat-lee-AYS*) **principle**, if we disturb an equilibrium—chemists refer to this as applying stress to the equilibrium—the rate of the forward or reverse reaction will change to offset the stress and regain equilibrium. Applying this principle, if we add more N_2 to our system (it appears on the reactant side of our equation), the rate of the forward reaction will increase, shifting the equilibrium to produce more products. This is because adding more N_2 molecules increases the chance that N_2 will collide with H_2 molecules, forming NH_3 more rapidly.

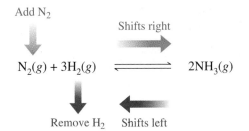

Alternatively, suppose we remove one of the substances, H_2. In order to regain equilibrium, the reverse reaction must be faster than the forward reaction, allowing the H_2 to be replenished. The equilibrium shifts to the left, forming more of the reactants. The lack of H_2 slows down the forward reaction, and to rebalance the rates, the reverse reaction speeds up, reestablishing a balance between the two rates.

In general, we can think of equilibrium as a balancing act between forward and reverse reactions. If one side of the reaction gains a substance, the reaction shifts to the other side to regain its equilibrium. If one side of the reaction loses a substance, the reaction will shift toward that side in order to regain its equilibrium.

Effect of Temperature on Equilibrium

What would happen to the equilibrium of our ammonia reaction if we change the temperature of the reaction? First we have to know whether the reaction itself is one that produces heat, an exothermic reaction, or one that absorbs heat from its surroundings, an **endothermic reaction**. (Determining whether a reaction is endothermic or exothermic is discussed further in Chapter 10.) The generation of ammonia is known to be an exothermic reaction. The chemical equation can be written as follows to signify that heat is produced:

$$N_2(g) + 3H_2(g) \rightleftharpoons 2NH_3(g) + \text{heat}$$

Because heat is a product of the reaction, if the temperature of the reaction is raised (heat added), the rate of the reverse reaction increases to offset the stress of adding heat. This causes the equilibrium to shift to the left. If the reaction were cooled down (heat removed), the rate of the forward reaction would increase to replenish the heat produced, shifting the equilibrium to the right.

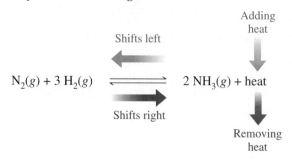

In an endothermic reaction (one that absorbs heat from its surroundings), like the reaction for production of NO gas shown, heat appears as a reactant, so the opposite shifts occur. Table 8.4 summarizes the effects of stress on a chemical equilibrium.

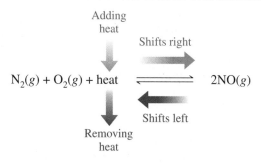

TABLE 8.4 EFFECTS OF CHANGES ON EQUILIBRIUM

Factor	Change (stress)	Reaction Shifts Toward
Concentration	Add more reactant	Right
	Remove reactant	Left
	Add product	Left
	Remove product	Right
Temperature	Raise temperature of endothermic reaction	Right
	Lower temperature of endothermic reaction	Left
	Raise temperature of exothermic reaction	Left
	Lower temperature of exothermic reaction	Right

SAMPLE PROBLEM 8.5

Factors Affecting Equilibrium

How does each of the following actions affect the equilibrium of the following reaction:

$$2SO_2(g) + O_2(g) \rightleftharpoons 2SO_3(g) + heat$$

a. add O_2 **b.** lower the temperature   **c.** add SO_3

SOLUTION

a. The addition of O_2 pushes the reaction toward products; the equilibrium shifts to the right.

b. Lowering the temperature removes heat, which favors the exothermic direction so the equilibrium shifts to the right.

c. The addition of SO_3 favors the reverse reaction toward reactants; the equilibrium shifts to the left.

PRACTICE PROBLEMS

8.13 What is meant by the term *reversible reaction*?

8.14 When does a reversible reaction reach equilibrium?

8.15 Write an equilibrium constant expression for the following reactions:

a. $CO(g) + H_2O(g) \rightleftharpoons H_2(g) + CO_2(g)$

b. $CH_3COOH(aq) + H_2O(l) \rightleftharpoons H_3O^+(aq) + CH_3COO^-(aq)$

8.16 Write an equilibrium constant expression for the following reactions:

a. $2N_2(g) + 3Br_2(g) \rightleftharpoons 2NBr_3(g)$

b. $C(s) + O_2(g) \rightleftharpoons CO_2(g)$

8.17 For the following values of K, indicate whether the products or reactants are present in larger amounts:

a. 1×10^{-5} b. 156   c. 1

8.18 For the following values of K, indicate whether the products or reactants are present in larger amounts:

a. 1×10^7 b. 0.0045   c. 0.00000079

8.19 Hydrogen chloride can be made by reacting hydrogen gas and chlorine gas.

$$H_2(g) + Cl_2(g) \rightleftharpoons 2HCl(g) + \text{heat}$$

What effect does each of the following changes have on the equilibrium?

a. add H_2 b. add heat (raise temperature)

c. remove HCl   d. remove Cl_2

8.20 Sulfur trioxide gas is produced by reacting sulfur dioxide gas with oxygen.

$$2SO_2(g) + O_2(g) \rightleftharpoons 2SO_3(g) + \text{heat}$$

What effect does each of the following changes have on the equilibrium?

a. add O_2 b. lower temperature   c. remove SO_2 d. remove O_2

8.21 In the lower atmosphere, oxygen is converted to ozone (O_3) by the energy provided from lightning.

$$3O_2(g) + \text{heat} \rightleftharpoons 2O_3(g)$$

What effect does each of the following changes have on the equilibrium?

a. add O_2 b. add O_3 c. raise temperature   d. lower temperature

8.22 When heated, carbon reacts with water to produce carbon monoxide and hydrogen.

$$C(s) + H_2O(g) + \text{heat} \rightleftharpoons CO(g) + H_2(g)$$

What effect does each of the following changes have on the equilibrium?

a. add heat   b. lower temperature

c. remove CO   d. add H_2O

8.23 After you open a bottle of soda, the drink eventually goes flat. How can Le Châtelier's principle explain this using the following reaction?

$$H^+(aq) + HCO_3^-(aq) \rightleftharpoons CO_2(g) + H_2O(l)$$

8.24 When you exercise, energy is produced by increasing the rate of the following reaction involving glucose. Why do you breathe faster when you exercise?

$$C_6H_{12}O_6(aq) + 6O_2(g) \rightleftharpoons 6CO_2(g) + 6H_2O(l)$$

8.4 Weak Acids and Bases

Equilibrium

All the principles of equilibrium from Section 8.3 apply to weak acids and bases because weak acids and bases only partially dissociate into ions, establishing an equilibrium in aqueous solution. For example, the dissociation of the weak acid acetic acid (CH_3COOH) into acetate anions (CH_3COO^-) and hydronium ions is as follows:

$$CH_3COOH(aq) + H_2O(l) \rightleftharpoons CH_3COO^-(aq) + H_3O^+(aq)$$

The equilibrium constant expression representing this reaction would be

$$K = \frac{[CH_3COO^-][H_3O^+]}{[CH_3COOH]}$$

Remember that pure liquids like $H_2O(l)$ are present in large amounts that do not change significantly as a reaction approaches equilibrium and so are considered constant and are not included in the equilibrium expression.

The Equilibrium Constant K_a

All weak acids dissociate the same way in water: by donating a proton to form hydronium ion. However, because they dissociate much less than 100%, each weak acid has an equilibrium constant called an **acid dissociation constant**, K_a. The K_a value for acetic acid is 1.75×10^{-5}. Notice that this number is less than 1, which indicates that more acetic acid molecules are present at equilibrium than acetate anions. Because all weak acids can set up an equilibrium in solution, they all have a defined K_a value at a given temperature. Each weak acid dissociates to a different extent so the K_a values are different from acid to acid. The strength of a weak acid can be determined from the K_a value. The larger the K_a value, the stronger the acid (the more protons dissociated). Two common organic functional groups that act as weak acids are carboxylic acids (acetic acid belongs to this family), which dissociate to form carboxylates, and protonated amines, which dissociate to form amines (see Figure 8.2).

▶ **FIGURE 8.2 Functional groups as acids and bases.** Two common organic functional groups act as acids in aqueous solution: carboxylic acids and protonated amines.

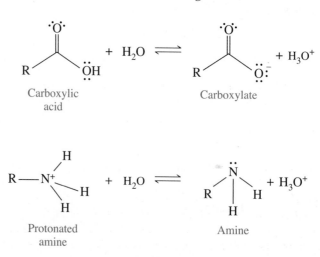

TABLE 8.5	K_a VALUES FOR SUBSTANCES ACTING AS WEAK ACIDS (25 °C)	
Name	Formula	K_a
Hydrogen sulfate ion	HSO_4^-	1.0×10^{-2}
Phosphoric acid	H_3PO_4	7.5×10^{-3}
Hydrofluoric acid	HF	6.5×10^{-4}
Nitrous acid	HNO_2	4.5×10^{-4}
Formic acid	$HCOOH$	1.8×10^{-4}
Acetic acid	CH_3COOH	1.75×10^{-5}
Carbonic acid	H_2CO_3	4.5×10^{-7}
Dihydrogen phosphate ion	$H_2PO_4^-$	6.6×10^{-8}
Ammonium ion	NH_4^+	6.3×10^{-10}
Hydrocyanic acid	HCN	6.2×10^{-10}
Bicarbonate ion	HCO_3^-	4.8×10^{-11}
Hydrogen phosphate ion	HPO_4^{2-}	1×10^{-12}

Increasing acid strength

SAMPLE PROBLEM 8.6

Strength of Weak Acids

Of the following acids—H_2CO_3, HF, and HCOOH—which is (a) the strongest? (b) the weakest?

SOLUTION

Use the K_a values in Table 8.5. The larger values (numbers closer to 1) are stronger acids.

a. The strongest acid of the group is HF, with a K_a value of 6.5×10^{-4}.

b. The weakest acid of the group is H_2CO_3, with a K_a value of 4.5×10^{-7}.

Conjugate Acids and Bases

According to the Brønsted–Lowry theory, the reaction between an acid and base involves a proton transfer, so if a weak acid is mixed with just water, water will act as a base. Consider the dissociation of acetic acid, CH_3COOH:

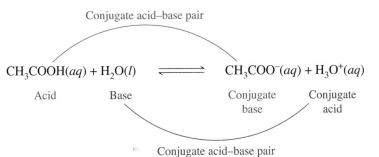

$$CH_3COOH(aq) + H_2O(l) \rightleftharpoons CH_3COO^-(aq) + H_3O^+(aq)$$

Acid — Base — Conjugate base — Conjugate acid

Conjugate acid–base pair

The acid CH_3COOH donates a proton to a molecule of water that accepts the proton, forming a hydronium ion, H_3O^+. What remains of the acid after the donation, CH_3COO^-—a carboxylate called acetate anion—is called the **conjugate base** of CH_3COOH. The term *conjugate base* comes from the fact that in the reverse reaction, the CH_3COO^- acts as a base and accepts the proton from the hydronium ion to form CH_3COOH. Likewise, when the water acts as a base, accepting a proton from acetic acid, a hydronium ion is formed. Hydronium ion is called the **conjugate acid** in this reaction because during the reverse reaction, hydronium ion acts as an acid donating its proton to the acetate anion. Molecules or ions related by the loss or gain of one H^+ are referred to as **conjugate acid–base pairs**. The functional groups carboxylic acid and carboxylate are conjugate acid–base pairs of the same functional group. Weak acids are generically designated as HA and their conjugate base as A^-.

A weak base such as an amine will accept a proton to form a protonated amine, in which case the water acts as an acid. The functional group's protonated amine and amine are conjugate acid–base pairs of the same functional group.

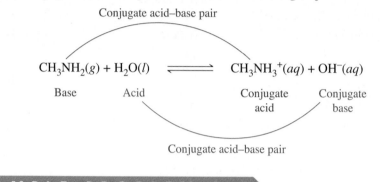

$$CH_3NH_2(g) + H_2O(l) \rightleftharpoons CH_3NH_3^+(aq) + OH^-(aq)$$

Base   Acid   Conjugate   Conjugate
                                               acid   base

Conjugate acid–base pair

SAMPLE PROBLEM 8.7

Identifying Conjugate Acid–Base Pairs

Label the acid, base, conjugate acid, and conjugate base in the following reaction:

$$HF(aq) + H_2O(l) \rightleftharpoons F^-(aq) + H_3O^+(aq)$$

SOLUTION

Compare HF on the reactant side to F^- on the product side. A proton is donated from the HF, leaving F^- as a product. So HF is acting as an acid. The product (F^-) is the conjugate base. Similarly, H_2O is accepting a proton to form the hydronium ion, so water is acting as a base, producing a conjugate acid, the hydronium ion.

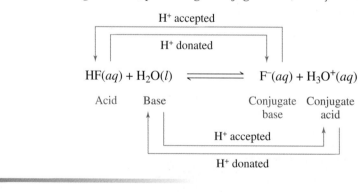

SAMPLE PROBLEM 8.8

Determining Formulas of Conjugates

Provide the following:

a. conjugate acid of OH^- **b.** conjugate acid of HPO_4^{2-}

c. conjugate base of H_2S **d.** conjugate base of HCO_3^-

SOLUTION

Keep in mind the definitions of Brønsted–Lowry acids and bases when trying to answer this type of problem.

a. To find the conjugate acid of OH^-, OH^- would be acting as a base by accepting a proton. The addition of H^+ to OH^- produces the neutral molecule H_2O as its conjugate acid because the $1+$ and $1-$ charges cancel each other out.

b. If HPO_4^{2-} acts as a base, it would be accepting a proton (H^+), thereby forming $H_2PO_4^-$ as its conjugate acid. (Note that when H^+ and HPO_4^{2-} combine, the negative charge changes from $2-$ to $1-$.)

c. To find the conjugate base of H_2S, H_2S would be acting as an acid by donating a proton, H^+. When a proton is donated, what is left over from the H_2S is HS^-. The $1-$ charge represents the pair of electrons that are left behind when H^+ is donated, leaving the electrons behind with the conjugate base.

d. If HCO_3^- acts as an acid, it would be donating a proton, forming CO_3^{2-} as its conjugate base. The donation of H^+ leaves behind a negative charge, making the total charge on conjugate base $2-$.

SAMPLE PROBLEM 8.9

Writing and Balancing Weak Acid–Base Equilibria

Complete the following reaction if $H_2PO_4^-$ acts as a base. Label the acid, conjugate acid, and conjugate base for the following reaction:

$$H_2PO_4^- + H_2O \rightleftharpoons$$

Base

SOLUTION

In order for you to identify, write, and balance an equilibrium reaction for a weak acid or base dissolving in water, a few simple rules can be followed:

- The conjugates will always appear on the product side of a chemical equilibrium.
- When a proton is donated, the conjugate base formed will have one more negative charge than the acid species from which it was formed.
- When a proton is accepted, the conjugate acid formed will have one more positive charge than the base from which it was formed.
- The total charge of the reactants equals the total charge of the products.

$$H_2PO_4^- + H_2O \rightleftharpoons H_3PO_4 + OH^-$$

Base   Acid   Conjugate   Conjugate
                   acid   base

PRACTICE PROBLEMS

8.25 Identify the stronger acid in each pair:

a. $H_2PO_4^-$ or HPO_4^{2-} b. HF or HCOOH   c. HBr or HNO_2

8.26 Identify the stronger acid in each pair:

a. HCN or H_2CO_3 b. H_3PO_4 or $H_2PO_4^-$ c. HCl or HSO_4^-

8.27 Identify the acid and base on the left side of the following equations and identify their conjugate species on the right side:

a. $HSO_4^-(aq) + H_2O(l) \rightleftharpoons H_3O^+(aq) + SO_4^{2-}(aq)$

b. $NH_4^+(aq) + H_2O(l) \rightleftharpoons H_3O^+(aq) + NH_3(aq)$

c. $HCN(aq) + NO_2^-(aq) \rightleftharpoons CN^-(aq) + HNO_2(aq)$

8.28 Identify the acid and base on the left side of the following equations and identify their conjugate species on the right side:

a. $H_3PO_4(aq) + H_2O(l) \rightleftharpoons H_3O^+(aq) + H_2PO_4^-(aq)$

b. $CO_3^{2-}(aq) + H_2O(l) \rightleftharpoons OH^-(aq) + HCO_3^-(aq)$

c. $H_3PO_4(aq) + NH_3(aq) \rightleftharpoons NH_4^+(aq) + H_2PO_4^-(aq)$

8.29 Write the formula and name of the conjugate base formed from each of the following acids:

a. HF b. H_2O c. H_2CO_3 d. HSO_4^-

8.30 Write the formula and name of the conjugate base formed from each of the following acids:

a. HCO_3^- b. H_3O^+ c. HPO_4^{2-} d. HNO_2

8.31 Write the formula and name of the conjugate acid formed from each of the following bases:

a. CO_3^{2-} b. H_2O c. $H_2PO_4^-$ d. Br^-

8.32 Write the formula and name of the conjugate acid formed from each of the following bases:

a. SO_4^{2-} b. CN^- c. OH^- d. ClO_2^-, chlorite ion

8.33 Complete the following reactions and label the conjugate acids and conjugate bases:

a. $HA(aq) + H_2O(l) \rightleftharpoons$

       Acid   Base

b. $H_2PO_4^-(aq) + H_2O(l) \rightleftharpoons$

       Acid   Base

c. $NH_3(aq) + H_2O(l) \rightleftharpoons$

       Base   Acid

8.34 Complete the following reactions and label the conjugate acids and conjugate bases:

a. $A^-(aq) + H_2O(l) \rightleftharpoons$

       Base   Acid

b. $HCO_3^-(aq) + H_2O(l) \rightleftharpoons$

       Base   Acid

c. $HCO_3^-(aq) + H_2O(l) \rightleftharpoons$

       Acid   Base

8.5 pH and the pH Scale

The Autoionization of Water, K_w

We have seen that water can act as a weak acid or base depending upon whether a base or acid is present in the solution. In pure water, the water molecules spontaneously react with each other, donating and accepting protons.

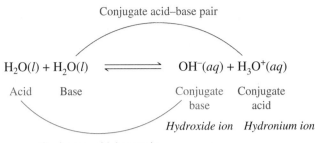

Conjugate acid–base pair

$$H_2O(l) + H_2O(l) \rightleftharpoons OH^-(aq) + H_3O^+(aq)$$

Acid   Base   Conjugate   Conjugate
                  base   acid

Hydroxide ion   Hydronium ion

Conjugate acid–base pair

This reaction, which is always present in water, is called the **autoionization of water**. If we were to write the equilibrium constant expression for water, K_w, (keeping in mind that pure liquid water will not appear), we would write

$$K_w = [OH^-][H_3O^+]$$

Wait, didn't we say in Chapter 7 that pure water does not conduct electricity because there are no dissolved ions present? Now we are saying there *are* ions found in pure water? Let's look at the measured *value* for K_w:

$$K_w = 1 \times 10^{-14} \text{ at } 25\,°C$$

From the small value of K_w and remembering the definition of K ([products]/[reactants]), we can see that the ionized products are found in pretty small amounts in pure water. In fact, K_w is 100 times smaller than the weakest acid found in Table 8.5. Pure water has *some* H_3O^+ and OH^- ions present, but because the amounts are so small, there are not enough ions present in pure water to conduct electricity.

Because the autoionization of water *always* occurs in aqueous solution, all aqueous solutions have small amounts of H_3O^+ and OH^- present.

$[H_3O^+]$, $[OH^-]$, and pH

In pure water, both the hydroxide and hydronium ion are being formed equally from the transfer of protons between water molecules, so in pure water $[H_3O^+] = [OH^-]$. At 25 °C both these values are 1×10^{-7} M. (Remember that "M" is the concentration unit molarity, or moles per liter.) When these concentrations are equal, the solution is said to be **neutral**. If an acid is added to water, there is an increase in $[H_3O^+]$ and a decrease in $[OH^-]$, which makes the solution **acidic**. If base is added, $[OH^-]$ increases and $[H_3O^+]$ decreases, making a **basic** solution (see Figure 8.3).

Most aqueous solutions are *not* neutral and have unequal concentrations of H_3O^+ and OH^-. The concentration of hydronium ion in an aqueous solution can range from about 18 M to 1×10^{-14} M (really close to zero). This range of numbers is huge! Because of this, it is more useful to compare $[H_3O^+]$ by the mathematical function

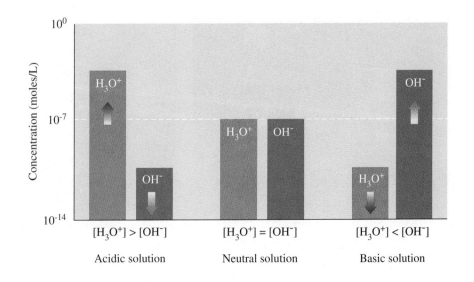

◀ **FIGURE 8.3 Relationships between hydronium and hydroxide in neutral, acidic, and basic solutions.** In a neutral solution, $[H_3O^+]$ and the $[OH^-]$ are equal. In acidic solutions, the $[H_3O^+]$ is greater than the $[OH^-]$. In basic solution, the $[OH^-]$ is greater than the $[H_3O^+]$.

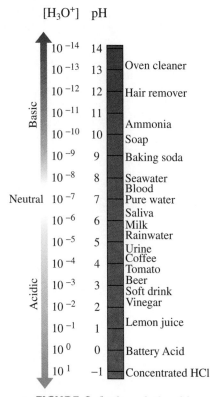

[H$_3$O$^+$] pH

[H$_3$O$^+$]	pH	
10^{-14}	14	
10^{-13}	13	Oven cleaner
10^{-12}	12	Hair remover
10^{-11}	11	Ammonia
10^{-10}	10	Soap
10^{-9}	9	Baking soda
10^{-8}	8	Seawater
10^{-7}	7	Blood / Pure water
10^{-6}	6	Saliva / Milk
10^{-5}	5	Rainwater
10^{-4}	4	Urine / Coffee / Tomato
10^{-3}	3	Beer / Soft drink
10^{-2}	2	Vinegar
10^{-1}	1	Lemon juice
10^{0}	0	Battery Acid
10^{1}	−1	Concentrated HCl

Basic / Neutral / Acidic

▲ **FIGURE 8.4 The relationship between [H$_3$O$^+$] and pH of some common substances.** Notice that the higher the pH, the smaller the [H$_3$O$^+$] and vice-versa.

"log" because it gives a set of numbers that usually falls between 0 and 14. This set of numbers describes the **pH** scale (see Figure 8.4). This scale was developed for the simple comparison of [H$_3$O$^+$] values. The letter "p" literally means the mathematical function negative log (−log), so a pH can be determined from the [H$_3$O$^+$] as

$$pH = -\log [H_3O^+]$$

Table 8.6 shows the relationship between pH and [H$_3$O$^+$].

TABLE 8.6   RELATIONSHIP BETWEEN ACIDITY, pH, AND [H$_3$O$^+$]

Solution Acidity	pH	[H$_3$O$^+$]
Basic	> 7	< 1.0 × 10^{-7} M
Neutral	= 7	= 1.0 × 10^{-7} M
Acidic	< 7	> 1.0 × 10^{-7} M

SAMPLE PROBLEM 8.10

The Relationship Between Acidity, pH, and [H$_3$O$^+$]

State whether solutions with the following conditions would be considered acidic, neutral, or basic.

a. pH = 5.0   **b.** [H$_3$O$^+$] = 2.3 × 10^{-9} M   **c.** pH = 12.0

d. [H$_3$O$^+$] = 4.3 × 10^{-4} M   **e.** [H$_3$O$^+$] = 1 × 10^{-7} M

SOLUTION

Refer to Table 8.6 or Figure 8.4.

a. acidic   **b.** basic   **c.** basic   **d.** acidic   **e.** neutral

Measuring pH

Living things prefer environments that are kept at a constant pH. For example, normal blood pH is strictly regulated between 7.35–7.45. The pH of a solution is commonly measured either electronically by using an instrument called a pH meter or visually using pH paper that is embedded with indicators that change color based on the pH of the test solution (see Figure 8.5).

▶ **FIGURE 8.5 pH measurement.** The pH of a solution can be determined using (a) a pH meter or (b) pH paper.

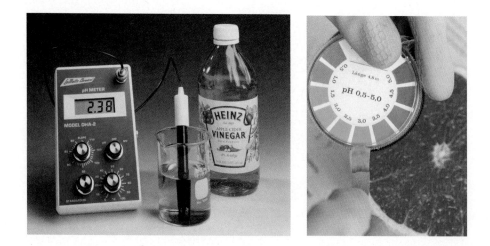

a.   b.

MATH MATTERS

Logarithms

The logarithm or *log* of 100 is 2. The log of 10^2 is therefore also 2. The log of 1000 or 10^3 is 3. Notice that the exponent (the 2 or the 3) is the log of these numbers. All positive numbers, even fractions of one, have a log value.

The purpose of the mathematical function log is to show the number of tens places included in a really large or really small positive number. Negative numbers do not have a log value. You may have seen the log function described as the *log of base 10* because the logs that are integers come from numbers that are whole-number multiples of 10 as we saw in the preceding examples.

Most scientific calculators have a log button on them so that log values of numbers that are not multiples of 10 can easily be determined. There is another logarithm function used in mathematics called the natural log, or *ln*, which is a completely different function and will not be discussed in this section.

SAMPLE PROBLEM 8.11

Calculating Logarithms of Base 10

Determine the log of the following numbers:

a. 6.0   b. 0   c. 10   d. 501

SOLUTION

Use a scientific calculator to find the log of the preceding numbers.

a. 0.78

b. Because all log values are greater than zero, the log 0 cannot produce a number. You may have gotten the message "error" when you tried this one on your calculator.

c. 1

d. 2.70

MATH MATTERS

Inverse Logarithms

Suppose we want to solve for *x* in the following equation:

$$4 = \log x$$

In this example, we know the log of a number but do not know the number that produced that log. To solve for *x*, the equation will have to be rearranged. To accomplish this, we need to take the inverse log of both sides of the equation. Many programmable calculators have an INV button that can be used to solve the preceding equation. Applying the inverse (INV) log function to both sides of the equation, we can solve for *x*:

$$\text{INV} \log 4 = \text{INV} \log (\log x)$$

$$\text{INV} \log 4 = x$$

$$1 \times 10^4 = x$$

All integers are logs of multiples of 10. Positive and negative numbers can have inverse logs because this number represents the exponent value. The actual inverse function of log *x* is 10^x, and many scientific calculators also have a 10^x button typically found as a second function above the log button. The y^x button can also be used if 10 is entered as *y*.

SAMPLE PROBLEM 8.12

Calculating Inverse Logarithms of Base 10

Find the number that gives the following log:

a. 5   b. −2.5   c. 0   d. −8.32

SOLUTION

If you have a programmable calculator, you can type in INV log and then the number in the problem. Alternatively, you can type the number then the function on a scientific calculator with a 10^x or a y^x button (type in 10 for y).

a. $10^5 = 1 \times 10^5$

b. $10^{-2.5} = 3.2 \times 10^{-3}$

c. It is possible to take the inverse log of zero; $10^0 = 1$

d. $10^{-8.32} = 4.8 \times 10^{-9}$

Calculating pH

Suppose we have a 0.050 M HCl solution, and we want to calculate its pH. Because strong acids fully ionize in solution, $[HCl] = [H_3O^+]$,

$$pH = -\log [H_3O^+]$$
$$pH = -\log 0.050$$
$$pH = -(-1.30) = 1.30$$

Let's take a quick look at the significant figures in the preceding calculation. In pH calculations with logarithms, the number of significant figures in the $[H_3O^+]$ will be the number of decimal places in the pH value.

$$[H_3O^+] = 0.050 \qquad\qquad pH = 1.30$$

Two significant figures   *Two decimal places*

Calculating [H₃O⁺]

Because we can easily measure the pH of most solutions, we are often more interested in finding the $[H_3O^+]$ from a measured pH value. Suppose we measured the pH of a solution to be 3.00 and we want to find the corresponding $[H_3O^+]$:

$$pH = -\log [H_3O^+]$$

To solve the pH equation for $[H_3O^+]$, we will have to multiply both sides of the equation by negative 1 and the inverse log function, INV log. In case you don't have an INV button on your calculator, keep in mind that the inverse log function is 10^x so to solve the equation for $[H_3O^+]$,

$$INV \log (-pH) = [H_3O^+]$$

or alternatively,

$$10^{-pH} = [H_3O^+]$$

From this equation we can solve for the $[H_3O^+]$ of a pH 3.00 solution as

$$INV \log (-3.00) \text{ or } 10^{-3.00} = [H_3O^+]$$
$$1.0 \times 10^{-3} \text{ M} = [H_3O^+]$$

In this case, the number of decimal places given in the pH measurement tells us the number of significant figures in the $[H_3O^+]$:

$$pH = 3.00 \qquad\qquad [H_3O^+] = 1.0 \times 10^{-3}$$

Two decimal places   *Two significant figures*

SAMPLE PROBLEM 8.13

Calculating pH from [H₃O⁺]

Determine the pH for the following solutions:
a. $[H_3O^+] = 1.0 \times 10^{-5}$ M   b. $[H_3O^+] = 5 \times 10^{-8}$ M

SOLUTION

To do these problems, you will be using the log function, log, on your calculator. Depending on the type of scientific calculator that you have, you will be inputting the numbers differently. Check with your instructor if you are having difficulty producing the correct answers using your calculator.
a. $pH = -\log [H_3O^+] = -\log (1.0 \times 10^{-5}) = -(-5.00) = 5.00$
b. $pH = -\log [H_3O^+] = -\log (5 \times 10^{-8}) = -(-7.3) = 7.3$

SAMPLE PROBLEM 8.14

Calculating [H₃O⁺] from a Measured pH

Determine the $[H_3O^+]$ for solutions having the following measured pH values:
a. pH = 7.35   b. pH = 11.0

SOLUTION

To do these problems, you will be using the inverse log function, either INV log or 10^x, on your calculator. Depending on the type of scientific calculator that you have, you will be inputting the numbers differently. Check with your instructor if you are having difficulty producing the correct answers using your calculator.
a. $[H_3O^+] = $ INV log $(-pH)$ or $10^{-pH} = $ INV log (-7.35) or $10^{-7.35} = 4.5 \times 10^{-8}$ M
b. $[H_3O^+] = $ INV log $(-pH)$ or $10^{-pH} = $ INV log (-11.0) or $10^{-11.0} = 1 \times 10^{-11}$ M

PRACTICE PROBLEMS

8.35 State whether each of the following solutions is acidic, basic, or neutral:
a. blood, pH 7.38   b. vinegar, pH 2.8
c. drain cleaner, pH 11.2   d. coffee, pH 5.5

8.36 State whether each of the following solutions is acidic, basic, or neutral:
a. soda, pH 3.2   b. shampoo, pH 5.7
c. laundry detergent, pH 9.4   d. rain, pH 5.8

8.37 State whether each of the following solutions is acidic, basic, or neutral:
a. $[H_3O^+] = 1.2 \times 10^{-8}$ M   b. $[H_3O^+] = 7.0 \times 10^{-3}$ M
c. $[H_3O^+] = 4.7 \times 10^{-11}$ M   d. $[H_3O^+] = 1.0 \times 10^{-7}$ M

8.38 State whether each of the following solutions is acidic, basic, or neutral:
a. $[H_3O^+] = 5.6 \times 10^{-10}$ M   b. $[H_3O^+] = 6.2 \times 10^{-8}$ M
c. $[H_3O^+] = 5 \times 10^{-2}$ M   d. $[H_3O^+] = 1.8 \times 10^{-6}$ M

8.39 Calculate the pH of each of the solutions in problem 8.37.

8.40 Calculate the pH of each of the solutions in problem 8.38.

8.41 Calculate the $[H_3O^+]$ for each of the following measured pH's:
a. drain cleaner, pH 12.10   b. coffee, pH 5.5
c. gastric juice, pH 2.00   d. toothpaste, pH 8.3

8.42 Calculate the $[H_3O^+]$ for each of the following measured pH's:
a. hand soap, pH 9.5   b. seawater, pH 8.2   c. apple juice, pH 3.5   d. beer, pH 4.5

8.6 pK_a

Which of the two acids, acetic acid or ammonium ion, is the stronger acid (see Figures 8.6 and 8.7)? We can determine this by comparing their K_a values as we did in Section 8.4. Notice from Table 8.7 that the K_a value for acetic acid, 1.75×10^{-5}, is a number closer to 1 than the K_a for ammonium, 6.3×10^{-10}, which tells us that acetic acid is the stronger of these two weak acids. In the previous section we learned that the mathematical function "p" means take the *negative log* of a number. As seen with the pH scale, it is easier to compare whole numbers than those in scientific notation. To make this comparison for acid strength, we can use **pK_a** values, which range from 0 to about 60. Notice that the pK_a value for acetic acid is a smaller number, 4.76, than that of ammonium, 9.20. This comparison illustrates the following rule for determining the strength of a weak acid using pK_a values: *The lower the pK_a, the stronger the acid.* Table 8.7 compares the pK_a and K_a values for some common weak acids.

▶ **FIGURE 8.6 Equilibrium reaction for the weak acid, acetic acid.** When acetic acid donates a proton to water, the carboxylate named acetate is formed. Recall that carboxylic acid and carboxylate are different forms of the same functional group.

▶ **FIGURE 8.7 Equilibrium reaction for the weak acid, ammonium.** When ammonium ion donates a proton to water, the amine ammonia is formed. Recall that protonated amine and an amine are different forms of the same functional group.

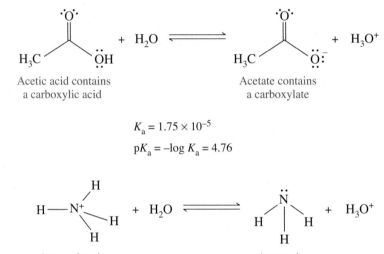

Acetic acid contains a carboxylic acid

Acetate contains a carboxylate

$$K_a = 1.75 \times 10^{-5}$$
$$pK_a = -\log K_a = 4.76$$

Ammonium ion, a protonated amine

Ammonia, an amine

$$K_a = 6.3 \times 10^{-10}$$
$$pK_a = -\log K_a = 9.20$$

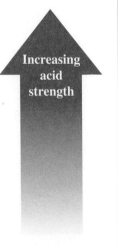

Increasing acid strength

TABLE 8.7	pK_a AND K_a VALUES FOR SUBSTANCES ACTING AS WEAK ACIDS (25 °C)		
Name	**Formula**	**pK_a**	**K_a**
Hydrogen sulfate ion	HSO_4^-	2.00	1.0×10^{-2}
Phosphoric acid	H_3PO_4	2.12	7.5×10^{-3}
Hydrofluoric acid	HF	3.19	6.5×10^{-4}
Nitrous acid	HNO_2	3.35	4.5×10^{-4}
Formic acid	HCOOH	3.74	1.8×10^{-4}
Acetic acid	CH_3COOH	4.76	1.75×10^{-5}
Carbonic acid	H_2CO_3	6.35	4.5×10^{-7}
Dihydrogen phosphate ion	$H_2PO_4^-$	7.18	6.6×10^{-8}
Ammonium ion	NH_4^+	9.20	6.3×10^{-10}
Hydrocyanic acid	HCN	9.21	6.2×10^{-10}
Bicarbonate ion	HCO_3^-	10.32	4.8×10^{-11}
Hydrogen phosphate ion	HPO_4^{2-}	12.00	1×10^{-12}

SAMPLE PROBLEM 8.15

Determining Strength of Weak Acids Using pK_a Values

Determine the stronger acid of the following pairs of weak acids:

a. ammonium ion or nitrous acid

b. phosphoric acid or dihydrogen phosphate ion

SOLUTION

Using the rule, the lower the pK_a, the stronger the weak acid, and Table 8.7,

a. Nitrous acid is the stronger acid.

b. Phosphoric acid is the stronger acid.

The Relationship Between pH and pK_a

We have seen two numbers associated with weak acid solutions, the pH and pK_a. Let's review what they tell us about a weak acid solution. The pK_a (and the K_a) tell us the fraction of acid molecules that dissociate. That is, the pK_a gives the ratio of conjugate base and hydronium ion (H_3O^+) to weak acid. The pH tells us how much hydronium ion is present in the solution. pK_a values are constant at a given temperature, that is, they do not change if acid or base is added to a weak acid solution. In contrast, the pH does change if acid or base is added to a weak acid solution.

Let's examine an acetic acid solution at three different pH's: a pH *below* the pK_a, a pH *equal* to pK_a, and a pH *above* the pK_a. If the pH of the solution is lower than the pK_a, extra H_3O^+ has been added to the equilibrium solution, and so the equilibrium has shifted to the left, meaning there is more acetic acid present than its conjugate base, acetate. When the ratio of acid to conjugate base is 1, in the equilibrium expression, the $K_a = [H_3O^+]$ and so the pH of the solution is equal to the pK_a. If the pH of the solution is higher than the pK_a, in effect less than equilibrium concentration of H_3O^+ is present and the equilibrium has shifted to the right. This produces more acetate, which will be present in higher concentration than the acetic acid. Table 8.8 summarizes these relationships.

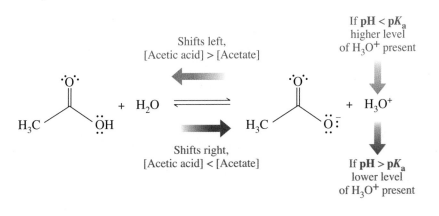

TABLE 8.8	RELATIONSHIP BETWEEN ACID (HA), CONJUGATE BASE (A⁻), pH, AND pK_a	
pH < pK_a	pH = pK_a	pH > pK_a
[HA] > [A⁻]	[HA] = [A⁻]	[HA] < [A⁻]

S A M P L E   P R O B L E M   8 . 1 6

Predominant Form at a Given pH and pK_a

Consider the weak acid methylammonium whose pK_a value is 10.7. Will the acid or the conjugate base predominate at the following pH's?

$$H_3C-\overset{\overset{\displaystyle H}{|}}{\underset{\underset{\displaystyle H}{|}}{\overset{+}{N}}}-H + H_2O \rightleftharpoons H_3C-\overset{\overset{\displaystyle }{|}}{\underset{\underset{\displaystyle H}{|}}{\ddot{N}}}-H + H_3O^+$$

      Methylammonium   *Methylamine*

a. 7.4   **b.** 12.1   **c.** 10.7

S O L U T I O N

The species that predominates (acid or conjugate base) depends upon the value of the pH versus the pK_a. Refer to Table 8.8.

a. A pH of 7.4 is below the pK_a for methylammonium (10.7), shifting the equilibrium to the left so the acid form, $CH_3NH_3^+$, predominates.

b. This pH is above the pK_a for methylammonium, shifting the equilibrium to the right, so the conjugate base form, CH_3NH_2, predominates.

c. This pH is equal to the pK_a for methylammonium, so there are equal concentrations of both $CH_3NH_3^+$ and CH_3NH_2 present.

P R A C T I C E   P R O B L E M S

8.43 Using Table 8.7, determine the stronger acid from the pairs of following acids:

a. $H_2PO_4^-$ or HPO_4^{2-} **b.** H_2SO_4 or acetic acid

c. formic acid or carbonic acid   **d.** ammonium ion or hydrocyanic acid

8.44 Using Table 8.7, determine the stronger acid from the following pairs of weak acids:

a. HCl or HCN   **b.** acetic acid or ammonium ion

c. carbonic acid or bicarbonate ion   **d.** ammonium ion or hydrogen phosphate ion

8.45 Consider the weak acid, benzoic acid, 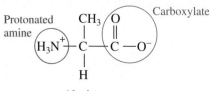 , whose pK_a value is

4.2. Will the acid or the conjugate base predominate at the following pH's?

a. 3.0   **b.** 4.2   **c.** 11.3

8.46 Consider the weak acid ethylammonium, $CH_3CH_2NH_3^+$, whose pK_a value is 10.8. Will the acid or the conjugate base predominate at the following pH's?

a. 8.0   **b.** 3.0   **c.** 13.5

8.7 Amino Acids: Common Biological Weak Acids

Consider the circled functional groups on the following molecule. These groups are identified as a protonated amine and a carboxylate.

Protonated amine   Carboxylate

$$\left(H_3\overset{+}{N}\right)-\overset{\overset{\displaystyle CH_3}{|}}{\underset{\underset{\displaystyle H}{|}}{C}}-\left(\overset{\overset{\displaystyle O}{\|}}{C}-O^-\right)$$

Alanine

This molecule is alanine and it belongs to a class of molecules called **amino** (from the amine functional group) **acids** (from the carboxylate functional group) that are the building blocks of proteins. Why is the amino acid alanine shown with a carboxylate group instead of a carboxylic acid and a protonated amine instead of an amine? Let's look at the pK_a values for the carboxylic acid and the protonated amine. The pK_a of alanine's carboxylic acid is 2.3, *below* physiological pH (~7.4), so, as we saw in Section 8.6, the conjugate base form (carboxylate) predominates. Likewise, the pK_a of the protonated amine group in alanine is 9.7, *above* physiological pH, so the acid form (protonated amine) predominates. The amino acid shown is in the form that predominates at a physiological pH of 7.4. This ionic form containing no net charge (+ and − cancel each other out) is called a **zwitterion**.

Amino acids and many other biological molecules contain more than one weak acid group, have more than one pK_a value, and so exist in different acid/conjugate base forms depending on the pH of the solution. Such molecules have a unique pH value at which the zwitterion is the predominant form. This point is called the **isoelectric point (pI)**. At the pI of alanine, the negative charge on the carboxylate is balanced by the positive charge on the ammonium ion, and the net charge of the amino acid is zero. The pI for the amino acid alanine is 6.0. Its pI is halfway between the pK_a values for the protonated amine and carboxylic acid as shown.

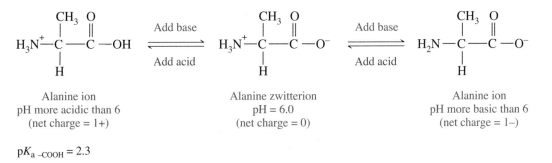

Alanine ion	Alanine zwitterion	Alanine ion
pH more acidic than 6	pH = 6.0	pH more basic than 6
(net charge = 1+)	(net charge = 0)	(net charge = 1−)

$pK_{a\ -COOH} = 2.3$

$pK_{a\ -\overset{+}{N}H_3} = 9.7$

SAMPLE PROBLEM 8.17

Amino Acids in Acid or Base

Write the zwitterion form of the amino acid cysteine shown in the following figure. If the pI for cysteine is 5.1, at what pH will the zwitterion be the predominant form?

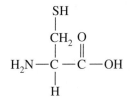

SOLUTION

The zwitterionic form contains a protonated amine and carboxylate group. This form will exist exclusively when pH = pI = 5.1 for cysteine.

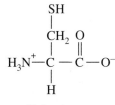

Zwitterion

PRACTICE PROBLEMS

8.47 Valine has the zwitterion structure shown in the following figure. The pK_a's and pI are listed. Given this information, what form of the molecule will predominate at the indicated pH values?

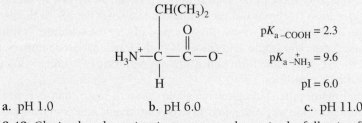

$$pK_{a-COOH} = 2.3$$
$$pK_{a-\overset{+}{N}H_3} = 9.6$$
$$pI = 6.0$$

a. pH 1.0   b. pH 6.0   c. pH 11.0

8.48 Glycine has the zwitterion structure shown in the following figure. The pK_a's and pI are listed. Given this information, what form of the molecule will predominate at the indicated pH values?

$$pK_{a-COOH} = 2.4$$
$$pK_{a-\overset{+}{N}H_3} = 9.6$$
$$pI = 6.0$$

a. pH 1.5   b. pH 12.0   c. pH 6.0

8.8 Buffers: An Important Property of Weak Acids and Bases

If your body temperature goes up by one or two degrees Fahrenheit, you have a fever and don't feel like yourself. Our bodies operate best under a very strict set of conditions that include not only temperature but also, concentration *and* pH. How do we keep the pH levels constant in our blood when we eat such a variety of foods with very different pH's? In our bodies, solutions of weak acids containing both acid and conjugate base help neutralize incoming bases and acids, thereby maintaining pH levels. A solution that contains both a weak acid and its conjugate base or a weak base and its conjugate acid is called a **buffer**. Strong acids and bases are not components of buffers. A buffer solution will resist a change in its pH if small amounts of acid or base are added.

Our blood is buffered mainly by the bicarbonate buffer system. Dissolved CO_2 produced during cellular respiration travels through the bloodstream for exhalation at the lungs. Although dissolved, this CO_2 is rapidly equilibrated through carbonic acid into bicarbonate ions. The intermediates (H_2CO_3 and H_2O) are often omitted since they are short lived in this reaction (see Figure 8.8).

Reaction p*K* is 6.1

$$H_2O(l) + CO_2(g) \underset{equilibrates}{\overset{Rapidly}{\rightleftharpoons}} H_2CO_3(aq) + H_2O(l) \rightleftharpoons H_3O^+(aq) + HCO_3^-(aq)$$

Carbonic   Bicarbonate
acid   ion
(acid)   (Conjugate base)

▲ **FIGURE 8.8 The bicarbonate buffer equilibrium.** CO_2 produced at the tissues rapidly equilibrates through carbonic acid to bicarbonate ion which is transported in the bloodstream to the lungs, reequilibrated to CO_2 and exhaled.

A buffer system like the bicarbonate buffer can be denoted by showing the acid and its conjugate base like this: H_2CO_3/HCO_3^-. Sometimes the conjugate base is represented as an ionic compound, a salt, like this: $H_2CO_3/NaHCO_3$.

Maintaining Physiological pH with Bicarbonate Buffer: Homeostasis

The bicarbonate buffer system helps to maintain optimal physiological pH in our bodies. The ability of an organism to regulate its internal environment by adjusting factors such as pH, temperature, and solute concentration is called **homeostasis**. Let's examine how the bicarbonate buffer system is able to regulate blood pH under adverse conditions.

Changes in Ventilation Rate

During normal breathing, appropriate amounts of CO_2 gas are removed from the bloodstream upon exhalation in the lungs, and blood pH is maintained. A person who **hypoventilates** may fail to exhale enough CO_2 from the lungs due to shallow breathing causing CO_2 gas to build up in the bloodstream. This can occur, for example, when a person is suffering from emphysema, a condition that blocks gas diffusion in the lungs. According to Le Châtelier's principle (Section 8.3), a buildup of CO_2 ultimately produces more H_3O^+ in the bicarbonate equilibrium, making the blood more acidic. This condition is known as **respiratory acidosis** and can occur in a number of health conditions (Table 8.9). Persons suffering from this condition must be treated to raise their blood pH back into the normal range. This can be done by administering a bicarbonate solution intravenously, which will have the effect of driving the

TABLE 8.9 ACIDOSIS AND ALKALOSIS: SYMPTOMS, CAUSES, AND TREATMENTS			
Respiratory Acidosis: CO_2 ↑ pH ↓		**Metabolic Acidosis: H_3O^+ ↑ pH ↓**	
Symptoms:	Failure to ventilate, suppression of breathing, disorientation, weakness, coma	Symptoms:	Increased ventilation, fatigue, confusion
Causes:	Lung disease blocking gas diffusion (e.g., emphysema, pneumonia, bronchitis, and asthma): depression of respiratory center by drugs, cardiopulmonary arrest, stroke, poliomyelitis, or nervous system disorders	Causes:	Renal disease, including hepatitis and cirrhosis: increased acid production in diabetes mellitus, hyperthyroidism, alcoholism, and starvation; loss of alkali in diarrhea: acid retention in renal failure
Treatment:	Correction of disorder, infusion of bicarbonate	Treatment:	Sodium bicarbonate given orally, dialysis for renal failure, insulin treatment for diabetic ketosis
Respiratory Alkalosis: CO_2 ↓ pH ↑		**Metabolic Alkalosis: H_3O^+ ↓ pH ↑**	
Symptoms:	Increased rate and depth of breathing, numbness, light-headedness, tetany	Symptoms:	Depressed breathing, apathy, confusion
Causes:	Hyperventilation because of anxiety, hysteria, fever, exercise; reaction to drugs such as salicylate, quinine, and antihistamines; conditions causing hypoxia (e.g., pneumonia, pulmonary edema, and heart disease)	Causes:	Vomiting, diseases of the adrenal glands, ingestion of excess alkali
Treatment:	Elimination of anxiety-producing state, rebreathing into a paper bag	Treatment:	Infusion of saline solution, treatment of underlying diseases, administer NH_4Cl

equilibrium shown back to the left when excess bicarbonate is present and reacting with the excess acid.

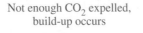

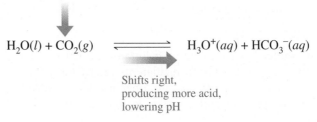

$$H_2O(l) + CO_2(g) \rightleftharpoons H_3O^+(aq) + HCO_3^-(aq)$$

Shifts right,
producing more acid,
lowering pH

Conversely, a person who **hyperventilates** (exhaling too much and not inhaling enough), which can occur under conditions of anxiety or overexercising, is exhaling too much CO_2 from the lungs. This has the effect of drawing H_3O^+ from the bloodstream, making the blood more basic. This condition is known as **respiratory alkalosis**. A person suffering from this condition needs to get more CO_2 back into their bloodstream by breathing a CO_2-rich atmosphere. This can easily be done by having the person breathe into a paper bag, which keeps some of the previously exhaled CO_2 available for the next breath. This will enrich the CO_2 present in the bloodstream, shifting the equilibrium back to the right, thereby producing more H_3O^+ and reestablishing the equilibrium.

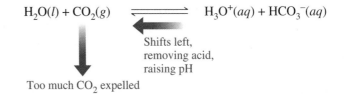

$$H_2O(l) + CO_2(g) \rightleftharpoons H_3O^+(aq) + HCO_3^-(aq)$$

Shifts left,
removing acid,
raising pH

Too much CO_2 expelled

Changes in Metabolic Acid Production

The chemical reactions that occur in our bodies can also directly change the pH of our blood by producing too much or too little H_3O^+. Diabetics use much less glucose for energy than a normal person because glucose is unable to get inside their cells for use. Because of this, diabetics often use fatty acids as a carbon source for energy production. A by-product of fatty acid chemical breakdown is acid production. In this case the imbalance is not caused by breathing, but by chemical reactions in the body so it is termed **metabolic acidosis**. As in the case of respiratory acidosis, a treatment is to administer bicarbonate to neutralize the excess acid, thereby forming more CO_2 that can then be exhaled.

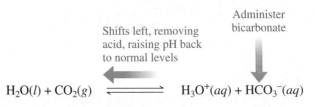

Shifts left, removing
acid, raising pH back
to normal levels

Administer
bicarbonate

$$H_2O(l) + CO_2(g) \rightleftharpoons H_3O^+(aq) + HCO_3^-(aq)$$

The body can also lose too much acid, thereby making the blood basic. This condition, known as **metabolic alkalosis**, can occur under conditions of excessive vomiting (the body is trying to replace the acid lost from the stomach, thereby making the blood basic). In order to lower the pH back to normal, ammonium chloride (NH_4Cl, a weak acid) can be administered to neutralize the excess base buildup.

PRACTICE PROBLEMS

8.49 Sometimes during stress or trauma, a person can start to hyperventilate. The person may then be instructed to breathe into a paper bag to avoid fainting. The bicarbonate equilibrium shown is affected.

$$H_2O(l) + CO_2(g) \rightleftharpoons H_3O^+(aq) + HCO_3^-(aq)$$

a. How does blood pH change during hyperventilation?

b. In which direction will the bicarbonate equilibrium shift during hyperventilation?

c. What is this condition called?

d. How does breathing into a paper bag help return the blood pH to normal?

e. In which direction will the bicarbonate equilibrium shift as the pH returns to normal?

8.50 A person who overdoses on antacids may neutralize too much stomach acid which causes an imbalance in the bicarbonate equilibrium in their blood.

$$H_2O(l) + CO_2(g) \rightleftharpoons H_3O^+(aq) + HCO_3^-(aq)$$

a. How does blood pH change when this occurs?

b. In which direction will the bicarbonate equilibrium shift during an Alka-Seltzer® overdose?

c. What is this condition called?

d. To treat severe forms of this condition, patients are administered NH_4Cl, which acts as a weak acid. How does this restore the bicarbonate equilibrium?

e. In which direction will the bicarbonate equilibrium shift as the pH returns to normal?

SUMMARY

8.1 Acids and Bases—Definitions

Chapter 8 describes properties of acids and bases, beginning with definitions. The Arrhenius definition states that acids produce H^+ and bases produce OH^- when dissolved in water. The Brønsted–Lowry definition builds on the Arrhenius definition by stating that an acid donates an H^+ and a base accepts an H^+. An H^+ or proton in solution exists in contact with a water molecule, thereby forming the hydronium ion, H_3O^+.

$$HCl(g) \xrightarrow[H_2O]{} H^+(aq) + Cl^-(aq)$$

Dissociates into ions

8.2 Strong Acids and Bases

Strong acids and bases completely ionize or dissociate in solution. There are six common strong acids: $HClO_4$, H_2SO_4, HI, HBr, HCl, and HNO_3. Water can act as either an acid or a base in a chemical reaction. A neutralization reaction occurs when a strong acid combines with a strong base. The products of neutralization are an ionic compound (salt) and water. Neutralization reactions are typically exothermic, meaning they produce heat. The use of an antacid provides a common example of a neutralization reaction. Antacids are basic compounds that neutralize excess stomach acid.

chapter eight

8.3 Chemical Equilibrium

Many chemical reactions are reversible operating in the forward direction up to a point when some of the products revert back to reactants. When forward and reverse reactions occur at the same rate, a chemical equilibrium is established. The equilibrium constant, K, defines the extent of a chemical reaction as a ratio of the concentration of the products to the concentration of the reactants. If a chemical reaction at equilibrium is disturbed, the reaction can regain its equilibrium according to Le Châtelier's principle by shifting to offset the disturbance. For example, if more reactants are added to a chemical reaction that is currently at equilibrium, the equilibrium will shift to the right forming more products to offset the addition. An equilibrium can also be offset by changes in temperature. Endothermic reactions require heat in order to occur. This is in contrast to exothermic reactions which give off heat when they occur.

$$N_2(g) + 3H_2(g) \rightleftharpoons 2NH_3(g) + heat$$

8.4 Weak Acids and Bases

Because weak acids partially ionize in solution ($\sim$5% ionized), they are reversible, can establish equilibrium, and have an equilibrium constant called the acid dissociation constant, K_a. When a weak acid dissociates, it produces a conjugate base and when a weak base dissociates, it produces a conjugate acid. These are referred to as conjugate acid–base pairs. The more a weak acid dissociates, the higher its K_a value, and the stronger the acid.

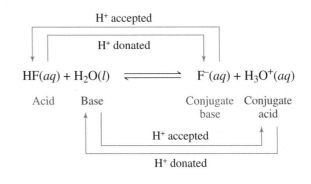

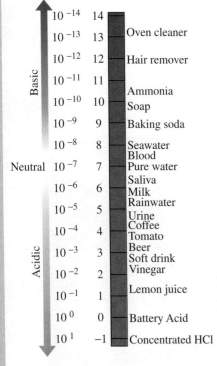

8.5 pH and the pH Scale

Water slightly ionizes, producing hydronium (H_3O^+) and hydroxide (OH^-) ions. When the concentrations of hydronium and hydroxide ions are equal, a neutral solution exists. This is the case for pure water. An excess of H_3O^+ in a solution makes a solution acidic. An excess of OH^- makes a solution basic. The pH scale is a simple measure of acidity with values typically falling between 0–14. Neutral solutions have a pH value of 7, acidic solutions have values less than 7, and basic solutions have pH values higher than 7. The pH of a solution can most easily be measured using an instrument called a pH meter, or by using pH paper. The pH is mathematically related to the concentration of H_3O^+ by the following equation:

$$pH = -\log [H_3O^+]$$

8.6 pK_a

The pH value changes with $[H_3O^+]$. In contrast, the pK_a value is constant for a specific weak acid at certain temperature. The pH value has an effect on which form of the weak acid predominates: the acid form or conjugate base form. If the pH of a solution is *the same as* the pK_a value of a weak acid, then the acid and conjugate base forms are present in equal amounts. If the pH of a solution is *higher* than the pK_a value of a weak acid, the conjugate base form predominates. If the pH of a solution is *lower* than the pK_a value of a weak acid, the acid form predominates in solution.

8.7 Amino Acids: Common Biological Weak Acids

An amino acid is an example of a biological molecule that contains acid–base functional groups. Amino acids are the building blocks of larger biomolecules called proteins. Amino acids contain both carboxylate and protonated amine functional groups at physiological pH (~7.4). This form containing a + and a − charge (net charge 0) is called a zwitterion. The isoelectric point of an amino acid solution has been reached when the pH of the solution is such that all the amino acid molecules are present as a zwitterion and the solution contains no net charge.

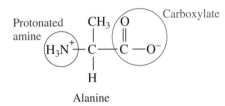

Alanine

8.8 Buffers: An Important Property of Weak Acids and Bases

Buffers resist changes in pH when acid or base is added to a solution. Buffer solutions consist of approximately equal amounts of a weak acid and its conjugate base. A strong acid cannot be part of a buffer system. The bicarbonate buffer system is an important buffer system in the blood. Blood pH is maintained in a narrow range of 7.35–7.45. If the blood pH drops below this range, a condition called acidosis occurs. This can be caused by changes in breathing (ventilation rate) or by changes in metabolism causing excess acid to build up in the bloodstream. If the blood pH becomes elevated, a condition called alkalosis exists. This condition can be caused by changes in breathing or by changes to metabolism causing acid to be removed from the bloodstream.

KEY TERMS

acid—A substance that dissolves in water and produces hydrogen ions (H^+), according to the Arrhenius theory. All acids are proton donors, according to the Brønsted–Lowry theory.

acid dissociation constant, K_a—A constant that establishes the ratio of products to reactants for a weak acid at equilibrium.

acidic—A term that describes an aqueous solution where the concentration of hydronium ions (H_3O^+) is greater than the concentration of hydroxide ions (OH^-).

amino acid—The building blocks of protein containing the functional groups protonated amine and carboxylate, which have weak acid–base forms.

autoionization of water—Spontaneous reaction of two water molecules to form a hydronium ion (H_3O^+) and a hydroxide ion (OH^-)

base—A substance that dissolves in water and produces hydroxide ions (OH^-), according to the Arrhenius theory. All bases are proton acceptors according to the Brønsted–Lowry theory.

basic—A term that describes an aqueous solution where the concentration of hydroxide ions (OH^-) is greater than the concentration of hydronium ions (H_3O^+).

buffer—A solution consisting of a weak acid and its conjugate base or a weak base and its conjugate acid that resists a change in pH when small amounts of acid or base are added.

chemical equilibrium—A state where the rate of formation of products and the rate of formation of reactants are balanced.

conjugate acid—The product after a base accepts a proton. It is capable of donating a proton in the reverse reaction.

conjugate acid–base pair—An acid and base that differ by one H^+.

conjugate base—The product after an acid donates a proton. It is capable of accepting a proton in the reverse reaction.

endothermic reaction—A reaction that absorbs energy in the form of heat from its surroundings.

equilibrium constant, K—A constant that is equal to the ratio of products to reactants at equilibrium.

exothermic reaction—A reaction that produces energy as heat.

homeostasis—The ability of an organism to regulate its internal environment by adjusting its physiological processes.

hydronium ion, H_3O^+—The ion formed by the attraction of a proton (H^+) to an H_2O molecule.

hyperventilation—A condition of expelling too much CO_2 from the lungs, thereby upsetting the bicarbonate equilibrium.

hypoventilation—A condition of expelling too little CO_2 from the lungs, thereby upsetting the bicarbonate equilibrium.

isoelectric point (pI)—The pH at which an amino acid has a net charge of zero.

Le Châtelier's principle—The principle stating that a chemical reaction at equilibrium that is disturbed (concentration change, temperature change) will shift its equilibrium to offset the disturbance.

metabolic acidosis—A physiological condition in which the blood pH is lower than 7.35. It is caused by cellular metabolism.

metabolic alkalosis—A physiological condition in which the blood pH is higher than 7.45. It is caused by cellular metabolism.

neutral—The term that describes a solution with equal concentrations of H_3O^+ and OH^-.

neutralization—The reaction between an acid and a base to form a salt and water.

pH—A measure of the H_3O^+ concentration in a solution. $pH = -\log [H_3O^+]$.

pK_a—A measure of the acidity of a weak acid. $pK_a = -\log K_a$.

respiratory acidosis—A physiological condition in which the blood pH is lower than 7.35. In this case, it is caused by changes in breathing.

respiratory alkalosis—A physiological condition in which the blood pH is higher than 7.45. In this case, it is caused by changes in breathing.

salt—An ionic compound containing a metal or a polyatomic ion (like NH_4^+) as the cation and a nonmetal or polyatomic ion as the anion (exception: OH^-).

strong acid—An acid that completely ionizes in water.

strong base—A base that completely ionizes in water.

weak acid—An acid that ionizes only slightly in solution.

weak base—A base that ionizes only slightly in solution.

zwitterion—The ionic form of an amino acid existing at physiological pH containing a $-\overset{+}{N}H_3$ and $-COO^-$ group and with no net charge.

ADDITIONAL PROBLEMS

8.51 Identify the following acids as strong or weak:
  a. H_2SO_4 **b.** H_2CO_3 **c.** HCl   **d.** HF

8.52 Identify the following acids as strong or weak:
  a. HNO_3 **b.** HNO_2 **c.** H_3PO_4 **d.** HBr

8.53 What are some similarities and differences between strong and weak acids?

8.54 What are some ingredients found in antacids? What do they do?

8.55 For the following reaction,

$$PbI_2(s) \rightleftharpoons Pb^{2+}(aq) + 2\,I^-(aq)$$

  a. write an equilibrium constant expression.
  b. If the value for $K = 7.1 \times 10^{-9}$, which substance predominates, the solid or the ions in solution?

8.56 For the following reaction,

$$N_2O_4(g) \rightleftharpoons 2\,NO_2(g)$$

  a. write an equilibrium constant expression.
  b. If the value for $K = 46$ at 500 K, which gas predominates, N_2O_4, or NO_2?

8.57 Consider the following reaction:

$$CO(g) + 2\,H_2(g) \rightleftharpoons CH_3OH(g) + heat$$

  a. Is the reaction exothermic or endothermic as written?
  b. Provide an equilibrium expression (K) for the reaction.
  c. How will the equilibrium shift if hydrogen gas is added to the reaction?
  d. How will the equilibrium shift if carbon monoxide is removed from the reaction?
  e. How will the equilibrium shift if the temperature is lowered?

8.58 Consider the following reaction:

$$Glycogen(aq)(\text{stored glucose}) + H_2O(l) \rightleftharpoons \text{blood glucose}(aq) + heat$$

  a. Is the reaction exothermic or endothermic as written?
  b. Provide an equilibrium expression (K) for the reaction.
  c. How will the equilibrium shift if blood glucose is added to the reaction?
  d. How will the equilibrium shift if the temperature is raised?

8.59 Determine the pH for the following solutions:
  a. $[H_3O^+] = 2.5 \times 10^{-8}$ M
  b. $[H_3O^+] = 7.0 \times 10^{-2}$ M

8.60 Are the solutions in problem 8.59 acidic, basic, or neutral?

8.61 Calculate the H_3O^+ concentration for a solution with the following pH:
  a. 3.52   **b.** 5.00   **c.** 9.25

8.62 Consider the buffer system of hydrofluoric acid, HF, and its salt, NaF:

$$HF(aq) + H_2O(l) \rightleftharpoons H_3O^+(aq) + F^-(aq)$$

  a. What is the purpose of a buffer system?
  b. Why is a salt of the acid needed?
  c. How does the buffer react when some acid (H_3O^+) is added?
  d. How does the buffer react when some base (OH^-) is added?

8.63 Consider the buffer system of nitrous acid, HNO_2, and its salt, $NaNO_2$:

$$HNO_2(aq) + H_2O(l) \rightleftharpoons H_3O^+(aq) + NO_2^-(aq)$$

  a. What is the purpose of a buffer system?

b. What is the purpose of NaNO$_2$ in the buffer?

c. How does the buffer react when some acid (H$_3$O$^+$) is added?

d. How does the buffer react when some base (OH$^-$) is added?

8.64 Adding a few drops of a strong acid to water will lower the pH appreciably. However, adding the same number of drops to a buffer does not appreciably alter the pH. Why?

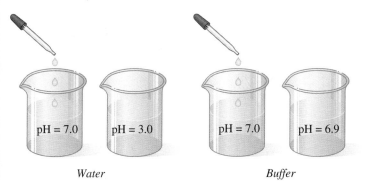

Water   Buffer

8.65 In blood plasma, pH is maintained by the carbonic acid–bicarbonate buffer equilibrium as follows:

$$H_2O(l) + CO_2(g) \rightleftharpoons H_3O^+(aq) + HCO_3^-(aq)$$

a. If excess acid is added to the bloodstream by a physiological condition other than breathing, what condition occurs?

b. Is the pH of the blood below or above normal?

c. How could such a condition be treated?

8.66 The pK_a's for carbonic acid and bicarbonate are 6.35 and 10.33, respectively. Which form of carbonic acid (H$_2$CO$_3$, HCO$_3^-$, or CO$_3^{2-}$) will predominate at physiological pH?

8.67 Consider the amino acid valine shown in its zwitterion form. The pI of valine is 6.0.

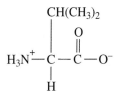

a. Draw the structure of the predominant form of valine at pH = 10.0

b. Draw the structure of the predominant form of valine at pH = 3.0

CHALLENGE PROBLEMS

8.68 Explain why the following amino acid cannot exist in the form shown in aqueous solution?

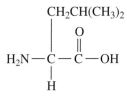

8.69 To determine the concentration of an unknown weak acid solution, a known amount of strong base can be titrated (added) to it until complete neutralization occurs (an endpoint is reached). If a 10.0-mL sample of vinegar (aqueous acetic acid, CH$_3$COOH) requires 16.5 mL of 0.500 M NaOH to reach the endpoint, what is the original molarity of the acetic acid solution?

HINT: *1 mole of NaOH will neutralize 1 mole of acetic acid*

ANSWERS TO ODD-NUMBERED PROBLEMS

Practice Problems

8.1 **a.** acid   **b.** base
       c. acid   **d.** base

8.3 Most hydrogen atoms contain one proton and one electron and no neutrons. Therefore, a positively charged hydrogen ion (H$^+$) is simply one proton.

8.5 **a.** HI is the acid (proton donor) and H$_2$O is the base (proton acceptor).
       b. H$_2$O is the acid (proton donor) and F$^-$ is the base (proton acceptor).

8.7 The strong acids are (a.) H$_2$SO$_4$, (b.) HCl, and (d.) HNO$_3$.

8.9 **a.** HNO$_3$(aq) + LiOH(s) $\longrightarrow$ H$_2$O(l) + LiNO$_3$(aq)
       b. H$_2$SO$_4$(aq) + Ca(OH)$_2$(s) $\longrightarrow$
                              2 H$_2$O(l) + CaSO$_4$(aq)

8.11 **a.** 3 HBr(aq) + Al(OH)$_3$(s) $\longrightarrow$
                              3 H$_2$O(l) + AlBr$_3$(aq)
        b. CaCO$_3$(s) + 2 HI(aq) $\longrightarrow$
                              CaI$_2$(aq) + CO$_2$(g) + H$_2$O(l)

8.13 Reversible reactions are reactions that can proceed in both the forward and reverse directions.

8.15 **a.** $K = \dfrac{[CO_2][H_2]}{[CO][H_2O]}$ **b.** $K = \dfrac{[CH_3COO^-][H_3O^+]}{[CH_3COOH]}$

8.17 a. reactants   **b.** products
    c. both reactants and products present in equal amounts

8.19 a. Products are favored.
    b. Reactants are favored.
    c. Products are favored.
    d. Reactants are favored.

8.21 a. Products are favored.
    b. Reactants are favored.
    c. Products are favored.
    d. Reactants are favored.

8.23 The escape of CO_2 gas removes CO_2; equilibrium favors the formation of the products. The soda no longer bubbles.

8.25 a. $H_2PO_4^-$ **b.** HF   **c.** HBr

8.27 a. acid HSO_4^-; conjugate base SO_4^{2-}
    base H_2O; conjugate acid H_3O^+
    b. acid NH_4^+; conjugate base NH_3
    base H_2O; conjugate acid H_3O^+
    c. acid HCN; conjugate base CN^-
    base NO_2^-; conjugate acid HNO_2

8.29 a. F^-, fluoride ion
    b. OH^-, hydroxide ion
    c. HCO_3^-, bicarbonate ion
    d. SO_4^{2-}, sulfate ion

8.31 a. HCO_3^-, bicarbonate ion
    b. H_3O^+, hydronium ion
    c. H_3PO_4, phosphoric acid
    d. HBr, hydrobromic acid

8.33

a. $HA(aq) + H_2O(l) \rightleftharpoons A^-(aq) + H_3O^+(aq)$

    Acid   Base   Conjugate   Conjugate
                          base   acid

b. $H_2PO_4^-(aq) + H_2O(l) \rightleftharpoons HPO_4^{2-}(aq) + H_3O^+(aq)$

    Acid   Base   Conjugate   Conjugate
                          base   acid

c. $NH_3(aq) + H_2O(l) \rightleftharpoons NH_4^+(aq) + OH^-(aq)$

    Base   Acid   Conjugate   Conjugate
                          acid   base

8.35 a. basic   **b.** acidic   **c.** basic   **d.** acidic

8.37 a. basic   **b.** acidic   **c.** basic   **d.** neutral

8.39 a. 7.92   **b.** 2.15   **c.** 10.33   **d.** 7.00

8.41 a. 7.9×10^{-13} M   **b.** 3×10^{-6} M
    c. 1.0×10^{-2} M   **d.** 5×10^{-9} M

8.43 a. $H_2PO_4^-$ **b.** H_2SO_4
    c. formic acid   **d.** ammonium ion

8.45 a. acid   **b.** neither   **c.** conjugate base

8.47

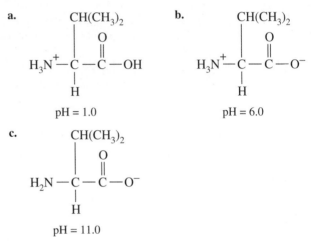

8.49 a. Hyperventilation will lower the CO_2 level in the blood, which decreases the H_3O^+ and increases the blood pH.
    b. The equilibrium will shift to the left.
    c. This condition is called respiratory alkalosis.
    d. Breathing into a bag will increase the CO_2 level, which increases H_3O^+ and lowers the blood pH.
    e. The equilibrium will shift to the right.

Additional Problems

8.51 a. strong   **b.** weak   **c.** strong   **d.** weak

8.53 Both strong and weak acids produce H_3O^+ in water. Weak acids are only slightly ionized, whereas a strong acid exists as ions in solution (fully ionizes).

8.55 a. $K = [Pb^{2+}][I^-]^2$ **b.** The solid is in excess.

8.57 a. exothermic   **b.** $K = \dfrac{[CH_3OH]}{[CO][H_2]^2}$

    c. It will shift to the right.   **d.** It will shift to the left.
    e. It will shift to the right.

8.59 a. 7.60   **b.** 1.15

8.61 a. 3.0×10^{-4} M   **b.** 1.0×10^{-5} M
    c. 5.6×10^{-10} M

8.63 a. A buffer system keeps the pH constant.
    b. The conjugate base, NO_2^-, neutralizes any acid added.
    c. Added H_3O^+ reacts with NO_2^-
    d. Added OH^- reacts with HNO_2.

8.65 a. metabolic acidosis
    b. below normal
    c. administer bicarbonate (HCO_3^-)

8.67

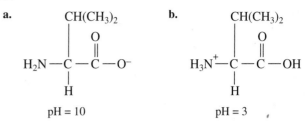

a. pH = 10   **b.** pH = 3

8.69

Number of moles of NaOH = number of moles of acetic acid

$$\frac{0.500 \text{ mole NaOH}}{1 \text{ L}} \times \frac{1 \text{ L}}{1000 \text{ mL}} \times 16.5 \text{ mL}$$

$$= 0.00825 \text{ mole NaOH}$$

$$\frac{0.00825 \text{ mole acetic acid}}{10.0 \text{ mL}} \times \frac{1000 \text{ mL}}{1 \text{L}} = 0.825 \text{ M acetic acid}$$

Guided Inquiry Activities **FOR CHAPTER 9**

SECTION 9.1

EXERCISE 1 Characteristics of Amino Acids

Information

There is a set of 20 amino acids required to make proteins in mammals. We can only manufacture 10 of them. The other 10 must be obtained in the diet. These are termed *essential* amino acids. Dietary proteins are complete or incomplete depending on whether they contain all essential amino acids. Plant sources are more likely to contain incomplete proteins, but when combined, they can provide complete protein sources. For example, protein from rice is deficient in the amino acid lysine, but rich in the amino acid methionine; beans are deficient in methionine and rich in lysine, so a meal of rice and beans can provide complete protein.

All 20 amino acids have some features in common:

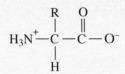

"R" is the part of the 20 amino acids that is different. In amino acids, R is often referred to as the "side chain." All the naturally occurring amino acids are L-*enantiomers*. The 20 different amino acids can be classified by the polarity of their side chain: polar neutral, polar charged (acidic or basic), or nonpolar.

At physiological pH, the amine group is in its protonated form, and the carboxyl group is in its conjugate base form as discussed in section 8.6. This state is called the *zwitterion* form.

Questions

1. What is an *essential* amino acid?

2. Gelatin is a protein produced from an animal protein called collagen that does not contain all 20 amino acids and lacks the essential amino acid tryptophan. Is gelatin considered a complete protein or an incomplete protein?

3. Identify the chiral center in the general structure of an amino acid shown in the preceding figure with an asterisk.

4. Name the two functional groups common to all 20 amino acids.

5. Locate the side chain (R) on each amino acid shown. Identify the functional group(s) found in the side chain (R). Is the R group polar or nonpolar?

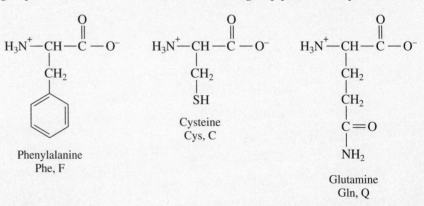

SECTION 9.2

EXERCISE 2 Condensation and Hydrolysis

Information

Amino acids combine to form peptides and ultimately proteins through a chemical reaction called a *condensation* (also known as *dehydration*) reaction. One water molecule is removed when a carboxylate and protonated amine react. The condensation of two amino acids is shown in the following figure.

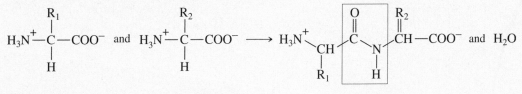

Peptide bond

When two amino acids combine as shown, a *dipeptide* is formed. A string of many amino acids bonded through peptide bonds is a *polypeptide*. A *protein* is a polypeptide that has a function.

Questions

1. Name the organic functional group represented in a peptide bond.

2. In the preceding condensation reaction, a dipeptide is formed. How many amino acids would be present in a tetrapeptide? _____ How many peptide bonds would be formed in a tetrapeptide? _____

3. The breaking of a peptide bond occurs through a hydrolysis reaction. In the body these reactions are catalyzed by enzymes called proteases. Draw the products of the following reaction if the dipeptide undergoes hydrolysis.

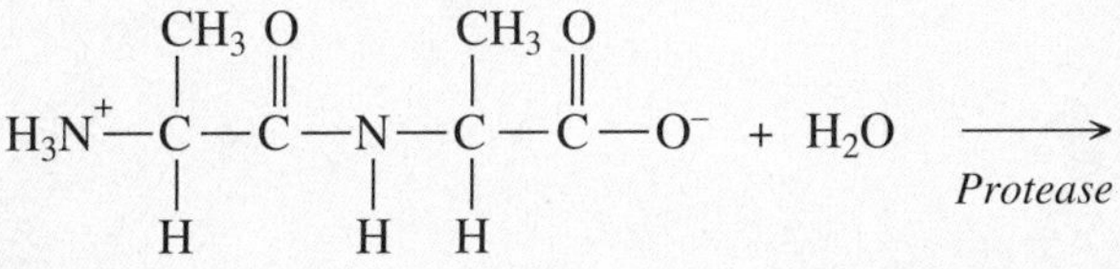

$$H_3N^+-\overset{\overset{\displaystyle CH_3}{|}}{\underset{\underset{\displaystyle H}{|}}{C}}-\overset{\overset{\displaystyle O}{\|}}{C}-\overset{}{\underset{\underset{\displaystyle H}{|}}{N}}-\overset{\overset{\displaystyle CH_3}{|}}{\underset{\underset{\displaystyle H}{|}}{C}}-\overset{\overset{\displaystyle O}{\|}}{C}-O^- \;+\; H_2O \quad \xrightarrow[Protease]{}$$

The dipeptide, Ala—Ala

Your hair is protein. Proteins are chains of amino acids linked together like beads on a string. These "beads" are held together through amide bonds. Proteins can get pretty big, so protein structure can be understood on several levels. To gain the basics of protein structure, read on.

Proteins
AMIDES AT WORK

9

We have seen amino acids acting as weak acids in Chapter 8. While their behavior as weak acids is significant, amino acids serve a biologically important role as the building blocks of life.

Much of the structure and function of our bodies is built on large biomolecules called **proteins**. Proteins transport oxygen in blood, serve as the primary components of skin and muscle, defend our bodies against infection, direct cellular chemistry as enzymes, and control metabolism as hormones. Proteins are polymers (*poly*—"many," *mers*—"parts") composed entirely of the building blocks of life, amino acids, covalently bonded in specific sequences. There are 20 different commonly occurring amino acids, and proteins can contain as few as 50 or as many as several hundreds or even thousands of these amino acids. The ordering of the amino acids in a protein determines both its structure and biological function.

In this chapter, we will discover that each amino acid has its own characteristics and chemical behavior. When the amino acids are linked together in a sequence like colored beads on a string, this chain can twist and fold as a result of intermolecular and ionic forces to form a unique three-dimensional structure that has a specific function.

9.1 Amino Acids—A Second Look

Amino Acid Structure

As mentioned in Section 8.6, the term *amino acid* gives us a way to remember its structure. "Amino" reminds us it contains a protonated amine ($-NH_3^+$), and "acid" reminds us it contains a carboxylic acid in the form of carboxylate ($-COO^-$). These two functional groups are bonded to a central carbon atom called the **alpha (α) carbon**. The protonated amine bonded to this carbon is often referred to as the **alpha (α) amino group** and the carboxylate ion as the **alpha (α) carboxylate group**. The α carbon is also bonded to a hydrogen atom and a larger side chain, which is responsible for the unique identity and characteristics of each amino acid.

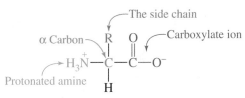

In all but one amino acid (glycine), the α carbon is a chiral carbon atom because it has four different groups bonded to it. As we saw in Chapter 4, compounds that contain a single chiral carbon atom can have two forms called enantiomers, which are nonsuperimposable mirror images. We can draw Fischer projections for amino acids similar to those drawn for the carbohydrates. As in the following figure, with the carbon chain drawn vertically, the carboxylate group at the top and the R group at the bottom, an amino acid has an enantiomer with the protonated amine on the left-hand side of the structure and one with the protonated amine on the right-hand side of the structure. The enantiomer with the protonated amine on the left is called the L-stereoisomer and its mirror image is the D-stereoisomer.

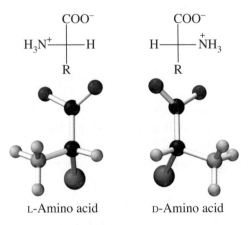

L-Amino acid   D-Amino acid

The L-amino acids are the building blocks of proteins. Some D-amino acids do occur in nature, but not in proteins.

The side chain or R group gives each amino acid its unique identity and characteristics. Twenty amino acids are found in most proteins. Nine different families of organic compounds are represented in the structures of the different amino acid side chains—alkanes, aromatics, thioethers, alcohols, phenols, thiols, amides, carboxylic acids (present as carboxylate ions), and amines (present as protonated amines). You may want to refer back to Table 4.6 to get reacquainted with the structures of these functional groups. The chemical properties and behaviors of the functional groups are present in the amino acids and are used to classify the amino acids into a few simple categories (nonpolar, polar, acidic, and basic). The structures of the amino acids, including their side chains, names, functional groups, and abbreviations are shown in Table 9.1.

TABLE 9.1 THE 20 COMMON AMINO ACIDS IN PROTEINS (COMMON FUNCTIONAL GROUPS NOTED IN SIDE CHAINS)

Nonpolar Amino Acids

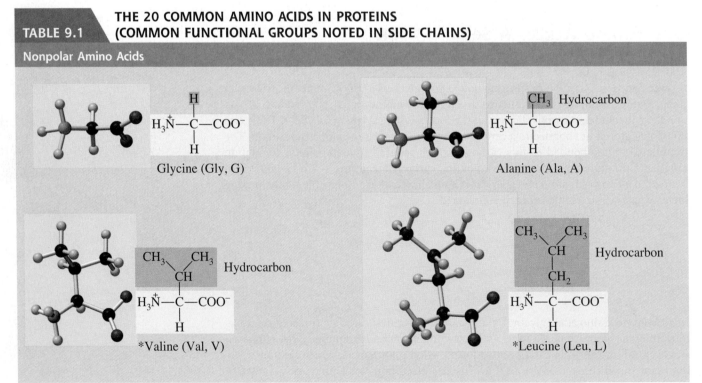

Glycine (Gly, G)

Alanine (Ala, A)

*Valine (Val, V)

*Leucine (Leu, L)

* Essential amino acids.

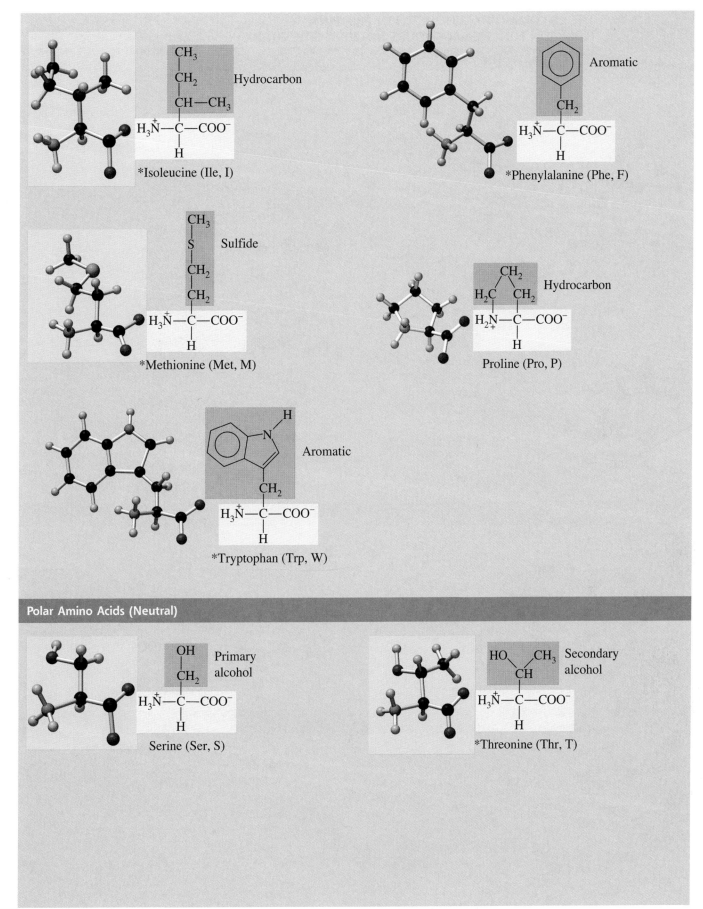

$$CH_3$$
$$CH_2$$
$$CH—CH_3$$
$$H_3\overset{+}{N}—C—COO^-$$
$$H$$
Hydrocarbon

*Isoleucine (Ile, I)

Aromatic

$$CH_2$$
$$H_3\overset{+}{N}—C—COO^-$$
$$H$$

*Phenylalanine (Phe, F)

$$CH_3$$
$$S$$
$$CH_2$$
$$CH_2$$
$$H_3\overset{+}{N}—C—COO^-$$
$$H$$
Sulfide

*Methionine (Met, M)

$$CH_2$$
$$H_2C \quad CH_2$$
$$H_2\overset{+}{N}—C—COO^-$$
$$H$$
Hydrocarbon

Proline (Pro, P)

$$H$$
$$N$$
$$CH_2$$
$$H_3\overset{+}{N}—C—COO^-$$
$$H$$
Aromatic

*Tryptophan (Trp, W)

Polar Amino Acids (Neutral)

$$OH$$
$$CH_2$$
$$H_3\overset{+}{N}—C—COO^-$$
$$H$$
Primary alcohol

Serine (Ser, S)

$$HO \quad CH_3$$
$$CH$$
$$H_3\overset{+}{N}—C—COO^-$$
$$H$$
Secondary alcohol

*Threonine (Thr, T)

* Essential amino acids.

| TABLE 9.1 | THE 20 COMMON AMINO ACIDS IN PROTEINS (COMMON FUNCTIONAL GROUPS NOTED IN SIDE CHAINS) (continued) |

Polar Amino Acids (Neutral)

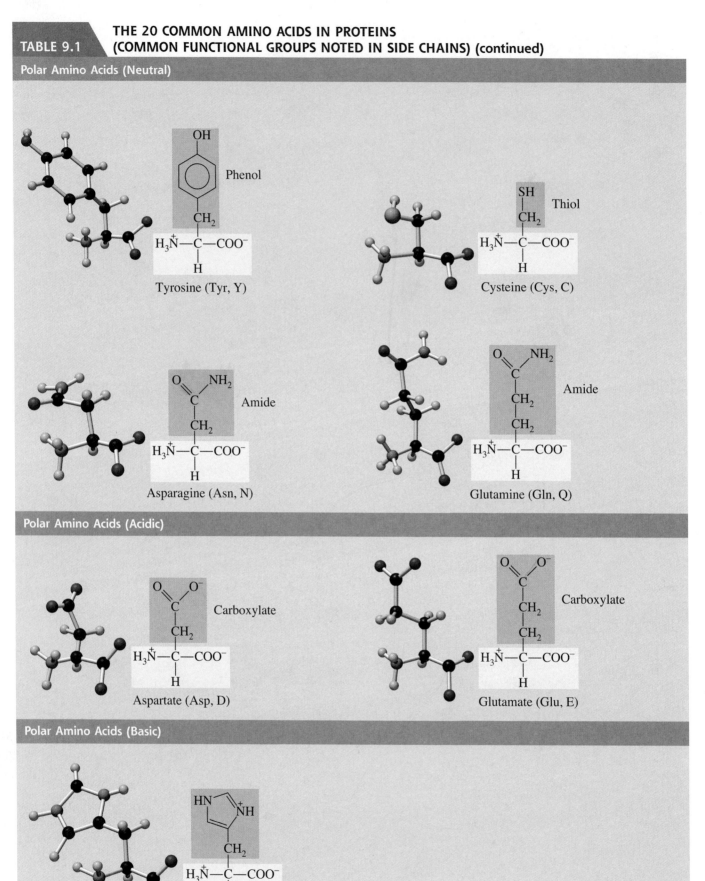

Phenol

Tyrosine (Tyr, Y)

Thiol

Cysteine (Cys, C)

Amide

Asparagine (Asn, N)

Amide

Glutamine (Gln, Q)

Polar Amino Acids (Acidic)

Carboxylate

Aspartate (Asp, D)

Carboxylate

Glutamate (Glu, E)

Polar Amino Acids (Basic)

*Histidine (His, H)

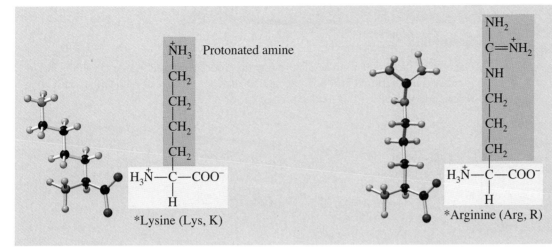

*Essential amino acids.

The 10 amino acids designated with an asterisk (*) in the table are called essential amino acids because they cannot be synthesized in the human body and must be obtained in the diet. Two of these amino acids, arginine and histidine, are essential in the diets of children, but not adults. The nonessential amino acids can be synthesized from the essential amino acids as needed by the body.

Proteins that contain all of the essential amino acids are known as *complete proteins*. Soybeans and most proteins found in animal products such as eggs, milk, fish, poultry, and meat are complete proteins. Some proteins from plant sources are *incomplete proteins* because they lack one or more of the essential amino acids. A complete protein meal can be obtained by combining foods like rice and beans or peanut butter on whole grain bread.

Classification of Amino Acids

The amino acids in Table 9.1 are separated into two broad classifications based on the characteristics of the side chains: **nonpolar amino acids** and **polar amino acids**. The polar amino acids are further divided into neutral, acidic, and basic.

If we look at the side chains of the amino acids that are classified as "nonpolar," we can see that, with a few exceptions, the side chains of these amino acids are composed entirely of carbon and hydrogen. Recall from Chapter 6 that the carbon–hydrogen bond is nonpolar and compounds composed of only carbon and hydrogen are nonpolar and hydrophobic (water-fearing). So, the side chains of this group of amino acids are nonpolar and hydrophobic. (In the case of tryptophan, the atoms other than carbon and hydrogen make up such a small part of the side chain that they contribute very little to the polarity of these amino acids. In the case of methionine, the C–S bonds present are nonpolar since C and S have the same electronegativity—see Chapter 3, Figure 3.13.)

The side chains of each of the amino acids classified as "polar" contain functional groups such as hydroxyl (—OH) and amide (—CONH$_2$) that create an uneven distribution of electrons in the side chain, making them polar. All but one of the polar side chains can form hydrogen bonds with water and, therefore, are hydrophilic (water-loving). The exception is cysteine, which has a polar thiol (—SH) group but does not form hydrogen bonds. Notice that the polar acidic and polar basic amino acids have charged side chains, allowing them to form ion–dipole interactions with water, so they are even more polar and hydrophilic than are the amino acids classified as polar neutral.

SAMPLE PROBLEM 9.1

Identifying Amino Acids

Name the following amino acids, provide the three-letter and one-letter abbreviations, identify the functional group in the side chain, identify any chiral centers with an asterisk, and indicate if the side chain is polar or nonpolar.

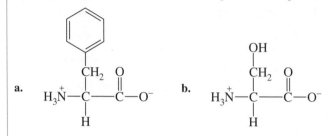

a.   b.

SOLUTION

a. Referring to Table 9.1, this amino acid is phenylalanine, abbreviated Phe or F. It contains one chiral center and an aromatic functional group. The side chain contains only C and H, so this amino acid is nonpolar (hydrophobic).

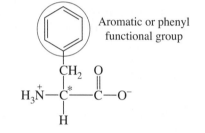

Aromatic or phenyl functional group

b. Referring to Table 9.1, this amino acid is serine, abbreviated Ser or S. It contains one chiral center. Because the side chain contains an OH, this is a polar (hydrophilic) neutral amino acid.

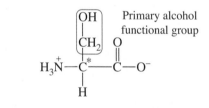

Primary alcohol functional group

PRACTICE PROBLEMS

9.1 Draw the structural formula for each of the following amino acids and put an asterisk (*) next to any chiral carbon atoms in your structure:

a. alanine   b. lysine   c. tryptophan   d. aspartate

9.2 Classify each of the amino acids in problem 9.1 as hydrophobic (nonpolar) or hydrophilic (polar).

9.3 Give the three-letter and one-letter abbreviations and identify the functional group in the side chain for each of the amino acids in problem 9.1.

9.4 Draw the structural formula for each of the following amino acids and put an asterisk (*) next to any chiral carbon atoms in your structure.

a. leucine   b. glutamate   c. methionine   d. threonine

9.5 Classify each of the amino acids in problem 9.4 as hydrophobic (nonpolar) or hydrophilic (polar).

9.6 Give the three-letter and one-letter abbreviations and identify the functional group in the side chain for each of the amino acids in problem 9.4.

9.7 Draw the structure for the amino acid represented by each of the following abbreviations:

a. G   b. His   c. Q   d. Ile

9.8 Draw the structure for the amino acid represented by each of the following abbreviations:

a. Pro   b. N   c. Val   d. Y

9.9 Give the names of the amino acids you drew in problem 9.7.

9.10 Give the names of the amino acids you drew in problem 9.8.

9.2 Protein Formation

With our amino acid building blocks in hand, we now need to understand how proteins form from these 20 different molecules. In Chapter 5, we showed how two monosaccharides bonded together through a condensation reaction to form a disaccharide. Condensations can also occur between amino acids, and the product formed is a dipeptide. In this case, the carboxylate ion ($-COO^-$) of one amino acid molecule reacts with the protonated amine ($-NH_3^+$) of a second amino acid. A water molecule is removed, and an **amide** functional group is formed.

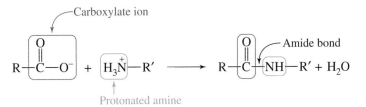

Amides are a family of organic compounds characterized by a functional group containing a carbonyl ($C=O$) bonded to a nitrogen atom. The nitrogen of the amide is bonded to two other atoms, which may be two hydrogen atoms, two carbon atoms, or one carbon atom and one hydrogen atom. When *amino acids* combine in a condensation reaction, the amide bond that is formed between them is called a **peptide bond**.

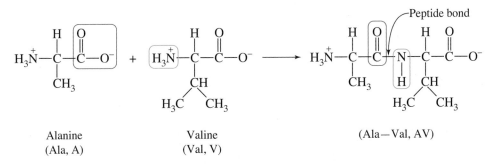

The product of the condensation reaction of alanine and valine, as shown in the preceding figure, is a **dipeptide**, which is represented as Ala—Val, or AV. In this dipeptide, alanine is called the **N-terminus** (or N-terminal amino acid) because it has an unreacted (free) α-amino group. Similarly, valine is called the **C-terminus** (or C-terminal amino acid) because it has an unreacted α-carboxylate group. By convention, dipeptides (and larger peptide-containing structures) are always written from the N-terminus to the C-terminus (left to right) whether using full structures or the one- or three-letter abbreviations.

Dipeptides can be formed from any two amino acids, and each pair of different amino acids can combine in two ways forming two different dipeptides. The two dipeptides, Ala—Val and Val—Ala, are structural isomers, different compounds, and have different properties. While the order of the amino acids may not seem that important, it is critical to both the structure and the function of the compound.

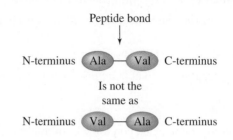

Dipeptides are the smallest members of the peptide class. Any compound containing amino acids joined by a peptide bond can be called a peptide. A compound with three amino acids is a tripeptide, one with four amino acids is a tetrapeptide, and so on. As the number of amino acids increases, the compounds are referred to as polypeptides. A biologically active polypeptide typically containing 50 or more amino acids joined together by peptide bonds is referred to as a protein.

SAMPLE PROBLEM 9.2

Identifying the Components of a Peptide

Consider the following dipeptide:

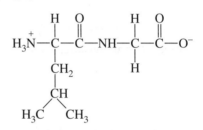

a. Circle the N-terminal amino acid and give its name.

b. Put a square around the C-terminal amino acid and give its name.

c. Locate the peptide bond by drawing an arrow to it.

d. Give the three-letter and one-letter abbreviations for this dipeptide.

SOLUTION

a., b., and c.

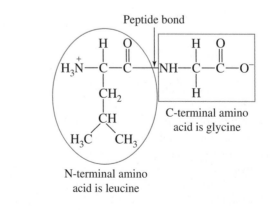

d. The dipeptide is Leu—Gly or LG.

PRACTICE PROBLEMS

9.11 Consider the following tripeptide:

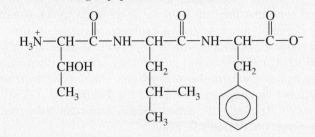

a. Draw a circle around the N-terminal amino acid and give its name. Draw a square around the C-terminal amino acid and give its name.

b. Give the one-letter and three-letter abbreviations of this tripeptide.

9.12 Consider the following tripeptide:

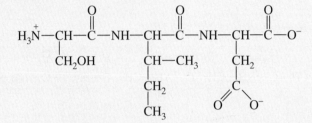

a. Draw a circle around the N-terminal amino acid and give its name. Draw a square around the C-terminal amino acid and give its name.

b. Give the one-letter and three-letter abbreviations of this tripeptide.

9.13 Draw the structural formula for each of the following peptides:

a. Phe—Glu   b. KCG

c. His—Met—Gln   d. WY

9.14 Draw the structural formula for each of the following peptides:

a. Ala—Asn—Thr   b. DS

c. Val—Arg   d. IYP

9.15 Label the N-terminus and the C-terminus for each of the peptides in problem 9.13.

9.16 Label the N-terminus and the C-terminus for each of the peptides in problem 9.14.

9.3 The Three-Dimensional Structure of Proteins

In Chapter 3, we saw that molecules are three-dimensional and that the shapes of simple molecules are determined by the charge clouds around a given atom using the VSEPR model. Peptides and proteins also have three-dimensional shapes or structures, but even the simplest dipeptide composed of two glycines has 17 atoms! It would be nearly impossible to determine its overall three-dimensional shape using the VSEPR model. Because proteins are more complex than simple organic molecules, a hierarchy was developed to describe the structure of proteins. This system organizes the structure of a protein into four levels: primary (1°), secondary (2°), tertiary (3°), and quaternary (4°). Each of these levels is a result of interactions between amino acids in the protein. Taken together, these four levels provide a complete description of the three-dimensional structure of a protein.

Primary Structure

Like different colored beads on a string, the **primary (1°) structure** of a protein is the order in which the amino acids (beads) are joined together by peptide bonds to form the **protein backbone** that lines up from the N-terminus to the C-terminus (left to right). The side chains of the amino acids are substituents dangling from this backbone.

The actual sequence of amino acids is important. To illustrate this, consider the word *protein*. Rearranging the letters in the word, we can make the word *pointer*, which has no meaning in common with the word *protein*. The same seven letters are used to make each word, but the meaning of each word is different. So it is with the primary structure of proteins. Arranging the amino acids in a different order creates a different peptide that no longer has the same function.

Figure 9.1 shows the primary structure of the eight amino acid peptide *angiotensin II*. This small peptide is involved in blood pressure regulation in humans. Its amino acid sequence is Asp—Arg—Val—Tyr—Ile—His—Pro—Phe (DRVYIHPF). Any other ordering of these same eight amino acids would not function as angiotensin II. The order of the amino acids determines the function of a protein.

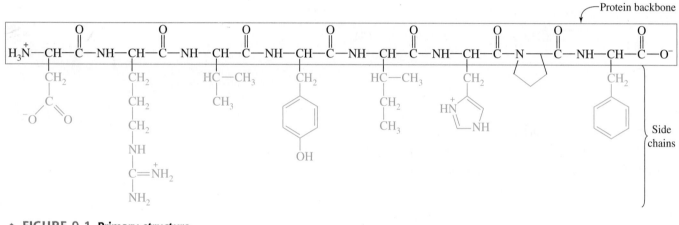

▲ **FIGURE 9.1 Primary structure of angiotensin II.** *Angiotensin II* is a peptide involved in the regulation of blood pressure in humans. Notice that the side chains (blue) are not part of the backbone.

Secondary Structure

Have you ever stretched a Slinky toy or a telephone cord and noticed the spiral shape of its coils? Like the coils of a slinky or a phone cord, the **secondary (2°) structure** of a protein describes repeating patterns of structure within the three-dimensional shape of a protein.

The two most common secondary structures are the **alpha helix (α helix)** and the **beta-pleated sheet (β-pleated sheet)**. The secondary structures in a protein are stabilized by hydrogen bonding along the protein backbone and involve amino acids that are near each other in the primary structure. (Recall from Chapter 6 that hydrogen bonding is not a covalent bond, but a strong intermolecular force.)

The α helix is a coiled structure in which the protein backbone adopts the form of a right-handed coiled spring. As shown in Figure 9.2, the helical structure is stabilized by hydrogen bonds formed between the carbonyl (C=O) oxygen atom (δ^-) of one amino acid and the N—H hydrogen atom (δ^+) of the amino acid located four amino acids away from it in the primary structure. The positioning of the hydrogen bonds along the axis of the helix allows it to stretch and recoil. Multiple hydrogen-bonding interactions along the backbone make the helix a strong structure. In the α helix, the side chains of the amino acids project outward away from the axis of the helix.

The β-pleated sheet is an extended structure in which segments of the protein chain align to form a zigzag structure much like a folded paper fan. As shown in Figure 9.3, strands called beta strands are held together side-by-side by hydrogen-bonding interactions between their backbones. In the β-pleated sheet, the side chains of the amino acids project above and below the sheet.

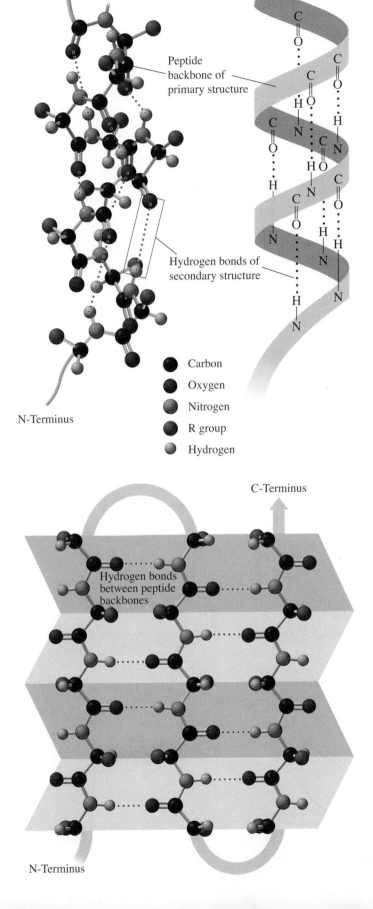

C-Terminus

Peptide
backbone of
primary structure

Hydrogen bonds of
secondary structure

- Carbon
- Oxygen
- Nitrogen
- R group
- Hydrogen

N-Terminus

◀ **FIGURE 9.2 The α helix.** The α helix has the hydrogen bonds positioned along the axis of the helix. There are three amino acids between the carbonyl and amine hydrogen that form the hydrogen-bonding interaction. Often the backbone is represented by a ribbon as shown on the right.

C-Terminus

Hydrogen bonds
between peptide
backbones

N-Terminus

◀ **FIGURE 9.3 β-pleated sheet.** Adjacent strands of the protein backbone are held together by hydrogen bonds in the β-pleated sheet. Alternating R groups (front and back) give the backbones a zig-zag or pleated shape.

In summary, secondary structures involve hydrogen bonding along the backbone of the protein chain. Interactions of the amino acid side chains within the structure of a protein lead to the next level of structure.

Tertiary Structure

Imagine a long phone cord lying on a desk with short lengths of the coils arranged next to or piled on top of each other. Like the coils of the phone cord, the α helices and β-pleated sheets of the polypeptide chain interact with each other and their environment to create an overall three-dimensional shape that is described by the **tertiary structure** (3°).

The folding and twisting of the polypeptide chain is caused by both hydrophobic and hydrophilic interactions between the side chains of the amino acids and the surrounding aqueous environment. Similar to the soap micelle that we saw in Chapter 6, in proteins the nonpolar side chains are repelled by the aqueous environment and end up in the interior of the protein while the polar side chains are attracted to the aqueous surroundings and appear on the surface. The tertiary structure is stabilized by the attractive forces between the side chains and the aqueous environment of the protein as well as the attractive forces between the side chains themselves. In a delicate balancing act that attempts to satisfy all the competing interactions, the protein folds into a specific three-dimensional shape.

The interactions involved when a protein folds up into its tertiary structure (see Figure 9.4) include the following:

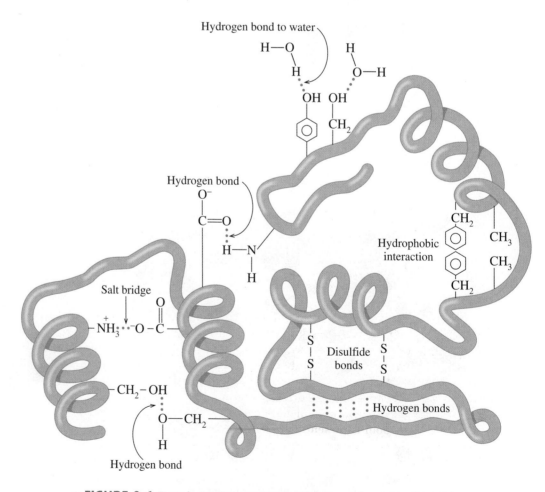

▲ **FIGURE 9.4 Protein tertiary structure.** The interactions of amino acid side chains with their environment and each other stabilize the tertiary structure of a protein.

1. **Nonpolar interactions** The nonpolar amino acid side chains are *unlike* the aqueous (polar) environment. They are repelled by the aqueous environment and aggregate in the interior of the protein. The ability of water to exclude the nonpolar side chains is referred to as the **hydrophobic effect** and drives the three-dimensional folding of proteins. The nonpolar side chains, gathered together in the interior of the protein, interact with each other through London forces, helping to stabilize the tertiary structure.

2. **Polar interactions** Polar amino acid side chains are *like* the surrounding aqueous environment and easily interact with water and each other through the following hydrophilic interactions: dipole–dipole, ion–dipole, and hydrogen bonding interactions. Because they are charged, polar acidic and polar basic side chains form ion–dipole interactions with water. The side chains of most of the polar amino acids (cysteine being an exception) can interact with each other and with water in the surrounding environment through hydrogen bonding.

3. **Salt bridges (ionic interactions)** Acidic and basic amino acid side chains exist in their ionized form in an aqueous environment. In other words, they exist as carboxylate ions ($-COO^-$), and the bases exist as the protonated amines ($-NH_3^+$). When a protein folds and a carboxylate ion on one amino acid side chain winds up near a protonated amine on a second amino acid side chain, the opposite charges attract, thereby forming a stabilizing interaction called a salt bridge or ionic interaction.

4. **Disulfide bonds** If two cysteine side chains are brought close together during the folding of a protein, the two $-SH$ groups (thiols) can react with each other through an oxidation reaction (losing hydrogens) to form a disulfide bond ($-S-S-$). Note that the disulfide bond is a covalent bond (electrons being shared between two atoms) in contrast to the other stabilizing attractive forces.

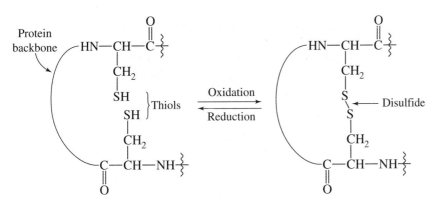

The twisting and folding of a protein chain results in the protein adopting a three-dimensional structure. Proteins can be loosely classified into groups based on how they fold into their resulting overall three-dimensional shape. **Globular proteins** are proteins that fold into a compact, spherical shape. Such proteins have the tertiary structure just described; that is, they have polar amino acid side chains on their surface, enhancing their solubility in water while the nonpolar amino acid side chains gather in the center of the spherical structure forming a nonpolar core. Myoglobin is an example of a globular protein (see Figure 9.5). It stores oxygen in skeletal muscle. While myoglobin is present in human muscle, it exists in much higher concentration in the muscles of sea mammals such as dolphins and whales and is the reason these animals can stay underwater for long periods of time without surfacing to breathe.

Fibrous proteins, on the other hand, have long, threadlike structures. Such proteins typically have high helical content. The association of multiple helices forms

▶ **FIGURE 9.5 Globular protein.**
Globular proteins like myoglobin fold into spherical structures with polar amino acid side chains on their surface and nonpolar amino acid side chains in the interior. The ribbon shown represents the protein backbone.

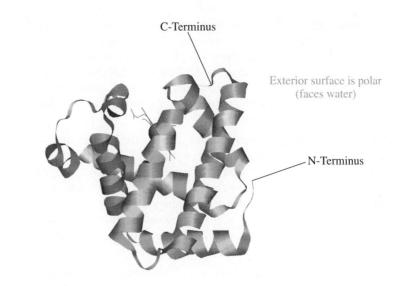

C-Terminus

Exterior surface is polar
(faces water)

N-Terminus

strong, durable structures. The most recognizable examples of fibrous proteins are keratins (see Figure 9.6), found in hair, nails, the scales of reptiles, and collagen, found in connective tissue, skin, tendons, ligaments, and cartilage (see Section 9.5).

▶ **FIGURE 9.6 Fibrous protein.**
The wool of sheep contains the fibrous protein alpha keratin. Such proteins are often composed of proteins with a high helical content.

α Helix

α-Keratin

Quaternary Structure

Imagine two or more of the balled-up phone cords mentioned earlier neatly arranged next to each other. The **quaternary (4°) structure** describes the interactions of two or more polypeptide chains to form a single biologically active protein. The individual polypeptide chains or subunits are held together by the same interactions that stabilize the tertiary structure of a single protein. The most well studied example of a protein with a quaternary structure is hemoglobin, which transports oxygen in blood. Hemoglobin consists of four polypeptide chains or subunits—two identical alpha subunits and two identical beta subunits (see Figure 9.9). In adult hemoglobin, all four subunits must be present in order for the protein to properly function as an oxygen carrier. It is important to note that not all biologically active proteins have a quaternary structure. Some proteins, myoglobin for example, are made up of a single polypeptide chain.

Figure 9.7 and Table 9.2 summarize the structural levels of proteins.

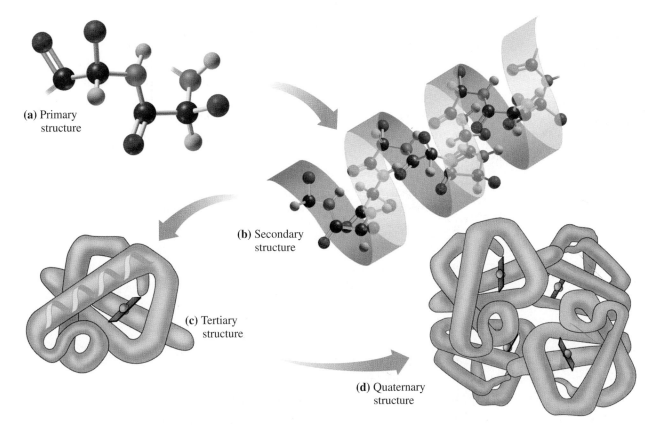

(a) Primary structure

(b) Secondary structure

(c) Tertiary structure

(d) Quaternary structure

▲ **FIGURE 9.7 The levels of protein structure.** All proteins have primary, secondary, and tertiary structures. Some also have a quaternary structure.

TABLE 9.2	SUMMARY OF LEVELS OF STRUCTURE AND STABILIZING FORCES IN PROTEINS
Level of Structure	**Forces Stabilizing Structure**
Primary (1°)	Peptide bonds
Secondary (2°)	Hydrogen bonding along the protein backbone between amino acids close together in sequence
Tertiary (3°)	London forces, hydrogen bonding, dipole–dipole and ion–dipole interactions, salt bridges, and disulfide bonds between amino acids far away from each other in sequence.
Quaternary (4°)	Same as tertiary structure but between subunits

SAMPLE PROBLEM 9.3

Side Chains in Tertiary Structure

Predict whether each of the following amino acid side chains would more likely be on the surface or in the interior of a protein after it has folded into its tertiary structure.

a. tyrosine **b.** leucine

SOLUTION

a. In Table 9.1, tyrosine is classified as a polar amino acid. Therefore, it would be on the surface of the protein. The —OH group of the tyrosine side chain would form hydrogen bonds with water.

b. In Table 9.1, leucine is classified as a nonpolar amino acid. This means that it would be repelled by the aqueous environment and would be found along with other nonpolar amino acids in the interior of the protein.

9.17 What type of bonding is present in the primary structure of a protein?

9.18 How many different tripeptides that contain one leucine, one glutamate, and one tryptophan are possible?

9.19 Name the stabilizing intermolecular force in the secondary structure of a protein.

9.20 When a protein folds into its tertiary structure, how does the primary structure change?

9.21 Describe the differences in the shape of an α helix and a β-pleated sheet.

9.22 What type of interaction would you expect between the side chains of each of the following pairs of amino acids in the tertiary structure of a protein?

a. histidine and aspartate
b. alanine and valine
c. two cysteines
d. serine and asparagine

9.23 What type of interaction would you expect between the side chains of each of the following pairs of amino acids in the tertiary structure of a protein?

a. lysine and glutamate
b. leucine and isoleucine
c. threonine and tyrosine
d. glutamine and arginine

9.24 Determine whether each of the following statements describes the primary, secondary, tertiary, or quaternary structure of a protein.

a. Side chains interact to form disulfide bonds.
b. Peptide bonds join amino acids in a polypeptide chain.
c. Two polypeptide chains are held together by hydrogen bonds.
d. Hydrogen bonding between amino acids in the same polypeptide gives a coiled shape to the protein.

9.25 Determine whether each of the following statements describes the primary, secondary, tertiary, or quaternary structure of a protein.

a. Three polypeptide chains interact to form a biologically active protein.
b. Hydrogen bonds form between adjacent segments of the backbone of the same protein to form a "folded-fan" type structure.
c. Nonpolar side chains are repelled by water and move to the interior of the protein.
d. Amino acids react in a condensation reaction to form a peptide bond.

9.26 Myoglobin is a protein containing 153 amino acids. Approximately half of the amino acids in myoglobin have polar side chains.

a. Where would you expect these amino acid side chains to be located in the tertiary structure of the protein?
b. Where would you expect the nonpolar side chains to be?
c. Would you expect myoglobin to be more or less soluble in water than a protein composed primarily of nonpolar amino acids? Why?

9.4 Denaturation of Proteins

Have you ever cracked open an egg and dropped it into a hot pan? When this happens, the clear part of the egg called the egg white quickly turns from clear to white in the hot pan. You are actually observing the **denaturation** of the proteins in the egg white.

Denaturation of a protein is a process that disrupts the stabilizing attractive forces in the secondary, tertiary, or quaternary structures. When a protein is denatured, its primary structure is not changed.

In the example of the egg white, the denaturing agent is heat. An increase in temperature disrupts intermolecular forces such as hydrogen bonding, London forces, and other polar interactions. A change in pH can also denature a protein. When the pH of a protein's environment changes, the side chains of the acidic and basic amino acids alter their charges and, therefore, their ability to form salt bridges. Certain organic compounds or heavy metal ions can also denature a protein by disrupting disulfide bridges. Table 9.3 gives several examples of denaturing agents and their effects on the various stabilizing forces of protein structure.

TABLE 9.3 EXAMPLES OF PROTEIN DENATURATION		
Denaturing Agent	**Disrupted Forces**	**Examples**
Heat above 50 °C	Hydrogen bonds and hydrophobic interactions	Cooking food
Acids and bases	Salt bridges and hydrogen bonds	Lactic acid from bacteria, which denatures milk proteins in the preparation of yogurt and cheese
Organic compounds	Disulfide bonds	Thiols, which are used in hairstyling for hair straightening or permanent waves
Heavy metal ions Ag^+, Pb^{2+}, Hg^{2+}	Disulfide bonds and salt bridges	Mercury and lead poisoning
Mechanical agitation	Hydrogen bonds and London forces	Whipped cream and meringue made from egg whites

Without the stabilizing interactions necessary to maintain its three-dimensional structure, a protein will unfold into a shapeless string of amino acids. When a protein loses its three-dimensional shape, it also loses its biological activity.

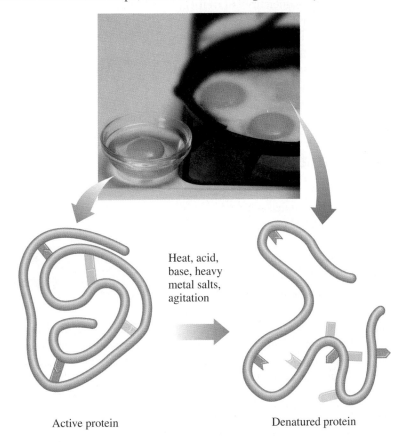

Heat, acid, base, heavy metal salts, agitation

Active protein   Denatured protein

Protein denaturation is applicable to hairstyling. If you have straight hair and you want it to be curly, you get a permanent wave. Alternatively, if you have curly hair and want it to be straight, you get it relaxed or straightened. Both these processes involve denaturing the proteins in your hair by disrupting the disulfide bonds found in the hair protein keratin, reshaping the hair, and forcing the disulfide bonds to reform at different points. Ammonium thioglycolate is one of the chemicals used to reduce the disulfides, and hydrogen peroxide is often used to oxidize the resulting thiols back to disulfide bonds.

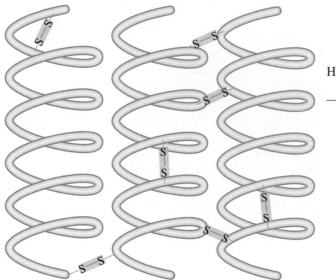 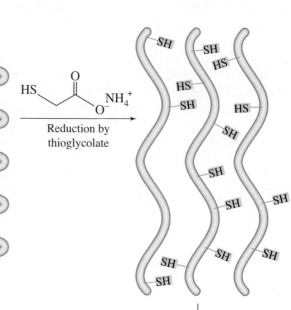

Proteins and their denaturation can also serve as an antidote for lead or mercury poisoning. A person who has accidentally ingested a heavy metal like lead or mercury is given egg whites to drink. The proteins in the egg whites are denatured by the mercury or lead and the combination forms a precipitate. Typically, an emetic is then administered to induce vomiting in order to discharge the metal–protein precipitate from the body.

9.5 Examples of Biologically Important Proteins

Next, we consider some biologically important proteins. In each example, notice how the structure of the protein dictates its function.

Collagen

About one-third of all the body's protein is collagen, making it the most abundant protein in the human body. It is found in connective tissues such as cartilage, skin, blood vessels, and tendons. The fibrous structure of collagen arises from three polypeptide strands woven together to form a special quaternary structure called a **triple helix** (see Figure 9.8). In contrast to the right-handed α helix, each of the polypeptides of the triple helix is a left-handed helix. The three left-handed helices wind together into a right-handed triple helix. This strong, ropelike structure allows collagen to perform its function of holding together bone, muscle, and vascular tissue in ligaments, tendons, and blood vessels. Collagen primarily contains the amino acids glycine, proline, and alanine along with a modified amino acid called hydroxyproline. The additional hydroxy group on hydroxyproline permits the polypeptide chains to form hydrogen bonds between the chains, stabilizing the quaternary structure, and adding extra strength to the collagen triple helix. The elongated and stabilized structure of the collagen helix facilitates its function as the primary component of connective tissues.

Triple helix   3 α Helix peptide chains

◀ **FIGURE 9.8 Collagen.** The collagen triple helix is composed of three left-handed helices woven together into a strong triple helix.

Scurvy, caused by a lack of vitamin C in the diet actually affects collagen formation. Vitamin C is critical to the conversion of proline to hydroxyproline, which functions to strengthen collagen. When hydroxyproline is absent, collagen becomes weakened and results in the common symptoms of spongy and bleeding gums, opening of healed scars, and nail loss. Scurvy can be reversed by resuming a diet containing normal amounts of vitamin C.

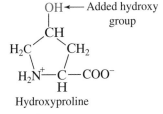

Hydroxyproline

Hemoglobin

As mentioned previously, the globular protein hemoglobin transports oxygen in the blood. Figure 9.9 shows the two α subunits in red and the β subunits in blue. Note that each of the chains of hemoglobin contains numerous α helices in its secondary structure (represented by the curly ribbons) folded and packed together. These four folded subunits of hemoglobin are attracted to each other through intermolecular forces including hydrogen bonds, London forces, and salt bridges to form a protein quaternary structure that is biologically active. Each subunit contains a nonprotein part called a **prosthetic group** that is vital to the protein's function. Hemoglobin's prosthetic group is called a heme (shown in green in Figure 9.9). Each heme group binds an Fe^{2+} which, in turn, binds oxygen (O_2). Each Fe^{2+} can bind one oxygen molecule for transport, and, therefore one molecule of hemoglobin can bind up to four molecules of oxygen.

The binding of O_2 to the Fe^{2+} of hemoglobin in the oxygen-rich environment of the lungs induces a change in the shape of the hemoglobin molecule. Biochemists call this a **conformational change**. The conformational change allows hemoglobin to hold on to the oxygen molecules long enough to deliver the bound oxygen to the tissues where oxygen levels are low. At the tissues, the oxygen dissociates from the hemoglobin, and the shape of the hemoglobin changes back to its de-oxygenated form. Because of hemoglobin's unique structure, it is able to interact with its environment and function as an oxygen transport and delivery system.

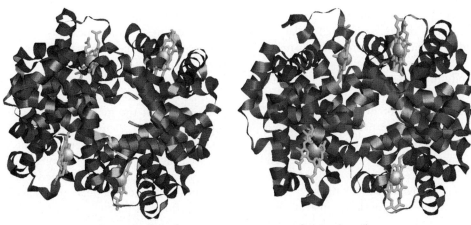

Oxygen unbound   Oxygen bound
(Deoxygenated state)   (Oxygenated state)

◀ **FIGURE 9.9 Hemoglobin.** Hemoglobin consists of four polypeptide subunits (two α chains shown in red and two β chains shown in blue) that are held together mainly by salt bridges to form the biologically active protein. The heme is shown in green. When oxygen binds to Fe^{2+} (orange ball), a conformational change occurs as salt bridges between the subunits collapse, closing the hole formed in the center of the protein.

Antibodies—Your Body's Defense System

When foreign substances like bacteria enter your body, your immune system produces proteins called antibodies (also known as immunoglobulins) to recognize and destroy the foreign substances. The foreign substance recognized by an antibody is called an **antigen**. Figure 9.10 shows the structure of an antibody. It consists of four polypeptide subunits, two identical heavy chains (higher mass, shown in blue and red) and two identical light chains (lighter mass, shown in yellow and green). The secondary structure contains β-pleated sheets (represented by the flat ribbons) that are stacked tightly together. As shown in the figure, the quaternary structure is held together through disulfide bridges between the polypeptide chains forming a unique "Y" shape, which is the biologically active form of the protein. The stem of the Y is similar in all antibodies and can bind to receptors on a variety of cells in the body. Antibodies bind antigens at the top of each arm of the Y. Each antibody has a unique primary sequence at the top of the Y that recognizes a single antigen. This ability to recognize a single antigen and the distinctive Y structure of an antibody is well suited to bind antigens and display them for future destruction by the immune system.

▶ **FIGURE 9.10 Structure of an antibody.** Antibodies contain four polypeptide chains held together in a "Y" shape by intermolecular forces and disulfide bridges.

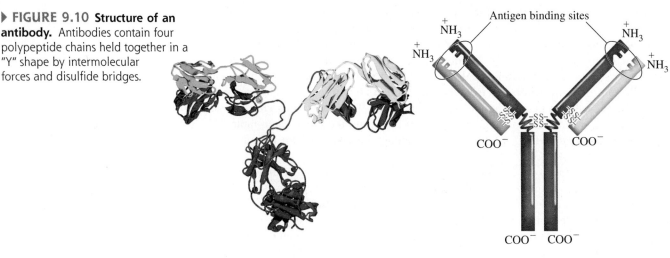

Integral Membrane Proteins

In Chapter 6, we saw that cells are surrounded by selectively permeable membranes consisting of a bilayer of phospholipids. The cell membrane has a nonpolar central region that inhibits the movement of polar substances into or out of the cell. In order to facilitate the movement of polar substances, membranes contain proteins, called **integral membrane proteins**, spanning the nonpolar region of the membrane.

An important integral membrane protein involved in electrolyte balance is the sodium potassium pump (Na^+/K^+ ATPase), shown in Figure 9.11. This protein is also composed of four polypeptide subunits held together by intermolecular forces. Because part of the protein is embedded in the nonpolar region of the membrane, the side chains on those parts of the protein are nonpolar, allowing them to interact with the nonpolar region of the membrane through London forces. These interactions help to anchor the protein within the membrane. The central cavity of the protein is lined with polar amino acids, which allow sodium and potassium ions to be pumped using energy provided by ATP through the channel. The positioning of the polar amino acids facing the inside of the channel allows the protein to perform its function of transporting ions through a nonpolar cell membrane.

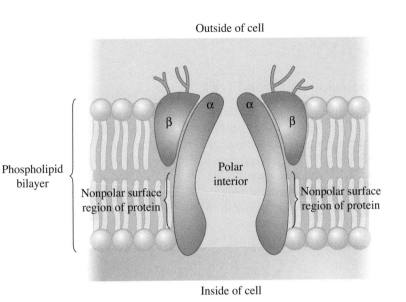

Outside of cell

Phospholipid bilayer

Nonpolar surface region of protein

Polar interior

Nonpolar surface region of protein

Inside of cell

◀ **FIGURE 9.11 An integral membrane protein.** The Na^+/K^+ ATPase is an integral membrane protein with a nonpolar surface and a polar interior. Alpha and beta represent separate polypeptides in the protein.

PRACTICE PROBLEMS

9.27 Match each protein in Column A with its function in Column B:

Column A	Column B
Collagen	Ion transporter
Hemoglobin	Structural connector
Antibody	Oxygen transporter
Na^+/K^+ ATPase	Bind foreign substances in body

9.28 Match each protein in Column A with its shape in Column B:

Column A	Column B
Collagen	Y-shaped
Hemoglobin	Ropelike
Antibody	Integral membrane protein
Na^+/K^+ ATPase	Roughly spherical

9.29 How does the structure of collagen uniquely suit it for its function as a structural protein found in tendons?

9.30 Hemoglobin is found in blood, which is composed mostly of water. Would you expect the amino acid side chains on the surface of hemoglobin to be polar or nonpolar? Explain.

chapter nine

SUMMARY

9.1 Amino Acids—A Second Look

Amino acids contain a central carbon atom, called the α carbon, bonded to four different groups—a protonated amine (amino) group, a carboxylate group, a hydrogen atom, and a side chain. Because of this structure, amino acids, with the exception of glycine, are chiral compounds. The L-enantiomers of amino acids are the building blocks of proteins. There are 20 different amino acids as identified by their side chains. The properties of the side chains determine whether the amino acids are classified as nonpolar or polar. Each amino acid has a unique name, as well as a one-letter and three-letter abbreviation.

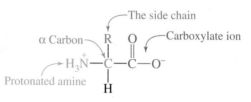

9.2 Protein Formation

Amino acids serve as the building blocks of proteins. Two amino acids join through a condensation reaction of the protonated amine group of one and the carboxylate group of the other. The bond that forms between the two amino acids is called a peptide bond and the new structure is a dipeptide. The newly formed dipeptide has an N-terminus with a free protonated amine group and a C-terminus with a free carboxylate group. In most cases, small structures (less than 50 amino acids) are referred to as peptides. A compound containing typically 50 or more amino acids linked by peptide bonds and which has biological activity is called a protein. In this way, proteins are polymers of amino acids.

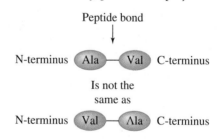

9.3 The Three-Dimensional Structure of Proteins

Proteins are large molecules with structures too complex to be described in the same way we describe the structures of small molecules. The three-dimensional structure of a protein is described using a system of four levels. The primary structure (1°) is the sequence of the amino acids that form the backbone of the protein. The bonding interaction in this level is the peptide bond. The secondary structure (2°) involves the interactions of amino acids that are near each other in the primary structure and describes patterns of regular or repeating structure. The most common secondary structures are the α helix and the β-pleated sheet. The secondary structure is stabilized by hydrogen bonding between atoms in the backbone. The tertiary structure (3°) is formed by the folding of the secondary structure onto itself and is driven by the hydrophobic interactions of the amino acid side chains with their aqueous environment. Nonpolar amino acids are forced to the interior of the protein, away from water, leaving polar amino acids on the surface to interact with the aqueous environment. This level of structure is stabilized by the attractive forces between the side chains and the aqueous environment of the protein as well as between two or more of the side chains and include nonpolar interactions, polar interactions (such as hydrogen bonding), salt bridges, and disulfide bonds. Proteins that fold into a roughly spherical shape are called globular proteins, whereas those that maintain elongated structures are referred to as fibrous proteins. Some proteins have a quaternary structure (4°), which involves the association of two or more peptides to form a biologically active protein. The forces that stabilize the quaternary structure are the same as in the tertiary structure.

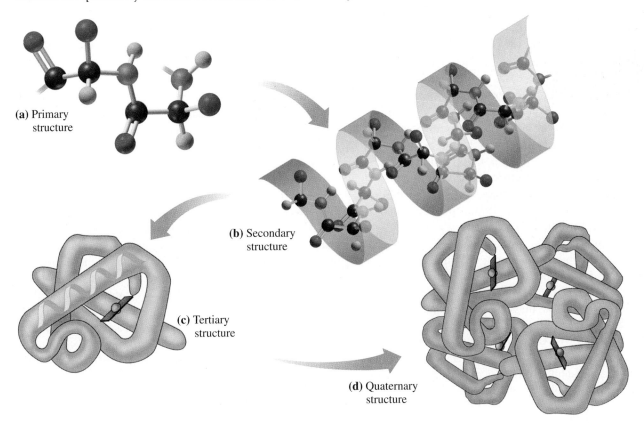

(a) Primary structure

(b) Secondary structure

(c) Tertiary structure

(d) Quaternary structure

9.4 **Denaturation of Proteins**

Denaturation of a protein is a process that disrupts the stabilizing attractive forces in the secondary, tertiary, or quaternary structure, often unfolding the protein. When a protein is denatured, its primary structure is not changed. Proteins can be denatured by heat, a change in the pH of their environment, reaction with small organic compounds and heavy metals such as lead or mercury, or by mechanical agitation. A denatured protein is no longer biologically active.

9.5 **Examples of Biologically Important Proteins**

In this section, several biologically important proteins are introduced including collagen, hemoglobin, antibodies, and the sodium potassium pump. These different proteins demonstrate that the biological function of a protein is dictated by its structure. Changes to the structure of a protein through mutation, denaturation, or other environmental conditions can adversely affect the function of a protein.

KEY TERMS

alpha (α) amino group—The protonated amine ($-NH_3^+$) bonded to the alpha carbon in an amino acid.

alpha (α) carbon—The carbon in an amino acid that is bonded to the carboxylate ion, the protonated amine, a hydrogen atom, and the side chain group.

alpha (α) carboxylate group—The carboxylate ion ($-COO^-$) bonded to the alpha carbon in an amino acid.

alpha (α) helix—A secondary protein structure resembling a right-handed coiled spring or a corkscrew.

amide—A family of organic compounds characterized by a functional group containing a carbonyl bonded to a nitrogen atom; amides are formed by a condensation reaction between a carboxylate ion and a protonated amine.

antigen—A molecule capable of eliciting the immune response of producing antibodies in the body.

beta (β)-pleated sheet—A secondary protein structure with a zigzag structure resembling a folded fan.

conformational change—A change in the tertiary or quaternary structure of a protein that occurs through bonds rotating, not bonds breaking.

C-terminus (also C-terminal amino acid or carboxy terminus)—In a peptide, the amino acid with the free carboxylate ion, always written to the far right of the structure.

denaturation—A process that disrupts the stabilizing attractive forces in the secondary, tertiary, or quaternary structure of a protein caused by heat, acids, bases, heavy metals, small organic compounds, or mechanical agitation.

dipeptide—Two amino acids joined by a peptide bond.

disulfide bond—Covalent bond formed between the —SH groups of two cysteines in a protein; involved in stabilizing the tertiary and quaternary structure of the protein.

fibrous protein—A protein with long, threadlike structures with high helical content typically found in fibers such as hair, wool, silk, nails, skin, and cartilage.

globular protein—A protein with a roughly spherical tertiary structure that typically has nonpolar amino acid side chains clustered in its interior and polar amino acid side chains on its surface.

hydrophobic effect—The movement of nonpolar amino acid side chains to the interior of a protein caused by an unfavorable interaction with the aqueous environment.

integral membrane protein—A protein that is found within or spanning the phospholipid bilayer of a membrane; many serve as passages for polar compounds to move across the membrane.

nonpolar amino acids—Amino acids whose side chains are composed almost entirely of carbon and hydrogen, resulting in an even distribution of electrons over the side chain portion of the molecule.

nonpolar interactions—The association of nonpolar amino acid side chains with each other; London forces

N-terminus (also N-terminal amino acid or amino terminus)—In a peptide, the amino acid with the free protonated amine, always written to the far left of the structure.

peptide bond—An amide bond that joins two amino acids.

polar amino acids—Amino acids whose side chains contain electronegative atoms, resulting in an uneven distribution of electrons over the side chain portion of the molecule.

polar interactions—The association of polar amino acid side chains with each other, includes hydrogen bonding, dipole–dipole interactions, and ion–dipole interactions.

primary structure—The first level of protein structure; simply the order of the amino acids bonded by peptide bonds forming a polypeptide chain.

prosthetic group—A non-amino acid portion of a protein required for protein function.

protein—Biologically active polymers typically containing 50 or more amino acids bonded together by peptide bonds.

protein backbone—The string of amino acids in a protein N-terminus to C-terminus. The side chains are pendant (hanging from) the main string.

quaternary structure—The highest level of protein structure that involves the association of two or more peptide chains to form a biologically active protein; can be stabilized by nonpolar interactions, polar interactions, ionic interactions, and disulfide bonds.

salt bridges (ionic interactions)—The attraction formed between a carboxylate ion on one amino acid's side chain with the protonated amine on a second amino acid's side chain.

secondary structure—The protein structural level that involves regular or repeating patterns of structure within the overall three-dimensional structure of a protein stabilized by hydrogen bonding along the protein backbone.

tertiary structure—The protein structural level that involves the folding of the secondary structure on itself driven by the hydrophobic interactions of nonpolar amino acids with the protein's aqueous environment; stabilized by polar interactions, nonpolar interactions, ionic interactions, and disulfide bonds between amino acid side chains and by polar interactions with water molecules in the environment.

triple helix—The quaternary structure found in collagen that involves three left-handed helical strands of protein woven together to form a single right-handed helix.

ADDITIONAL PROBLEMS

9.31 What functional groups are common to all α-amino acids?

9.32 Give the name and three-letter abbreviation for the amino acid described by each of the following:

  a. the nonpolar amino acid with a sulfur atom in its side chain

  b. a polar amino acid with a single nitrogen atom in its side chain

  c. the nonpolar amino acid with only one carbon in its side chain

9.33 Give the name and three-letter abbreviation for the amino acid described by each of the following:

  a. the polar amino acid with a benzene ring in its side chain

  b. the nonpolar amino acid whose side chain forms a ring with its α-amino group

  c. the polar amino acid with a sulfur atom in its side chain

9.34 Give the name and three-letter abbreviation for the amino acid that does not have a chiral center.

9.35 Two of the amino acids have two chiral centers. Give the name and three-letter abbreviation of each.

9.36 Aspartame, which is commonly known as Nutrasweet, contains the following dipeptide:

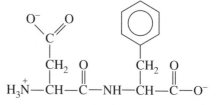

a. What are the amino acids in aspartame?

b. Give the three-letter abbreviation of the N-terminal amino acid.

c. Circle the peptide bond.

d. Draw the structure of the isomer of this dipeptide where the C-terminal and N-terminal amino acids are switched.

9.37 Vegetables, nuts, and seeds are often deficient in one or more of the essential amino acids. From the Protein Source column in the following table, choose combinations of two or more foods you could eat to get a complete protein.

AMINO ACID DEFICIENCY IN SOME FOODS	
Protein Source	Amino Acid Missing
Rice	Lys
Oatmeal	Lys
Peas	Met
Beans	Trp, Met
Almonds	Lys, Trp
Corn	Lys, Trp

9.38 Draw the structure of the possible dipeptides formed from one alanine combining with one cysteine.

9.39 Consider the amino acids glycine, proline, and lysine.

a. How many tripeptides can be formed from these three amino acids if each is used only once in the structure?

b. Using three-letter abbreviations for the amino acids, give the sequence for each of the possible tripeptides.

c. Draw the structure of the tripeptide that has proline as its N-terminal amino acid and glycine as its C-terminal amino acid. Circle each peptide bond.

9.40 a. Draw the structure of Val—Ala—Leu.

b. Would you expect to find this segment at the center or on the surface of a globular protein? Why?

9.41 a. Draw the structure of Ser—Lys—Asp.

b. Would you expect to find this segment at the center or on the surface of a globular protein? Why?

9.42 Name the covalent bond that helps to stabilize the tertiary structure of a protein.

9.43 Identify some differences between the following pairs:

a. primary and secondary protein structures

b. complete and incomplete proteins

c. fibrous and globular proteins

9.44 Identify some differences between the following pairs:

a. α helix and β-pleated sheet

b. ionic interactions (salt bridge) and polar interactions

c. polar and nonpolar amino acids

9.45 Identify the level of protein structure associated with each of the following:

a. more than one polypeptide

b. hydrogen bonding between backbone atoms

c. the sequence of amino acids

d. intermolecular forces between R groups

9.46 Identify the level of protein structure associated with each of the following:

a. α helix

b. disulfide bridge

c. peptide bond

d. salt bridges between polypeptides

9.47 Briefly describe the structure of collagen and discuss how it is different from other helix-containing proteins.

9.48 A piece of a polypeptide is folded so that a serine side chain is located on the surface of the polypeptide. Describe the effect on the protein of changing that amino acid to a isoleucine.

9.49 Indicate what type(s) of intermolecular forces are disrupted and what level of protein structure is changed by the following denaturing treatments:

a. an egg placed in water at 100 °C and boiled for 10 minutes

b. acid added to milk during the preparation of yogurt

c. egg whites whipped in a mixing bowl to make meringue

9.50 Describe the changes that occur in the primary structure when a protein is denatured.

9.51 What types of covalent bonds can be disrupted when a protein is denatured? Name a denaturing agent that could accomplish this.

9.52 Describe what happens to the protein structure of hair when it is curled with a heated curling iron.

9.53 Collagen contains an amino acid that is a modified form of the naturally occurring amino acid.

a. Name the natural amino acid.

b. How is the structure of the side chain of this amino acid modified?

c. What is the purpose of the modification in terms of the structure and function of the collagen?

9.54 Hemoglobin requires a prosthetic group in order to transport oxygen. Explain what a prosthetic group is and name the one found in hemoglobin.

9.55 Integral membrane proteins are located across a cell membrane and have nonpolar amino acids on their surface and polar amino acids in the interior. This is in contrast to proteins like hemoglobin that are found in the bloodstream. Explain why the surface of the integral membrane protein is nonpolar.

CHALLENGE PROBLEMS

9.56 When lemon juice is added to milk, the milk curdles. Why does lemon juice have this effect? What is happening to the proteins in milk?

9.57 How is the structure of a soap micelle (Chapter 6) similar to the structure of a globular protein?

HINT: *Both of these are found in aqueous solution.*

9.58 Insulin is a protein hormone that functions as two polypeptide chains whose amino acid sequences are as follows:

A chain: GIVEQCCTSICSLTQLENYCN

B chain: FVNQHLCGDHLVEALYLVCGERGFFYTPKT

a. The highlighted region in chain B forms an α helix. In an α helix the R groups are found on the outside of the helix, protruding out from the center, and every fourth amino acid appears on the same side of the helix. Considering the polarity of the R groups, is this helix amphipathic (having a polar part and nonpolar part)? Which side do you think faces the exterior?

b. Considering the amino acid sequences, suggest how these two polypeptide chains might be held together in an active insulin molecule.

ANSWERS TO ODD-NUMBERED PROBLEMS

Practice Problems

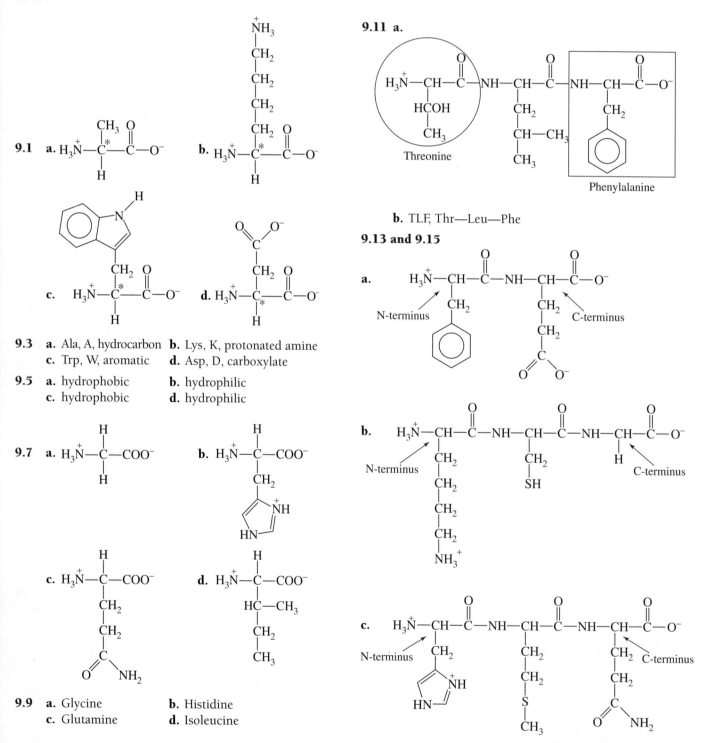

9.1 a.

b.

c.

d.

9.3 a. Ala, A, hydrocarbon **b.** Lys, K, protonated amine
c. Trp, W, aromatic **d.** Asp, D, carboxylate

9.5 a. hydrophobic **b.** hydrophilic
c. hydrophobic **d.** hydrophilic

9.7 a. **b.**

c. **d.**

9.9 a. Glycine **b.** Histidine
c. Glutamine **d.** Isoleucine

9.11 a.

Threonine

Phenylalanine

b. TLF, Thr—Leu—Phe

9.13 and 9.15

a. N-terminus   C-terminus

b. N-terminus   C-terminus

c. N-terminus   C-terminus

d.

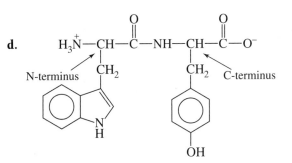

N-terminus   C-terminus

9.17 covalent bonding, peptide bond

9.19 hydrogen bonding

9.21 An α helix is a coiled structure with the amino acid side chains protruding outward from the helix. The β-pleated sheet is an open, extended, zigzag structure with the side chains of the amino acids oriented above and below the sheet.

Additional Problems

9.31 carboxylate and protonated amine

9.33 a. Tyrosine, Tyr   **b.** Proline, Pro
   c. Cysteine, Cys

9.35 Isoleucine, Ile, and Threonine, Thr

9.37 Several possibilities including rice and beans, peas and corn.

9.39 a. 6
   b. Gly—Pro—Lys, Gly—Lys—Pro, Pro—Gly—Lys,
   Pro—Lys—Gly, Lys—Pro—Gly, Lys—Gly—Pro

c.

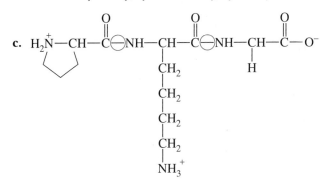

9.41

a.

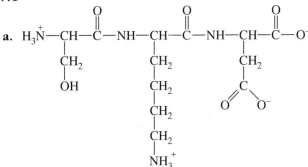

   b. This segment would be on the surface in the aqueous environment. The side chains are hydrophilic and would interact with the water in the surrounding environment.

9.23 a. salt bridge
   b. London force (nonpolar interaction)
   c. hydrogen bonding
   d. hydrogen bonding

9.25 a. quaternary   **b.** secondary
   c. tertiary   **d.** primary

9.27 Collagen, Structural connector
   Hemoglobin, Oxygen transporter
   Antibody, Bind foreign substances in body
   Na^+/K^+ ATPase, Ion transporter

9.29 Each of the individual helices of the triple helix is strong and flexible. Winding three helices together gives a much stronger structure but maintains the flexibility. This creates the necessary strength for the tendons to maintain attachments but provides the flexibility required for motion.

9.43 a. Primary structures are held together by covalent bonding, and secondary structures are held together by intermolecular forces.
   b. Complete proteins are those that contain all essential amino acids and incomplete proteins do not.
   c. Fibrous proteins have elongated structures and globular proteins have roughly spherical structures.

9.45 a. quaternary structure
   b. secondary structure
   c. primary structure
   d. tertiary and quaternary structure

9.47 Collagen's structure consists of three left-handed helices wrapped around each other to form a right-handed helix called a triple helix. Most of the single helices found in proteins are right-handed, not left-handed as in collagen.

9.49 a. hydrogen bonds and nonpolar attractions; secondary and tertiary
   b. hydrogen bonds and salt bridges; secondary and tertiary
   c. hydrogen bonds and nonpolar attractions; secondary and tertiary

9.51 Disulfide bonds, Thioglycolate.

9.53 a. Proline
   b. An OH group is added to the side chain to form hydroxyproline.
   c. The OH on the side chain allows for the formation of more hydrogen bonds between side chains, which increases the strength of the collagen.

9.55 The middle portion of these proteins span the nonpolar portion of the membrane. This surface interacts with the nonpolar environment and therefore must be nonpolar itself.

9.57 Both have nonpolar interiors and polar surfaces. On a micelle the polar heads of the fatty acid salts face outward into the aqueous environment, similar to the polar amino acid side chains on the globular protein. The nonpolar tails of the fatty acid salts gather together in the interior of the micelle just as the nonpolar amino acid side chains gather in the interior of the globular protein.

Guided Inquiry Activities **FOR CHAPTER 10**

EXERCISE 1 **Thermodynamics of Chemical Reactions**

Information

The diagrams shown are called reaction energy diagrams and graphically show the progress of two different chemical reactions (reactions I and II) on the horizontal axis and the amount of energy needed as the reaction moves forward from reactants to products on the vertical axis. The activation energy is the amount of energy needed to get the reactants into position so they can actually react.

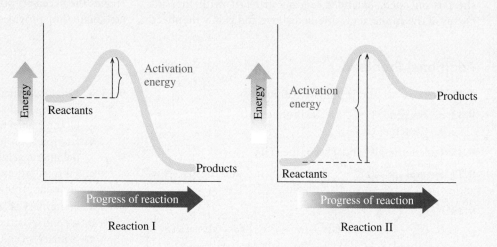

Reaction I   Reaction II

Questions

1. Which reaction has the larger activation energy?
2. Based on the information, which reaction can form products more quickly?
3. A catalyst speeds up a chemical reaction by lowering the activation energy. It does not change the energy of the reactants or products. Sketch the reaction energy diagram for Reaction I and draw a second curve on the same diagram for Reaction I if it uses a catalyst. Label the two traces "catalyzed" and "uncatalyzed."
4. Reactions that release energy (as heat) are called *exothermic* and reactions that require energy are called *endothermic*.
   a. Which reaction (I or II) has more energy in the reactants than in the products?
   b. Which reaction is exothermic?
5. Examine your sketch from question 3. Does a catalyst change the amount of energy produced or required in a chemical reaction? How did you decide?
6. Draw a reaction energy diagram for a slow, exothermic reaction and a reaction energy diagram for a fast, endothermic reaction. What is different about the two diagrams?

EXERCISE 2 **Factors That Affect Enzyme Activity**

Information

The activity of an enzyme (a measure of how fast it catalyzes a chemical reaction) can be affected by changes in the enzyme's environment. Enzymes are most efficient at an optimal pH and temperature.

Questions

1. Pepsin and trypsin are enzymes that catalyze the digestion of proteins. Pepsin has an optimal pH of 2 and trypsin has an optimal pH of 7 to 8. Which of these two enzymes is likely to digest proteins in the stomach?

2. Lactase, the enzyme that hydrolyzes the disaccharide lactose, operates at an optimal pH of ~6.5 and an optimal temperature of 37 °C. How would the following changes affect the rate of catalysis—increase, decrease, or stay the same?
   a. lowering the pH to 2
   b. raising the temperature to 50 °C
   c. increasing the amount of lactose available

Enzyme Inhibition

An enzyme's activity can be reversibly decreased or inhibited if (a) a molecule structurally similar to the normal substrate competes for the active site or (b) if a molecule binds to a second site on the enzyme, changing the shape of the active site.

Questions

1. Which of the two scenarios, (a) or (b), would be called *competitive inhibition*?

2. Which of the two scenarios, (a) or (b), would be called *noncompetitive inhibition*?

3. Succinate is the substrate for the enzyme succinate dehydrogenase, one of the eight enzymes found in the citric acid cycle. Malonate is a reversible inhibitor of succinate dehydrogenase. What type of inhibitor is malonate likely to be? Explain your reasoning.

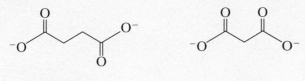

Succinate   Malonate

4. The enzyme hexokinase catalyzes the first step in glycolysis; glucose $\longrightarrow$ glucose-6-phosphate. The activity of this enzyme can be inhibited by the buildup of the product glucose-6-phosphate. Binding glucose-6-phosphate to a second site on the enzyme changes the shape of the active site. This is an example of which type of inhibition?

5. Which type of inhibition do you think could be overcome by adding more substrate? Explain your answer.

6. Which type of inhibition does not allow the enzyme to reach top speed, even if more substrate is added? Explain your answer.

7. The analgesic aspirin acts as a type of inhibitor called a "suicide" inhibitor. Aspirin forms a covalent bond to the amino acid serine whose side chain is found in the active site of the enzyme cyclooxygenase. Provide an explanation for the name of this inhibition.

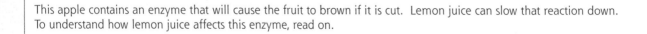

This apple contains an enzyme that will cause the fruit to brown if it is cut. Lemon juice can slow that reaction down. To understand how lemon juice affects this enzyme, read on.

Enzymes
NATURE'S CHEMISTS

Have you ever wondered how the food you eat is transformed into the energy that you need to live? The process involves a group of remarkable molecules known as enzymes. Enzymes are typically large globular proteins and they are present in every cell of your body. Enzymes act as *catalysts*, compounds that accelerate the reactions of metabolism (chemical reactions in the body) but are not consumed or changed by those reactions. Like all other catalysts, an enzyme cannot force a reaction to occur that would not normally occur. Instead, an enzyme simply makes a reaction occur faster.

In this chapter, we will discover how catalysts work to speed up reactions and how the structure of an enzyme makes it uniquely suited to function as a catalyst in cellular chemistry. We will also consider the production or consumption of energy, specifically heat, during chemical reactions. This study of energy is called **thermodynamics**.

10.1 Enzymes and Their Substrates

Like most other proteins, enzymes are large molecules with complex three-dimensional shapes. Because many enzymes in your body function in an aqueous environment, they fold so that polar amino acids are on their surface. The final folded (tertiary) structure of an enzyme plays an important role in its function. Let's look at the features of an actual enzyme, hexokinase. This enzyme's job is to catalyze the transfer of a phosphate group from the energy molecule adenosine triphosphate, ATP, to a six-carbon sugar (hexose) like D-glucose. It functions in the first step of glycolysis (the process by which glucose is broken down in the body) by adding a phosphate to the sixth carbon of glucose forming the product, glucose-6-phosphate.

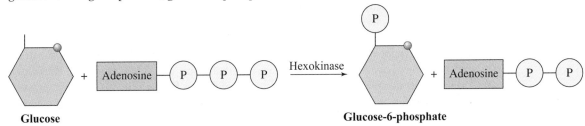

Glucose　　　　　　　　　　　　　　　　**Glucose-6-phosphate**

In the chemical equation, the enzyme name usually appears above or below the reaction arrow. The phosphate group is represented by a P in a circle. It contains a phosphorus surrounded by four oxygen atoms. In Section 3.2, we saw phosphate as one of the polyatomic ions. The Lewis structure of phosphate under physiological conditions is the anion hydrogen phosphate, HPO_4^{2-}, also called inorganic phosphate and abbreviated P_i. The structure of P_i is

$$\overset{\displaystyle \ddot{\text{O}}}{\underset{\displaystyle \ddot{\text{O}}\text{H}}{\overset{\displaystyle \|}{-}\text{P}-}}$$

The Active Site

Figure 10.1 shows the folded structure of the enzyme hexokinase. In this representation, the protein backbone of hexokinase is represented as a ribbon. As hexokinase folds into its tertiary structure, an indentation forms on one part of its surface. This pocket is known as the **active site**, and it is lined with amino acid side chains. As its name implies, the active site is the functional part of the enzyme where catalysis occurs. The active site of hexokinase is labeled in Figure 10.1.

▶ **FIGURE 10.1 The active site of hexokinase.** Hexokinase's protein structure is shown as a ribbon. Glucose fits in the pocket of the active site between the upper and lower lobes of the protein during catalysis. It is held in place by multiple hydrogen bonding interactions (red dashes) with amino acid side chains.

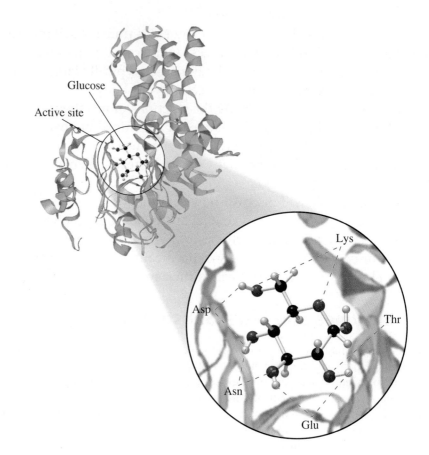

Like a left foot fitting into a left shoe, glucose, the reactant for hexokinase, fits snugly into its active site. In an enzyme-catalyzed reaction, the *reactant* is called the **substrate**. The active site of hexokinase fits D-glucose but L-glucose, its enantiomer, or even D-galactose, its epimer (one chiral center difference), will not fit into the active site. Because the shape of an enzyme's active site is complementary to the shape of its substrate, most enzymes have only a few substrates, and many only one. This property of enzymes is called **substrate specificity**. Just as we know which shoe fits on which foot, many enzymes are specific for one enantiomer of the same compound (see Figure 10.2). The opposite enantiomer will not fit perfectly in the active site and cannot undergo catalysis.

Hexokinase also has a nonprotein "helper" in the form of a magnesium ion (Mg^{2+}) in its active site. Mg^{2+} assists with catalysis. There are two categories of nonprotein helpers—**cofactors** and **coenzymes**. Cofactors are inorganic substances like the magnesium ion that can be obtained in a normal diet from minerals. Coenzymes are small organic molecules, and many of them are derived from vitamins such as riboflavin (vitamin B_2) found in the coenzyme flavin adenine dinucleotide (FAD).

Enzyme–Substrate Models

How does a substrate like glucose find the active site of an enzyme like hexokinase and situate itself for catalysis? Glucose is drawn to the active site of hexokinase by intermolecular attractions like hydrogen bonding. Five polar amino acid side chains—asparagine (Asn),

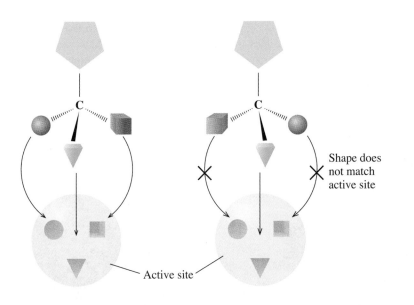

◀ **FIGURE 10.2 Enzymes and enantiomers.** The pocket of an enzyme's active site (yellow) has a specific shape that can distinguish between the mirror image shapes of enantiomers.

asparatate (Asp), glutamate (Glu), lysine (Lys), and threonine (Thr)—form hydrogen bonds to glucose, orienting it correctly within hexokinase's active site for catalysis (see Figure 10.1). This initial interaction of an enzyme with a substrate is called the **enzyme–substrate complex (ES)**. The formation of this complex occurs before catalysis can begin. Notice that there are no covalent bonds formed when the substrate and enzyme come into first contact.

There are several different models that describe how the enzyme interacts with the substrate to form the ES. Two will be described here. The first model introduced was the **lock-and-key model**. In this model, the active site is thought of as a rigid, inflexible shape that is an exact complement to the shape of its substrate. According to the lock-and-key model, each enzyme accommodates one and only one substrate, much as a lock has only one key (see Figure 10.3a).

Now we know that many enzymes can react with two or more similar substrates so a newer model called the **induced-fit model** was developed. In this model, the enzyme's active site is flexible and has a shape that is roughly complementary to the shape of its substrate. As the substrate interacts with the enzyme, the active site undergoes a conformational change, adjusting to fit the shape of the substrate (see Figure 10.3b). The shape of the substrate may change slightly as well.

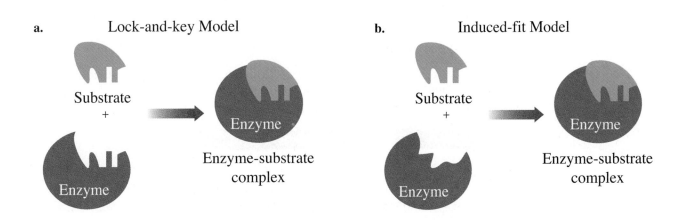

▲ **FIGURE 10.3 Models for enzyme–substrate interaction.** (a) In the lock-and-key model, the enzyme and substrate have rigid complementary shapes that fit together to form ES. (b) In the induced-fit model, the enzyme and substrate have flexible, but similar shapes that adjust when the substrate gets closer to the enzyme, thereby forming a unique Enzyme-substrate complex.

Hexokinase and glucose form an enzyme-substrate complex that is a good example of the induced-fit model. In Figure 10.4 the shape of hexokinase is shown in a ball-and-stick representation. Notice how the shape changes when the enzyme interacts with glucose. The two lobes of the enzyme close snugly around the substrate to form the ES.

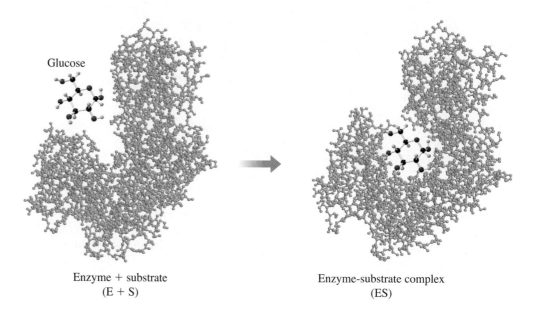

Glucose

Enzyme + substrate
(E + S)

Enzyme-substrate complex
(ES)

▲ **FIGURE 10.4 Interaction of hexokinase and glucose through induced fit.** When glucose enters the active site of hexokinase (left), the enzyme's two lobes come together, wrapping glucose snugly into its active site forming the enzyme–substrate complex (right).

Now we have seen how an enzyme's structure is perfectly suited to allow it to interact with its substrate. In Section 10.2, we will consider the energy changes that take place during various chemical reactions and then apply them to our study of enzymes.

SAMPLE PROBLEM 10.1

Enzymes and Substrates

Describe the function of the active site of an enzyme.

SOLUTION

The active site of an enzyme is a pocket uniquely fitted to the substrate and contains the amino acid side chains that catalyze the reaction.

PRACTICE PROBLEMS

10.1 What level of protein structure is involved in the formation of an enzyme's active site?

10.2 Describe how a substrate is drawn to an enzyme to form ES.

10.3 Distinguish between a cofactor and a coenzyme.

10.4 Describe the key difference in the lock-and-key and induced-fit models.

10.5 Which model for enzyme–substrate interaction describes the action of hexokinase? Describe how the enzyme changes when the enzyme and glucose form ES.

10.2 Thermodynamics of Chemical Reactions

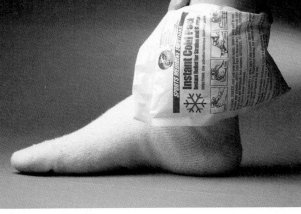

If you have ever sprained your ankle on a hiking trip or at an athletic event, you may have used an instant cold pack to help control the swelling. These packs typically contain two compartments, one filled with water and the other containing a white solid (often ammonium nitrate). When the pack is squeezed, the plastic membrane between the chambers breaks, allowing the materials to combine, and the whole pack gets very cold.

When chemical reactions occur, some bonds are broken and new bonds are formed, and in the process the overall amount of energy changes. As discussed in Chapter 8, reactions that release (or produce) energy as heat are *exothermic*, and reactions that absorb (or consume) energy as heat are *endothermic*. In this case, the mixing of the two chemicals, while not a chemical reaction in the cold pack is an endothermic process.

For a chemical reaction to occur, the molecules of the reactant(s) must collide with each other with enough energy and the proper orientation. The energy necessary to align the reactant molecules and to cause them to collide with enough energy to form products is known as the **activation energy**. If the energy available is less than the activation energy, the molecules will simply bounce off each other without forming product.

We can think of activation energy as the energy hill that molecules must overcome before products can form. If enough energy is available to allow the molecules to get over the energy hill (also known as the activation barrier), they can react to form products.

Figure 10.5 shows reaction energy diagrams for exothermic and endothermic reactions. These diagrams are a visual way to represent the energy changes that occur in a reaction. Notice that each reaction has an activation energy hill to climb before proceeding to products.

In Figure 10.5, you can also see that each reaction has the **heat of reaction** labeled. The heat of reaction is the energy difference between the products and the reactants. Notice that the reactants are higher in energy than the products in an exothermic reaction; therefore, energy is released into the surroundings as the reaction progresses. In an endothermic reaction, the products

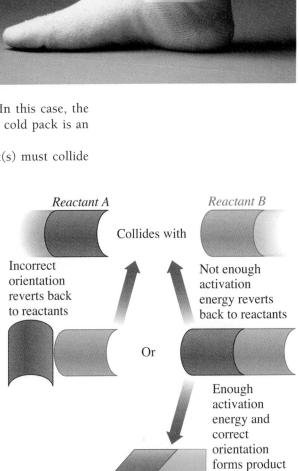

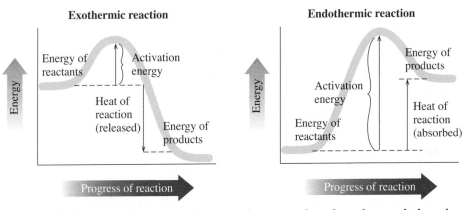

▲ **FIGURE 10.5 Reaction energy diagrams for an exothermic and an endothermic reaction.** Reaction energy diagrams are a visual method for representing the energy changes in a chemical reaction.

are higher in energy than the reactants, indicating that energy is absorbed from the surroundings.

The activation energy also has an effect on how quickly a reaction proceeds. We can liken this to an experience of a tour group visiting the Empire State Building. Suppose the tour group is given two options for the hour that they are visiting: They can visit the gift shop on the first floor of the building or climb the stairs and view the New York City skyline. Visiting the gift shop would take less energy and less time to accomplish than climbing the stairs and viewing the skyline. The same is true in chemical reactions. Reactions with a low activation energy (like visiting the gift shop) proceed more quickly (have a faster rate) than reactions with a high activation energy (like climbing the stairs to the top of the building). We can measure the **rate of reaction** by determining the amount of product formed (or reactant used up) in a certain period of time. The rate of a reaction is affected by several factors including temperature and the concentration of reactants. However, here we are especially interested in how catalysts, such as enzymes, speed up reactions.

A catalyst speeds up a reaction by *lowering* the activation energy. With a lower energy hill to climb, the reaction progresses to form products more quickly. A catalyst does not affect the energy (or the amount) of products or reactants in a reaction (whether the reaction is exothermic or endothermic); it only speeds up formation of the products. An enzyme-catalyzed reaction does this by first forming ES, before proceeding to form product (see Figure 10.6).

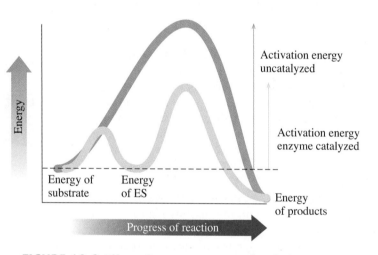

▲ **FIGURE 10.6 Effect of an enzyme on a chemical reaction.** Like all other catalysts, an enzyme lowers the activation energy of a reaction, thereby increasing the rate of the reaction. Enzymes accomplish this by first forming ES.

As an example, consider hydrogen peroxide, H_2O_2, often used as an antiseptic to clean cuts. When hydrogen peroxide is put on a cut, the area bubbles because oxygen gas is produced by the reaction of hydrogen peroxide with the enzyme catalase found in the blood. The balanced chemical reaction is

$$2H_2O_2(l) \xrightarrow{\text{Catalase}} 2H_2O(l) + O_2(g)$$

In the absence of the enzyme, this reaction occurs very slowly. In fact, if you pour hydrogen peroxide on your uncut skin, nothing will happen. Hydrogen peroxide works as an antiseptic by reacting with catalase to produce a highly concentrated oxygen environment at the wound site that kills germs.

SAMPLE PROBLEM 10.2

Reaction Energy Diagrams

Which of the following reaction energy diagrams is a(n)

a. exothermic reaction?

b. endothermic reaction?

c. faster reaction?

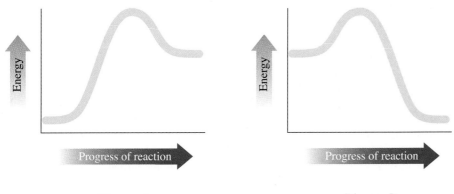

Diagram A   Diagram B

SOLUTION

a. The energy of the reactants is greater than the energy of the reactants in Diagram B. This represents an exothermic reaction.

b. The energy of the reactants is less than the energy of the products in Diagram A. This represents an endothermic reaction.

c. The diagram that has the lower activation energy will be the faster reaction. This is the exothermic reaction in Diagram B.

PRACTICE PROBLEMS

10.6 Fill in the blanks in the following sentence: In order for a reaction to occur, the reactants must collide with each other with enough _____ and proper _____.

10.7 When water and solid ammonium nitrate (NH_4NO_3) are combined in a test tube, the solid dissolves in an endothermic process. Would the test tube feel hot or cold? Explain your reasoning.

10.8 a. In your own words, define heat of reaction.

b. How does the heat of reaction differ in exothermic and endothermic reactions?

c. In an endothermic reaction, are the products or the reactants higher in energy?

10.9 a. Describe activation energy for a chemical reaction.

b. How does a catalyst increase the rate of a chemical reaction?

c. In an exothermic reaction, are the products or the reactants lower in energy?

10.10 Classify the following as exothermic or endothermic reactions:

a. When two solids are combined in a test tube, the test tube gets hot.

b. On a reaction energy diagram, the reactants are lower energy than the products.

10.11 Classify the following as exothermic or endothermic reactions:

a. The reaction of hexokinase produces energy.

b. On a reaction energy diagram, the products are lower in energy than the reactants.

10.12 Describe how the rate of a reaction can be measured.

10.3 Enzymes and Catalysis

Enzymes catalyze reactions by lowering the activation energy, but how? In the hexokinase reaction from Section 10.1, glucose and ATP react in the active site of the enzyme and a phosphate is transferred from ATP to glucose. The products of the reaction are glucose-6-phosphate and adenosine diphosphate, ADP. If hexokinase is not present, the reaction between glucose and ATP would occur much more slowly. What is it about the environment of the active site of hexokinase that causes these two molecules to react more quickly to form products?

Thus far, we have seen that the first step in an enzyme-catalyzed reaction is the formation of ES. We have also seen that catalysts lower the activation energy of a reaction. Putting these two pieces of information together, it makes sense that the formation of ES is key to lowering the activation energy and allowing a biochemical reaction to proceed and quickly form products.

This is accomplished during the formation of ES through the interactions between the enzyme and the substrate. Each interaction releases a small amount of energy, stabilizing the complex. These small interactions combine to lower the activation energy for an enzyme-catalyzed reaction versus its uncatalyzed reaction by a lot! Let's look at some interactions that help to lower the activation energy.

Proximity

The active site of an enzyme has a small volume. When the ES forms, the active site is "filled" with substrate. This means that the reacting molecules (for hexokinase, glucose, and ATP) are in close proximity to each other and to the catalytic side chains of the amino acids lining the active site. The closer they are, the more likely they will react. The activation energy is lowered by this effect. The substrates don't have to find each other as they would in solution; they are already close together in the active site.

The amino acid side chains in the active site of an enzyme are the "tools" used by the enzyme to help facilitate the reaction. These side chains are often the functional groups of the acidic and basic amino acids.

Orientation

In the active site of an enzyme, the substrate molecules are held at the appropriate distance and in correct alignment to each other to allow the reaction to occur. The arrangement of the amino acid side chains in the active site creates interactions that orient the substrates so they will react. Without an enzyme to orient the substrates, every time the substrates might bump into each other in solution, they may not react because of incorrect positioning. The enzyme guarantees the substrates are lined up correctly. This lowers the activation energy needed.

In the absence of an enzyme, it is more difficult for substrates to get close enough and orient themselves to react.

In the presence of an enzyme, it is easier for substrates to get close enough and orient themselves to react.

Bond Energy

Think of a bond as a rubber band. Which is easier to cut with scissors: one that is stretched or one that is unstretched? The stretched one is easier because the rubber molecules are weakened or strained and it takes less energy to break the molecules with scissors. Likewise, when an enzyme interacts with its substrate to form ES, the bonds of the substrate molecule(s) are weakened (strained). The weakening of the bonds means that they will more readily react. In other words, weaker bonds break more easily and the activation energy is therefore lowered by this effect.

Looking back at our hexokinase reaction, the coenzyme Mg^{2+} holds the ATP molecule in one area of the active site and the glucose interacts with another area. The amino acid side chains in the active site of hexokinase form multiple hydrogen bonds with the glucose (refer to Figure 10.1). Each hydrogen-bonding interaction formed stabilizes ES and lowers the activation energy.

When glucose enters the active site, the enzyme undergoes a significant conformational change causing the two lobes of the enzyme to close around the substrate (following the induced-fit model—Section 10.1). With the two lobes of the enzyme closer together, the ATP is moved closer to the glucose and is in proper alignment for the reaction.

When in the active site of hexokinase, glucose is positioned so that the hydroxyl group on C6, activated by an aspartate side chain, is able to remove a phosphate from ATP to form glucose-6-phosphate + ADP (see Figure 10.7). Because the enzyme is less attracted to the products than the substrates, the lobes of the enzyme move apart and the products are released.

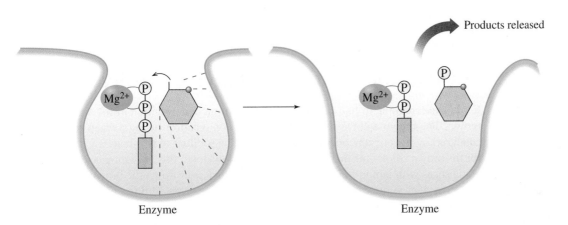

▲ **FIGURE 10.7 Catalysis by hexokinase.** The coenzyme Mg^{2+} holds ATP in place (blue lines) and hydrogen bonds (red lines) hold glucose in the active site. The O on C6 is perfectly positioned to form a bond to the end phosphate of ATP. The product does not fit the active site as snugly as the substrate and is released.

SAMPLE PROBLEM 10.3

List three factors that contribute to lowering the activation energy when ES is formed.

SOLUTION

Three factors are (1) close proximity of reactants to each other and to interactions with amino acid side chains, (2) favorable orientation of reactants, and (3) weakening of bond energy in reactants.

PRACTICE PROBLEMS

10.13 What kind of interaction attracts the cofactor Mg^{2+} and ATP to each other? **HINT:** *Look at the structure of the phosphate group.*

10.14 In this section, we saw that glucose is held in the active site of hexokinase by hydrogen bonding. The outline of the active site of a hypothetical enzyme and its substrate glyceraldehyde is represented in the following figure. Draw dotted lines to show the possible hydrogen bonds between glyceraldehyde and the hypothetical enzyme's active site.

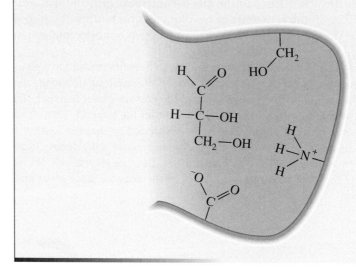

10.4 Factors That Affect Enzyme Activity

If you slice an apple in half and allow it to sit untouched, the flesh of the apple will turn brown. The brown color is caused by an oxidation reaction (using oxygen from the air) catalyzed by the enzyme polyphenol oxidase. The product of this reaction is responsible for the brown color. One method often used to slow down this browning is to sprinkle the apple slices with lemon juice. How does lemon juice slow down the browning of apples?

First, lemon juice contains vitamin C, which can react with the product, temporarily removing the brownish color. Second, lemon juice contains citric acid that slows down or *inhibits* the enzymatic reaction by changing the pH environment of the enzyme. Enzyme-catalyzed reactions, like most other chemical reactions, are affected by the reaction conditions. These include changes in substrate concentration, pH, and temperature and the presence of inhibitors. Changing reaction conditions affects how fast an enzyme converts its substrate to product. Measuring how fast an enzymatic reaction occurs is a measure of an enzyme's **activity**.

Substrate Concentration

How does substrate concentration affect enzyme activity? Imagine a factory that assembles bicycles. The supplier brings in enough parts to produce 10 bicycles, and the workers with their tools build 10 bicycles in one hour. The manager of the factory asks the supplier to bring in enough parts to make 15 bicycles, and the workers also complete that task in one hour. Impressed, the manager asks the supplier to bring in enough parts to make 50 bicycles. In one hour, the workers assemble only 25 bicycles. The manager then asks the supplier to bring in enough parts to make 100 bicycles, but the workers still assemble only 25 in one hour. The workers are working at their maximum capacity by assembling 25 bicycles in one hour. Regardless of the number of extra parts the manager orders, if the resources at the factory do not increase (number of workers, number of tools, etc.), only 25 bicycles can be built in one hour. Enzymes also operate at a particular rate, just like the workers with their tools, no matter how many substrate molecules (or bicycle parts) are present.

As we saw earlier, the first step in an enzyme-catalyzed reaction is the formation of ES. If the amount of enzyme remains unchanged, an increase in the substrate concentration

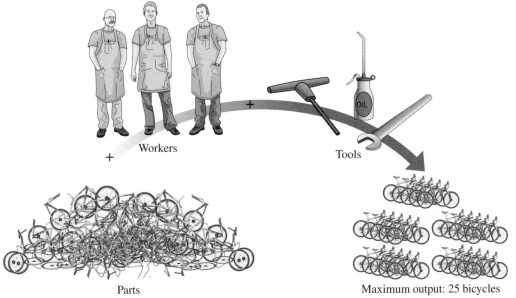

increases the enzyme's activity up to the point, where the enzyme becomes saturated with its substrate. That is, all of the enzyme contains substrate, leaving no more free active sites. At this point the enzyme is working at maximum activity. Increasing the amount of substrate will not increase the activity further. At maximum activity, the conditions under which the enzyme is operating are considered to be in a **steady state**. Under steady-state conditions, substrate is being converted to product as efficiently as possible (see Figure 10.8).

pH

In addition to substrate concentration, the pH of an enzyme's environment is another condition that affects enzyme activity. Let's go back to the workers at the bicycle factory for an explanation. Suppose that it is flu season and several workers come to work with early stages of the flu, breathe on their coworkers, and eventually infect most of the workforce. With fewer workers at the factory, fewer bicycles can be assembled in an hour. Like the environment at the factory, when an enzyme's environment is changed (in this case the pH), its tertiary structure is disrupted, altering the active site and causing the enzyme's activity to decrease.

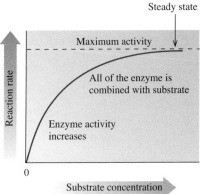

▲ **FIGURE 10.8 Effect of substrate concentration on enzyme activity.** Increasing the amount of substrate increases the enzyme's activity until the enzyme becomes saturated with substrate and a steady state is reached.

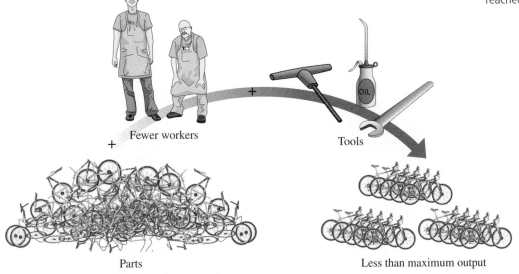

Enzymes are most active at a pH known as their **optimum pH**. At this pH, the enzyme maintains its tertiary structure and, therefore, its active site. Changes in the pH not only affect the structure of the enzyme but may also change the nature of an amino acid side chain. For example, if an enzyme requires a carboxylate ion ($-COO^-$) to function properly, lowering the pH could convert the carboxylate ion to its carboxylic acid form ($-COOH$). This change would cause the enzyme's activity to decrease.

In the body, an enzyme's optimal pH is based on the location of the enzyme. For example, enzymes in the stomach are designed to function at much lower pH's because of the acidity. Optimum pH values for several common enzymes are shown in Table 10.1.

TABLE 10.1	OPTIMUM pH FOR SELECTED ENZYMES		
Enzyme	**Location**	**Substrate**	**Optimum pH**
Pepsin	Stomach	Peptide bonds	2
Sucrase	Small intestine	Sucrose	6.2
Urease	Liver	Urea	7.4
Hexokinase	All tissues	Glucose	7.5
Trypsin	Small intestine	Peptide bonds	8
Arginase	Liver	Arginine	9.7

Temperature

As with pH, enzymes have an **optimum temperature** at which they are most active. If we consider the bicycle factory workers, if the factory air conditioning broke down on a hot day in the middle of August, it is likely that the workers would not be able to work as efficiently to assemble 25 bicycles in one hour! The optimal temperature for most human enzymes is normal body temperature, 37 °C. Above their optimum temperature, enzymes lose activity due to the disruption of the intermolecular forces stabilizing the tertiary structure. At high temperatures, an enzyme denatures, which in turn modifies the structure of the active site. At low temperatures, enzyme activity is radically reduced due to the lack of energy present for the reaction to take place at all. Because enzymes are the major culprits in food spoilage, we store foods in a refrigerator or freezer to slow the spoilage process. The enzymes present in bacteria can also be destroyed by high temperatures, in processes like boiling contaminated drinking water or sterilizing instruments and other equipment in hospitals and laboratories.

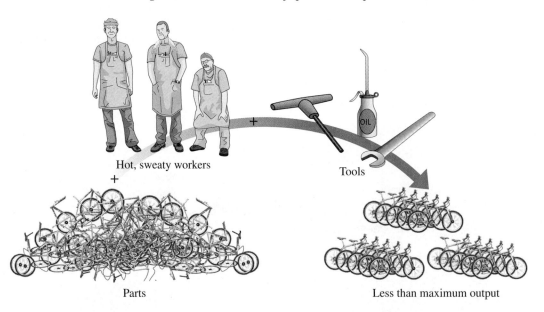

Hot, sweaty workers

Tools

+

Parts

Less than maximum output

Inhibitors

Certain types of molecules known as **inhibitors** can also cause enzymes to lose activity. If we consider the bicycle factory, suppose some of the workers' tools became rusted and could not be used to assemble the bicycles. This would definitely inhibit bicycle production. Enzyme inhibitors function in different ways, but they all prevent the active site from interacting with the substrate to form ES. Some inhibitors cause enzymes to lose catalytic activity only temporarily, while others function to cause enzymes to lose catalytic activity permanently.

In **reversible inhibition**, the inhibitor causes the enzyme to lose catalytic activity; however, if the inhibitor is removed, the enzyme becomes functional again. Reversible inhibitors can be competitive or noncompetitive.

Competitive inhibitors are molecules that compete with the substrate for the active site. A competitive inhibitor has a structure that resembles the substrate of the enzyme. The competitive inhibitor will interact with the active site to form an enzyme-inhibitor complex, but usually no reaction will take place. As long as the inhibitor remains in the active site, the enzyme cannot interact with its substrate and form product. This lowers the enzyme's activity. Figure 10.9 diagrams how a competitive inhibitor works.

A medical therapy based on competition involves the enzyme liver alcohol dehydrogenase (LAD) which oxidizes ethanol, the alcohol found in wine and spirits, as well as the compounds ethylene glycol and methanol, both found in antifreeze. All three are substrates of LAD and compete for the active site. Notice that the structures of the three are very similar. In fact, one remedy for ethylene glycol poisoning in pets and methanol poisoning in humans is the slow intravenous infusion of ethanol maintaining a controlled concentration in the bloodstream over several hours. This slows the production of the toxic metabolic products of ethylene glycol or methanol, giving the kidneys time to filter out the excess substrates in the urine.

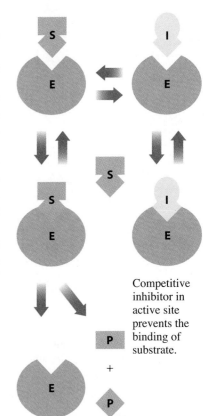

Competitive inhibitor in active site prevents the binding of substrate.

▲ **FIGURE 10.9 Competitive inhibition.** The structure of a competitive inhibitor is similar to substrate, which allows it to compete for the active site.

$$CH_3CH_2OH \qquad HOCH_2CH_2OH \qquad CH_3OH$$
Ethanol $\qquad$ Ethylene glycol $\qquad$ Methanol

Noncompetitive inhibitors typically do not resemble the substrate, and so they do not compete for the enzyme's active site. Instead, these inhibitors bind to another site on the enzyme that is usually remote to the active site. When a noncompetitive inhibitor binds to the enzyme, it changes the shape of the enzyme and the active site loses its shape or is distorted so it is no longer able to interact effectively with the substrate. Again, as long as the inhibitor remains bound to the enzyme, the enzyme cannot function. Figure 10.10 diagrams how a noncompetitive inhibitor works.

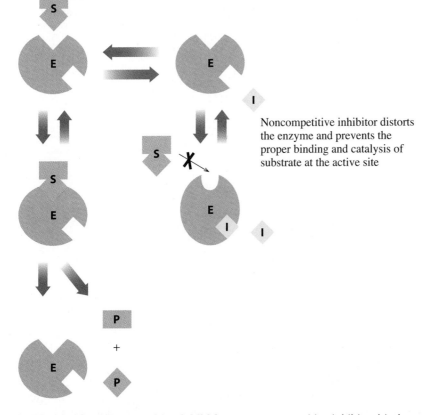

Noncompetitive inhibitor distorts the enzyme and prevents the proper binding and catalysis of substrate at the active site

▲ **FIGURE 10.10 Noncompetitive inhibition.** A noncompetitive inhibitor binds to a site other than the active site, which distorts the shape of the active site, causing the enzyme to lose catalytic activity.

As suggested in the diagrams, the inhibition caused by competitive and noncompetitive inhibitors can be reversed. Inhibition caused by a competitive inhibitor can be reversed by adding more substrate to the reaction. Since the inhibitor and the substrate are competing for the active site, the higher the concentration of the substrate, the more likely it is that it will win the competition for the active site. In the case of the noncompetitive inhibitor, adding more substrate has no effect because regardless of the amount of enzyme, a certain portion of the enzyme is inactivated by the inhibitor. Reversing the effect of a noncompetitive inhibitor typically requires a special chemical reagent to remove the inhibitor and restore the catalytic activity of the enzyme. Going back to the bicycle factory, if a tool technician removed the rust from all the workers' tools so they were again in working order, the number of bicycles produced by the factory would increase.

In **irreversible inhibition**, the inhibitor forms a covalent bond with an amino acid side chain in the enzyme's active site. With the inhibitor covalently bonded in the active site, the substrate is excluded or the catalytic reaction is blocked. Regardless of the method, irreversible inhibitors permanently inactivate enzymes (see Figure 10.11). Likewise, if a fire destroyed a major portion of the bicycle factory, bicycle production would cease.

▶ **FIGURE 10.11 Irreversible inhibition.** An irreversible inhibitor forms a covalent bond with the enzyme's active site, rendering it permanently inactive.

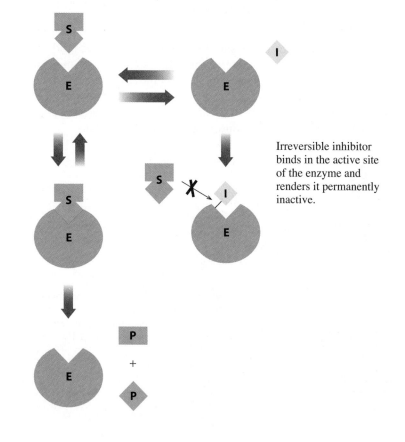

Irreversible inhibitor binds in the active site of the enzyme and renders it permanently inactive.

Heavy metals like silver, mercury, and lead can inactivate enzymes irreversibly by binding to the thiol ($-SH$) functional groups of the amino acid cysteine if found in the active site. As we saw in Chapter 9, this can also denature the proteins.

Antibiotics Inhibit Bacterial Enzymes

Enzyme inhibitors have long been used in the battle against diseases. The well-known antibiotic penicillin is another example of an irreversible inhibitor. Penicillin binds to the active site of an enzyme that bacteria use in the synthesis of their cell walls, structures present in bacterial cells but not human cells. When the bacterial enzyme bonds to penicillin, the enzyme loses its catalytic activity, and the growth of the bacterial cell wall slows. Without a proper cell wall for protection, bacteria cannot survive and the infection stops.

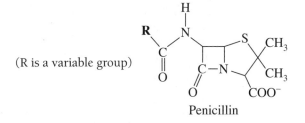

(R is a variable group)

Penicillin

SAMPLE PROBLEM 10.4

Factors Affecting Enzyme Activity

Lactase catalyzes the hydrolysis of the disaccharide lactose into the monosaccharides galactose and glucose. It is active in the small intestine where the pH is around 6.8. Supplements of lactase can help people who are lactose intolerant digest lactose.

Lactose $\xrightarrow{\text{Lactase}}$ galactose + glucose

a. Would the enzyme also be just as active in the stomach with a pH of 1–3? Explain.

b. Would a lactase supplement act just as well on lactose if it is first stirred into cold milk prior to drinking? Explain.

SOLUTION

a. No. The enzyme would be less active in the stomach since it operates at an optimum pH of 6.8.

b. No. The enzyme has an optimum pH of body temperature, 37 °C. Its activity would be much lower in cold milk.

PRACTICE PROBLEMS

10.15 How would the following changes affect enzyme activity for an enzyme whose optimal conditions are normal body temperature and physiological pH?

a. raising the temperature from 37 °C to 60 °C

b. lowering the pH from 7.5 to 4.0

10.16 Chymotrypsin is an enzyme located in the small intestine that catalyzes peptide bond hydrolysis in proteins. Based on the examples given in Table 10.1, would you expect chymotrypsin to be most active at a pH of 4.5, 7.8, or 10.0?

10.17 The enzyme urease catalyzes the formation of ammonia and carbon dioxide from urea as shown:

$$H_2N-\overset{\overset{\displaystyle O}{\|}}{C}-NH_2 + H_2O \xrightarrow{\text{Urease}} 2NH_3 + CO_2$$

Describe what effect the following changes would have on the rate of this reaction:

a. adding excess urea

b. lowering the temperature to 0 °C

10.18 Indicate whether each of the following describes a competitive or a noncompetitive inhibitor.

a. The structure of the inhibitor is similar to that of the substrate.

b. Adding more substrate to the reaction has no effect on the enzyme activity.

c. The inhibitor competes with the substrate for the active site.

d. The structure of the inhibitor has no resemblance to the structure of the substrate.

e. Adding more substrate to the reaction restores the enzyme activity.

10.19 Penicillin is used to treat bacterial infections by destroying bacteria cells. Why does the antibiotic kill bacteria, but not humans?

SUMMARY

10.1 Enzymes and Their Substrates

Enzymes are large, globular proteins that serve as catalysts in biological systems. The functional part of an enzyme is the active site, which is a small groove or cleft on the surface of the molecule where catalysis occurs. Substrates are the reactants in the reactions catalyzed by enzymes. Because of the three-dimensional shape of the active site, enzymes have few (many only one) substrates that will bind and react. The lock-and-key and induced-fit theories explain how an enzyme interacts with its substrate to form ES.

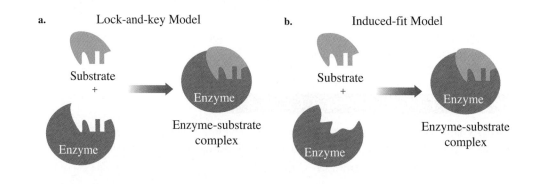

a. Lock-and-key Model   b. Induced-fit Model

Substrate + Enzyme → Enzyme-substrate complex

Substrate + Enzyme → Enzyme-substrate complex

10.2 Thermodynamics of Chemical Reactions

Thermodynamics is the study of the energy changes that occur during a chemical reaction. Activation energy is the energy required to get a reaction started. The activation energy also plays a role in the rate of the reaction, how fast it proceeds from reactants to products. The heat of reaction is a measure of the production or consumption of energy during a reaction and is the difference in the energy of the products compared to the reactants. An endothermic reaction absorbs (or consumes) heat energy from its environment as it proceeds. An exothermic reaction releases (or produces) heat energy as it proceeds. A catalyst speeds up a reaction by lowering the activation energy. Enzymes form ES first before catalysis, which assists in lowering the activation energy.

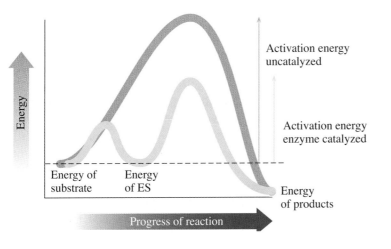

10.3 Enzymes and Catalysis

The active site of an enzyme is designed to help a reaction proceed to products. The formation of ES lowers the activation energy for the catalyzed reaction in several ways. The atoms undergoing the reaction are brought close together in space and aligned with amino acid side chains in the active site for catalysis. The active site also aligns the reactants with the optimal orientation for the reaction to occur, and the interactions of the substrate with the active site weaken the bonds between atoms in the substrate so that the bonds are easier to break during the reaction.

10.4 Factors That Affect Enzyme Activity

Enzyme activity is measured by how fast an enzyme catalyzes a reaction. Environmental factors such as substrate concentration, pH, temperature, and the presence of inhibitors can affect the activity of an enzyme. An increase in substrate concentration increases the rate of an enzyme-catalyzed reaction until the enzyme is saturated with substrate. At this point, the enzyme is operating in a steady state and as efficiently as possible. Enzymes have an optimum pH and optimum temperature at which they function best. Inhibitors decrease or eliminate an enzyme's catalytic abilities. The effect of an inhibitor can be reversible or irreversible. Reversible inhibitors can be competitive inhibitors, which compete with the substrate for the enzyme's active site or noncompetitive inhibitors, which bind to the enzyme at a site other than the active site, changing the shape of the active site.

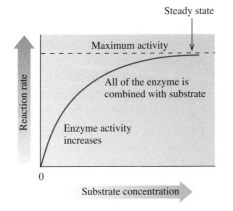

KEY TERMS

activation energy—The energy necessary to align the reactant molecules and to cause them to collide with enough energy to form products.

active site—A pocket on the surface of an enzyme formed during the folding of the tertiary structure and lined with amino acid side chains; portion of the enzyme involved in the catalysis of chemical reactions.

activity—A measure of how fast an enzyme converts its substrate to product.

coenzyme—A small organic molecule necessary for the function of some enzymes; many are obtained from a normal diet as vitamins.

cofactor—An inorganic substance, like a metal ion, necessary for the function of some enzymes; often obtained in the diet as minerals.

competitive inhibitor—A molecule structurally similar to the substrate that competes for the active site.

enzyme–substrate complex (ES)—The complex that is formed when an enzyme binds its substrate in its active site.

heat of reaction—In a chemical reaction, the energy difference between the products and the reactants.

induced-fit model—A model describing the initial interaction of an enzyme to its substrate; an enzyme's active site and its substrate have flexible shapes that are roughly complementary and adjust to allow the formation of ES.

inhibitor—A molecule that causes an enzyme's catalytic activity to decrease.

irreversible inhibition—The loss of enzyme catalytic activity that is permanent.

lock-and-key model—A model describing the initial interaction of an enzyme with its substrate; each enzyme has only one substrate whose shape is complementary to the shape of the active site of the enzyme.

noncompetitive inhibitor—A molecule that inhibits an enzyme's catalytic activity by interacting with a site other than the active site and distorting the shape of the active site.

optimum pH—The pH at which an enzyme functions most efficiently.

optimum temperature—The temperature at which an enzyme functions most efficiently.

rate of reaction—A measure of how much product is formed (or reactant used up) in a certain period of time in a chemical reaction.

reversible inhibition—The lowering of enzyme catalytic activity that is caused by a competitive or noncompetitive inhibitor that can be reversed.

steady state—A state describing optimal enzyme activity where substrate is being converted to product as efficiently as possible.

substrate—The reactant in a chemical reaction catalyzed by an enzyme.

substrate specificity—The property that explains why enzymes have very few substrates and many only one substrate.

thermodynamics—The study of the relationship between a chemical reaction and the heat energy that it produces or consumes.

ADDITIONAL PROBLEMS

10.20 Describe the role enzymes serve for chemical reactions in the body.

10.21 What occurs at the active site of an enzyme?

10.22 Match the terms (1) ES, (2) enzyme, and (3) substrate with the following descriptions:
   a. has a tertiary structure that recognizes the substrate
   b. is the combination of an enzyme with the substrate
   c. has a structure that fits the active site of an enzyme

10.23 Match the terms (1) active site, (2) lock-and-key model, and (3) induced-fit model with the following descriptions:
   a. the portion of an enzyme where catalytic activity occurs
   b. the active site adapts to the shape of a substrate
   c. the active site has a rigid shape

10.24 Do the amino acids that are in the active site of an enzyme have to be near each other in the enzyme's primary structure?

10.25 The enzyme trypsin catalyzes the breakdown of many structurally diverse proteins in foods. Does the induced-fit or lock-and-key model explain the action of trypsin best? Explain.

10.26 The enzyme sucrase catalyzes the hydrolysis of the disaccharide sucrose, but not the disaccharide lactose. Does the induced-fit or lock-and-key model explain the action of sucrase better? Explain.

10.27 A diet deficient in thiamine results in a condition called beriberi, which is characterized by fatigue, weight loss, and poor appetite among several other symptoms. Patients with beriberi do not properly metabolize glucose and other sugars. If thiamine is not an enzyme, what kind of compound do you suppose it is? Explain.

10.28 What type of interactions between an enzyme and its substrate help to stabilize ES?

10.29 Does each of the following statements describe a simple enzyme (no cofactor or coenzyme necessary), an enzyme that requires a cofactor, or an enzyme that requires a coenzyme?
   a. contains Ca^{2+} in the active site
   b. consists of one polypeptide chain in its active form
   c. contains vitamin B_6 in its active site

10.30 When calcium chloride is dissolved in water, the process is moderately exothermic. Would this reaction be used in a hot pack or a cold pack?

10.31 Which reaction occurs at a faster rate, an exothermic reaction with a low activation energy or an endothermic reaction with a high activation energy? Explain.

10.32 What is measured by the heat of reaction?

10.33 Use the following reaction energy diagram to answer the questions:

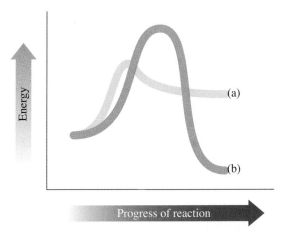

a. Which curve represents the faster reaction, and which represents the slower?
b. Which curve represents an endothermic reaction, and which curve represents an exothermic reaction?

10.34 Two curves for the same reaction are shown in the following reaction energy diagram. Which curve represents the uncatalyzed reaction, and which curve represents the reaction with a catalyst present?

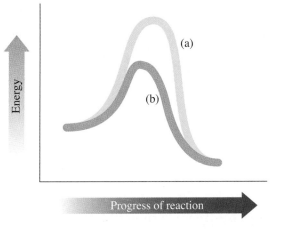

10.35 In the following reaction energy diagram, label 1,2,3, and 4 as one of the following: (a) energy of reactants (b) energy of products (c) activation energy (d) heat of reaction

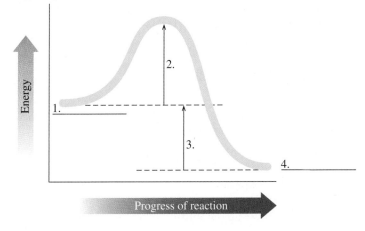

10.36 Is the reaction represented by the diagram in problem 10.35 endothermic or exothermic?

10.37 Draw and label a reaction energy diagram for an endothermic reaction whose activation energy is two times greater than its heat of reaction.

10.38 Draw and label a reaction energy diagram for an exothermic reaction whose heat of reaction is three times greater than its activation energy.

10.39 The following hydrogenation reaction is moderately exothermic, but it occurs at a very slow rate in the absence of a catalyst. When a platinum catalyst is added to the reaction, the reaction rate increases dramatically. Sketch and label reaction energy diagrams for this reaction with (a) no catalyst present and (b) with a platinum catalyst.

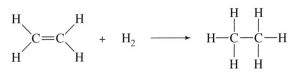

10.40 A substrate is held in the active site of an enzyme by intermolecular forces between the substrate and the amino acid side chains. For the outlined regions A, B, and C on the following substrate molecule:
   a. name one possible intermolecular force it could form in the active site.
   b. name a possible amino acid whose side chain could be involved in the enzyme's active site.

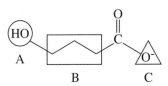

10.41 If each of the following amino acid side chains is present in the active site of an enzyme, indicate whether it would (a) serve a catalytic function, (b) serve to hold the substrate, or (c) both.
   a. aspartate

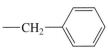

   b. phenylalanine

   c. valine

   d. lysine

$$-CH_2CH_2CH_2CH_2NH_3^+$$

10.42 Chymotrypsin, an enzyme that hydrolyzes peptide bonds in proteins, functions in the small intestine at an optimum pH of 7.7 to 8.0. How is the rate of the chymotrypsin-catalyzed reaction affected by each of the following conditions?
   a. decreasing the concentration of proteins
   b. changing the pH to 3.0
   c. running the reaction at 75 °C

10.43 Pepsin, an enzyme that hydrolyzes peptide bonds in proteins, functions in the stomach at an optimum pH of 1.5 to 2.0. How is the rate of a pepsin-catalyzed reaction affected by each of the following conditions?
   a. increasing the concentration of proteins
   b. changing the pH to 5.0
   c. running the reaction at 0 °C

10.44 Problems 10.42 and 10.43 both mention enzymes that hydrolyze peptide bonds. How do you account for the fact that pepsin has a high catalytic activity at pH 1.5, but chymotrypsin has very little activity at pH 1.5?

10.45 After the reaction catalyzed by an enzyme is complete, why do the products of the reaction migrate away from the active site?

10.46 Describe the key difference between a competitive inhibitor and a noncompetitive inhibitor.

10.47 How does an irreversible inhibitor function differently than a reversible inhibitor?

10.48 When lead acts as a poison, it can do so by either replacing another ion (such as zinc) in the active site of an enzyme or it can react with cysteine side chains to form covalent bonds. Which of these is irreversible and why?

10.49 Increasing the substrate concentration of an enzyme-catalyzed reaction increases the rate of the reaction until the enzyme becomes saturated and reaches its maximum activity. If the amount of *enzyme* in the reaction is increased under saturating conditions, would the rate of the reaction increase, decrease, or remain unchanged? Explain.

10.50 A hypothetical enzyme has an optimum temperature of 40 °C. If we wish to double the rate of the reaction catalyzed by this enzyme, can we do so by increasing the temperature to 80 °C? Explain your answer.

10.51 Cadmium is a poisonous metal used in industries that produce batteries and plastics. Cadmium ions (Cd^{2+}) are inhibitors of hexokinase. Increasing the concentration of glucose or ATP, the substrates of hexokinase, or Mg^{2+}, the cofactor of hexokinase, does not change the rate of the cadmium-inhibited reaction. Is cadmium a competitive or noncompetitive inhibitor? Explain.

10.52 Meats spoil due to the action of enzymes that degrade the proteins. Fresh meats can be preserved for long periods of time by freezing them. Explain how freezing meats works to prevent spoilage.

10.53 Many drugs are competitive inhibitors of enzymes. When scientists design inhibitors to serve as drugs, why do you suppose they choose to design competitive inhibitors instead of noncompetitive inhibitors?

10.54 Fresh pineapple contains the enzyme bromelain which degrades proteins.
   a. The directions on a package of Jello® (a protein-containing food product) say to add canned pineapple (which is heated to high temperatures to preserve it), not fresh pineapple. Why?
   b. Fresh pineapple is used in a marinade to tenderize tough meat. Why?

CHALLENGE PROBLEMS

10.55 The active site of a hypothetical enzyme contains a pocket of nonpolar amino acid side chains next to a cleft containing acidic and basic amino acid side chains.

   a. Which amino acids are likely responsible for catalysis? Explain.
   b. Which amino acids are present primarily for holding the substrate in the active site?

   c. Would you predict that the substrate is polar or non-polar based on your answers to part a and part b? Why?

10.56 Contact lens wearers often soak their lenses in a solution containing enzymes to remove protein deposits from the lenses. Why is it essential to thoroughly rinse the lenses before placing them in the eyes?

ANSWERS TO ODD-NUMBERED PROBLEMS

Practice Problems

10.1 Tertiary

10.3 Both are nonprotein portions of an enzyme, a coenzyme is organic, a cofactor is not.

10.5 The induced-fit model. When glucose is fit into the active site, the enzyme undergoes a conformational change closing around the substrate.

10.7 Cold. If the reaction is endothermic, it absorbs heat from its surroundings which cools the reaction.

10.9 a. The activation energy is the amount of energy it takes for the reactants in a chemical reaction to react to form products.
   b. A catalyst increases the rate of a chemical reaction by lowering the activation energy.
   c. Products

10.11 a. exothermic   **b.** exothermic

10.13 Because the phosphates on ATP have negative charges and Mg^{2+} has positive charges, the attraction is ionic.

10.15 a. The enzyme activity will decrease if the temperature is raised above the optimum temperature.

b. The activity will be lowered if the pH is changed.

10.17 a. no effect   **b.** rate decreases

10.19 Bacterial cells have a cell wall, human cells do not.

Additional Problems

10.21 The active site is the location on an enzyme where catalysis occurs.

10.23 a. (1) active site

b. (3) induced-fit model

c. (2) lock-and-key model

10.25 Because it can have more than one substrate, trypsin's action is better described by the induced-fit model.

10.27 Since thiamine is an organic molecule involved in enzymatic reactions but is not itself an enzyme then it must be a coenzyme.

10.29 a. requires a cofactor

b. describes a simple enzyme

c. requires a coenzyme

10.31 An exothermic reaction with a low activation energy. Reactions with lower activation energy occur more quickly.

10.3 a. Curve (a) is faster, curve (b) is slower.

b. Curve (a) is endothermic, curve (b) is exothermic.

10.35 1—Energy of reactants, 2—Activation energy, 3—Heat of reaction, 4—Energy of products

10.37

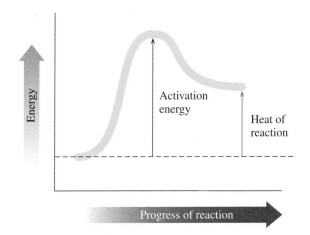

10.39

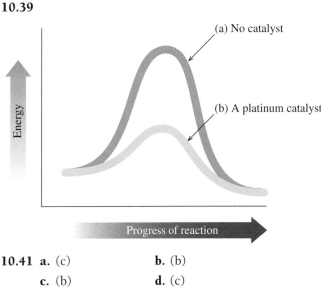

10.41 a. (c)   **b.** (b)
c. (b)   **d.** (c)

10.43 a. The rate would increase until a steady state is reached.

b. The rate would decrease.

c. The rate would decrease.

10.45 The products do not fit snugly into the active site so are released.

10.47 An irreversible inhibitor forms covalent bonds to the enzyme and permanently inactivates it.

10.49 If the amount of enzyme were increased in a reaction occurring at a steady state, the maximum rate of the reaction would increase since more enzyme is present to convert substrate into product.

10.51 Cadmium would be a noncompetitive inhibitor because increasing the substrate and cofactor has no effect on the rate. This implies that the cadmium is binding to another site on the enzyme.

10.53 When designing an inhibitor, usually the substrate of an enzyme is known even if the structure of the enzyme is not. It is easier to design a molecule that resembles the substrate (competitive) than to find an inhibitor that binds to a second site on an enzyme (noncompetitive).

10.55 a. The polar amino acids are likely involved in catalysis.

b. The nonpolar amino acids are likely involved in aligning the substrate correctly.

c. The substrate is likely nonpolar with a small polar portion.

Guided Inquiry Activities **FOR CHAPTER 11**

EXERCISE 1 Components of Nucleotides

Information

▼ **SCHEME 1 Component molecules found in nucleic acids.**

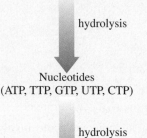

Nucleic Acid
(DNA, RNA)

↓ hydrolysis

Nucleotides
(ATP, TTP, GTP, UTP, CTP)

↓ hydrolysis

Nitrogenous Base + Sugar + Phosphate

Questions

1. Name the three component molecules found in a nucleotide.
2. What chemical reaction occurs when a nucleotide is *formed* from its component molecules?

Nitrogenous Bases

Five different aromatic rings with nitrogen known as nitrogenous bases fit into two classes, purines and pyrimidines. The general structure of these rings is

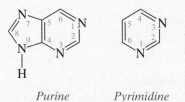

Purine   *Pyrimidine*

The five nitrogenous bases found in nucleotides and nucleic acids are

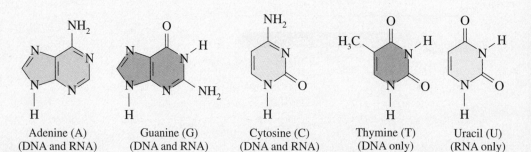

Adenine (A)   Guanine (G)   Cytosine (C)   Thymine (T)   Uracil (U)
(DNA and RNA)   (DNA and RNA)   (DNA and RNA)   (DNA only)   (RNA only)

Questions

3. Which of the nitrogenous bases are purines? Which are pyrimidines?

4. Which are found in DNA? Which in RNA?

5. Compare the structures of uracil and thymine. How are they different?

Pentose Sugars

There are two aldopentose sugars found in nucleic acids, ribose and deoxyribose. They are shown in the following Haworth projection. The sugars in nucleic acids are numbered with a prime (′) to distinguish their carbons from the carbons of the nitrogenous bases.

Questions

6. a. Compare the structures of ribose and 2-deoxyribose. How are they different?
   b. Which one do you think is found in DNA? RNA?

Phosphate

A phosphate is a functional group that contains a phosphorus surrounded by four oxygen atoms as described in Section 10.1.

Putting It All Together

As implied in Scheme 1, the condensation of a phosphate, sugar, and nitrogenous base produces a *nucleotide*. The phosphate's OH will condense with OH on C5′. A purine's N9 will condense with the OH on C1′. A pyrimidine's N1 will condense with the OH on C1′.

Questions

7. Draw the structure of the nucleotide adenosine monophosphate (AMP) using the ribose sugar shown in the following figure:

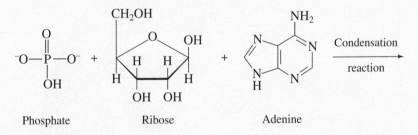

Phosphate   Ribose   Adenine

8. If AMP is adenosine monophosphate, how would the structure of ATP, adenosine triphosphate, differ?

Pentose sugars in RNA and DNA

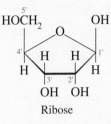

Ribose

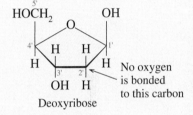

Deoxyribose

No oxygen is bonded to this carbon

SECTION 11.2

EXERCISE 2 Nucleic Acids

Information

Nucleotides (phosphate, sugar, and nitrogenous base) can be combined through a condensation reaction to form the nucleic acids RNA or DNA. When combined, one phosphate is connected between the 5′ and 3′ carbons of sugar molecules.

The *backbone* of a nucleic acid is made up of alternating adjacent sugar and phosphate groups. The nitrogenous bases are dangling off the backbone (similar to how the side chains of a protein strand dangle from its backbone). When writing nucleic acids, their structure is abbreviated by the base abbreviation since the sugar and the phosphate are the same in each nucleotide. The convention for designating the order of the nucleic acids is from the 5′ ⟶ 3′ end. On the string shown in the following figure, (GTCA), the 5′ end is on the left and the 3′ end is on the right.

▶ **FIGURE 1 A nucleic acid has directionality.** The arrangements of the component parts can be abbreviated down to a set of base letters.

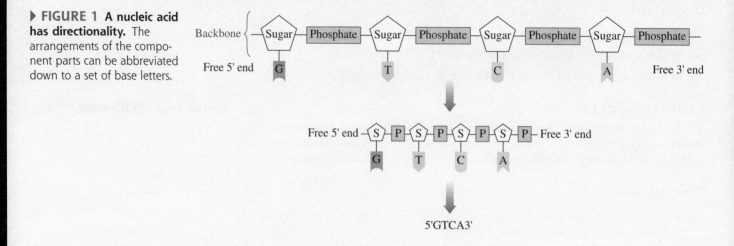

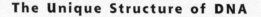

5'GTCA3'

The Unique Structure of DNA

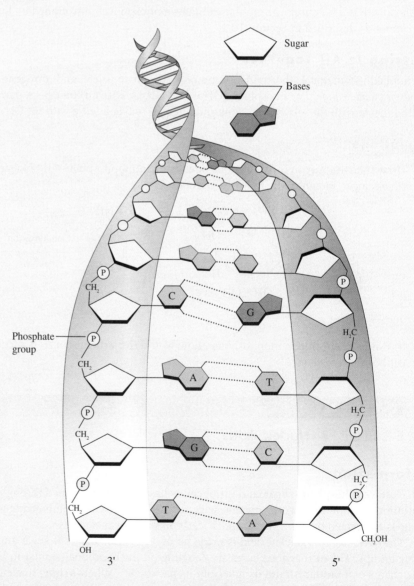

378

Characteristics of DNA

- Double stranded—two strands of nucleic acid line up in opposite directions (antiparallel) with the bases in the center.

- Hydrogen bonded—The bases H-bond across the strands like rungs in a ladder: G≡C, A═T. These pairings are referred to as *complementary base pairs*.

- Double helix—Due to the chirality of the sugars, the two antiparallel strands have a natural twist to them, forming a helical shape called a double helix.

A strand of double stranded DNA can be abbreviated as

5′ATTCGG3′
3′TAAGCC5′

Questions

1. **a.** How would you designate the following molecule in a one-letter abbreviation?

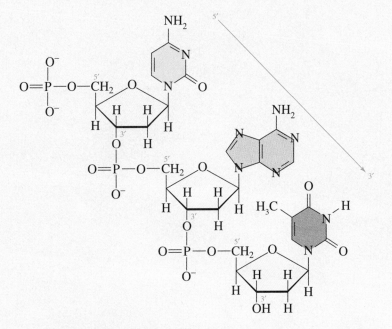

b. Is this a strand of DNA or RNA? How do you know?

c. Provide an abbreviation for the strand of DNA with complementary bases to the one shown in part a. Indicate the 5′ and 3′ ends.

2. Provide the complementary DNA strand to the following strand:

5′AATTCCGCTAACG3′

3. From what you have seen in this section and your own experience, have your group describe the structure and function of DNA in a brief paragraph.

The nucleic acid called DNA defines our hereditary characteristics like eye and hair color. These large molecules are made up of smaller component parts. Chapter 11 also describes how DNA is used to make proteins in the cell along with many other aspects of nucleic acids.

Nucleic Acids
BIG MOLECULES WITH A BIG ROLE

The nucleic acid deoxyribonucleic acid, or DNA, appears regularly on crime shows and in courtrooms in the context of DNA testing. What is so unique and important about this molecule? What does it look like? **DNA** is the molecule in our cells that stores and directs information responsible for cell growth and reproduction. DNA, found in the nucleus of the cell, contains genetic information and is called the **genome**. One part of the genome, the **gene**, contains information to make a particular protein for the cell. The human genome contains an estimated 20,000 to 25,000 genes. Every cell in your body contains the same DNA, so why is a skin cell different from a liver cell? Different cell types use different parts of the genome to express their characteristics. Together the DNA in your tissues governs your physical traits like your eye and hair color, your growth rate, and your metabolic rate.

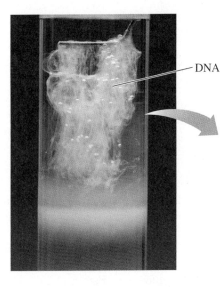

DNA

DNA molecules are so long and stringy that when they are isolated, we can actually see them with the naked eye. Nucleic acids like DNA and its counterpart ribonucleic acid (**RNA**) are big molecules made up of repeating building blocks called nucleotides. Exploring the structure of the nucleotide building blocks is a starting point for understanding the more complex structure of the nucleic acids and genomics, the study of genes and their interactions.

11.1 Components of Nucleic Acids

Just as proteins are strings of amino acids, nucleic acids are strings of molecules called **nucleotides**. While most proteins contain 20 different amino acids in a given sequence, most nucleic acids are created from a set of nucleotides in a given sequence. The differences between DNA and RNA lie in slight variations of the components found in these nucleotides. Each nucleotide has three basic components: a nitrogenous base, a five-carbon sugar (pentose), and a phosphate functional group.

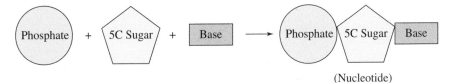

(Nucleotide)

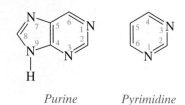

Purine *Pyrimidine*

Nitrogenous Bases

There are four different nitrogenous bases in nucleic acids, and each of the bases contains one of two nitrogen-containing aromatic rings, either the purine ring or the pyrimidine ring.

DNA contains two purines, adenine (A) and guanine (G), and two pyrimidines, thymine (T) and cytosine (C). RNA contains the same bases, except thymine (5-methyluracil) is replaced with uracil (U) (see Figure 11.1).

▶ **FIGURE 11.1 Purine and pyrimidine bases.** DNA contains the nitrogenous bases A, G, C, and T; RNA contains A, G, C, and U. Notice the structural difference between T and U is a methyl group.

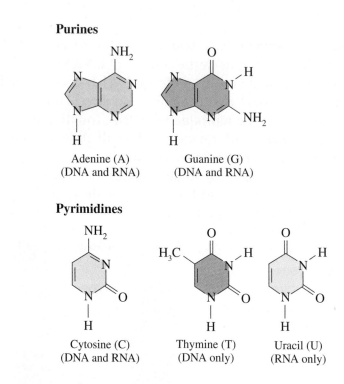

Purines

Adenine (A)
(DNA and RNA)

Guanine (G)
(DNA and RNA)

Pyrimidines

Cytosine (C)
(DNA and RNA)

Thymine (T)
(DNA only)

Uracil (U)
(RNA only)

Ribose and Deoxyribose

Nucleotides also contain five-carbon pentose sugars. To distinguish the carbons in the nitrogenous bases from the carbons in the sugar rings, a prime symbol (′) is added to the carbon numbering of the sugars (1′, 2′, 3′, 4′, and 5′). RNA contains the pentose *ribose* (the "R" in RNA) and DNA contains the pentose *deoxyribose* (the "D" in DNA). Deoxyribose lacks an oxygen on carbon 2′ of the pentose, but otherwise its structure is identical to ribose (see Figure 11.2).

▶ **FIGURE 11.2 Five-carbon sugars.** The five-carbon pentose sugar found in RNA is ribose and is deoxyribose in DNA. Notice that the carbon numbering includes the prime (′) symbol.

Pentose sugars in RNA and DNA

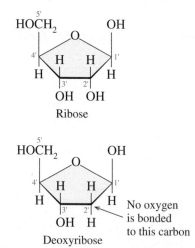

Ribose

Deoxyribose

No oxygen is bonded to this carbon

SAMPLE PROBLEM 11.1

Components of Nucleic Acids

Identify each of the following bases as a purine or a pyrimidine:

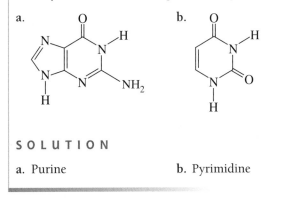

a.   b.

SOLUTION

a. Purine   b. Pyrimidine

Condensation of the Components

In Chapter 5 we saw that sugars could bond to each other through a condensation reaction. In that condensation reaction, a molecule of water is removed and a glycosidic bond forms between two sugars. Similarly, a pentose and a nitrogenous base can join by this reaction when a nitrogen in the base (N1 of pyrimidines or N9 of purines) bonds to C1′ of the pentose, forming a carbon-to-nitrogen glycosidic bond. The condensation reaction between ribose and the base adenine is shown forming a molecule called adenosine.

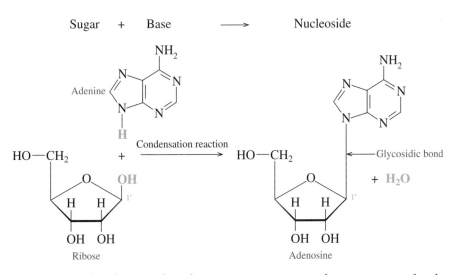

We noted earlier that a nucleotide contains a nitrogenous base, a sugar, and a phosphate, and that these nucleotides make up nucleic acids. Where does adenosine fit in? It does not contain the phosphate component. Adeno*sine* is a **nucleoside**, formed when just the pentose sugar and the nitrogenous base components combine in a condensation reaction. It is the first step in the synthesis of a nucleotide.

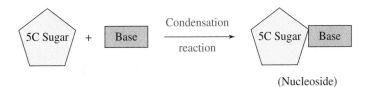

Suppose a hydrogen phosphate (HPO_4^{2-}) reacts with the —OH group on C5′ of an adenosine molecule. The molecule formed in this condensation reaction is a *nucleotide*. In this case, the nucleotide *adenosine monophosphate* (AMP) is produced.

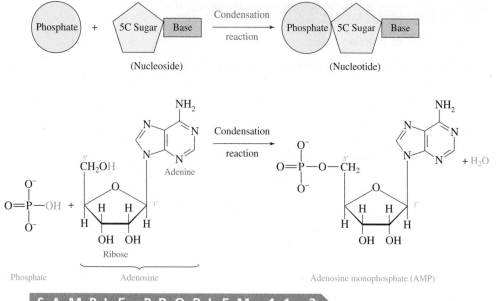

Phosphate   Adenosine   Adenosine monophosphate (AMP)

SAMPLE PROBLEM 11.2

Condensation of the Components

Provide the products for each of the following condensation reactions:

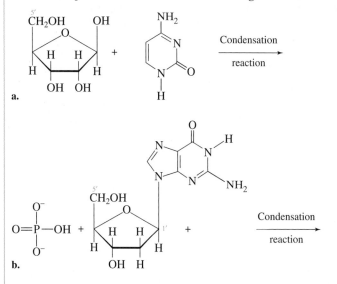

a.

b.

SOLUTION

a. This is a condensation reaction producing a nucleoside. Nitrogen 1 of the pyrimidine condenses with C1′ of the ribose, forming the nucleoside cytidine and a molecule of water.

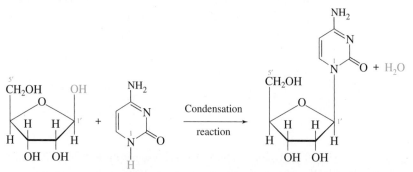

b. In this reaction, a water molecule and the nucleotide GMP are produced. C5′ of deoxyribose condenses with the phosphate—OH group.

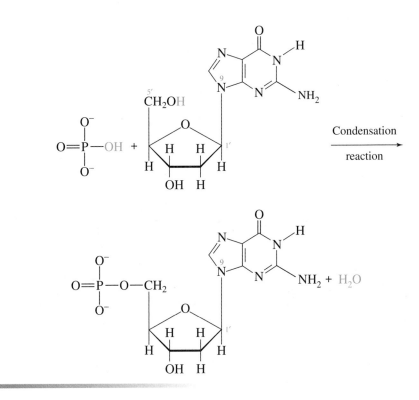

Naming Nucleotides

Nucleotides include the nucleoside name and the number of phosphates present. Because the names can get quite long, nucleotide names are commonly abbreviated. This abbreviation indicates the type of sugar (ribose or deoxyribose) and the nitrogenous base. If deoxyribose is found in the nucleotide a lowercase *d* is inserted at the beginning of the abbreviation. The names and abbreviations of the nucleotides found in nucleic acids are shown in Figure 11.3 and Table 11.1

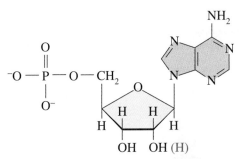

Adenosine 5'-monophosphate (AMP)
Deoxyadenosine 5'-monophosphate (dAMP)

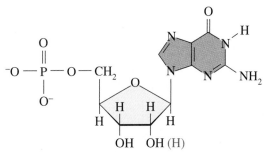

Guanosine 5'-monophosphate (GMP)
Deoxyguanosine 5'-monophosphate (dGMP)

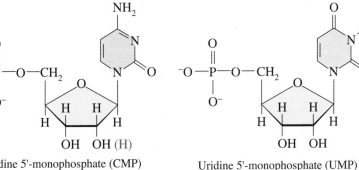

Cytidine 5'-monophosphate (CMP)
Deoxycytidine 5'-monophosphate (dCMP)

Uridine 5'-monophosphate (UMP)

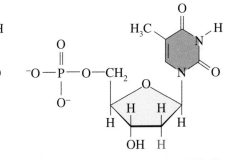

Deoxythymidine 5'-monophosphate (dTMP)

▲ **FIGURE 11.3 Naming nucleotides.** The nucleotides of RNA (black names) are identical to those of DNA (red names) except in DNA the sugar is deoxyribose and deoxythymidine replaces uridine.

TABLE 11.1	NUCLEOSIDES AND NUCLEOTIDES IN DNA AND RNA	
Base	Nucleosides	Nucleotides
RNA		
Adenine (A)	Adenosine (A)	Adenosine 5′-monophosphate (AMP)
Guanine (G)	Guanosine (G)	Guanosine 5′-monophosphate (GMP)
Cytosine (C)	Cytidine (C)	Cytidine 5′-monophosphate (CMP)
Uracil (U)	Uridine (U)	Uridine 5′-monophosphate (UMP)
DNA		
Adenine (A)	Deoxyadenosine (A)	Deoxyadenosine 5′-monophosphate (dAMP)
Guanine (G)	Deoxyguanosine (G)	Deoxyguanosine 5′-monophosphate (dGMP)
Cytosine (C)	Deoxycytidine (C)	Deoxycytidine 5′-monophosphate (dCMP)
Thymine (T)	Deoxythymidine (T)	Deoxythymidine 5′-monophosphate (dTMP)

SAMPLE PROBLEM 11.3

Nucleotides

Identify which nucleic acid (DNA or RNA) contains each of the following nucleotides and state the components of each nucleotide:

a. deoxyguanosine 5′-monophosphate (dGMP)

b. adenosine 5′-monophosphate (AMP)

SOLUTION

a. This is a DNA nucleotide because it contains deoxyribose. It also contains guanine and one phosphate.

b. This is an RNA nucleotide because it contains ribose (does not have deoxy pre-fix). It also contains adenine and one phosphate.

PRACTICE PROBLEMS

11.1 Identify each of the following as being a purine or a pyrimidine:

a. thymine

b.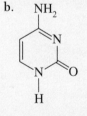

11.2 Identify each of the following as being a purine or a pyrimidine:

a. guanine

b.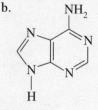

11.3 Identify the bases in problem 11.1 as present in RNA, DNA, or both.

11.4 Identify the bases in problem 11.2 as present in RNA, DNA, or both.

11.5 List the names and abbreviations of the four nucleotides in DNA.

11.6 List the names and abbreviations of the four nucleotides in RNA

11.7 Provide the products for each of the following condensation reactions:

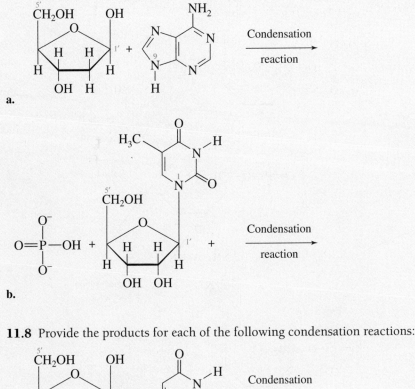

a.

b.

11.8 Provide the products for each of the following condensation reactions:

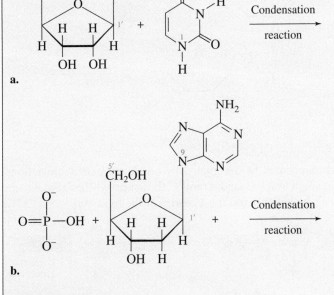

a.

b.

11.2 Nucleic Acid Formation

Many nucleotides linked together form **nucleic acids**. Analogous to the amino acids in proteins being linked together by peptide bonds, nucleotides are linked together through **phosphodiester bonds** where the phosphate oxygens are connected between the 3′ and 5′ C's of adjacent sugar molecules (see Figure 11.4).

Primary Structure: Nucleic Acid Sequence

Analgous to a protein backbone, the backbone of a nucleic acid consists of alternating sugar and phosphate with the bases dangling from the sugar. A protein's primary structure is the sequence of amino acids, and a nucleic acid's primary structure is indicated by its nucleotide sequence. The backbone of a nucleic acid runs directionally with the phosphates connected between the 3′ carbon of one sugar and the 5′ carbon of the

▶ **FIGURE 11.4 Condensation of nucleotides.** Nucleotides bond to each other through a condensation reaction between the 3′—OH of one nucleotide and the phosphate group (bonded to the 5′—OH of the nucleotide) of a second nucleotide forming a dinucleotide.

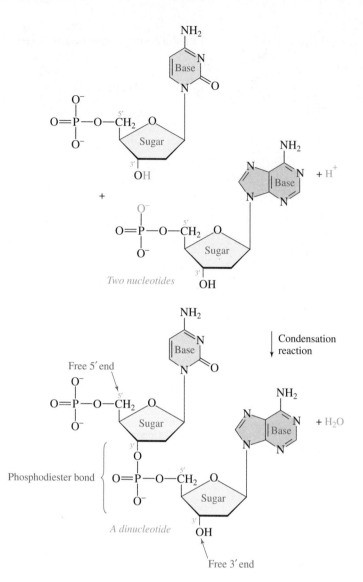

neighboring sugar. The sequence can be designated by one-letter base abbreviations. Convention states that if a single nucleic acid strand is shown drawn horizontally, the 5′ end of a nucleic acid is at the left end and the 3′ end is at the right end.

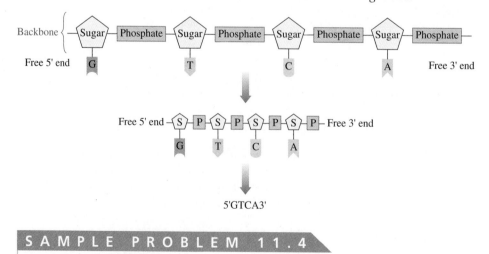

Bonding of Nucleotides

Draw the RNA dinucleotide (two nucleotides joined by a phosphodiester) formed if two adenosine monophosphates undergo a condensation reaction. Label the 5′ and 3′ ends of your structure. Identify the phosphodiester bond.

SOLUTION

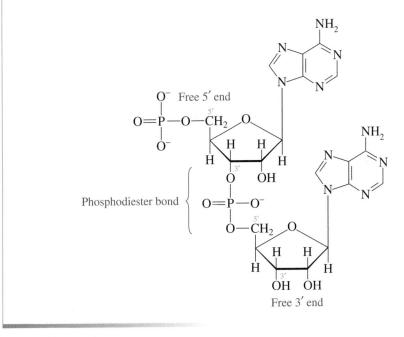

O⁻ Free 5′ end

Phosphodiester bond

Free 3′ end

Sequence Abbreviation

Provide an abbreviation for the following nucleic acid sequence:

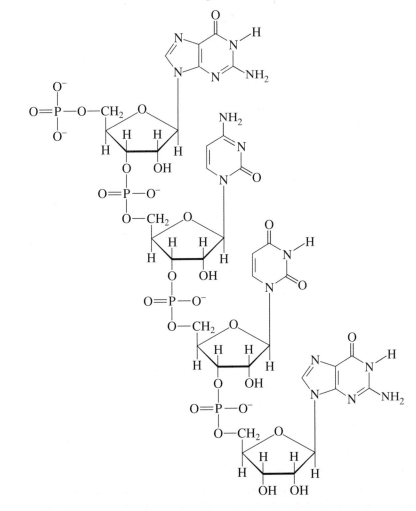

SOLUTION

The 5′ end of this molecule is at the top. The sequence is abbreviated by identifying the nitrogenous base present in each nucleotide beginning at the 5′ end. This nucleic acid is abbreviated GCUG.

PRACTICE PROBLEMS

11.9 How are nucleotides bonded together in a nucleic acid chain?

11.10 Describe the differences in the two ends of a nucleic acid.

11.11 Draw the dinucleotide GC that would be found in RNA. Label the 5′ and 3′ ends of your structure. Identify the phosphodiester bond.

11.12 Draw the dinucleotide AT that would be found in DNA. Label the 5′ and 3′ ends of your structure. Identify the phosphodiester bond.

11.3 DNA

Nucleic acid–base sequences stored as DNA in the nucleus of the cell hold the code for cellular production of proteins. A few key discoveries beginning in the late 1940s led to what we now understand as the structure of DNA. Erwin Chargaff noted that the amount of adenine (A) is always equal to the amount of thymine (T) (A=T), and the amount of guanine (G) is always equal to the amount of cytosine (C) (G=C). This also implies that the number of purines equals the number of pyrimidines in DNA.

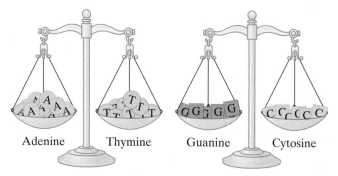

Adenine   Thymine   Guanine   Cytosine

Secondary Structure: Complementary Base Pairing

DNA's secondary structure is described by the interaction of two nucleic acids to form a **double helix**. This was first proposed in 1953 by James Watson and Francis Crick. A double helix can be envisioned as a twisted ladder (see Figure 11.5). Imagine that the sugar–phosphate backbones make up the rails of the ladder and the bases dangling off the backbone interact as the rungs in the center of the ladder. The two strands both have the bases in the center and the backbones run in opposite directions, or as biochemists describe them, they are *antiparallel* to each other. One strand goes in the 5′ to 3′ direction and the other strand goes in the 3′ to 5′ direction.

Each of the rungs in DNA's helical ladder contains one base from each of the strands. The two bases in a ladder rung associate with each other through hydrogen bonding. Watson and Crick realized that all the rungs are the same length, so they must contain one purine and one pyrimidine. The pairs A–T and G–C are called **complementary base pairs** (see Figure 11.6). Adenine forms two hydrogen bonds to thymine, and guanine forms three hydrogen bonds to cytosine. The DNA in one human cell contains about 3 billion of these base pairs!

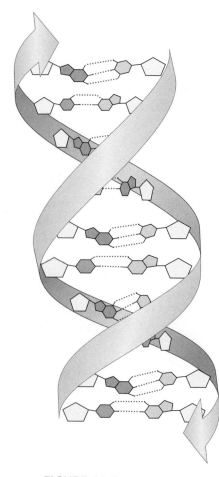

▲ **FIGURE 11.5 A DNA double helix.** The two backbones run antiparallel to each other (noted by arrows on ends) and the bases line up in the interior space.

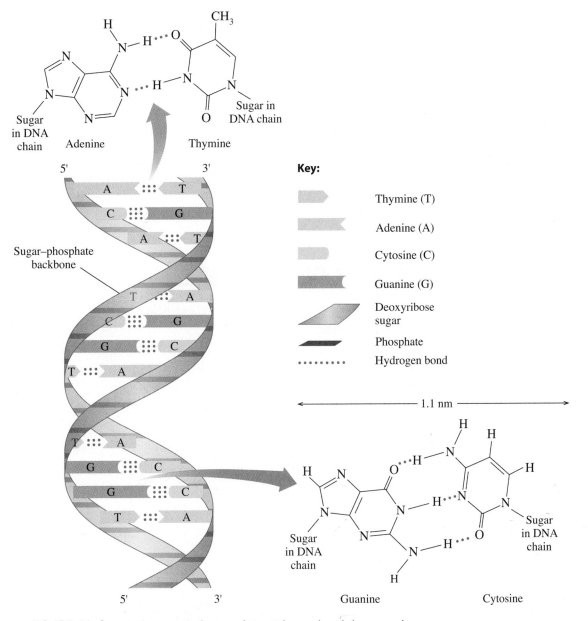

▲ **FIGURE 11.6 Complementary base pairs.** Hydrogen bonds between the complementary base pairs hold the nucleic acid strands together in DNA's double helix.

Complementary Base Pairs

Write the base sequence and indicate the 3′ and 5′ ends of the complementary strand for a segment of DNA with the base sequence 5′ACGATCT3′.

S O L U T I O N

In the complementary strand, A pairs with T and G pairs with C. The two strands run opposite each other so

Given strand: 5′ACGATCT3′
Complement: 3′TGCTAGA5′

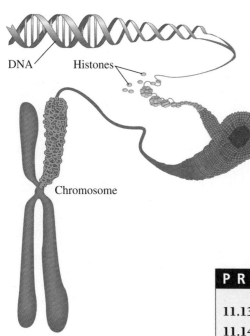

DNA

Histones

Chromosome

▲ FIGURE 11.7 **Chromosome structure.** DNA is supercoiled into tight structures called chromosomes. The DNA is wound around proteins called histones which then further wind into fibers that are organized into the chromosome structure.

Tertiary Structure: Chromosomes

Try this. Hold a rubber band vertically at the top and bottom and twist it. There are only so many times that you can twist it before it starts to double up on itself. As in proteins, the tertiary structure refers to a large biomolecule's overall shape when it folds onto itself. Because DNA has a helical twist (the double helix), any further twisting (doubling it up on itself) that makes the DNA more compact constitutes its tertiary structure. This further twisting is called **supercoiling**. The 3 billion base pairs of DNA in one human cell would stretch out as a double helix to about 6 feet in length. The DNA in a human cell is separated into 46 pieces (23 from mother, 23 from father) that are supercoiled around proteins called histones. These pieces of DNA wound about histones pack into **chromosomes** (see Figure 11.7). Chromosomes are an efficient package for lots of DNA information.

PRACTICE PROBLEMS

11.13 How are the two strands of nucleic acid in DNA held together?

11.14 Define complementary base pairing.

11.15 Write the base sequence and indicate the 3′ and 5′ ends of the complementary strand for a segment of DNA with the following base sequences:

a. 5′AAAA3′

b. 5′CCCCTTTT3′

c. 5′ACATTGG3′

d. 5′TGTGAACC3′

11.16 Write the base sequence and indicate the 3′ and 5′ ends of the complementary strand for a segment of DNA with the following base sequences:

a. 5′AAAAAACC3′

b. 5′GGGGGAT3′

c. 5′AAAATTTT3′

d. 5′CGCGATATTA3′

11.17 Fill in the following table with the analogous nucleic acid structures:

	Protein	Nucleic Acid
Primary structure	Sequence of amino acids	
Secondary structure	Local hydrogen bonding between backbone atoms	
Tertiary structure	Folding of backbone associating amino acids far away in sequence	

11.18. List the similarities and differences in the secondary structure of a protein and the secondary structure of DNA.

11.4 RNA and Protein Synthesis

RNA can be considered the "middleman" in the process of creating a protein from a gene in DNA. Like DNA, RNA is a string of nucleotides; however, some important differences exist. These differences are listed in the Table 11.2. One of the main differences is that RNA does not contain the base thymine. In RNA, the base uracil is substituted and it is complementary to adenine, forming two hydrogen bonds (A = U).

TABLE 11.2	DIFFERENCES BETWEEN RNA AND DNA	
Differences	**RNA**	**DNA**
Sugar	Ribose	Deoxyribose
Base	Uracil	Thymine
Stranding	Single strand of nucleic acid	Double strand of nucleic acid
Size	Much smaller	Larger

RNA Types and Where They Fit In

The three main types of RNA found in the cell are involved in transforming a DNA sequence into a protein sequence. The names of the three RNAs describe their role in this transformation: *messenger RNA, ribosomal RNA*, and *transfer RNA*.

Messenger RNA and Transcription

The process of making a protein from DNA has two key steps, the first of which is to make a gene copy from the DNA found in the cell nucleus. This step is called transcription. In **transcription**, DNA's double helix temporarily unwinds so that a complementary copy can be made from one of the strands. This complementary copy is the messenger RNA (**mRNA**). Messenger RNA is a single-stranded piece of RNA containing the complementary bases of the original DNA strand (see Figure 11.8). This gene copying is catalyzed by **RNA polymerase**, a complex enzyme containing several subunits that binds to the DNA. Once completed, the mRNA travels from the nucleus to an organelle called the ribosome where the mRNA sequence is processed into protein.

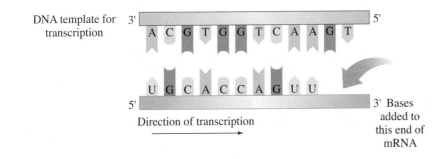

◀ **FIGURE 11.8 Transcription.** In transcription, a messenger RNA is transcribed from one strand of genetic DNA (the template) as a complementary copy.

Ribosomal RNA and the Ribosome

The cell organelle called the ribosome is made up of ribosomal RNA (**rRNA**) and protein. A ribosome can be thought of as a protein factory. This is where the nucleotide sequence delivered from the mRNA is interpreted into an amino acid sequence. The ribosome has two rRNA/protein subunits called the small subunit and the large subunit. The general shape of each is shown in Figure 11.9. The mRNA strand fits into a groove on the small subunit with the bases pointing toward the large subunit.

▶ **FIGURE 11.9 Ribosome structure.** A typical ribosome consists of a small subunit and a large subunit. Both subunits contain rRNA and protein. The mRNA fits in a groove between them.

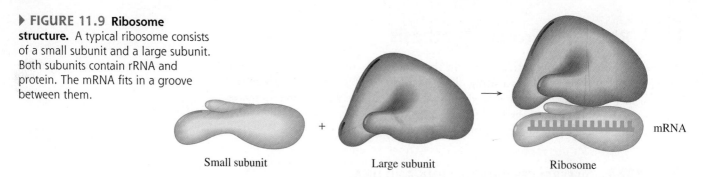

Small subunit   Large subunit   Ribosome   mRNA

Transfer RNA and Translation

The second step in the process of making a protein from DNA occurs in the ribosome and is called **translation**. The mRNA sequence must be translated into a protein sequence. The facilitator for this process is the transfer RNA (**tRNA**). There are several areas on the tRNA sequence where complementary bases can hydrogen bond with each other. This gives the tRNA a tightly compacted T-shaped structure (see Figure 11.10). The tRNA has a three-base sequence (triplet) called an **anticodon** at its *anticodon loop*. When in the ribosome, the anticodon of a tRNA can hydrogen bond to three complementary bases on the mRNA. The tRNA also has a place at the opposite end called the *acceptor stem* where it can bind an amino acid. The amino acid is joined to the tRNA through ester bond formation in a reaction called esterification. (We saw this reaction previously in triglyceride formation in Section 6.2.) The only way to get an amino acid incorporated into a growing protein chain is by bringing it to the ribosome bonded to the tRNA. Each of the 20 amino acids has one or more different tRNAs available for this.

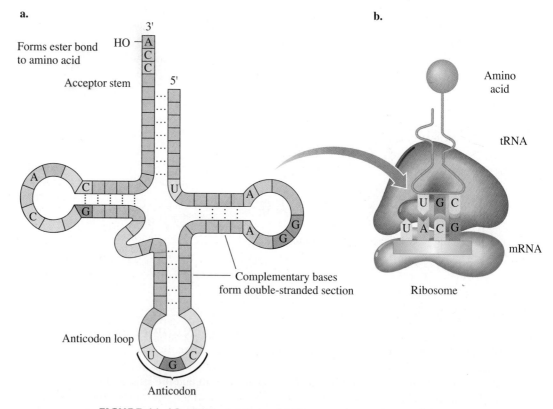

a.

Forms ester bond to amino acid

Acceptor stem

Complementary bases form double-stranded section

Anticodon loop

Anticodon

b.

Amino acid

tRNA

mRNA

Ribosome

▲ **FIGURE 11.10 tRNA structure.** (a) The secondary structure of the tRNA showing complementary base pairing. (b) The tRNA with amino acid and mRNA in the ribosome. A typical tRNA molecule has an anticodon loop that complements three bases on mRNA and has an acceptor stem at the 3′ end of the nucleic acid where an amino acid attaches.

Transcription

The sequence of bases in a DNA template strand is 5′CGATCA3′. What is the corresponding mRNA that is produced from this DNA?

SOLUTION

To form the mRNA, the bases in the DNA template are paired with their complementary bases: G with C, C with G, T with A, and A with U.

DNA template	5′CGATCA3′
Complementary mRNA	3′GCUAGU5′

PRACTICE PROBLEMS

11.19 Name the three types of RNA.

11.20 List the function of each of the three types of RNA.

11.21 In your own words, define the term *transcription*.

11.22 In your own words, define the term *translation*.

11.23 List the mRNA bases that complement the bases A, T, G, and C in DNA.

11.24 The sequence of bases in a DNA template strand is 5′CCGAAGGTTCAC3′. What is the corresponding mRNA produced?

11.25 The sequence of bases in a DNA template strand is 5′TACGGCAAGCTA3′. What is the corresponding mRNA produced?

11.5 Putting It Together: The Genetic Code and Protein Synthesis

If you were on a beach at night and saw a light in the ocean giving off three short flashes of light followed by three long flashes of light and three more short flashes of light, what would you do? Many of us would do nothing because we may not recognize that this is the Morse code distress signal for SOS. Someone who could translate Morse code would immediately get help if they saw this signal. Similarly, the genetic code from our gene sequences of DNA is a code transcribed into mRNA and decoded as a protein sequence. This *translation* involves tRNA and takes place at the ribosome.

The Genetic Code

The mRNA transcribed from the DNA contains a sequence of bases specifying the protein to be made. A given triplet called a **codon** in the mRNA translates to a specific amino acid. For example, the sequence UUU in a mRNA specifies the amino acid phenylalanine. The **genetic code** (see Table 11.3) assigns all 20 amino acids to codons of mRNA. Sixty-four codon combinations are possible from the four bases A, G, C, and U. The three codons UGA, UAA, and UAG are stop signals. When they appear in the mRNA sequence, this is a signal to stop adding amino acids to the growing protein chain. The triplet AUG has two roles in protein synthesis. If at the 5′ end of a mRNA, the codon AUG represents the start codon initiating protein synthesis. If it is found anywhere else in the mRNA, it codes for the amino acid methionine.

Codons in mRNA	5′ UUU|GGG|CGC 3′	
Translation	↓ ↓ ↓	
Amino acid sequence	Phe – Gly – Arg	(three-letter abbreviation)
	F G R	(one-letter abbreviation)

TABLE 11.3 mRNA CODONS: THE GENETIC CODE FOR AMINO ACIDS

First Letter from 5′ End	Second Letter				Third Letter
	U	**C**	**A**	**G**	
U	UUU ⎱ Phe UUC ⎰ UUA ⎱ Leu UUG ⎰	UCU ⎱ UCC ⎱ Ser UCA ⎰ UCG ⎰	UAU ⎱ Tyr UAC ⎰ UAA STOP UAG STOP	UGU ⎱ Cys UGC ⎰ UGA STOP UGG Trp	U C A G
C	CUU ⎱ CUC ⎱ Leu CUA ⎰ CUG ⎰	CCU ⎱ CCC ⎱ Pro CCA ⎰ CCG ⎰	CAU ⎱ His CAC ⎰ CAA ⎱ Gln CAG ⎰	CGU ⎱ CGC ⎱ Arg CGA ⎰ CGG ⎰	U C A G
A	AUU ⎱ AUC ⎱ Ile AUA ⎰ ªAUG Met/ start	ACU ⎱ ACC ⎱ Thr ACA ⎰ ACG ⎰	AAU ⎱ Asn AAC ⎰ AAA ⎱ Lys AAG ⎰	AGU ⎱ Ser AGC ⎰ AGA ⎱ Arg AGG ⎰	U C A G
G	GUU ⎱ GUC ⎱ Val GUA ⎰ GUG ⎰	GCU ⎱ GCC ⎱ Ala GCA ⎰ GCG ⎰	GAU ⎱ Asp GAC ⎰ GAA ⎱ Glu GAG ⎰	GGU ⎱ GGC ⎱ Gly GGA ⎰ GGG ⎰	U C A G

ªCodon that signals the start of a peptide chain.
STOP codons signal the end of a peptide chain.

SAMPLE PROBLEM 11.8

Codons

What is the sequence of amino acids coded by the following codons in mRNA?

5′ GUC|AGC|CCA 3′

SOLUTION

According to the genetic code in Table 11.3, GUC codes for valine, AGC for serine, and CCA for proline. The amino acid sequence is Val–Ser–Pro or VSP.

Protein Synthesis

Let's pause to review the process of protein synthesis beginning with transcription (see Figure 11.11).

Transcription

DNA found in the nucleus unwinds at the site of a gene, and a complementary copy of this DNA template, called mRNA is created. *RNA polymerase* is the name of the enzyme complex (many subunits) that binds to the DNA and links nucleotides to create mRNA. The mRNA then travels out of the nucleus to the ribosome.

tRNA Activation

Before the tRNA can be used in the ribosome, an amino acid must be attached to its acceptor stem. An enzyme called tRNA synthetase attaches the correct amino acid to the acceptor stem of the tRNA so it is ready for use in protein synthesis (see Figure 11.12).

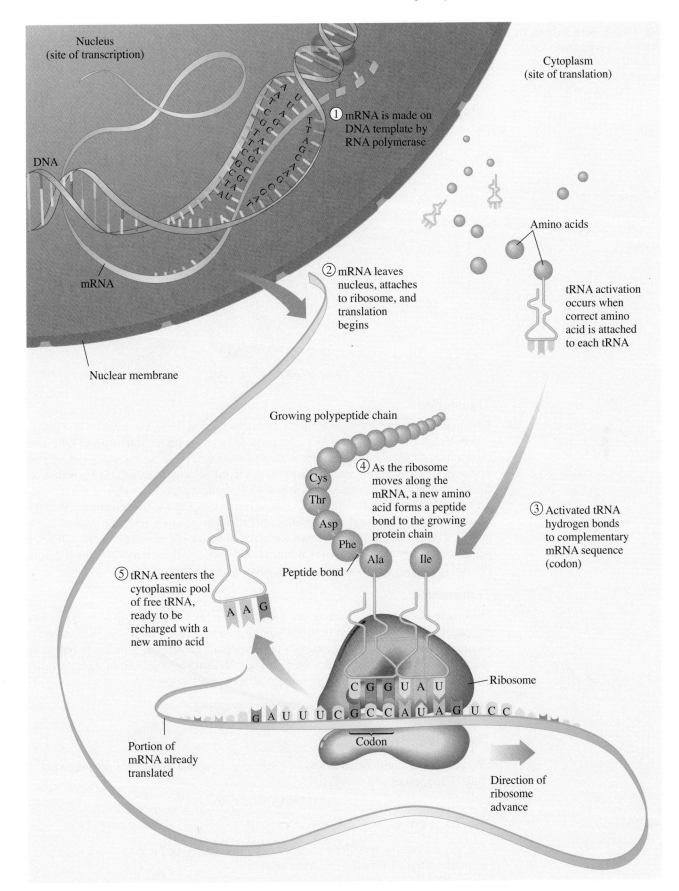

**Nucleus
(site of transcription)**

**Cytoplasm
(site of translation)**

DNA

mRNA

① mRNA is made on
DNA template by
RNA polymerase

② mRNA leaves
nucleus, attaches
to ribosome, and
translation
begins

Nuclear membrane

Amino acids

tRNA activation
occurs when
correct amino
acid is attached
to each tRNA

Growing polypeptide chain

Cys

Thr

Asp

Phe

Ala   Ile

Peptide bond

④ As the ribosome
moves along the
mRNA, a new amino
acid forms a peptide
bond to the growing
protein chain

③ Activated tRNA
hydrogen bonds
to complementary
mRNA sequence
(codon)

⑤ tRNA reenters the
cytoplasmic pool
of free tRNA,
ready to be
recharged with a
new amino acid

A A G

C G G U A U

Ribosome

G A U U U C G C C A U A G U C C

Codon

**Portion of
mRNA already
translated**

**Direction of
ribosome
advance**

▲ **FIGURE 11.11 Protein synthesis.** Protein synthesis begins with transcription of gene
DNA to the mRNA (upper left) which moves to the ribosome for translation via activated
tRNA molecules. Amino acids are linked until a stop codon on the mRNA is reached.

▶ **FIGURE 11.12 Activated tRNA.**
An activated tRNA with anticodon AGU bonds a serine at the acceptor stem.

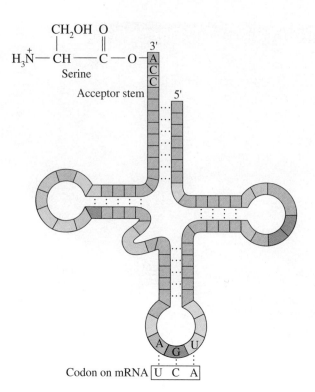

Codon on mRNA U C A

Translation

Protein synthesis begins when a mRNA positions itself at the ribosome. The first codon in a mRNA is the start codon, AUG. An activated tRNA with an anticodon of UAC and methionine attached enters the ribosome and hydrogen bonds to the mRNA. A second activated tRNA matching the next codon on the mRNA enters an adjacent position on the ribosome. The two amino acids then join, forming a peptide bond, and the methionine detaches from the first tRNA. The deactivated tRNA in the first position leaves the ribosome, and the second tRNA shifts into the first position with the dipeptide attached. This shifting is called *translocation*. As the ribosome moves along the mRNA, a peptide bond joins each subsequent amino acid that enters with a tRNA anticodon to the mRNA codon. In this way a growing polypeptide chain emerges.

Termination

Eventually, the ribosome encounters a stop codon, and protein synthesis ends. The polypeptide chain is released from the ribosome. The initial amino acid methionine is often removed from the beginning of the polypeptide chain. The growing polypeptide chain folds into its tertiary structure, forming any disulfide links, salt bridges, or other interactions that make the polypeptide a biologically active protein.

SAMPLE PROBLEM 11.9

Translation

What order of amino acids would you expect in a peptide made from the following mRNA sequence?

5′ UCA|AAA|GCC|CUU| 3′

SOLUTION

Each of the codons specifies a particular amino acid. Using Table 11.3, we write a peptide with the following one-letter sequence:

mRNA codons	5′ UCA|AAA|GCC|CUU 3′
Three-letter amino acid sequence	Ser – Lys – Ala – Leu
One-letter sequence	S K A L

PRACTICE PROBLEMS

11.26 List the difference between a codon and an anticodon.

11.27 Why are there at least 20 tRNAs?

11.28 Provide the three-letter amino acid sequence expected from each of the following mRNA segments:

a. 5′ AAA|AAA|AAA 3′

b. 5′ UUU|CCC|UUU|CCC 3′

c. 5′ UAC|GGG|AGA|UGU 3′

11.29 Provide the three-letter amino acid sequence expected from each of the following mRNA segments:

a. 5′ AAA|CCC|UUG|GCC 3′

b. 5′ CCU|CGA|AGC|CCA|UGA 3′

c. 5′ AUG|CAC|AAA|GAA|GUA|CUU 3′

11.30 In your own words describe how a peptide chain is extended.

11.31 In your own words define translocation.

11.32 The following portion of DNA is in the template DNA strand:

3′ GCT|TTT|CAA|AAA 5′

a. Write the corresponding mRNA section. Show the nucleic acid sequence as triplets and label the 5′ and the 3′ ends.

b. Write the anticodons corresponding to the codons on the mRNA.

c. Write the amino acids that will be placed in a peptide chain.

11.33 The following portion of DNA is in the template DNA strand:

3′ TGT|GGG|GTT|ATT 5′

a. Write the corresponding mRNA section. Show the nucleic acid sequence as triplets and label the 5′ and the 3′ ends.

b. Write the anticodons corresponding to the codons on the mRNA.

c. Write the amino acids that will be placed in a peptide chain.

11.6 Genetic Mutations

What will happen to protein synthesis if the DNA sequence is changed? Any change in a DNA nucleotide sequence is called a **mutation**. Let's examine the possibilities.

- No change in protein sequence. Sometimes a change in a DNA base will have no effect. Only about 2.5% of the DNA in your chromosomes actually encodes for proteins. The rest of your DNA is nongene or what is called "junk" DNA. This DNA is likely not junk, in that it contains recognition sites for protein binding. Also, recall that there is more than one codon that codes for each amino acid. For example, if the codon UUU were changed to UUC, the amino acid phenylalanine would still be placed in the growing polypeptide chain. These types of mutations are called **silent mutations**.

- A change in protein sequence occurs, but it has no effect on protein function. If, for example, the codon AUU is mutated to GUU, then the amino acid isoleucine would be changed to valine. These two amino acids are similar in polarity and size, and it is likely that such a substitution would not have much of an effect on the protein function. This is another type of silent mutation.

- A change in protein sequence occurs and affects protein function. If, for example, the codon AUU were mutated to AAU, then isoleucine would be changed to asparagine. In this case, a nonpolar amino acid is replaced with a polar amino acid, which could affect the structure and the function of the protein. Other mutations that can have a negative effect on protein synthesis include mutating a codon into a stop codon, or inserting or deleting a base into the DNA. The latter has the effect of shifting the triplets that are read in the mRNA and would change the identity of all subsequent amino acids after the insertion or deletion.

SAMPLE PROBLEM 11.10

Mutations

A mRNA has the sequence of codons 5′ CCC|AGA|GCC 3′. If a base substitution in the DNA changes the mRNA codon of AGA to GGA, how is the amino acid sequence affected in the resulting protein? Can you predict whether this might have an effect on the protein function?

SOLUTION

The initial mRNA sequence of 5′ CCC|AGA|GCC 3′ codes for the amino acids proline, arginine, and alanine. When the mutation occurs, the new sequence of mRNA codons, 5′ CCC|GGA|GCC 3′, codes for proline, glycine, and alanine. Arginine is replaced by glycine, which changes a polar charged amino acid into a small uncharged amino acid. This change could have an effect on protein structure.

Sources of Mutations

How do mutations occur? Sometimes when DNA replicates itself, errors occur at random. This is considered a **spontaneous mutation**. Environmental agents like chemicals or radiation that produce mutations in DNA are referred to as **mutagens**. Many mutagens can cause cancer and are also designated **carcinogens**. Viruses can also cause mutations (Section 11.7).

One common chemical mutagen is sodium nitrite ($NaNO_2$), which is used as a preservative in processed meats like hot dogs and bologna. In the presence of amines (found in proteins), sodium nitrite forms a compound called a nitrosamine, a known carcinogen in animals. Nitrosamines assist in the conversion of cytosine into uracil, which effectively converts a C–G base pair into a U–A base pair. As a known mutagen, why is sodium nitrite still used to preserve meat? The amounts used in the preservation process are small, and the benefits to meat preservation outweigh the risk of mutation.

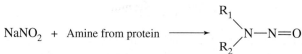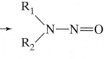

$NaNO_2$ + Amine from protein ⟶

Sodium nitrite, food preservative used in meats

Nitrosamine

If a mutation occurs in a **somatic cell** (any cell type other than egg or sperm), it affects only the individual organism and can cause conditions like cancer. But mutations that occur in **germ cells** (sperm or egg cells) can be passed on to future generations. A germ cell mutation can lead to malformation (and subsequent malfunction) of a particular protein. In this way, germ cell mutations cause **genetic diseases**. Over 4,000 genetic diseases have been identified. Several of these are listed in Table 11.4.

DNA $\xrightarrow{\text{Mutation}}$ Alteration ⟶ Defective ⟶ Genetic disease (germ cells)
of DNA protein or cancer (somatic cells)

TABLE 11.4	SOME GENETIC DISEASES	
Genetic Disease	**Result**
Galactosemia	The transferase enzyme required for the metabolism of galactose-1-phosphate is absent. Accumulation leads to cataracts and mental retardation.
Cystic fibrosis	The most common inherited disease. Thick mucus secretions make breathing difficult and block pancreatic function.
Down syndrome	The leading cause of mental retardation, occurring in about 1 of every 800 live births. Mental and physical problems including heart and eye defects are the result of the formation of three chromosomes, usually number 21, instead of a pair of chromosomes.
Familial hypercholesterolemia	A mutation of a gene on chromosome 19 results in high cholesterol levels that lead to early coronary heart disease in people 30–40 years old.
Muscular dystrophy (Duchenne)	One of 10 forms of MD. A mutation in the X chromosome results in the low or abnormal production of *dystrophin*. This muscle-destroying disease appears at about age 5, with death by age 20, and it occurs in about 1 of 10,000 males.
Huntington's disease (HD)	Appearing in middle age. HD affects the nervous system, leading to total physical impairment. It is the result of a mutation in a gene on chromosome 4, which can now be mapped to test people in families with HD.
Sickle-cell anemia	Defective hemoglobin from a mutaion in a gene on chromosome 11 decreases the oxygen-carrying ability of red blood cells, which take on a sickled shape, causing anemia and plugged capillaries from red blood cell aggregation.
Hemophilia	One or more defective blood-clothing factors lead to poor coagulation, excessive bleeding, and internal hemorrhages.
Tay-Sachs disease	Hexosaminidase A is defective, causing an accumulation of lipids in the brain, resulting in mental retardation, loss of motor control, and early death.

PRACTICE PROBLEMS

11.34 In your own words define a silent mutation.

11.35 In your own words define mutagen.

11.36 Consider the following portion of mRNA produced by the normal order of DNA nucleotides:

5′ ACA|UCA|CGG|GUA 3′

a. Write the amino acid sequence that would be produced from this mRNA.
b. Write the amino acid sequence if a mutation changes UCA to ACA. Is this likely to affect protein function?
c. Write the amino acid sequence if a mutation changes CGG to GGG. Is this likely to affect protein function?
d. What happens to protein synthesis if a mutation changes UCA to UAA?
e. What happens if a G is added to the beginning of the chain?
f. What happens if the A is removed from the beginning of the chain?

11.37 Consider the following portion of mRNA produced by the normal order of DNA nucleotides:

5′ CUU|AAA|CGA|GUU 3′

a. Write the amino acid sequence that would be produced from this mRNA.
b. Write the amino acid sequence if a mutation changes CUU to AUU. Is this likely to affect protein function?
c. Write the amino acid sequence if a mutation changes CGA to AGA. Is this likely to affect protein function?
d. What happens to protein synthesis if a mutation changes AAA to UAA?

11.7 Viruses

Viruses are small particles containing from 3 to 200 genes that can infect any cell type. Viruses are not considered cells because they cannot make their own proteins or their own energy. They contain only parts needed to infect a cell. Viruses have their own nucleic acid but use the infected cell or the **host** cell's ribosomes and RNA to make its proteins. Some infections caused by viruses invading human cells are listed in Table 11.5.

TABLE 11.5	SOME DISEASES CAUSED BY VIRAL INFECTION
Disease	**Virus**
Common cold	Coronavirus (over 100 types)
Influenza	Orthomyxovirus
Warts	Papovavirus
Herpes	Herpesvirus
Leukemia, cancers, AIDS	Retroviruses
Hepatitis	Hepatitis A virus (HAV), hepatitis B virus (HBV), hepatitis C (HCV)
Mumps	Paramyxovirus
Epstein–Barr	Epstein–Barr virus (EBV)

Viruses are very simple, and even though they have a variety of shapes, they all contain a nucleic acid (either DNA or RNA) enclosed in a protein coat called a **capsid**. Many viruses also contain an additional protective coat called an **envelope** surrounding the capsid. (See Figure 11.13.) The function of all viruses is the same: to monopolize the functions of the host cell for the benefit of the virus.

▶ **FIGURE 11.13 Virus structure.** All viruses contain nucleic acid protected by a protein coat called a capsid. Many animal viruses also contain exterior spikes made of protein.

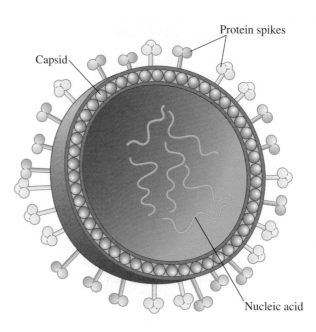

Capsid
Protein spikes
Nucleic acid

A virus infects a cell when an enzyme in the protein coat makes a hole in the host cells, allowing the viral nucleic acid to enter and mix with host cell material. If the virus contains DNA, the host cell begins to replicate the viral DNA in the same way it would replicate normal DNA. Viral DNA produces viral RNA, which proceeds to make the proteins for the virus. The completed virus particles are assembled and then released from the cell to infect more cells. This often occurs by a process called budding (see Figure 11.14).

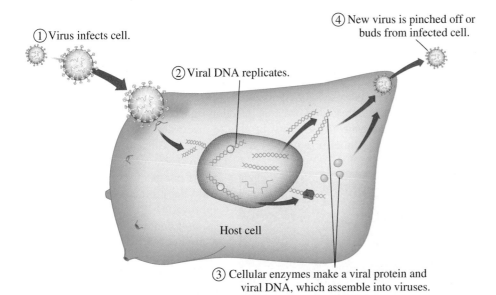

① Virus infects cell.

④ New virus is pinched off or buds from infected cell.

② Viral DNA replicates.

Host cell

③ Cellular enzymes make a viral protein and viral DNA, which assemble into viruses.

◀ **FIGURE 11.14 The life cycle of a virus.** After a virus attaches to the host cell, it injects its viral DNA and uses the host cell's amino acids to synthesize its viral proteins. It uses the host cell's nucleic acids, enzymes, and ribosomes to make viral mRNA, new viral DNA, and viral proteins. The newly assembled viruses are released to infect other cells.

Vaccines are often inactive forms of viruses that boost the immune response by causing the body to produce antibodies to fight the virus. Several childhood diseases such as polio, mumps, chicken pox, and measles can be prevented through the use of vaccines.

Retroviruses

A virus that contains RNA as the nucleic acid is called a **retrovirus**. Once retroviral RNA gets into the cell, it must first make viral DNA through a process known as reverse transcription. Retroviruses contain an enzyme called reverse transcriptase that uses the viral RNA to make complementary strands of viral DNA using the host nucleotides. The viral DNA joins the DNA of the host cell and uses the cell's enzymes and ribosomes to replicate virus particles (see Figure 11.15).

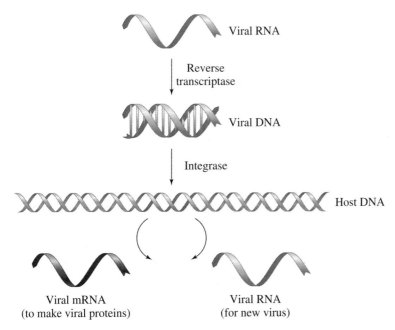

Viral RNA

Reverse transcriptase

Viral DNA

Integrase

Host DNA

Viral mRNA (to make viral proteins)

Viral RNA (for new virus)

▲ **FIGURE 11.15 Retroviral nucleic acid replication.** Once the viral RNA is injected into the cell, it is first transcribed into DNA using the viral enzyme reverse transcriptase. The viral DNA gets incorporated into the host DNA by a viral enzyme integrase and is then transcribed to make both mRNA for viral protein synthesis and new viral RNA.

HIV-1 and AIDS

*H*uman *I*mmunodeficiency *V*irus type 1, or HIV-1, is a retrovirus responsible for the disease AIDS (*A*cquired *I*mmune *D*eficiency *S*yndrome). HIV-1 infects a type of white blood cell known as a T4 lymphocyte that is part of the human immune system. The depletion of these immune cells caused by the HIV-1 infection reduces a person's ability to fight off other infections, so AIDS patients are very susceptible to skin cancers, sarcomas, and bacterial infections like pneumonias.

Therapies to inactivate viruses must be able to inactivate unique parts of the virus life cycle to minimize damage to the host cell. Current AIDS therapies involve several drugs that attack HIV-1 at the points of reverse transcription and viral protein synthesis. The first approved AIDS drug, called AZT (azidothymidine) and sold under the name Retrovir®, is similar to the nucleoside deoxythymidine, but it lacks a 3′—OH group. If reverse transcriptase incorporates this nucleoside into the viral DNA, transcription will halt. Several other nucleoside analogs have been developed since AZT. This

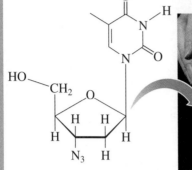
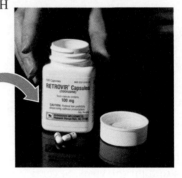

AZT
Azidothymidine

variety has allowed doctors to prescribe a "cocktail" of medicines that can extend the remission periods for AIDS patients.

A second set of drugs used to treat HIV-1 is called protease inhibitors and includes ritonavir, indinavir, and saquinavir. Once one of the unique proteins made by HIV's mRNA was identified and characterized, researchers set to work attempting to inactivate it. The targeted protein is called HIV protease, and it is responsible for clipping the viral proteins down to size for viral assembly. Many of the HIV protease inhibitors are competitive inhibitors that block the active site, lowering enzyme activity.

Other areas where drugs are being developed and approved to combat HIV-1 infection include cell entry drugs such as maraviroc and enfuvirtide, which block insertion of RNA into the host cell, and integrase inhibitors like raltegravir. Integrase is the enzyme involved in incorporating viral DNA into host DNA. Researchers are hopeful that combination therapies can keep the HIV infection in check.

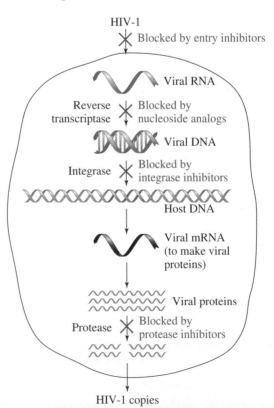

SAMPLE PROBLEM 11.11

Viruses

Why are viruses unable to replicate on their own?

SOLUTION

Viruses contain only DNA or RNA and not the necessary replication machinery including enzymes, nucleosides, and ribosomes.

PRACTICE PROBLEMS

11.38 Name two components common to all viruses.

11.39 Why do viruses need to enter a host cell?

11.40 If a virus contains viral RNA,

a. name the first step that must occur before DNA replication.

b. name the class of virus.

11.41 What is the purpose of a vaccine?

11.42. How do nucleoside analogs disrupt the life cycle of the HIV-1 virus?

11.43 How do protease inhibitors disrupt the life cycle of the HIV-1 virus?

11.8 Recombinant DNA Technology

Since the discovery of DNA as the genetic material in all cells, scientists have been finding ways to manipulate DNA to produce proteins needed as medicines, proteins helpful in crop resistance to pests, and proteins capable of positive identification of individuals (paternity testing) to name a few.

As it sounds, the term **recombinant DNA** involves recombining DNA from two different sources. Often called *genetic engineering* or *gene cloning*, the genome of one organism is altered by splicing in a section of DNA containing a gene from a second organism. Why would anyone want to do that? Inserting a higher organism's gene into smaller organism (like bacteria) whose life cycle is shorter produces larger amounts of the desired protein more quickly. The idea of combining genes from two different organisms is not really that new. Humans have been crossbreeding plants and animals for centuries, exchanging DNA to produce desired traits. The recombinant DNA techniques developed in the mid-1970s just work more quickly, expand the utility, and are more predictable.

There are a few basic steps to recombining DNA and producing a protein in another organism (see Figure 11.16).

- *Identify and isolate a gene of interest.* A gene must be located on a chromosome and removed. This gene of interest is referred to as the *donor DNA*. The removal is done with enzymes called **restriction enzymes** that recognize specific DNA sequences four to eight bases in length. One of these enzymes is called *Eco*RI (pronounced "eco–R–one") and recognizes the following DNA sequence cutting the DNA as shown.

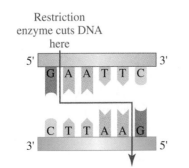

▶ **FIGURE 11.16 Basics of recombinant DNA and protein expression.** A DNA segment containing the gene to be cloned is isolated, incorporated into a plasmid vector, and the protein is then expressed and isolated from bacteria.

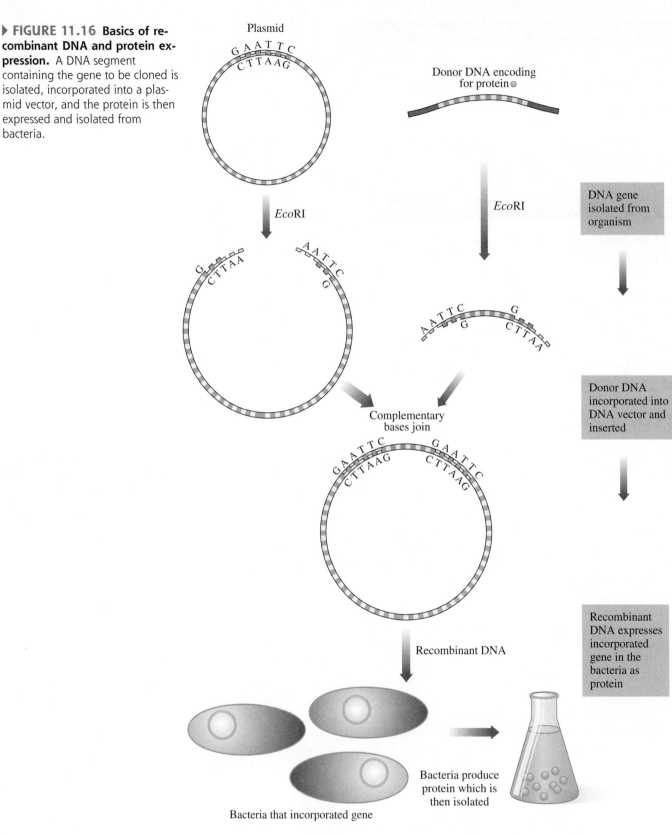

- *Insert donor DNA into the organism DNA using a vector.* A **vector** is a transporter for the donor DNA. It can incorporate donor DNA into the genome of the organism. One common vector found in bacteria is a piece of circular DNA called a **plasmid**. Using the same restriction enzymes, the plasmid is opened and the isolated gene DNA can be inserted. The vector DNA can be incorporated into the bacterial DNA as the bacteria grow and divide.

- *Express the incorporated gene in the new organism.* Bacteria that have the donor DNA incorporated will produce the protein from the gene of interest as they undergo normal protein synthesis. This protein production of a nonnative gene is called **expression**. Because this organism has no use for the produced protein, it can subsequently be isolated and purified for its intended use.

Therapeutic Proteins

One of the first applications of recombinant DNA technology incorporated the human insulin gene into bacteria called *Escherichia coli (E. coli)*, thereby allowing these bacteria to produce human insulin. Prior to 1982, pig or cow insulin was harvested for use by diabetics. The amino acid sequences of the pig and cow insulin have two or three amino acid differences from human insulin, respectively. Although pig and cow insulin sequences are similar, impurities in the extraction process produced rejection in some people. Almost all insulin used today is expressed human insulin causing fewer side effects.

Genetically Modified Crops

Recombinant DNA technology has also allowed the insertion of genes into crop and food plants offering crops a growth advantage. In the late 1990s Monsanto introduced Roundup Ready® cotton to farmers. This cotton seed has a gene inserted into it that gives it resistance to Roundup®, a weed killer. When farmers apply Roundup to a Roundup Ready crop, weeds are selectively killed and the crop survives. There are also many food crops—like the papaya—that have been genetically modified to resist viral infection (see Figure 11.17).

◀ **FIGURE 11.17 Genetically modified crops.** The genetically modified papaya (right) is resistant to the papaya ringspot virus that affects normal papaya (left), providing higher yields and larger fruits.

Genetic Testing

In the 1990s, U.S. government agencies in conjunction with private industry took on the task of sequencing the entire human genome in what was called the Human Genome Project (HGP). The last chromosome was finally completed in May 2006. The human genome contains over 3 billion base pairs and an estimated 20,000 to 25,000 genes. Because of the HGP we can now identify genes responsible for many genetic diseases. Through genetic testing, a person's DNA can be screened for mutated genes. Such tests can assist couples planning to have children, identify paternity, and predict certain cancers, Alzheimer's disease, and Huntington's diseases.

Nuclear Transplantation— Cloning an Organism

Is cloning a gene the same thing as cloning an organism? No. The term **clone** means to make an exact copy. In one case you are making an exact copy of a single gene to express a single protein, in the other, you are making an exact copy of an entire organism. Cloning an organism creates a genetic copy of the original organism. This is done by taking the nuclear DNA from an adult cell (somatic cell) and transplanting it into an egg cell whose DNA has been removed. In some cases such an egg cell with a full set of chromosomes behaves like a fertilized egg and will begin to divide forming an embryo. The embryo can then be transplanted into a surrogate until it fully develops. The first cloned mammal, Dolly the sheep, was born in 1996 and was the sole survivor of 276 attempts to create and implant an embryo. Since that time a number of other animals have been successfully cloned, including the cow, horse, monkey, deer, and cat (see Figure 11.18).

▶ **FIGURE 11.18 Cloned animals.** (a) Dolly the sheep was the first cloned animal. (b) In 2001, the first kitten—named Cc—was cloned.

S A M P L E   P R O B L E M   1 1 . 1 2

Recombinant DNA Technology

Describe how a gene from one organism can be incorporated into a second organism using recombinant DNA technology.

S O L U T I O N

The steps in recombinant DNA include cutting the gene of interest (donor DNA) from an organism's genome using a restriction enzyme, inserting that gene into a vector, and incorporating the donor DNA into the genome of the host organism.

P R A C T I C E   P R O B L E M S

11.44 In your own words define recombinant DNA.

11.45 In your own words define gene expression.

11.46 Describe the structure of a plasmid.

11.47 Describe the function of a vector.

SUMMARY

11.1 Components of Nucleic Acids

Deoxyribonucleic acid (DNA) and ribonucleic acid (RNA) are nucleic acids. They consist of strings of nucleotides. A nucleotide has three components: a nitrogenous base, a five-carbon sugar, and a phosphate. A nucleoside consists of only the nitrogenous base and the five-carbon sugar. The sugar deoxyribose is found in DNA, and the sugar ribose is found in RNA. The bases adenine (A), guanine (G), and cytosine (C) are found in both DNA and RNA. Thymine (T) is a fourth base found in DNA and uracil (U) is a fourth base found in RNA. Nucleotides are named as the nucleoside + the number of phosphates (up to three) bonded to it. For example, ATP, a metabolically relevant nucleotide, stands for adenosine (nucleoside) triphosphate (three phosphates). The components of nucleosides, as well as those of nucleotides, link together through condensation reactions.

DNA

11.2 Nucleic Acid Formation

Each nucleic acid has a unique sequence that is its primary structure. Nucleic acids form when nucleotides undergo condensation linking sugar to phosphate. The 3′—OH on the sugar of one nucleotide condenses with the phosphate on the 5′ end of a neighboring nucleotide. The backbone of a nucleic acid is formed from alternating sugar-phosphate-to-sugar-phosphate groups. The nitrogenous bases dangle from the backbone. Each nucleic acid has a single free 5′ and 3′ end.

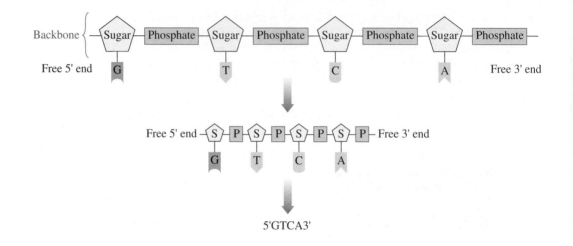

5'GTCA3'

11.3 DNA

A DNA molecule resembles a twisted ladder. It consists of two strands of nucleic acid, one upside down from the other (running oppositely or antiparallel) with their bases facing inward toward each other. The two strands are held together through the bases (rungs of the ladder), which hydrogen bond to complementary bases on the other strand, giving DNA its secondary structure. The base A forms two hydrogen bonds to T, and G forms three hydrogen bonds to C. In most plants and animals, DNA is found in the cell nucleus and is compacted into a tertiary structure called a chromosome. Chromosomes also contain a protein component called a histone, around which the DNA supercoils itself. Humans have 23 pairs of chromosomes in their cells.

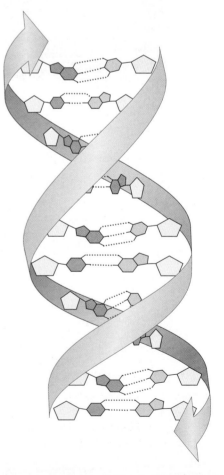

11.4 RNA and Protein Synthesis

RNA differs from DNA in that it contains a ribose sugar instead of a deoxyribose. It contains a uracil (instead of thymine) that can hydrogen bond to an adenine. RNA is also smaller than DNA and is single stranded. The three types of RNA are involved in transforming the DNA sequence into a protein sequence in the cell. These types are messenger RNA (mRNA), ribosomal RNA (rRNA), and transfer RNA (tRNA). The messenger RNA is involved in transcribing a complementary copy from gene DNA in the cell nucleus and taking that copy to the ribosome. Ribosomes are cell organelles where protein synthesis takes place. They consist of rRNA and protein. The tRNA is a compact RNA structure that acts as a conduit between the code on the messenger RNA and an amino acid sequence. In this regard it has two features: the anticodon at one end that is complementary to a sequence on the mRNA and the acceptor stem where an amino acid can attach.

11.5 Putting It Together: The Genetic Code and Protein Synthesis

The genetic code is a series of base triplet sequences on mRNA specifying the order of amino acids in a protein. The codon AUG signals the start of transcription and the codons UAG, UGA, and UAA signal it to stop. Protein synthesis begins with transcription where a mRNA creates a complementary copy of a DNA gene. tRNA activation involves binding an amino acid to the tRNA. Activation is catalyzed by the enzyme tRNA synthetase. During translation, tRNAs bring the appropriate amino acids to the ribosome and peptide bonds form until termination when a stop codon is reached. The polypeptide becomes a functional protein upon release.

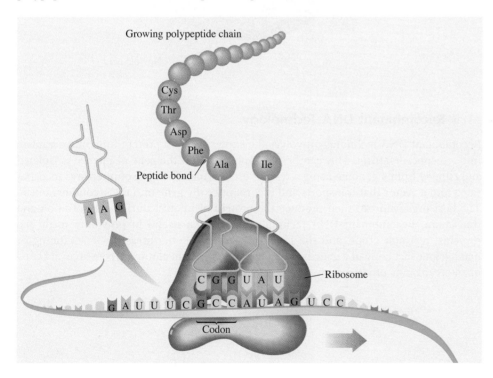

11.6 Genetic Mutations

Base alterations from normal cell DNA sequences are called mutations. Mutations can have different effects on protein synthesis depending upon their location. Some mutations are silent and have no effect on protein synthesis, and some may change the sequence but not alter protein function. Others can affect protein sequence, structure, and function. Mutations can be random during DNA replication or they can be caused by mutagens like chemicals or radiation. A mutation in a germ cell can be inherited. If such a mutation results in a defective protein, a genetic disease results.

11.7 Viruses

Viruses are particles containing DNA or RNA and a protein coat called a capsid. They invade a host cell and use the host cell's machinery (nucleotides, ribosomes, etc.) to replicate more virus particles. Viruses containing RNA are called retroviruses. These viruses must go through an initial step of reverse transcription to make viral DNA from viral RNA. HIV-1 is a retrovirus. Several areas of retroviral replication have been studied and promising drugs have been developed in recent years to slow down HIV-1 infection in AIDS patients. These include the development of nucleoside analogs, protease inhibitors, integrase inhibitors, and most recently, entry point inhibitors.

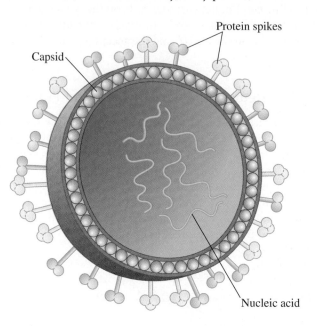

11.8 Recombinant DNA Technology

Recombinant DNA technology involves the expression of a protein from one organism into a second organism. This can be accomplished after the gene of interest is isolated and clipped from a genome using a restriction enzyme. The gene of interest is incorporated into a vector that transports and incorporates the gene into the second organism. The host organism can then produce the protein of interest during transcription and translation. Recombinant DNA techniques have been useful in providing medicinal proteins, resistant crops, and the identification of genetic diseases. Gene cloning or making an exact copy of a gene is different from organism cloning where an exact copy of an organism is produced.

KEY TERMS

anticodon—The triplet of bases in the center loop of tRNA that is complementary to a codon on mRNA.

capsid—The protein coat on a virus enclosing nucleic acid.

carcinogen—A mutagen that causes cancer.

chromosome—A single piece of DNA wound around histones found in the nucleus of the cell.

clone—An exact copy.

codon—A sequence of three bases in mRNA that specifies a certain amino acid to be placed in a protein. A few codons signal the start or stop of transcription.

complementary base pair—Base pairing first described by Watson and Crick. In DNA, adenine and thymine are always paired (A–T or T–A) and guanine and cytosine are always paired (G–C or C–G). In RNA, adenine pairs with uracil (A–U or U–A).

DNA—Deoxyribonucleic acid; the genetic material of all cells containing nucleotides with deoxyribose sugar, phosphate, and the four nitrogenous bases adenine, thymine, guanine, and cytosine.

double helix—A shape that describes the double stranded helical twist of DNA. It is like a twisted ladder with rails as the sugar–phosphate backbone and base pairs as the rungs.

envelope—An additional protein coat found on some viruses surrounding the capsid.

expression—In recombinant DNA technology, the synthesis of a protein from one organism cloned into a second organism.

gene—The part of a piece of DNA that encodes for a particular protein.

genetic code—The information in DNA that is transferred to mRNA as a sequence of codons for the synthesis of protein.

genetic disease—Disease caused by an inherited mutation from a germ cell.

genome—The set of DNA in one cell.

germ cell—A sperm or egg cell.

host—The cell that a virus particle uses to reproduce itself.

mRNA—Messenger RNA; produced in the nucleus by DNA to carry the genetic information to the ribosomes for the construction of a protein.

mutagen—An environmental agent that causes a mutation.

mutation—A change in DNA sequence.

nucleic acid—Large molecules composed of nucleotides, found in cells as double stranded helical DNA and single stranded RNA.

nucleoside—A pentose sugar condensed with one of the five nitrogenous bases (adenine, thymine, guanine, cytosine, or uracil) at C1.

nucleotide—A pentose sugar condensed with one of the five nitrogenous bases (adenine, thymine, guanine, cytosine, or uracil) at C1$'$ and up to three phosphates at C5$'$.

phosphodiester bond—Bond linking adjacent nucleotides in a nucleic acid. This bond connects the 3$'$ —OH of one nucleotide to the phosphate linked to the 5$'$ C of a second nucleotide.

plasmid—A piece of circular DNA found in bacteria commonly used as a vector.

recombinant DNA—DNA that has been recombined from two organisms.

restriction enzyme—Enzymes that cut DNA at specific DNA sequences.

retrovirus—A virus that contains RNA and uses the enzyme reverse transcriptase to make viral DNA from its RNA.

RNA—Ribonucleic acid; a type of nucleic acid that is a single strand of nucleotides containing adenine, cytosine, guanine, and uracil.

RNA polymerase—The enzyme complex that catalyzes the construction of mRNA from a DNA template.

rRNA—Ribosomal RNA; a major component of the organelle called the ribosome.

silent mutation—A change in DNA sequence that does not result in a change to protein function.

spontaneous mutation—Mutations arising from random errors during DNA replication.

somatic cell—All cells except the reproductive cells, sperm or egg. In humans these cells have 23 pairs of (46) chromosomes.

supercoiling—The doubling up of a helical molecule to form a compact coiled structure.

transcription—The transfer of a gene copy from DNA via the formation of an mRNA.

translation—The interpretation of the codons in the mRNA as amino acids in a peptide.

tRNA—Transfer RNA; an RNA that places a specific amino acid into a peptide chain at the ribosome. There is one or more tRNA for each of the 20 different amino acids.

vector—A delivery agent for a gene from one organism for incorporation into another organism.

ADDITIONAL PROBLEMS

11.48 List the three components of a nucleotide. What type of chemical reaction takes place when the components you listed combine to form nucleotide?

11.49 Identify each of the following bases as a pyrimidine or a purine:
    a. cytosine   **b.** adenine
    c. uracil   **d.** thymine
    e. guanine

11.50 Indicate if each of the bases in problem 11.49 are found in DNA only, RNA only, or both DNA and RNA.

11.51 Identify the base and sugar in each of the following nucleotides:
    a. CMP   **b.** dAMP
    c. dGMP   **d.** UMP

11.52 Identify the base and sugar in each of the following nucleotides:
    a. dTMP   **b.** AMP
    c. dCMP   **d.** GMP

11.53 How do the bases thymine and uracil differ?

11.54 How do the sugars ribose and deoxyribose differ?

11.55 Fill in the following table:

	Protein	Nucleic Acid
Repeating unit	Amino acid	
Backbone repeat	$N–C_\alpha–C–N–C_\alpha–C$, etc.	
One-letter abbreviation	Name of amino acid	
Free left end	Amino or N terminus	
Free right end	Carboxy or C terminus	

11.56 Write the complementary base sequence for each of the following DNA segments. Indicate the 5′ and the 3′ ends.
    a. 5′GACTTAGGC3′
    b. 5′TGCAAACTAGCT3′
    c. 5′ATCGATCGATCG3′

11.57 Write the complementary base sequence for each of the following DNA segments. Indicate the 5′ and the 3′ end.
    a. 5′TTACGGACCGC3′
    b. 5′ATAGCCCTTACTGG3′
    c. 5′GGCCTACCTTAACGACG3′

11.58 Match the following statements with mRNA, rRNA, or tRNA:
    a. combines with proteins to form ribosomes
    b. carries the genetic information from the nucleus to the ribosome
    c. carries amino acids to ribosome for protein synthesis

11.59 Match the following statements with mRNA, rRNA, or tRNA:
    a. contains codons for protein synthesis
    b. contains anticodons
    c. found in the small subunit and the large subunit of the ribosome

11.60 List the possible codons for each of the following amino acids:
    a. threonine   **b.** serine   **c.** cysteine

11.61 List the possible codons for each of the following amino acids:
    a. valine   **b.** proline   **c.** histidine

11.62 Provide the amino acid corresponding to each of the following codons:
    a. ACG   **b.** CCA   **c.** GCA

11.63 Provide the amino acid corresponding to each of the following codons:
    a. UUG   **b.** CGG   **c.** AUC

11.64 Endorphins are polypeptides that reduce pain. What is the amino acid sequence formed from the following mRNA that codes for a pentapeptide that is an endorphin called Leu-enkephalin?

    5′ AUG|UAC|GGU|GGA|UUU|CUA|UAA 3′

11.65 What is the one-letter amino acid sequence formed from the following mRNA that codes for a pentapeptide that is an endorphin called Met-enkephalin?

    5′ AUG|UAC|GGU|GGA|UUU|AUG|UAA 3′

11.66 What is the anticodon on tRNA for each of the following codons in a mRNA?
    a. AGC   **b.** UAU   **c.** CCA

11.67 What is the anticodon on tRNA for each of the following codons in a mRNA?
    a. GUG   **b.** CCC   **c.** GAA

11.68 a. A base substitution changes a codon for an enzyme from GCC to GCA. Why is there no change in the amino acid order in the protein?

    b. In sickle-cell anemia, a base substitution in the hemoglobin gene replaces glutamate (a polar amino acid) with valine. Why does the replacement of one amino acid cause such a drastic change in biological function?

11.69 a. A base substitution for an enzyme replaces leucine (a nonpolar amino acid) with alanine. Why does this change in amino acids have little effect on the biological activity of the enzyme?

b. A base substitution replaces cytosine in the codon UCA with adenine. How might this substitution affect the amino acids in the protein?

11.70 Discuss whether each of the following is a viable area for drug development to inactivate a retrovirus:
   a. reverse transcriptase inhibitors
   b. tRNA synthetase inhibitors
   c. cell entry blockers

11.71 Discuss whether each of the following is a viable area for drug development to inactivate a retrovirus:
   a. nucleoside analogs

b. integrase inhibitors
c. RNA polymerase inhibitors

11.72 How do integrase inhibitors disrupt the life cycle of the HIV-1 virus?

11.73 How do entry point inhibitors disrupt the life cycle of the HIV-1 virus?

11.74 List two societal benefits to recombinant DNA technology.

11.75 Distinguish between gene cloning and the cloning of an organism.

CHALLENGE PROBLEMS

11.76 Oxytocin is a small peptide containing nine amino acids. How many nucleotides would be found in the mRNA for this protein?

11.77 A protein contains 35 amino acids. How many nucleotides would be found in the mRNA code for this protein?

11.78 a. If the DNA chromosomes in salmon contain 28% adenine, what is the percent of thymine, guanine, and cytosine?

   b. If the DNA chromosomes in humans contain 20% cytosine, what is the percent of guanine, adenine, and thymine?

11.79 The DNA double helix can unwind or denature at temperatures between 90 °C–99 °C. Denaturing occurs when H bonds are broken. Which of the following strands of DNA would be expected to denature at a higher temperature? Provide an explanation.

DNA strand 1	5′ATTTTCCAAAGTATA3′
	3′TAAAAGGTTTCATAT5′
DNA strand 2	5′CGCGAGGGTCCACGC3′
	3′GCGCTCCCAGGTGCG5′

ANSWERS TO ODD-NUMBERED PROBLEMS

Practice Problems

11.1 a. pyrimidine
   b. pyrimidine
11.3 a. DNA
   b. both DNA and RNA
11.5 deoxyadenosine 5′-monophosphate (dAMP), deoxythymidine 5′-monophosphate (dTMP), deoxycytidine 5′-monophosphate (dCMP), deoxyguanosine 5′-monophosphate (dGMP)

11.7 a.

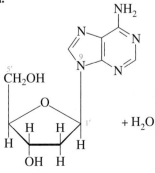

b.

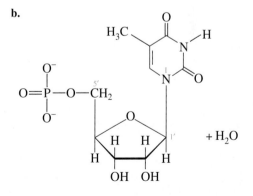

11.9 The nucleotides in nucleic acids are held together by phosphodiester bonds between the 3′-OH of a sugar (ribose or deoxyribose) and a phosphate group on the 5′-carbon of another sugar.

11.11

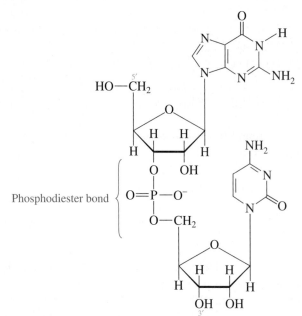

Phosphodiester bond

11.13 The two DNA strands are held together by hydrogen bonds between the bases in each strand.

11.15 a. 3′TTTT5′
    b. 3′GGGGAAAA5′
    c. 3′TGTAACC5′
    d. 3′ACACTTGG5′

11.17

	Protein	Nucleic Acid
Primary structure	Sequence of amino acids	Sequence of nucleotides
Secondary structure	Local hydrogen bonding between backbone atoms	Hydrogen bonding between bases in nucleic acid strands
Tertiary structure	Folding of backbone associating amino acids far away in sequence	Supercoiling of DNA into tight, compact structures like chromosomes

11.19 Messenger RNA, ribosomal RNA, and transfer RNA

11.21 In transcription, the sequence of nucleotides on one strand of a DNA template (the double helix is temporarily unwound) is used to produce a complementary mRNA copy using the enzyme RNA polymerase.

Additional Problems

11.49 a. pyrimidine
    b. purine
    c. pyrimidine
    d. pyrimidine
    e. purine

11.23 A in DNA complements with U in mRNA, T in DNA complements with A in mRNA, G in RNA complements with C in mRNA, and C in DNA complements with G in mRNA.

11.25 3′AUGCCGUUCGAU5′

11.27 There are 20 amino acids that are brought into the ribosome, each bonded to a different tRNA.

11.29 a. Lys–Pro–Leu–Ala
    b. Pro–Arg–Ser–Pro–Stop
    c. Met–His–Lys–Glu–Val–Leu

11.31 Translocation is the movement of the second tRNA to the spot vacated by the first tRNA, allowing for the next tRNA to bind to the ribosome. The amino acid is added to the growing peptide.

11.33 a. 5′ ACA | CCC | CAA | UAA 3′
    b. 3′ UGU | GGG | GUU | AUU 5′
    c. three-letter code: Thr–Pro–Gln–Stop, one-letter code: TPQ-Stop

11.35 A mutagen is an environmental agent that produces a mutation (change in base sequence) in DNA.

11.37 a. three-letter code: Leu–Lys–Arg–Val, one-letter code: LKRV
    b. Changing leucine to isoleucine, which are both nonpolar, will not have an effect on the protein function.
    c. Nothing, both codons code for arginine.
    d. Leucine will be the last amino acid incorporated into the growing protein chain because UAA is a stop codon.

11.39 Viruses cannot replicate themselves without a host cell. They do not have the cell machinery (enzymes, organelles, etc.).

11.41 A vaccine allows the body to mount an immune response, by producing antibodies to a less active form of a virus so that if an active form is encountered later, the body can fight against it more effectively.

11.43 Protease inhibitors inactivate an HIV-1 specific protease that is needed to process viral proteins.

11.45 Gene expression is producing a protein from a DNA gene sequence (usually of a non-native gene).

11.47 A vector is a transporter for donor DNA. It enables incorporation of the donor DNA into the organism DNA.

11.51 a. cytosine, ribose
    b. adenine, deoxyribose
    c. guanine, deoxyribose
    d. uracil, ribose

11.53 Thymine contains a methyl group ($-CH_3$) that uracil lacks.

11.55

	Protein	Nucleic Acid
Repeating unit	Amino acid	Nucleotide
Backbone repeat	$N-C_\alpha-C-N-C_\alpha-C$, etc.	sugar–phosphate–sugar–phosphate, etc.
One-letter abbreviation	Name of amino acid	Name of base
Free left end	Amino or N terminus	5′ end
Free right end	Carboxy or C terminus	3′ end

11.57 a. 3′AATGCCTGGCG5′
   b. 3′TATCGGGAATGACC5′
   c. 3′CCGGATGGAATTGCTGC5′

11.59 a. mRNA   **b.** tRNA   **c.** rRNA

11.61 a. GUU, GUC, GUA, GUG
   b. CCU, CCC, CCA, CCG
   c. CAU, CAC

11.63 a. leucine   **b.** arginine   **c.** isoleucine

11.65 YGGFM

11.67 a. CAC   **b.** GGG   **c.** CUU

11.69 a. Both alanine and leucine have small, nonpolar R groups. Both amino acids are found in similar environments in proteins so this substitution will unlikely affect protein structure or function.
   b. If a serine codon (UCA) is replaced with a stop codon (UAA), any remaining amino acids will not be linked onto the growing polypeptide chain. This can affect both the structure and function of the protein.

11.71 a. Nucleoside analogs that can be incorporated into viral DNA to stop reverse transcription are a viable development area.
   b. Integrase is a viral-only enzyme so this would be a viable method for inactivating the virus.
   c. Because RNA polymerase is used to make all proteins in the cell, this would not be a viable method to inactivate the virus.

11.73 Entry point inhibitors block sites on the outside of a host cell where a virus would attach and eventually enter. If the virus cannot enter the cell, the virus cannot replicate.

11.75 In gene cloning, a copy of a single gene is created. In organism cloning, many genes are copied and an exact copy of an organism is created.

11.77 3 nucleotides/amino acid + 1 start codon + 1 stop codon = 35 × 3 + 3 + 3 = 111 nucleotides

11.79 The more hydrogen bonds present, the more energy (as heat) must be applied to denature the DNA. Because strand 2 has more G–C base pairs with three hydrogen bonds each, it is predicted to have the higher denaturation temperature.

Guided Inquiry Activities **FOR CHAPTER 12**

EXERCISE 1 Reaction Pathways

Information

In the body, reactions occur in a series called a pathway. Each reaction in the pathway is catalyzed by a different enzyme. Most pathways have at least one irreversible reaction. These tend to be control points for pathways.

$$A \xrightarrow[I]{E_1} B \underset{II}{\overset{E_2}{\rightleftharpoons}} C \underset{III}{\overset{E_3}{\rightleftharpoons}} D \underset{IV}{\overset{E_4}{\rightleftharpoons}} products$$

▲ **SCHEME 1 A reaction pathway.** This pathway contains four reactions labeled *I – IV*.

Questions

1. Label the items in Scheme 1 with the following terms if the reactions are occuring left to right.

   reactant   product   enzyme   equilibrium arrow   nonequilibrium arrow

2. Do you think any of the reactions (*I–IV*) are reversible? Irreversible? List them.

3. Can you tell the direction of an equilibrium reaction from Scheme 1? If yes, in which direction is that reaction occurring?

4. What would you need to know in order to decide question 3?

5. Think back to Le Châtelier's principle (Chapter 8).
   a. If more A were added to the pathway, would more products be produced?
   b. If products were added to the pathway, would more A be produced?
   c. If products were added to the pathway, would more B be produced?
   d. If products were removed from the pathway, what would happen to the amounts of A, B, C, and D?

6. Considering your answers to the previous questions, do you think that reaction pathways are reversible?

EXERCISE 2 Carbohydrate Metabolism in the Body

Carbohydrate Catabolism Overview   Carbohydrate Anabolism Overview

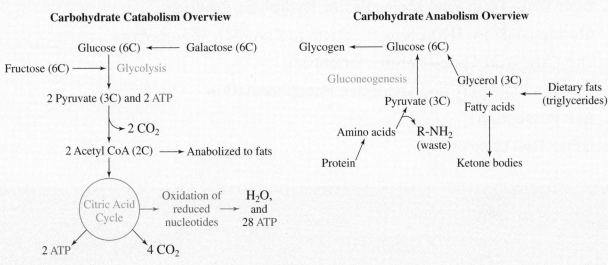

Questions

Answer the questions using the preceding diagrams.

1. Which of the overviews describes the breakdown of glucose?

2. Which of the overviews describes the synthesis of glucose?

Glucose Catabolism

3. Provide the name of the process that describes the breakdown of glucose to pyruvate.

4. What polysaccharide stored in the body can provide glucose?

5. At the end of glucose catabolism, where did the six carbons from glucose go?

6. How many ATP are generated by the catabolism of one molecule of glucose?

7. Only glucose goes through the entire glycolytic pathway (glycolysis). Explain how it is possible that we can get energy from the monosaccharides galactose and fructose.

8. Common culture says "too much sugar makes you fat." If carbohydrates and dietary fats are structurally different molecules, explain how eating too much sugar could make you fat.

Glucose Anabolism

9. Provide the name of the process that describes the synthesis of glucose from pyruvate.

10. Name two molecules containing three carbons each that can be used by the body to make glucose.

11. Of polysaccharides (complex carbohydrates), proteins, and dietary fats, which do you think would provide glucose the quickest when the body needs it? Explain your reasoning.

12. If glucose is not readily available in the diet, the body will manufacture it from protein or fat. What other by-products are made when glucose is synthesized from protein? From fat?

Our bodies use the biomolecules in food to produce energy. Now that we have explored their chemical structures and basic reactivity, we are ready to put all of this together to see how these molecules are used in the body to produce and store energy. Read Chapter 12 to find out.

Food as Fuel
A METABOLIC OVERVIEW

Have you ever heard someone claim they have a high metabolism? When they say this, they usually mean they can eat more than the average person and still maintain their body weight. Likewise, many nutrition products and weight-loss plans promise to "boost your metabolism." In this context, if your metabolism is "boosted," you will lose weight. What exactly is metabolism? **Metabolism** refers to the chemical reactions occurring in the body that break down or build up molecules. Most of the time in biological systems, the chemical reactions of metabolism occur in a series of steps called a **metabolic pathway**. In Chapter 10 we saw that chemical reactions can absorb or give off energy. In the body, this chemical energy is captured in the nucleotide ATP, which then serves as the fuel for the processes of life. Among other things, ATP is used to contract muscles, transport molecules across membranes, and transmit nerve impulses. One of the ways our bodies produces ATP is by metabolizing the carbohydrate glucose. In this chapter we explore the absorption of biomolecules from food, metabolic pathways involved in glucose metabolism, controlling glucose levels in the blood, and the ways in which other biomolecules, for example, proteins and lipids, can feed into the metabolic pathways.

12.1 Overview of Metabolism

All animals, including humans, must ingest food for fuel to carry out daily functions like circulation, respiration, and movement. How does eating food produce energy and how do our bodies convert food into useful energy? Animals can get energy (measured as calories) from the covalent bonds contained in carbohydrates, fats, and proteins. As an example, let's look at what happens when someone eats a bean and cheese burrito.

In the first stage of metabolism, the large biomolecules in food are digested or broken down into their smaller units through hydrolysis reactions. Polysaccharides, like starch in the tortilla, are hydrolyzed into monosaccharide units, the triglycerides in the cheese are ultimately broken down to glycerol and fatty acids, and the proteins in the beans and cheese are hydrolyzed into their amino acid units. These hydrolysis products, the smaller molecules produced by the breakdown of the large biomolecules, are absorbed through the intestinal wall into the blood stream and are eventually transported to different tissues for use by the cells. Once in the cells the hydrolysis products are broken down even further into a few common metabolites containing two or three carbons (see Figure 12.1). **Metabolites** are chemical intermediates formed by enzyme-catalyzed reactions in the body. At this point, as long as the cells have oxygen and there is a need for energy production, the two-carbon metabolites, known as acetyl groups, can be broken down further to carbon dioxide (one carbon) through a series of chemical reactions called the citric acid cycle. This cycle works in conjunction with the electron transport and oxidative phosphorylation pathways to produce a lot of energy that is

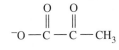

a. *Pyruvate*

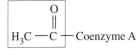

b. *Acetyl group*

▲ **FIGURE 12.1**
Common metabolites. (a) Pyruvate (a three-carbon molecule) and (b) acetyl groups (containing two carbons) found in acetyl coenzyme A are common intermediates in many biochemical pathways where carbon is metabolized.

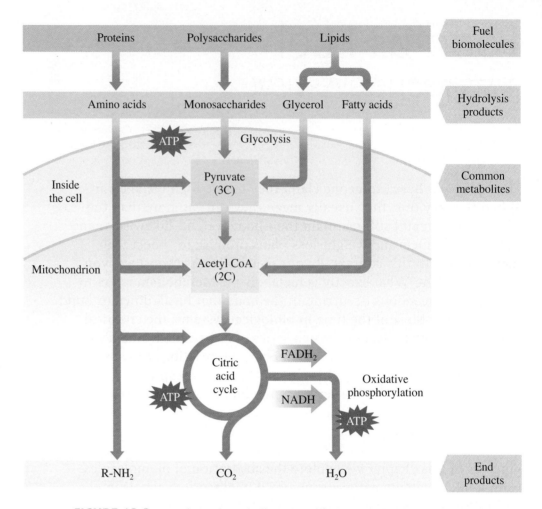

▲ **FIGURE 12.2 Overview of catabolism.** Food fuels are hydrolyzed to their hydrolysis products and further oxidized to a few common metabolites, as they move from the intestine to the interior regions of the cell. These fuels ultimately produce CO_2 and H_2O and store the energy produced as ATP.

transferred through the cells by the molecules ATP, reduced nicotinamide adenine dinucleotide (NADH), and reduced flavin adenine dinucleotide ($FADH_2$). One burrito contains a lot of carbon atoms, but we have a lot of cells that need fuel (see Figure 12.2).

Metabolism is the term used to describe the sum of all chemical reactions occurring in living systems. These reactions work together in the body to digest food, produce energy, and build the biomolecules needed to maintain healthy cells and tissues. As stated in the introduction, chemical reactions in the body are grouped together into sets called metabolic pathways. For example, glucose (containing six carbons) is broken down to two molecules of pyruvate (three carbons each) through a series of chemical reactions collectively referred to as *glycolysis*.

Metabolism can be considered in two parts: catabolism and anabolism. **Catabolism** refers to chemical reactions where larger molecules are broken down into a few common smaller molecules called metabolites. These reactions tend to generate energy and are oxidative, they involve the loss of hydrogen or the addition of oxygen. **Anabolism** refers to chemical reactions where metabolites combine to form larger molecules. These reactions tend to require energy and are reductive, they involve the removal of oxygen or the addition of hydrogen. The energy released during catabolic reactions is stored in molecules such as ATP until it is needed to drive an anabolic reaction. ATP and other energy-rich molecules couple the reactions of catabolism and anabolism together. In this way, the energy produced in catabolism is transferred to the energy-requiring reactions in anabolism (see Figure 12.3).

▲ **FIGURE 12.3 Catabolism and anabolism.** Catabolic and anabolic reactions together define metabolism. Catabolism produces the energy-rich molecules and hydrolysis products that can be used as building blocks for anabolism (production of large biological molecules).

Metabolic Pathways in the Animal Cell

Metabolic pathways occur in different parts of the cell. It is therefore important to know some of the major parts of a common animal cell (see Figure 12.4) to place the main pathways in metabolism.

In animals, a cell membrane separates the materials inside the cell from the exterior aqueous environment. The structure of a cell membrane was discussed in Section 6.5. The *nucleus* contains DNA that controls cell replication and protein synthesis for the cell. The **cytoplasm** consists of all the material between the nucleus and the cell membrane. The **cytosol** is the fluid part of the cytoplasm. It is the aqueous solution of electrolytes and enzymes that catalyzes many of the cell's chemical reactions.

Within the cytoplasm are specialized structures called organelles that carry out specific functions in the cell. We have already seen (Chapter 11) that the ribosomes are the sites of protein synthesis. The **mitochondria** are the energy-producing factories of the cells. A mitochondrion consists of an outer membrane and an inner membrane, with an intermembrane space between them. The fluid section encased by the inner membrane is called the matrix. Enzymes located in the matrix and along the inner membrane catalyze the oxidation of carbohydrates, fats, and amino acids. All of the oxidation pathways produce CO_2, H_2O, and energy which are used to form energy-rich compounds. Table 12.1 provides a summary of the functions of the cellular components in animal cells.

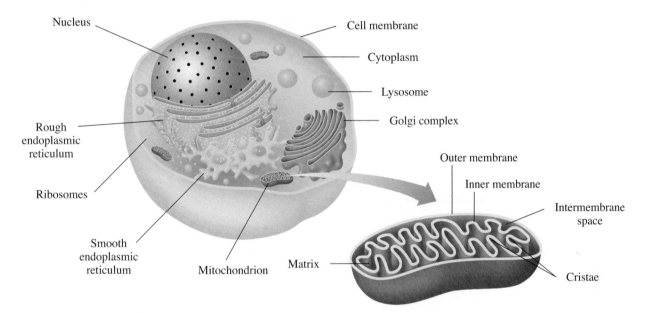

▲ **FIGURE 12.4 The cell.** The major components of an animal cell.

TABLE 12.1   LOCATIONS AND FUNCTIONS OF COMPONENTS IN ANIMAL CELLS

Component	Description and Function
Cell membrane	Separates the contents of a cell from the external environment and contains structures that communicate with other cells
Cytoplasm	Consists of all the cellular contents between the cell membrane and nucleus
Cytosol	Is the fluid part of the cytoplasm that contains enzymes for many of the cell's chemical reactions
Endoplasmic reticulum	Rough type processes proteins for secretion and synthesizes phospholipids; smooth type synthesizes fats and steroids
Golgi complex	Modifies and secretes proteins from the endoplasmic reticulum and synthesizes glycoproteins and cell membranes
Lysosomes	Contain hydrolytic enzymes that digest and recycle old cell structures
Mitochondria	Contain the structures for the synthesis of ATP from energy-producing reactions
Nucleus	Contains genetic information for the replication of DNA and the synthesis of protein
Ribosomes	Are the sites of protein synthesis using mRNA templates

> ### S A M P L E   P R O B L E M   1 2 . 1
>
> **Metabolism and Cell Structure**
>
> Identify the following as catabolic or anabolic reactions:
>
> a. digestion of polysaccharides
> b. synthesis of proteins
> c. oxidation of glucose to CO_2 and H_2O
>
> ### S O L U T I O N
>
> a. The breakdown of large molecules is a catabolic reaction.
> b. The synthesis of large molecules requires energy and involves anabolic reactions.
> c. Oxidation of molecules such as glucose involves catabolic reactions.

> ### P R A C T I C E   P R O B L E M S
>
> **12.1** How could you identify a catabolic reaction?
>
> **12.2** How could you identify an anabolic reaction?
>
> **12.3** Name a chemical reaction that breaks down biomolecules into their component parts.
>
> **12.4** Pyruvate and acetyl groups are common chemical intermediates called _____.
>
> **12.5** The ultimate end product for the catabolism of glucose (six carbons) is _____.
>
> **12.6** Name a molecule that transfers energy within the cell.
>
> **12.7** Identify the cell organelle considered the energy-producing factory for the cell.

12.2 Metabolically Relevant Nucleotides

In Chapter 11 we saw that nucleotides are the building blocks of the nucleic acids DNA and RNA. Nucleotides also have important metabolic functions in cells. They act as energy exchangers and can also be coenzymes. All these nucleotides have two forms: a high-energy form and a low-energy form. They consist of some basic components introduced in Chapter 11: the nucleoside adenosine, a phosphate, and a five-carbon sugar. Many of these molecules also have a vitamin within their structure. Because the structures of the basic components were shown in Chapter 11, they are illustrated here using their component names. Table 12.2 shows the structures of the metabolic nucleotides described in the following sections in both their high- and low-energy forms. The nucleotide portion (base, sugar, and at least one phosphate) and the vitamin portion are also indicated.

ATP/ADP

The nucleotide ATP is often referred to as the energy currency of the cell. ATP can undergo hydrolysis to adenosine diphosphate (ADP) as shown in the following equation. The nucleotide components are shown beneath the chemical structure for simplicity. During hydrolysis, energy is released as a product, so in this case, ATP is the high-energy form and ADP is the low-energy form. The energy given off during the hydrolysis of ATP can be coupled to drive a chemical reaction that requires energy during anabolism.

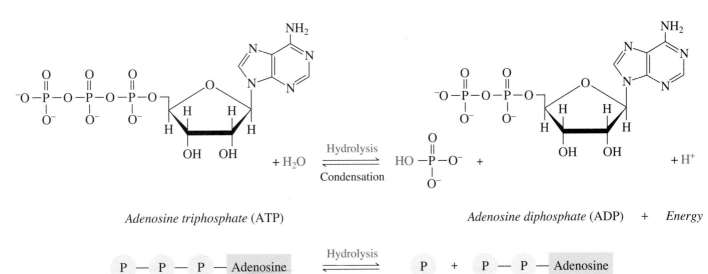

Adenosine triphosphate (ATP) $+ H_2O$ $\underset{\text{Condensation}}{\overset{\text{Hydrolysis}}{\rightleftharpoons}}$ $HO-\overset{O}{\underset{O^-}{\overset{|}{P}}}-O^-$ + Adenosine diphosphate (ADP) + Energy $+ H^+$

$$P - P - P - Adenosine \underset{\text{Condensation}}{\overset{\text{Hydrolysis}}{\rightleftharpoons}} P \quad + \quad P - P - Adenosine$$

NADH/NAD$^+$ and FADH$_2$/FAD

Nicotinamide adenine dinucleotide (NAD$^+$) and flavin adenine dinucleotide (FAD) are two energy-transferring compounds whose high-energy form is reduced (hydrogen added) and whose low-energy form is oxidized (hydrogen removed). Recall that in reductions, hydrogens are added. The abbreviations for these forms are NADH (reduced form)/NAD$^+$ (oxidized form) and FADH$_2$ (reduced form)/FAD (oxidized form). The active end of each molecule, the part that is being oxidized or reduced, contains a vitamin component. Nicotinamide is derived from the vitamin niacin (B$_3$), and riboflavin (B$_2$) is found in FAD.

Acetyl Coenzyme A and Coenzyme A

Another important energy exchanger containing a nucleotide is coenzyme A (CoA). The two forms of this compound are acetyl coenzyme A (high energy) and coenzyme A (low energy). Energy is released from acetyl coenzyme A when the C–S bond in the thioester functional group is hydrolyzed, producing an acetyl group and coenzyme A. CoA contains adenosine, three phosphates, and a pantothenic acid (vitamin B$_5$) derived portion.

TABLE 12.2 METABOLICALLY RELEVANT NUCLEOTIDES

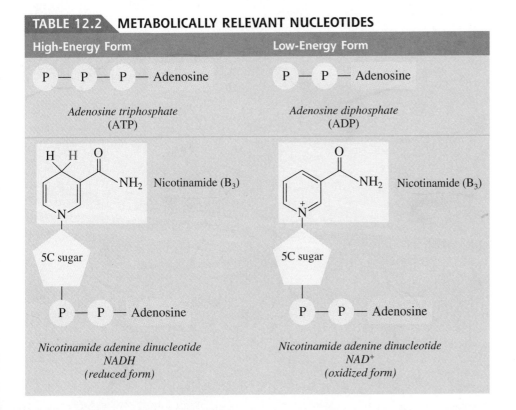

High-Energy Form	Low-Energy Form
P — P — P — Adenosine	P — P — Adenosine
Adenosine triphosphate (ATP)	Adenosine diphosphate (ADP)
Nicotinamide adenine dinucleotide NADH (reduced form)	Nicotinamide adenine dinucleotide NAD$^+$ (oxidized form)

TABLE 12.2 METABOLICALLY RELEVANT NUCLEOTIDES (CONTINUED)

High-Energy Form	Low-Energy Form

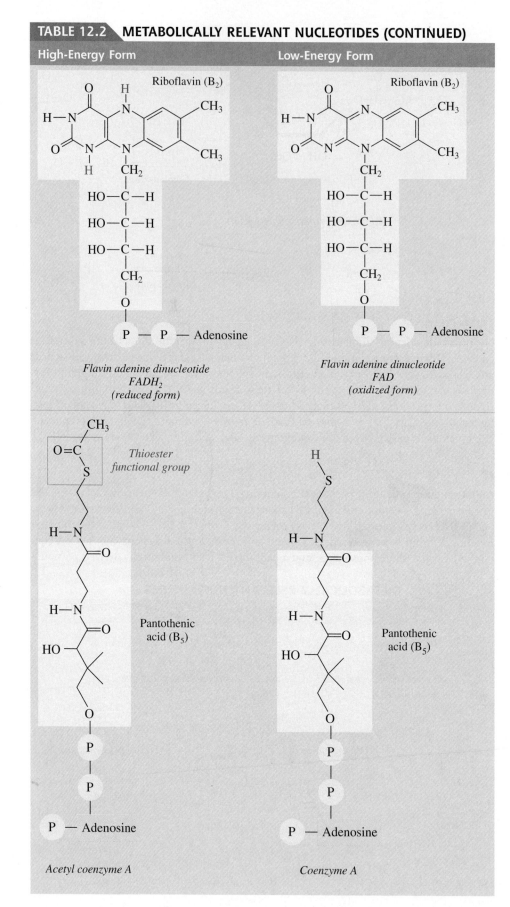

Riboflavin (B$_2$)

Flavin adenine dinucleotide
FADH$_2$
(reduced form)

Riboflavin (B$_2$)

Flavin adenine dinucleotide
FAD
(oxidized form)

Thioester functional group

Pantothenic acid (B$_5$)

Pantothenic acid (B$_5$)

Acetyl coenzyme A

Coenzyme A

SAMPLE PROBLEM 12.2

Metabolic Nucleotides

Provide the abbreviation for the high-energy form of the following:

a. nicotinamide adenine dinucleotide

b. flavin adenine dinucleotide

SOLUTION

The high-energy forms of these nucleotides are the reduced forms (contain H).

a. NADH   **b.** $FADH_2$

PRACTICE PROBLEMS

12.8 Identify the metabolic nucleotide described by the following:

a. contains a form of the vitamin niacin

b. the main energy currency in the body

c. contains a thioester functional group

12.9 Match one of the following metabolic nucleotides to its description:

NADH	ATP
NAD^+	ADP
$FADH_2$	Acetyl CoA
FAD	CoA

a. the reduced form of nicotinamide adenine dinucleotide

b. exchanges energy when a C – S bond is hydrolyzed

c. the oxidized form of flavin adenine dinucleotide

12.10 Match one of the following metabolic nucleotides to its description:

NADH	ATP
NAD^+	ADP
$FADH_2$	Acetyl CoA
FAD	CoA

a. exchanges energy when a phosphate bond is hydrolyzed

b. the high energy form of coenzyme A

c. the oxidized form of nicotinamide adenine dinucleotide

12.3 Digestion—From Fuel Molecules to Hydrolysis Products

When food enters the body, it begins to break down in a process called **digestion**. In digestion, large biomolecules are hydrolyzed into smaller hydrolysis products. These hydrolysis products are then able to be absorbed by the body and delivered to the cells where they can be further catabolized.

Carbohydrates

Have you ever eaten a saltine cracker and, after swallowing most of it, had a sweet taste in your mouth? This happens because the starch (amylose and amylopectin) in the cracker actually begins to be digested in your mouth through the action of the enzyme

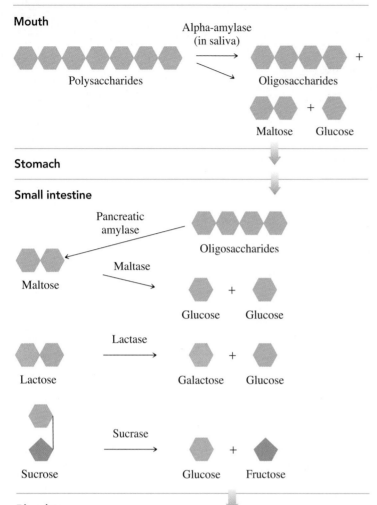

Mouth

Stomach

Small intestine

Bloodstream

▲ **FIGURE 12.5 Digestion of carbohydrates.** The digestion of carbohydrates begins in the mouth and is completed in the small intestine.

alpha-amylase secreted in saliva. This salivary amylase hydrolyzes some of the α-glycosidic bonds in the starch molecules, producing some glucose, the disaccharide maltose, and oligosaccharides. The slightly sweet taste is due to the presence of glucose in the hydrolysis products.

Only monosaccharides are small enough to be transported through the intestinal wall and into the bloodstream. To complete the digestion of starch, enzymes in the small intestine hydrolyze starch into monosaccharides. The disaccharides lactose, sucrose, and maltose are also hydrolyzed in the small intestine. These hydrolysis reactions were discussed in Section 5.5 and are shown in Figure 12.5.

We eat other carbohydrates like cellulose that cannot be digested since we lack the enzyme cellulase which hydrolyzes its β-glycosidic bonds. These indigestible fibers are referred to as insoluble fibers and although not useful as fuel for the body, they are important for a healthy digestive tract. Insoluble fiber stimulates the large intestine to help the body excrete waste.

Fats

Dietary fats such as the triglycerides and cholesterol are nonpolar molecules, so their digestion in aqueous digestive juices is a little tricky. To assist in the digestion of dietary fats, a substance called bile is excreted from the gall bladder into the stomach during digestion. Bile contains soap-like molecules called bile salts. They are soap-like because bile salts are *amphipathic*: They contain a nonpolar part and a polar part. The amphipathic bile salts place their nonpolar face toward the dietary fats and their polar face toward the water, forming micelles very similar to the soap micelles that break up a greasy dirt stain on clothing. This process of breaking up larger nonpolar globules into smaller droplets (micelles) is called **emulsification**. The smaller micelles are able to move the dietary fats closer to the intestinal cell wall so cholesterol can be absorbed and triglycerides can be hydrolyzed via pancreatic lipase into free fatty acids and monoglycerides, which are then absorbed in the small intestine. Once across the intestinal wall, free fatty acids and monoglycerides are reassembled as triglycerides while the cholesterol is linked to another free fatty acid forming a cholesterol ester. These are both repackaged with protein as a lipoprotein called a **chylomicron**. Chylomicrons ultimately transport triglycerides through the bloodstream to the tissues where they are used for energy production or are stored in the cells (see Figure 12.6).

Proteins

As shown in Figure 12.7, protein digestion begins in the stomach where proteins are denatured (unfolded) by the acidic digestive juices. Digestive enzymes like pepsin, trypsin, and chymotrypsin hydrolyze peptide bonds as the denatured proteins begin their journey through the stomach and into the intestine. The resulting amino acids are absorbed through the intestinal wall into the bloodstream for delivery to the tissues.

Small intestine

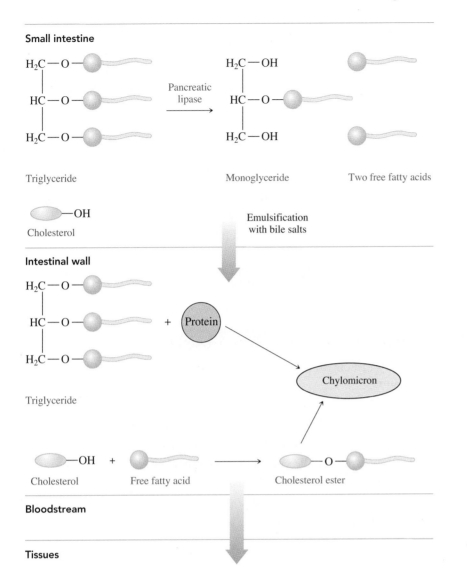

FIGURE 12.6 Digestion of dietary fats. Triglycerides are hydrolyzed to monoglycerides and free fatty acids for transport across the intestinal wall. They are reassembled along with cholesterol ester and protein into droplets called chylomicrons that are delivered to the tissues.

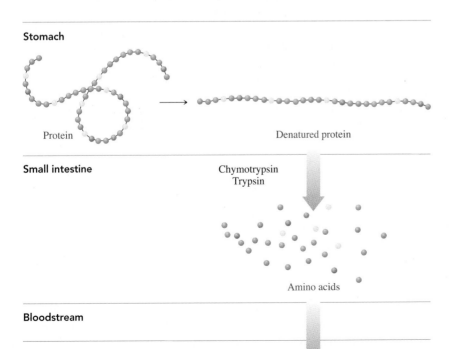

FIGURE 12.7 Digestion of proteins. Proteins are denatured in the stomach and hydrolyzed to amino acids in the small intestine.

S A M P L E   P R O B L E M   1 2 . 3

Digestion

Indicate the enzyme(s) involved in the digestion of each of the following:

a. lactose

b. triglyceride

c. protein

S O L U T I O N

a. The disaccharide lactose is broken into the monosaccharides galactose and glucose by the enzyme lactase (see Figure 12.5).

b. A triglyceride is broken down into two free fatty acids and a monoglyceride by an enzyme called a lipase (see Figure 12.6).

c. A protein is broken up into component amino acids by digestive enzymes, some of which are pepsin, trypsin, and chymotrypsin (see Figure 12.7).

P R A C T I C E   P R O B L E M S

12.11 Name a carbohydrate (if any) that undergoes digestion in each of the following sites:

    a. mouth   **b.** stomach   **c.** small intestine

12.12 α-Amylase is produced in the _____ and it catalyzes _____.

12.13 Name the chemical reaction occurring during most digestive processes.

12.14 Explain the role of bile salts in the digestion of fats.

12.15 Describe how cholesterol is packaged after absorption in the intestine.

12.16 Name the end products for digestion of starch.

12.17 Name the end products for digestion of proteins.

12.4 Glycolysis—From Hydrolysis Products to Common Metabolites

The main source of fuel for the body is glucose, which is catabolized through a chemical pathway called *glycolysis*. Glucose is such an important fuel source that when there is not enough glucose entering the body during periods of sleeping or fasting, the body makes its own glucose through the anabolic process called **gluconeogenesis**.

Glucose from the blood stream enters the cells through a glucose transporter protein in the cell membrane. In the cell, glycolysis occurs in the cytosol when the six-carbon monosaccharide glucose is broken down into two three-carbon molecules of pyruvate. Other hydrolysis products like fructose, amino acids, and free fatty acids can be used as fuel by the cells by entering glycolysis later in the pathway or by chemical conversion to one of the common metabolites.

The Chemical Reactions in Glycolysis

In the body, energy must be transferred in small amounts to minimize the heat released in the process. Reactions that produce energy are coupled with reactions that require energy, thereby helping to maintain a constant body temperature. In glycolysis, energy is transferred through phosphate groups undergoing condensation and hydrolysis reactions. There are 10 chemical reactions in glycolysis that result in the formation of two molecules of pyruvate from one molecule of glucose. As shown in Figure 12.8, the first five reactions require an energy investment of two molecules of ATP, which are used to add two phosphate groups to the sugar molecule. This newly formed molecule is then split into two sugar phosphates. Reactions 6 through 10 generate two high-energy NADH molecules during the addition of two more phosphates and four ATP molecules when the four phosphates are removed from the sugar phosphates.

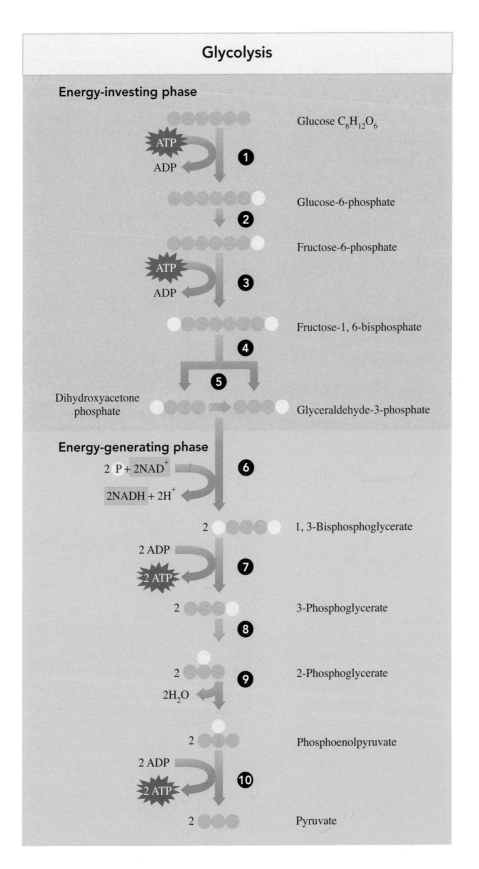

◀ **FIGURE 12.8 Schematic of Glycolysis.** In glycolysis, a six-carbon glucose is catabolized into two three-carbon pyruvate molecules. One molecule of glucose produces two ATP (net) and two NADH molecules.

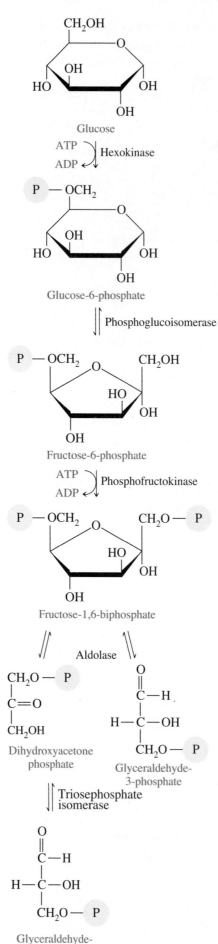

Glucose

Glucose-6-phosphate

Fructose-6-phosphate

Fructose-1,6-biphosphate

Dihydroxyacetone
phosphate

Glyceraldehyde-
3-phosphate

Triosephosphate
isomerase

Glyceraldehyde-
3-phosphate

Reaction 1 Phosphorylation: First ATP invested
Glucose is converted to glucose-6-phosphate when
ATP is hydrolyzed to ADP. The reaction is catalyzed
by the enzyme *hexokinase*

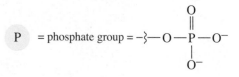

Reaction 2 Isomerization
The enzyme *phosphoglucoisomerase* converts
glucose-6-phosphate, an aldose, to its isomer,
fructose-6-phosphate, a ketose.

Reaction 3 Phosphorylation: Second ATP invested
A second ATP is hydrolyzed to ADP and the
phosphate is transferred to fructose-6-phosphate
forming fructose-1,6-bisphosphate. The word
bisphosphate indicates that the phosphates are on
different carbons in fructose. This reaction is
catalyzed by the enzyme *phosphofructokinase*.

Reaction 4 Cleavage: Two trioses are formed
Fructose-1,6-bisphosphate is split or cleaved into
two triose phosphates – dihydroxyacetone
phosphate and glyceraldeyhhye-3-phosphate –
catalyzed by the enzyme *aldolase*.

Reaction 5 Isomerization of a triose
In reaction 5, *triosephosphate isomerase* converts
one of the triose products, dihydroxyacetone
phosphate, to the other, glyceraldehydes-3-
phosphate. Now all 6 carbon atoms from glucose
are in two identical 3 carbon triose phosphates.

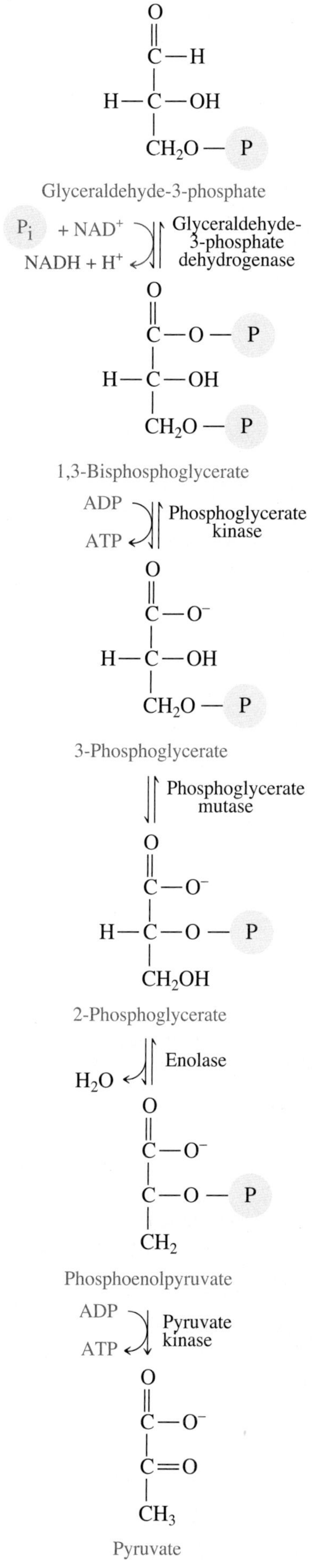

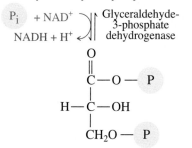

Glyceraldehyde-3-phosphate

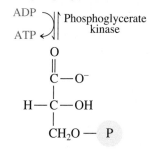

Reaction 6 First energy production yields NADH
The aldehyde group of glyceraldehyde-3-phosphate is oxidized and phosphorylated by *glyceraldehyde-3-phosphate dehydrogenase*. The coenzyme NAD$^+$ is reduced to the high energy compound NADH and H$^+$ in the process.

1,3-Bisphosphoglycerate

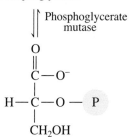

Reaction 7 Next energy production yields ATP
The energy rich 1,3-bisphosphoglycerate now drives the formation of ATP when *phosphoglycerate kinase* transfers one phosphate from 1,3-bisphosphoglycerate to ADP.

3-Phosphoglycerate

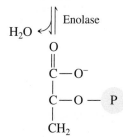

Reaction 8 Formation of 2-phosphoglycerate
A *phosphoglycerate mutase* transfers the phosphate group from carbon 3 to carbon 2 to yield 2-phosphoglycerate.

2-Phosphoglycerate

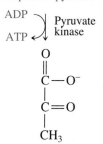

Reaction 9 Removal of water makes a high energy enol
An *enolase* catalyzes the removal of water to yield phosphoenolpyruvate, a high-energy compound that can transfer its phosphate in the next step.

Phosphoenolpyruvate

Reaction 10 Third energy production yields a second ATP
ATP is generated in this final reaction when the phosphate from phosphoenolpyruvate is transferred. The reaction is catalyzed by *pyruvate kinase*.

Pyruvate

Through the 10 reactions in glycolysis, one molecule of glucose is converted into two molecules of pyruvate. Because two molecules of ATP were originally required in the energy-investment phase and four are produced in the energy-generation phase, the net energy output for one molecule of glucose is two NADH and two ATP. The net chemical reaction is shown.

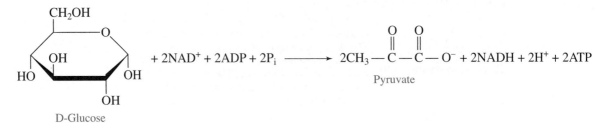

D-Glucose
$+ 2NAD^+ + 2ADP + 2P_i \longrightarrow 2CH_3 - \overset{O}{\underset{\|}{C}} - \overset{O}{\underset{\|}{C}} - O^- + 2NADH + 2H^+ + 2ATP$

Pyruvate

Regulation of Glycolysis

As a dam regulates water output from a reservoir, most metabolic pathways have at least one reaction step that regulates the flow of reactants to final products through the pathway. In glycolysis this regulation occurs in step 3 of the pathway. The enzyme phosphofructokinase, which catalyzes the phosphorylation of fructose-1-phosphate to fructose-1,6-bisphosphate, is heavily regulated by the cells. ATP acts as an inhibitor of phosphofructokinase. If cells have plenty of ATP, glycolysis slows down.

> ### S A M P L E   P R O B L E M   1 2 . 4
>
> ### Glycolysis
>
> If one NADH is generated in step 6 of glycolysis, how is a total of two NADH generated from one molecule of glucose?
>
> ### S O L U T I O N
>
> Two molecules of glyceraldhyde-3-phosphate (three carbon molecules) are produced from one molecule of glucose so this reaction occurs two times for every molecule of glucose going through the pathway.

The Fates of Pyruvate

The metabolite pyruvate generated from glycolysis can still be catabolized further, producing more energy for the body. How this occurs depends on the availability of oxygen in the cell. When ample oxygen is available (referred to as **aerobic** conditions), pyruvate is oxidized further to acetyl coenzyme A. When oxygen is in short supply (referred to as **anaerobic** conditions), pyruvate is reduced to lactate.

Aerobic Conditions

When oxygen is readily available in the cell, pyruvate produces more energy for the cell. When pyruvate breaks down further, the carboxylate functional group of pyruvate is liberated as CO_2, producing a two-carbon acetyl group and a reduced NADH. The acetyl group binds to coenzyme A through a sulfur atom, creating a thioester functional group and producing acetyl CoA, another common metabolite. This reaction occurs in the mitochondria in animals.

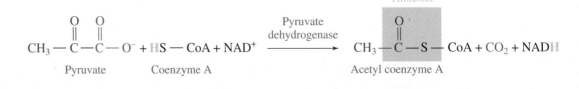

Anaerobic Conditions

During times of strenuous exercise, oxygen is in short supply in the muscles.

Under these anaerobic conditions, the middle carbonyl in pyruvate is reduced (hydrogen added) to an alcohol functional group, and lactate is formed. The hydrogen (and energy) required for this reaction is supplied by NADH and H^+, producing NAD^+. Because no ATP is generated by pyruvate under anaerobic conditions, the NAD^+ produced funnels back into glycolysis to oxidize more glyceraldehyde-3-phosphate (step 6), providing a small but necessary amount of ATP. This reaction occurs in the cytosol.

Figure 12.9 provides a summary of the aerobic and anaerobic fates of pyruvate.

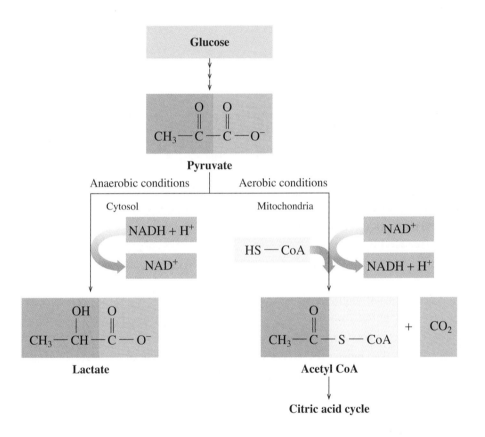

◀ **FIGURE 12.9 The fates of pyruvate.** Pyruvate is converted to acetyl CoA under aerobic conditions and lactate under anaerobic conditions.

Another single-celled organism, yeast, converts pyruvate to ethanol under anaerobic conditions. This process is called **fermentation**. In the preparation of alcoholic beverages like wine, yeast produces pyruvate from glucose in grape juices and under low oxygen conditions transforms pyruvate into ethanol.

Fructose and Glycolysis

Can fructose enter glycolysis to produce energy? As we saw in Chapter 5, fructose tastes sweeter than an equivalent amount of glucose, so less can be used to produce a sweet taste. In fact, refined sugars like table sugar and high fructose corn syrup do just this. Fructose is readily taken up in the muscle and liver. In the muscles it is converted to fructose-6-phosphate, entering glycolysis at step 3. In the liver, it is enyzmatically converted to the trioses dihydroxyacetone phosphate and glyceraldehyde-3-phosphate used in step 5 of glycolysis. Because glycolysis is regulated earlier in the pathway at step 3, the triose products created by fructose in the liver provide an excess of reactants that funnel into step 6, creating excess pyruvate and acetyl CoA that is not needed in the cells and is ultimately converted to fat.

SAMPLE PROBLEM 12.5

Fates of Pyruvate

Is each of the following products from pyruvate produced under anaerobic or aerobic conditions?

a. acetyl CoA   b. lactate   c. ethanol

SOLUTION

a. aerobic conditions   b. anaerobic conditions   c. anaerobic conditions

PRACTICE PROBLEMS

12.18 Name the starting reactant of glycolysis.

12.19 Name the end product of glycolysis.

12.20 How many ATP molecules are invested in the initial steps of glycolysis for one molecule of glucose?

12.21 How many ATP molecules are produced when one molecule of glucose undergoes glycolysis?

12.22 How many NADH molecules are produced when one molecule of glucose undergoes glycolysis?

12.23 In terms of high-energy molecules, what is the net output for one molecule of glucose undergoing glycolysis?

12.24 Name the coenzyme produced during the conversion of pyruvate to acetyl CoA.

12.25 Name the coenzyme produced during the conversion of pyruvate to lactate.

12.26 The formation of lactate permits glycolysis to continue under anaerobic conditions. Explain.

12.27 Name the products of fermentation.

12.28 Explain how the catabolism of fructose differs from that of glucose.

12.29 What is the main regulation point of glycolysis? How does this explain why fructose is readily converted into fat?

12.5 The Citric Acid Cycle—Central Processing

We have just seen how glucose begins its catabolism in the cell. What about amino acids and fatty acids? During aerobic catabolism when O_2 is present, all three of these biomolecules funnel into and out of the citric acid cycle. The **citric acid cycle** is a series of reactions that degrades the two-carbon acetyl groups from the metabolite acetyl CoA into CO_2 and generates the high-energy reduced molecules NADH and $FADH_2$. This cycle is also known as the Kreb's cycle or the tricarboxylic acid cycle. We will use the term citric acid cycle in this book.

The name citric acid cycle originated because the first step in the cycle involves the formation of the six-carbon molecule citrate, the conjugate base of citric acid. This initial reaction is a condensation reaction between an entering acetyl CoA molecule and the four-carbon molecule oxaloacetate. The six-carbon molecule citrate loses first one and then a second carbon as CO_2, forming the four-carbon molecule succinyl CoA. These carbon–carbon bond-breaking reactions transfer energy and produce NADH from the coenzyme NAD^+. Succinyl CoA then runs through a set of reactions regenerating oxaloacetate, and the cycle begins again (see Figure 12.10).

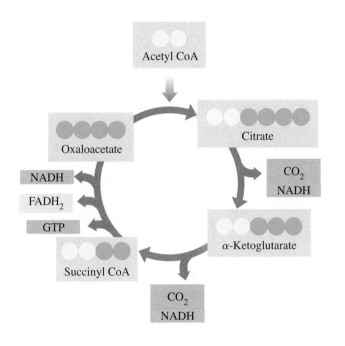

◀ **FIGURE 12.10 Schematic of the citric acid cycle.** The two carbons of acetyl CoA condense with oxaloacetate to form citrate (yellow and green circles). Two carbons are sequentially removed from citrate as CO_2, ultimately reforming oxaloacetate. During the cycle the high-energy nucleotides NADH, $FADH_2$, and GTP are produced.

Reactions of the Citric Acid Cycle

There are eight reactions in the citric acid cycle, each of which is catalyzed by an enzyme. These reactions occur in the mitochondrial matrix, deep within the mitochondria (see Figure 12.4). The eight reactions are shown in Figure 12.11 and described here beginning with the formation of citrate.

Reaction 1 Formation of Citrate
In reaction 1, the acetyl group from acetyl CoA (two carbons) combines with oxaloacetate (four carbons), forming citrate (six carbons) and CoA.

Reaction 2 Isomerization to Isocitrate
If you examine the reactant and product of this reaction, you will see that the two molecules are structural isomers. The —OH and one of the H atoms have been swapped in citrate to form isocitrate. Citrate contains a tertiary alcohol whereas isocitrate contains a secondary alcohol (see Section 5.2). A secondary alcohol can be further oxidized to a carbonyl, whereas a tertiary alcohol cannot. This rearrangement is necessary because isocitrate is oxidized in the next reaction.

Reaction 3 First Oxidative Decarboxylation (Release of CO_2)
The term **oxidative decarboxylation** details the reaction that is taking place. "Oxidative" tells us that this reaction is an oxidation. Here an alcohol is being

oxidized (two hydrogens removed) to a ketone called α-ketoglutarate. When this occurs, a corresponding reduction reaction must take place (remember, oxidation and reduction always occur together). The coenzyme NAD^+ is reduced to NADH, accepting the proton and electrons removed during the oxidation. "Decarboxylation" tells us that a carboxylate (COO^-) is removed as CO_2.

Reaction 4 Second Oxidative Decarboxylation

A second decarboxylation occurs in this step, and a four-carbon molecule is produced. In this reaction a second oxidation-reduction reaction also takes place. The thiol group of CoA is oxidized (loses a hydrogen), and another NAD^+ is reduced to NADH. The CoA is bonded to the four-carbon molecule, thus producing succinyl-CoA.

Reaction 5 Hydrolysis of Succinyl CoA

In this step, succinyl CoA undergoes hydrolysis to succinate and coenzyme A. The energy produced during the hydrolysis of the thioester is transferred to produce the high-energy nucleotide guanosine triphosphate or GTP from GDP and P_i. In the cell, GTP is readily converted to ATP through the transfer of a phosphate group.

$$GTP + ADP \longrightarrow GDP + ATP$$

Reaction 6 Dehydrogenation of Succinate

In this oxidation, one hydrogen is eliminated from each of the two central carbons of succinate, forming a trans $C=C$ bond and producing fumarate. These two hydrogens fuel the reduction of the coenzyme FAD to $FADH_2$.

Reaction 7 Hydration of Fumarate

Water adds to the trans double bond of fumarate as $-H$ and $-OH$ in this step, forming malate.

Reaction 8 Oxidation of Malate

As in reaction 3, the secondary alcohol of malate is oxidized to a ketone, oxaloacetate in this case, providing protons and electrons to reduce the coenzyme NAD^+ to NADH.

Citric Acid Cycle Summary

One turn of the citric acid cycle produces a net energy output of three NADH, one $FADH_2$, and one GTP (forms ATP). Two CO_2 and one CoA are also produced. Because the reactants in the cycle are regenerated, the net reaction for one turn of this eight-step cycle is

$$Acetyl\ CoA + 3\ NAD^+ + FAD + GDP + P_i + 2\ H_2O \longrightarrow$$
$$2\ CO_2 + 3\ NADH + 2\ H^+ + FADH_2 + CoA + GTP$$

SAMPLE PROBLEM 12.6

Citric Acid Cycle

When one acetyl CoA completes the citric acid cycle, how many of each of the following are produced?

a. NADH   b. CO_2 c. $FADH_2$

SOLUTION

a. One turn of the citric acid cycle produces three molecules of NADH.

b. Two molecules of CO_2 are produced by the decarboxylation of isocitrate and α-ketoglutarate in steps 3 and 4.

c. One turn of the citric acid cycle produces one molecule of $FADH_2$.

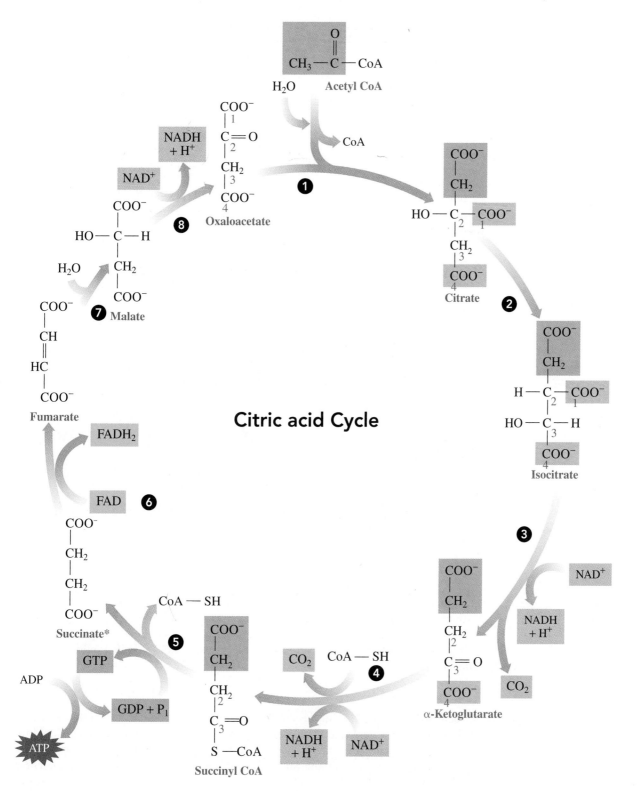

* Succinate is a symmetrical compound so we can no longer track the acetyl group which entered the cycle as acetyl CoA.

▲ **FIGURE 12.11 The reactions of the citric acid cycle.** In the citric acid cycle the oxidative catabolism produces two CO_2 molecules, reduced coenzymes NADH and $FADH_2$, and a GTP molecule.

PRACTICE PROBLEMS

12.30 Name the products of one turn of the citric acid cycle.

12.31 Name the compounds needed to start the citric acid cycle.

12.32 Name the reaction in the citric acid cycle that is a condensation.

12.33 Name the reactions in the citric acid cycle that involve oxidative decarboxylation.

12.34 Name the reactions of the citric acid cycle that involve hydration (addition of H_2O).

12.35 Name the reactions of the citric acid cycle that reduce NAD^+.

12.36 Name the reactions of the citric acid cycle that reduce FAD.

12.6 Electron Transport and Oxidative Phosphorylation

Let's review the number of ATPs produced from the metabolism of a single molecule of glucose in glycolysis and the citric acid cycle. Two ATPs are produced in glycolysis, and two ATPs are produced in the citric acid cycle (as GTP) from the two pyruvate molecules that enter as acetyl CoA. Organisms undergoing aerobic catabolism can produce more energy, but so far we have not seen a substantial number of ATPs produced. Where is all the energy? Remember that high-energy reduced forms of the nucleotides NADH and $FADH_2$ are also produced in glycolysis (two NADH per glucose), from pyruvate oxidation to acetyl CoA (two NADH per glucose), and in the citric acid cycle (six NADH and two $FADH_2$ per glucose).

Glycolysis	2 NADH and 2 ATPs
Oxidation of 2 pyruvate	2 NADH
Citric acid cycle (2 acetyl CoA)	6 NADH, 2 $FADH_2$, and 2 ATPs

As we will see, these high-energy reduced forms of the nucleotides transfer their electrons and hydrogens through the inner mitochondrial membrane and combine with oxygen to form H_2O. The energy generated as a result of this process is captured and used to drive the reaction of ADP + P_i to form ATP. The process of producing ATP using the energy from the oxidation of reduced nucleotides is called **oxidative phosphorylation**.

Electron Transport

Mitochondria are the ATP factories of the cell. Reduced nucleotides from the citric acid cycle are produced here, and their energy upon oxidation is used to generate ATP. Mitochondria (see Figure 12.4) have two membranes that isolate their contents from the rest of the cell: the outer membrane (exterior) and the inner membrane (interior). These areas are separated by what is called the inner membrane space. The central area of the mitochondria bounded by the inner membrane is called the matrix. The reactions of the citric acid cycle occur in the matrix, and the reduced nucleotides, NADH and $FADH_2$, begin their journey through the inner membrane here.

A set of enzyme complexes, commonly called complexes I through IV, are embedded in the inner membrane of the mitochondria and contain a set of electron carriers that transport the electrons and protons of NADH and $FADH_2$ through the inner mitochondrial membrane. Two of the electron carriers, coenzyme Q and cytochrome *c*, are not firmly attached to any one complex and serve to shuttle electrons between the complexes (see Figure 12.12).

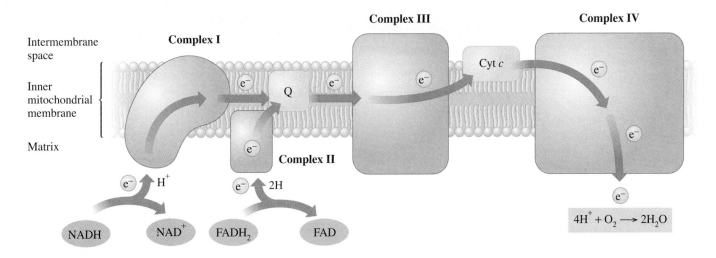

▲ FIGURE 12.12 **The electron transport chain.** Electrons flow to molecular oxygen through a series of electron carriers embedded in proteins in the inner mitochondrial membrane.

Complex I NADH Dehydrogenase
At complex I, NADH enters electron transport. During its oxidation, two electrons and two protons are transferred to the electron transporter coenzyme Q, reducing its two ketone groups to alcohols (see figure at right). NAD^+ is regenerated and returns to a catabolic pathway as in the citric acid cycle. The overall reaction at complex I is

$$NADH + H^+ + Q \longrightarrow NAD^+ + QH_2$$

Complex II Succinate Dehydrogenase
$FADH_2$ enters electron transport at complex II after the reduced nucleotide is produced in the conversion of succinate to fumarate in the citric acid cycle by complex II. Two electrons and two protons from $FADH_2$ are also transferred to a coenzyme Q to yield QH_2. The overall reaction at complex II is

$$FADH_2 + Q \longrightarrow FAD + QH_2$$

Complex III Coenzyme Q—Cytochrome c Reductase
At complex III, the reduced coenzyme Q (QH_2) molecules formed in complex I or II are reoxidized to ubiquinone (Q), and the electrons pass through a series of electron acceptors until they arrive as single electrons in the mobile protein cytochrome c, which moves the electron from complex III to complex IV.

Complex IV Cytochrome c Oxidase
At complex IV, electrons are transferred from cytochrome c through another set of electron acceptors to combine with hydrogen ions and oxygen (O_2) to form water. This is the final stop for the electrons.

$$4H^+ + 4e^- + O_2 \longrightarrow 2\,H_2O$$

$$Q + 2H^+ + 2\,e^- \rightleftharpoons QH_2$$

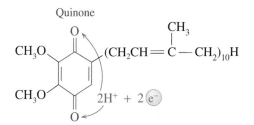

Quinone

Oxidized coenzyme Q (Q)

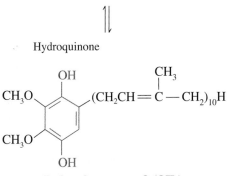

Hydroquinone

Reduced coenzyme Q (QH_2)

Electron Transport

Give the abbreviation for each of the following electron transporters:

a. the reduced form of flavin adenine dinucleotide

b. the reduced form of coenzyme Q

S O L U T I O N

a. $FADH_2$ b. QH_2

Oxidative Phosphorylation

How does the movement of electrons through a set of enzyme complexes in the inner membrane generate ATP for the cell? Biochemist Peter Mitchell first proposed the **chemiosmotic model** linking electron transport to the generation of a proton (H^+) difference or *gradient* on either side of the inner membrane and the resulting production of ATP. In this model, three of the complexes (I, III, and IV) span the inner membrane and pump (relocate) protons out of the matrix and into the intermembrane space as electrons are shuttled through the complexes. These protons are not generated directly from the oxidation of NADH and $FADH_2$, but are simply present in the matrix. A difference in both charge (electrical) and concentration (chemical) of protons on either side of the membrane results in an **electrochemical gradient**. The formation of the proton gradient across the inner mitochondrial membrane provides the energy for ATP synthesis.

The inner mitochondrial membrane is a very tight membrane that is impermeable to even a small proton. The only way to move H^+ back into the matrix is through a protein complex, called complex V, or ATP synthase. The movement of protons from an area with many protons to an area with fewer protons releases energy. As protons flow back into the matrix through complex V, the resulting release of energy drives the synthesis of ATP.

$$ADP + P_i + energy \longrightarrow ATP$$

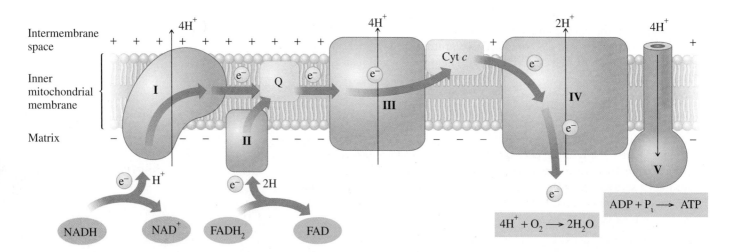

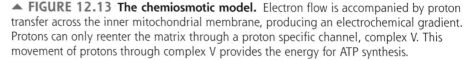

▲ **FIGURE 12.13 The chemiosmotic model.** Electron flow is accompanied by proton transfer across the inner mitochondrial membrane, producing an electrochemical gradient. Protons can only reenter the matrix through a proton specific channel, complex V. This movement of protons through complex V provides the energy for ATP synthesis.

Thermogenesis—Uncoupling ATP Synthase

If the protons pumped during electron transport and their return to the matrix during ATP production is disrupted, ATP is not produced. If ATP cannot be produced, the energy that would have been harnessed as ATP is simply released as heat. This generation of heat in the body is called **thermogenesis**. Some animals adapted to cold climates produce small organic molecules called uncouplers, which *uncouple* electron transport and oxidative phosphorylation and thereby assist in regulating their body temperature through thermogenesis. These animals have a higher amount of a tissue called *brown fat* that appears brown due to the high concentrations of mitochondria present. The cytochrome molecules present in mitochondria contain an iron ion that is responsible for the brown color. Newborn babies have higher levels of brown fat than do adults because newborns do not have much stored fat. Brown fat deposits are located near major blood vessels that carry the warmed blood through the body, allowing a newborn to generate more heat to warm its body surface.

SAMPLE PROBLEM 12.8

Chemiosmotic Model

Provide another name for complex V and describe its function.

SOLUTION

Complex V is also known as ATP synthase. It provides a channel for H^+ to flow back into the matrix (after being pumped out by complexes I, III, and IV) and is the site for ATP synthesis, which occurs in the matrix.

PRACTICE PROBLEMS

12.37 Name the electron carrier that transports electrons from complex I to complex III.

12.38 Name the electron carrier that transports electrons from complex III to complex IV.

12.39 Name the complex where water is produced.

12.40 Name the complex where $FADH_2$ enters electron transport.

12.41 According to the chemiosmotic theory, how does the proton gradient provide energy to synthesize ATP?

12.42 How is the proton gradient established?

12.7 ATP Production

Oxidative phosphorylation couples the energy of electron transport to proton pumping and finally to ATP synthesis. How many ATPs are synthesized for each reduced nucleotide entering electron transport? Because NADH enters the electron transport chain at complex I and $FADH_2$ enters at complex II, the number of protons pumped for these two molecules is different. Ten H^+ are pumped into the inner membrane space for every NADH entering electron transport and six H^+ are pumped into the inner membrane space for each $FADH_2$ transported (see Figure 12.13). The most widely accepted value for the number of H^+ needed to drive the synthesis of one ATP is four. So, the number of ATPs synthesized per NADH is 2.5 and the number of ATP synthesized per $FADH_2$ is 1.5.

Nucleotide Input	Protons (H^+) Pumped	ATP Output
NADH	10	2.5
$FADH_2$	6	1.5

> ## SAMPLE PROBLEM 12.9
>
> ### ATP Production
>
> Why does the oxidation of NADH provide energy for the formation of 2.5 ATP molecules whereas $FADH_2$ produces 1.5 ATPs?
>
> ### SOLUTION
>
> Electrons from the oxidation of NADH enter electron transport earlier through Complex I than do those from $FADH_2$. For every 4 protons pumped through the inner membrane, one ATP can be synthesized. One NADH pumps 10 protons into the intermembrane which can synthesize 2.5 ATPs, whereas one $FADH_2$ pumps 6 protons, forming 1.5 ATPs.

Counting ATP from One Glucose

How many ATP molecules can be generated from one molecule of glucose undergoing complete catabolism? As noted in Section 12.1, the end products for the oxidative catabolism for any fuel (foodstuff) are CO_2 and H_2O. For glucose ($C_6H_{12}O_6$), the overall balanced oxidation equation is

$$C_6H_{12}O_6 + 6O_2 \longrightarrow 6CO_2 + 6H_2O$$

This chemical equation does not account for the amount of energy that can be generated from oxidative catabolism. Let's review the pathways of the biochemical oxidation of glucose and count the total number of ATPs produced.

Glycolysis

In glycolysis, the oxidation of glucose produces two NADH molecules and two ATP molecules. Recall that glycolysis occurs in the *cytosol*, and electron transport draws NADH from the matrix inside the *mitochondria*. The two NADH from glycolysis must be shuttled into the matrix in order to enter electron transport. The direct shuttling of two NADH into the matrix results in the production of five ATPs.

$$\text{Glucose} \longrightarrow 2\,\text{Pyruvate} + 2\,\text{ATP} + 2\,\text{NADH (5 ATP)}$$

Oxidation of Pyruvate

After glycolysis, the two pyruvates enter the mitochondria, where they are further oxidized to produce a total of two acetyl CoA, two CO_2, and two NADH. The oxidation of two pyruvates thus leads to the production of five ATPs.

$$2\,\text{Pyruvate} \longrightarrow 2\,\text{Acetyl CoA} + 2\,\text{NADH (5 ATP)}$$

Citric Acid Cycle

The two acetyl CoAs produced from the two pyruvate next enter the citric acid cycle. Each acetyl CoA entering the cycle yields two CO_2, three NADH, one $FADH_2$, and one GTP (converted directly to ATP). Because two acetyl CoA enter the cycle from one glucose, a total of six NADH, two $FADH_2$, and two ATPs are produced.

6 NADH	$\longrightarrow$	15 ATP
2 $FADH_2$	$\longrightarrow$	3 ATP
Directly in pathway	$\longrightarrow$	2 ATP

Total ATP for two acetyl CoAs: 20 ATP

Total ATP from Glucose Oxidation

By summing the ATPs produced from glycolysis, oxidation of pyruvate, and the citric acid cycle, a net number of ATPs produced from one glucose can be estimated (see Figure 12.14).

TABLE 12.3 ATP PRODUCED BY THE COMPLETE OXIDATION OF GLUCOSE

Pathway	Reduced Nucleotides Produced	ATP Yield
Glycolysis	$2NADH_{cytosol} \longrightarrow 2NADH_{matrix}$	5 ATP
(Produced directly in pathway)		2 ATP
2 Pyruvate $\longrightarrow$ 2 Acetyl CoA	2NADH	5 ATP
Citric acid cycle Two turns of the cycle accommodates 2 acetyl CoA	6 NADH 2FADH$_2$	15 ATP 3 ATP
(Produced as GTP in pathway)		2 ATP
	TOTAL ATP	32 ATP

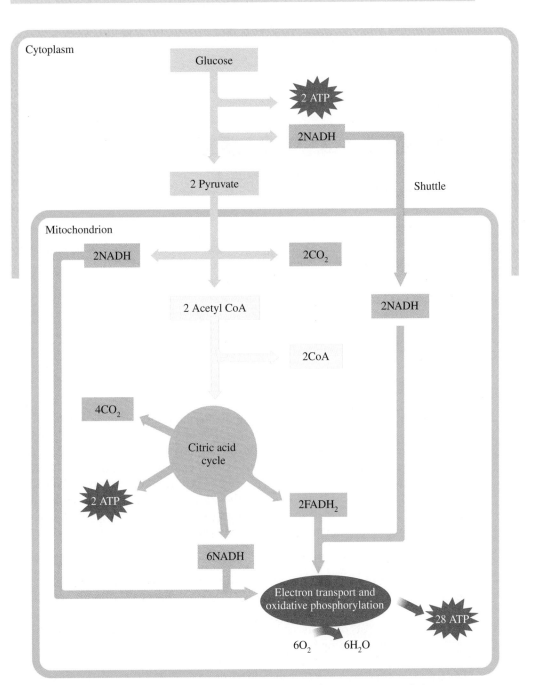

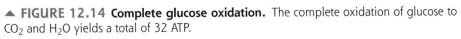

▲ **FIGURE 12.14 Complete glucose oxidation.** The complete oxidation of glucose to CO_2 and H_2O yields a total of 32 ATP.

SAMPLE PROBLEM 12.10

ATP Production

Indicate the total number of ATP produced by the following oxidations:

a. one pyruvate to one acetyl CoA
b. one acetyl CoA turning through the citric acid cycle

SOLUTION

a. The oxidation of pyruvate to acetyl CoA produces one NADH, which yields 2.5 ATP.
b. One turn of the citric acid cycle produces

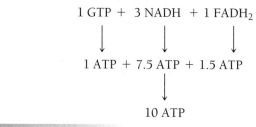

PRACTICE PROBLEMS

12.43 List the energy yield in ATP molecules for each of the following:
a. NADH $\longrightarrow$ NAD$^+$
b. glucose $\longrightarrow$ 2 pyruvate
c. 2 pyruvate $\longrightarrow$ 2 acetyl CoA + 2CO$_2$
d. acetyl CoA $\longrightarrow$ 2CO$_2$

12.44 List the energy yield in ATP molecules for each of the following:
a. FADH$_2$ $\longrightarrow$ FAD
b. glucose + 6O$_2$ $\longrightarrow$ 6CO$_2$ + 6H$_2$O
c. glucose $\longrightarrow$ 2 lactate
d. pyruvate $\longrightarrow$ lactate

12.8 Other Fuel Choices

Glucose is our primary source of fuel. When there is more than enough to take care of the energy needs of the cell, it is stored in the liver and muscles as glycogen. Hydrolysis of glycogen (called *glycogenolysis*) produces glucose quickly when glucose concentrations are low (during sleeping and fasting), thereby maintaining glucose levels. Once glycogen stores are depleted, glucose can be synthesized from noncarbohydrate sources (originating from proteins and triglycerides) that feed into *gluconeogenesis*. Amino acids and fatty acids can also feed into the catabolic oxidation at different points, allowing ATP production to continue in the absence of glucose.

Energy from Fatty Acids

If glycogen and glucose are not available, cells that still need ATP can oxidize fatty acids to acetyl CoA through a degradation pathway known as **beta oxidation (β oxidation)**. In β oxidation carbons are removed two at a time from an activated fatty acid. An activated fatty acid consists of a fatty acid bonded to a CoA, called a **fatty acyl CoA**. The removal of two carbons during β oxidation produces an acetyl CoA and a fatty acyl CoA shortened by two carbons. The cycle repeats until the original fatty acid is completely degraded to two-carbon acetyl CoA units. The acetyl CoAs can then enter the citric acid cycle in the same way as acetyl CoA derived from glucose. This reaction cycle occurs in the mitochondrial matrix.

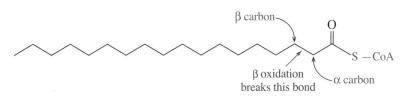

Fatty acyl CoA containing 18 carbons

The Beta Oxidation Cycle

β Oxidation (see Figure 12.15) includes a cycle of four reactions that convert the
$-CH_2-$ of the beta carbon to a beta ketone. Once this is formed, the two-carbon
acetyl group splits from the fatty acyl carbon chain, shortening the fatty acid. One cycle
of β oxidation yields one $FADH_2$ and one NADH.

Reaction 1 Oxidation (Dehydrogenation)
The first reaction removes one hydrogen from each of the alpha and beta carbons, and
a double bond is formed. These hydrogens are transferred to FAD to form $FADH_2$.

Reaction 2 Hydration
In reaction 2, water is added to the α and β carbon double bond as $-H$ and
$-OH$, respectively.

Reaction 3 Oxidation (Dehydrogenation)
The alcohol formed on the β carbon is oxidized to a ketone. As we have seen before
in the citric acid cycle, the hydrogen from the alcohol reduces NAD^+ to NADH.

Reaction 4 Removal of Acetyl CoA
In the fourth reaction of the cycle, the bond between the α and β carbon is
broken and a second CoA is added, forming an acetyl CoA and a fatty acyl CoA
shorter by two carbons. The fatty acyl CoA can be run through the cycle again.

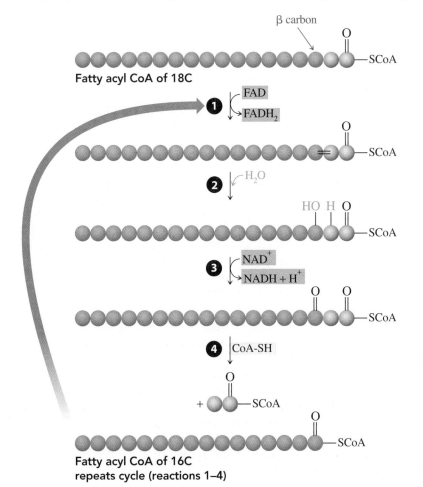

◀ **FIGURE 12.15 β oxidation.**
A fatty acyl CoA is broken up two
carbons at a time into acetyl CoA units
through a four-step cycle known as β
oxidation.

The net reaction for one cycle of β oxidation is summarized as

Fatty acyl CoA$_{n\text{ carbons}}$ + NAD$^+$ + FAD + H$_2$O + CoA $\longrightarrow$

Fatty acyl CoA$_{n-2\text{ carbons}}$ + Acetyl CoA + NADH + H$^+$ + FADH$_2$

S A M P L E   P R O B L E M   1 2 . 1 1

β Oxidation

Match each of the following descriptions to the reaction of β oxidation:

(1) oxidation one   (2) hydration

(3) oxidation two   (4) removal of acetyl CoA

a. Water is added to a double bond.

b. A fatty acyl CoA of two fewer carbons is produced.

c. FAD is reduced to FADH$_2$.

d. NAD$^+$ is reduced to NADH.

S O L U T I O N

a. (2) hydration   b. (4) removal of acetyl CoA

c. (1) oxidation one   d. (3) oxidation two

Cycle Repeats and ATP Production

How many cycles of β oxidation does one fatty acid go through and how many acetyl CoA are produced? Let's consider the saturated fatty acid stearic acid (18 carbons) as it undergoes β oxidation. Every 2 carbons will produce an acetyl CoA, so nine acetyl CoA are produced. This will take eight turns of the β oxidation cycle because the last turn produces two acetyl CoA.

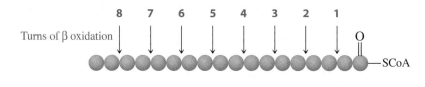

Fatty acyl CoA of 18C

Next, let's calculate the total number of ATP produced from one stearic acid molecule (18C) that completes eight turns of β oxidation. Nine acetyl CoAs are produced that can produce ATP through the citric acid cycle. In Section 12.7 we calculated that each turn of the citric acid cycle ultimately produces 10 ATP. In addition, one NADH and one FADH$_2$ are produced per turn of β oxidation, producing a total of eight NADH and eight FADH$_2$ per stearic acid molecule.

ATP PRODUCTION FROM β OXIDATION FOR A STEARIC ACID (18C) MOLECULE	
9 Acetyl CoA 9 Acetyl CoA × 10 ATP/acetyl CoA	90 ATP
8 turns of β oxidation 8 NADH × 2.5 ATP/NADH	20 ATP
8 FADH$_2$ × 1.5 ATP/FADH$_2$	12 ATP
Total	**122 ATP**

SAMPLE PROBLEM 12.12

ATP Production from β Oxidation

Calculate the number of ATPs produced from the β oxidation of one myristic acid (saturated C14) entering β oxidation as an acyl CoA.

SOLUTION

Myristic acid will go through six turns of β oxidation, producing seven acetyl CoA molecules. Each acetyl CoA going into the citric acid cycle ultimately produces 10 ATP. The six turns of β oxidation produce six NADH and six $FADH_2$. Each NADH produces 2.5 ATP and each $FADH_2$ produces 1.5 ATP.

7 Acetyl CoA		
7 Acetyl CoA × 10 ATP/acetyl CoA		70 ATP
6 turns of β oxidation		
6 NADH × 2.5 ATP/NADH		15 ATP
6 $FADH_2$ × 1.5 ATP/$FADH_2$		9 ATP
	Total	94 ATP

Too Much Acetyl CoA—Ketosis

In the absence of carbohydrates, the body breaks down its body fat through β oxidation to continue ATP production. This may seem efficient to the dieter considering a low-carb diet; however, some tissue types like the brain preferentially use glucose for energy. Even in the absence of carbohydrates, the liver will produce glucose from pyruvate through gluconeogenesis for tissues like the brain. The oxidation of large amounts of fatty acids can cause acetyl CoA molecules to accumulate in the liver. When this occurs, the two carbon acetyl units condense in the liver, forming the four-carbon ketone molecules β-hydroxybutyrate and acetoacetate and the molecule acetone. These are collectively referred to as **ketone bodies**. Their formation, termed *ketogenesis*, is outlined in Figure 12.16. The condition known as **ketosis** occurs when an excessive amount of ketone bodies is present in the body (and cannot be efficiently metabolized). This condition is often seen in diabetics who have difficulty getting glucose into the cells for energy, individuals on low-carbohydrate or high-fat diets, or individuals undergoing starvation.

Because two of the ketones are carboxylic acids, the excessive formation of ketone bodies can lower the blood pH and cause *metabolic acidosis*. Acetone vaporizes easily, giving someone suffering from ketosis an odd, sweet-smelling breath upon exhalation similar to that of someone who has been drinking alcohol.

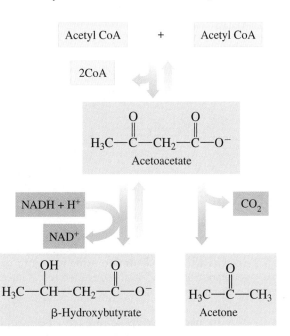

▲ **FIGURE 12.16 Ketogenesis.** Excess acetyl CoA molecules not needed for ATP production combine to produce the ketone bodies acetoacetate, β-hydroxybutyrate, and acetone.

Energy from Amino Acids

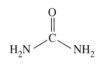

Urea

Amino acids from proteins produce nitrogen when metabolized in the body. When excess protein is ingested, amino acids must be degraded. The α-amino group of an amino acid is removed, thereby yielding an α-keto acid through a process called **transamination**. The α-keto acid produced can then be converted into intermediates for other metabolic pathways. The excess ammonium ions produced in this process must be excreted from the body because they are toxic if allowed to accumulate. A series of reactions called the **urea cycle** converts ammonium ions (NH_4^+) into urea, which can then be excreted in the urine.

Various amino acids offer a way to replenish the intermediates in the citric acid cycle and therefore have the ability to generate ATP. Which intermediate they replenish depends on the number of carbons in the amino acid structure. Amino acids like alanine, containing three carbons, can enter the pathways as pyruvate. Amino acids with four carbons are converted to oxaloacetate, and five-carbon amino acids are converted to α-ketoglutarate. Some amino acids can enter at more than one point depending on need (see Figure 12.17).

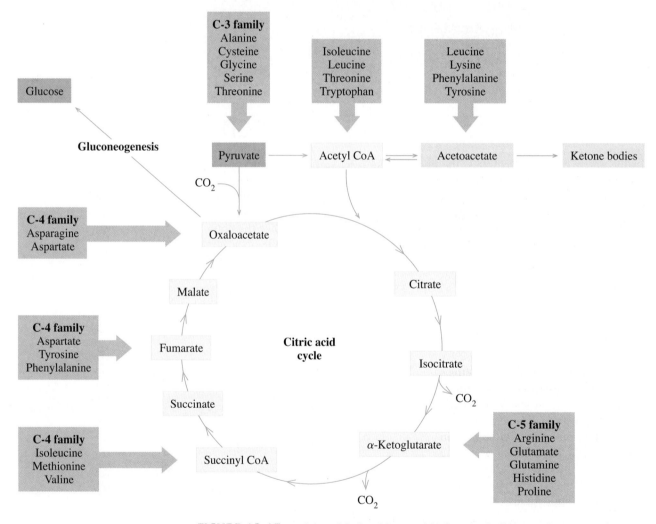

▲ **FIGURE 12.17 Amino acids feed into oxidative catabolism.** Carbon atoms from degraded amino acids are converted to the intermediates of the citric acid cycle and other pathways.

In this way we get some energy from amino acids, but it is only about 10% of needed energy under normal conditions. More energy is extracted from amino acids in conditions like fasting or starvation when carbohydrate and fat stores are depleted. If amino acids remain the only energy source for a long time, proteins in body tissues are degraded for fuel.

Carbon Atoms from Amino Acids

Which amino acids provide carbon atoms that enter the citric acid cycle as α-ketoglutarate?

SOLUTION

Those amino acids with five carbons feed into α-ketoglutarate. These are arginine, glutamate, glutamine, histidine, and proline.

PRACTICE PROBLEMS

12.45 Name the location in the cell where β oxidation takes place.

12.46 Name the coenzymes needed for β oxidation.

12.47 Capric acid is a saturated C10 fatty acid.
a. Draw fatty acyl capric acid activated for β oxidation.
b. Identify the α and β carbons in fatty acyl capric acid.
c. Write the overall equation for the complete β oxidation of capric acid.

12.48 Arachidic acid is a saturated C20 fatty acid.
a. Draw fatty acyl arachidic acid activated for β oxidation.
b. Identify the α and β carbons in fatty acyl arachidic acid.
c. Write the overall equation for the complete β oxidation of arachidic acid.

12.49 Under what conditions are ketone bodies produced in the body?

12.50 Explain why diabetics produce high levels of ketone bodies.

12.51 Why does the body convert NH_4^+ into urea?

12.52 Draw the structure of urea.

12.53 Name the metabolic substrate(s) that can be produced from the carbon atoms of each of the following amino acids:
a. alanine   b. aspartate
c. valine   d. glutamine

12.54 Name the metabolic substrate(s) that can be produced from the carbon atoms of each of the following amino acids:
a. leucine   b. asparagine
c. cysteine d. arginine

Putting It Together: Linking the Pathways

This chapter introduced several catabolic pathways and the high-energy molecules they produce. Degradation of food biomolecules begins with digestion, and when the cell needs energy and oxygen is plentiful, larger molecules are metabolized into smaller metabolites that ultimately funnel into the citric acid cycle, electron transport, and oxidative phosphorylation. Through anabolic pathways, larger molecules can be synthesized from the smaller metabolites when necessary. The biological hydrolysis products can be shifted into anabolic or catabolic pathways depending on the needs of the cell. Glucose can be degraded to acetyl CoA entering the citric acid cycle to produce energy or be converted to glycogen for storage in the cells. Amino acids provide nitrogen for anabolism of nitrogen compounds, but their carbons can enter the citric acid cycle as α-keto acids if necessary. Figure 12.18 gives a visual summary of the material discussed in this chapter.

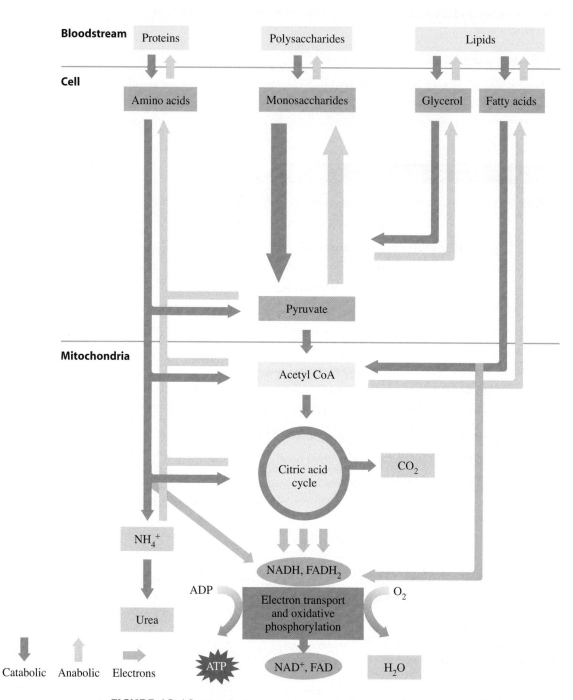

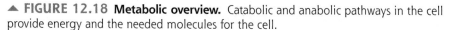

▲ **FIGURE 12.18 Metabolic overview.** Catabolic and anabolic pathways in the cell provide energy and the needed molecules for the cell.

SUMMARY

12.1 Overview of Metabolism

Metabolism refers to chemical reactions occurring in the body. Catabolism refers to reactions that break down larger molecules into smaller ones. These reactions usually produce energy and the processes are oxidative. Anabolism refers to reactions that synthesize larger biological molecules from smaller ones. These reactions typically require energy and are overall reductive. In the body, chemical reactions are usually grouped in metabolic pathways. Biochemical reactions that produce energy tend to be coupled to reactions requiring energy. Metabolic pathways tend to be compartmentalized in different parts of the cell. The mitochondria are the energy-producing factories of the cells.

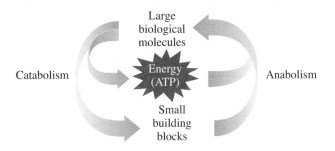

12.2 Metabolically Relevant Nucleotides

Nucleotides are used in nucleic acids. Their components are also found in the coenzymes of metabolism and are used to transfer energy throughout the cell. ATP is considered the main energy currency of the cell and produces energy when hydrolyzed to ADP + P_i. NADH and $FADH_2$ contain a nucleotide portion and a vitamin portion and are important coenzymes that transport hydrogens and electrons in the cell. They have a high-energy reduced form (NADH and $FADH_2$) and a low-energy oxidized form (NAD^+ and FADH). Coenzyme A (CoA) also contains a nucleotide and vitamin portion. The high energy form of Co A is called acetyl CoA and has a two-carbon acetyl group attached to Co A through a high-energy thioester bond.

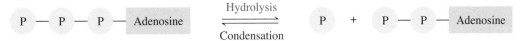

12.3 Digestion—From Fuel Molecules to Hydrolysis Products

Carbohydrates are digested in a series of reactions as the glycosidic bonds of polysaccharides are hydrolyzed, yielding oligo- and disaccharides. These disaccharides and those from other food sources undergo hydrolysis to monosaccharides. These monosaccharides can be absorbed into the bloodstream and carried into the cells for metabolism. Because they are nonpolar and do not mix in water, dietary fats like triglycerides must be emulsified by bile salts so they can be enzymatically hydrolyzed to monoglycerides for absorption. In the blood, triglycerides are repackaged into lipoproteins called chylomicrons for delivery to the cells. Protein digestion begins in the stomach and continues in the small intestine. It involves the hydrolysis of peptide bonds, yielding amino acids that can be absorbed into the bloodstream and transported to the cells for metabolism.

12.4 Glycolysis—From Hydrolysis Products to Common Metabolites

Glycolysis describes the series of 10 reactions that break down a six-carbon glucose molecule to two three-carbon pyruvate molecules. These 10 reactions yield two ATP and two NADH molecules per glucose. Under aerobic conditions, pyruvate is further oxidized in the mitochondria to acetyl CoA. In the absence of oxygen, pyruvate is reduced to lactate and regenerates NAD^+ so that glycolysis can continue. Some organisms like yeast undergo fermentation, producing ethanol instead of lactate under anaerobic conditions. Glucose can be stored as glycogen in the liver and muscle for later use. Fructose can undergo glycolysis; however, in the liver its catabolism is unregulated and can produce excess products that become stored in the body as fat.

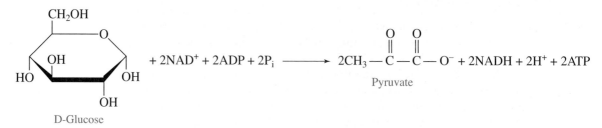

12.5 The Citric Acid Cycle—Central Processing

The citric acid cycle occurs in the mitochondrial matrix and combines acetyl CoA (two carbons) with oxaloacetate (four carbons) producing citric acid, which undergoes oxidation and decarboxylation to yield two CO_2, GTP, three NADH, and one $FADH_2$. The cycle ultimately regenerates oxaloacetate to begin again. GTP readily converts to ATP in the cell.

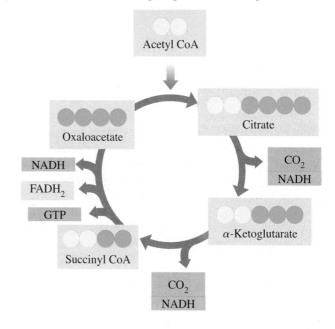

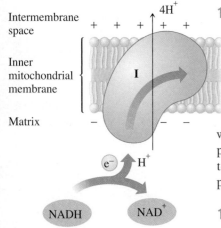

12.6 Electron Transport and Oxidative Phosphorylation

The reduced nucleotides NADH and $FADH_2$ become oxidized, transporting H^+ and electrons through a series of enzyme complexes in the inner mitochondrial membrane. The final electron acceptor in this process is O_2, which combines with H^+ to form H_2O. The complexes act as proton pumps, moving protons from the matrix to the inner membrane space, producing a proton gradient during electron transport. The protons can return to the matrix through complex V, ATP synthase, which generates energy for the formation of ATP. This process is known as oxidative phosphorylation. When ATP and proton transport are uncoupled, the energy in the proton gradient is released as heat, which assists in maintaining body temperature in a process called thermogenesis.

12.7 ATP Production

For every four H^+ pumped into the inner membrane space and returned to the matrix, one ATP can be synthesized. The oxidation of one NADH provides enough energy to synthesize 2.5 ATP. One $FADH_2$ provides energy to synthesize 1.5 ATP. Under aerobic conditions, the complete oxidation of one molecule of glucose produces a total of 32 ATP.

12.8 Other Fuel Choices

Fatty acids can be used for fuel when glucose supplies are low. Fatty acids link to coenzyme A, forming activated fatty acyl CoA which is transported to the mitochondria for catabolism in a reaction cycle called β oxidation. Fatty acyl CoA is oxidized, thereby producing a new fatty acyl CoA that is two carbons shorter and one molecule of acetyl CoA. Each turn of β oxidation produces one NADH and one $FADH_2$. High levels of acetyl CoA in the cell activate the ketogenesis pathway, thereby forming ketone bodies that can lead to ketosis and acidosis. Amino acids can also be used for fuel when other fuel supplies are low and they are not needed to produce nitrogen-containing compounds in the cell. When amines are removed from amino acids as ammonium ions, they are converted to urea for excretion. The carbons from amino acids can feed into oxidative catabolism as different intermediates depending on the amino acid.

KEY TERMS

aerobic—In the presence of oxygen.

anabolism—Metabolic chemical reactions in which smaller molecules are combined to form larger ones. These reactions tend to be reductive and require energy.

anaerobic—In the absence of oxygen.

beta (β) oxidation—The degradation of fatty acids by removing two-carbon segments from a fatty acid at the oxidized β carbon.

catabolism—Metabolic chemical reactions in which larger molecules are broken down into smaller ones. These reactions tend to be oxidative and produce energy.

chemiosmotic model—The conservation of energy from the transfer of electrons in the electron transport chain by pumping protons into the intermembrane space to produce a proton gradient that provides the energy to synthesize ATP.

chylomicrons—Droplets of dietary fat and protein that circulate from the intestine to the tissues for absorption.

citric acid cycle—A series of reactions that degrades two-carbon acetyl groups from acetyl CoA into CO_2, generating the high-energy molecules NADH and $FADH_2$ in the process.

cytoplasm—The material in a cell between the nucleus and the cell membrane.

cytosol—The aqueous solution of the cytoplasm.

digestion—The breakdown of large molecules into smaller components for absorption, usually through hydrolysis.

electrochemical gradient—A difference in both concentration and electrical charge of an ion across a membrane; the driving force for oxidative phosphorylation.

emulsification—Breaking up large insoluble globules into smaller droplets via an amphipathic molecule.

fatty acyl CoA—A fatty acid bonded to coenzyme A through a thioester bond.

fermentation—The anaerobic production of ethanol from pyruvate.

gluconeogenesis—The anabolic process of synthesizing glucose.

ketone bodies—Ketone compounds formed from the condensation of excess acetyl CoA in the liver.

ketosis—A condition where ketone bodies are not all metabolized, leading to low blood pH.

metabolic pathway— A series of chemical reactions leading to a common product.

metabolism— The sum of all chemical reactions occurring in an organism.

metabolite—A chemical intermediate formed during enzyme-catalyzed metabolism. Some common ones are pyruvate and the acetyl group.

mitochondria—Cell organelle where energy-producing reactions take place.

oxidative decarboxylation—An oxidation where a carboxylate is released as CO_2. The corresponding reduction forms a reduced nucleotide like NADH.

oxidative phosphorylation—The production of ATP from ADP and P_i using energy generated by the oxidation reactions of electron transport.

thermogenesis—The generation of heat. In the mitochondria it can be caused by the uncoupling of ATP synthase from the electron transport chain.

transamination—The transfer of an amino group from an amino acid, thereby producing an α-keto acid.

urea cycle—The process by which excess ammonium ions from the degradation of amino acids are converted to urea for excretion.

SUMMARY OF REACTIONS

Hydrolysis of ATP

$$ATP \longrightarrow ADP + P_i + energy$$

Formation of ATP

$$ADP + P_i + Energy \longrightarrow ATP$$

Digestion of Proteins

$$Protein + H_2O \xrightarrow[Enzyme]{H^+} Amino\ acids$$

Glycolysis

$$+ 2NAD^+ + 2ADP + 2P_i \longrightarrow 2CH_3-\overset{O}{\overset{||}{C}}-\overset{O}{\overset{||}{C}}-O^- + 2NADH + 2H^+ + 2ATP$$

D-Glucose   Pyruvate

Hydrolysis of Disaccharides

$$Lactose + H_2O \xrightarrow{Lactase} Galactose + glucose$$

$$Sucrose + H_2O \xrightarrow{Sucrase} Glucose + fructose$$

$$Maltose + H_2O \xrightarrow{Maltase} Glucose + glucose$$

Digestion of Triglycerides

$$Triglycerides + 3\,H_2O \xrightarrow{Lipases} Glycerol + 3\ Fatty\ acids$$

Oxidation of Pyruvate to Acetyl CoA

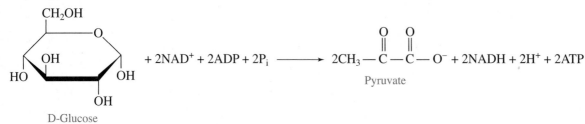

Pyruvate   Coenzyme A   Acetyl coenzyme A

Reduction of Pyruvate to Lactate

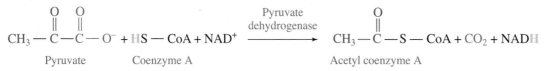

Pyruvate   Lactate

Citric Acid Cycle

$$Acetyl\ CoA + 3NAD^+ + FAD + GDP + P_i + 2H_2O \longrightarrow$$
$$2CO_2 + 3NADH + 2H^+ + FADH_2 + CoA + GTP$$

Complete Oxidation of Glucose

$$C_6H_{12}O_6 + 6O_2 \longrightarrow 6CO_2 + 6H_2O$$

β Oxidation of Fatty Acid

$$Fatty\ acyl\ CoA_{n\ carbons} + NAD^+ + FAD + H_2O + CoA \longrightarrow$$
$$Fatty\ acyl\ CoA_{n-2\ carbons} + Acetyl\ CoA + NADH + H^+ + FADH_2$$

ADDITIONAL PROBLEMS

12.55 Name the location in the cell where the following catabolic processes take place:
  a. glycolysis
  b. citric acid cycle
  c. ATP synthesis

12.56 Name the location in the cell where the following catabolic processes take place:
  a. pyruvate oxidation
  b. electron transport
  c. β-oxidation

12.57 Identify the type of food—carbohydrate, fat, or protein—that gives each of the following digestion products:

a. glucose **b.** fatty acid
c. maltose **d.** glycerol
e. amino acids

12.58 How and where does sucrose undergo digestion in the body? Name the products.

12.59 How and where does lactose undergo digestion in the body? Name the products.

12.60 If glycolysis occurs in the cytosol and the citric acid cycle occurs in the mitochondrial matrix, how do the products of glycolysis get inside the mitochondrial matrix?

12.61 Which of the following reactions in glycolysis produce ATP or NADH?

a. 1,3-bisphosphoglycerate to 3-phosphoglycerate
b. glucose-6-phosphate to fructose-6-phosphate
c. phosphoenolpyruvate to pyruvate

12.62 Which of the following reactions in glycolysis produce ATP or NADH?

a. glucose to glucose-6-phosphate
b. glyceraldehyde-3-phosphate to 1,3-bisphosphoglycerate
c. dihydroxyacetone phosphate to glyceraldehyde-3-phosphate

12.63 Which of the reactions given in problem 12.61 represent isomerizations where the reactants and products are structural isomers?

12.64 Which of the reactions given in problem 12.62 represent isomerizations where the reactants and products are structural isomers?

12.65 After running a marathon, a runner has muscle pain and cramping. What might have occurred in the muscle cells to cause this?

12.66 After eating a large, starchy meal, what does the body do with all the glucose from starch?

12.67 Under what condition is pyruvate converted to lactate in the body?

12.68 When pyruvate is used to form acetyl CoA, the product has only two carbon atoms. What happened to the third carbon?

12.69 Refer to the diagram of the citric acid cycle to answer each of the following:

a. Name the six-carbon compounds.
b. Name the five-carbon compounds.
c. Name the oxidation reactions.
d. Name the reactions where secondary alcohols are oxidized to ketones.

12.70 Refer to the diagram of the citric acid cycle to answer each of the following:

a. Name the four-carbon compounds.
b. Name the reactant that undergoes a hydration reaction.
c. Name the reaction that is coupled to GTP formation.
d. Provide the number of CO_2 molecules produced per turn of the citric acid cycle.

12.71 If there are no reactions in the citric acid cycle that use oxygen, O_2, why does the cycle operate only in aerobic conditions?

12.72 What products of the citric acid cycle are used in electron transport?

12.73 Identify the following as the reduced or oxidized form:

a. NAD^+ **b.** $FADH_2$ **c.** QH_2

12.74 Identify the following as the reduced or oxidized form:

a. NADH   **b.** FAD   **c.** Q

12.75 During electron transport, H^+ are pumped out of the _____, across the inner membrane, and into the _____.

12.76 During ATP synthesis, H^+ move from the _____, across the inner membrane, and into the _____.

12.77 In the chemiosmotic model, how is energy provided to synthesize ATP?

12.78 What is the effect of proton accumulation in the intermembrane space?

12.79 How many ATPs are produced when glucose is oxidized to pyruvate compared to when glucose is oxidized to CO_2 and H_2O?

12.80 What metabolic substrate(s) can be produced from the carbon atoms of each of the following amino acids?

a. histidine   **b.** isoleucine
c. methionine   **d.** phenylalanine

12.81 Consider the complete oxidation of capric acid, a saturated C10 fatty acid.

a. How many acetyl CoA units are produced?
b. How many turns of β oxidation are needed ?
c. How many ATPs are generated from the complete oxidation of capric acid?

12.82 Consider the complete oxidation of arachidic acid, a saturated C20 fatty acid.

a. How many acetyl CoA units are produced?
b. How many turns of β oxidation are needed?
c. How many ATPs are generated from the complete oxidation of arachidic acid?

CHALLENGE PROBLEMS

12.83 Identify all steps in the oxidative catabolism of glucose that can be considered oxidative decarboxylations.

12.84 How many turns of the citric acid cycle does it take for the two carbons from an acetyl group entering the citric acid cycle to be liberated as CO_2?

12.85 Lauric acid, a saturated C12 fatty acid, is found in coconut oil.

a. Draw fatty acyl lauric acid activated for β oxidation.
b. Identify the α and β carbons in fatty acyl lauric acid.
c. Write the overall equation for the complete β oxidation of lauric acid.

d. How many acetyl CoA units are produced?

e. How many cycles of β oxidation are needed?

f. Account for the total ATP yield from β oxidation of lauric acid (C12) by completing the following table:

_____Acetyl CoA × 10 ATP/acetyl CoA	_____ATP
_____NADH × 2.5 ATP/NADH	_____ATP
_____FADH$_2$ × 1.5 ATP/FADH$_2$	_____ATP
Total	

12.86 A person is brought to the emergency room in what appears to be a drunken stupor. On closer examination she seems to be breathing very rapidly and you notice a sweet-smelling odor on her breath. You find out this person has not been drinking alcohol. What is her likely condition?

12.87 Can acetyl CoA feed into gluconeogenesis producing glucose for the body? Can fatty acids be used to produce glucose?

ANSWERS TO ODD-NUMBERED PROBLEMS

Practice Problems

12.1 In metabolism, a catabolic reaction breaks apart molecules, releases energy, and is an oxidation.

12.3 hydrolysis

12.5 CO_2 and H_2O

12.7 mitochondria

12.9 a. NADH   **b.** Acetyl CoA   **c.** FAD

12.11 a. starch
     b. none
     c. oligosaccharides, disaccharides

12.13 hydrolysis

12.15 Cholesterol is esterified and packaged into lipoproteins.

12.17 amino acids

12.19 pyruvate

12.21 4 ATP (net 2 ATP)

12.23 2 NADH and 2 ATP

12.25 NAD^+

12.27 ethanol, CO_2, and NAD^+

12.29 The main regulation point of glycolysis is phosphofructokinase, the enzyme involved in step 3. Because the products of the metabolism of fructose enter glycolysis at step 5, they bypass the main regulation point, which leads to the production of excess pyruvate and acetyl CoA that is not needed in the cells and is ultimately converted to fat.

12.31 acetyl CoA and oxaloacetate

12.33 isocitrate $\longrightarrow$ α-ketoglutarate (reaction 3) and α-ketoglutarate $\longrightarrow$ succinyl CoA (reaction 4)

12.35 isocitrate $\longrightarrow$ α-ketoglutarate (reaction 3), α-ketoglutarate $\longrightarrow$ succinyl CoA (reaction 4), and malate $\longrightarrow$ oxaloacetate (reaction 8)

12.37 Coenzyme Q, also called ubiquinone

12.39 complex IV

12.41 As protons flow through ATP synthase, energy is released to produce ATP.

12.43 a. 2.5 ATP   **b.** 7 ATP
      c. 5 ATP   **d.** 10 ATP

12.45 mitochondrial matrix

12.47

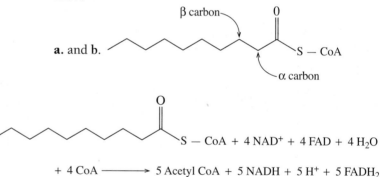

12.49 Ketone bodies form in the body when excess acetyl CoA results from the breakdown of large amounts of fat.

12.51 NH_4^+ is toxic if it accumulates in the body.

12.53 a. pyruvate   **b.** oxaloacetate, fumarate
      c. succinyl CoA   **d.** α-ketoglutarate

Additional Problems

12.55 a. cytosol
  b. mitochondrial matrix
  c. mitochondrial matrix

12.57 a. carbohydrate   **b.** fat
  c. carbohydrate   **d.** fat
  e. protein

12.59 Lactose undergoes digestion in the small intestine to yield galactose and glucose.

12.61 a. produces ATP   **b.** neither   **c.** produces ATP

12.63 b.

12.65 The runner's muscles may be switching over to anaerobic catabolism to keep ATP production going. The muscles produce lactate and H^+ during this process causing soreness.

12.67 anaerobic (low oxygen) conditions

12.69 a. citrate and isocitrate
  b. α-ketoglutarate
  c. isocitrate $\longrightarrow$ α-ketoglutarate (reaction 3), α-ketoglutarate $\longrightarrow$ succinyl CoA (reaction 4), succinate $\longrightarrow$ fumarate (reaction 6), and malate $\longrightarrow$ oxaloacetate (reaction 8)
  d. isocitrate $\longrightarrow$ α-ketoglutarate (reaction 3) and malate $\longrightarrow$ oxaloacetate (reaction 8)

12.71 O_2 is used during electron transport. The coenzymes NAD^+ and FAD needed in the citric acid cycle are only available if they are oxidized during electron transport.

12.73 a. oxidized   **b.** reduced   **c.** reduced

12.75 matrix, intermembrane space

12.77 Energy released as protons flow down a concentration gradient through ATP synthase back to the matrix is utilized for the synthesis of ATP.

12.79 The oxidation of glucose to pyruvate produces 7 ATP whereas 32 ATP are produced from the complete oxidation of glucose to CO_2 and H_2O.

12.81 a. 5   **b.** 4   **c.** 66 ATP

12.83 Oxidation of pyruvate: pyruvate $\longrightarrow$ acetyl CoA

Reaction 3 of citric acid cycle:
isocitrate $\longrightarrow$ α-ketoglutarate

Reaction 4 of the citric acid cycle:
α-ketoglutarate $\longrightarrow$ succinyl CoA

12.85

a. and **b.**

c.

$+ 5\ NAD^+ + 5\ FAD + 5\ H_2O + 5\ CoA \longrightarrow$

$6\ Acetyl\ CoA + 5\ NADH + 5\ H^+ + 5\ FADH_2$

d. 6   **e.** 5

f.

_____6_____ Acetyl CoA × 10 ATP/acetyl CoA	60	ATP
_____5_____ NADH × 2.5 ATP/NADH		12.5 ATP
_____5_____ FADH$_2$ × 1.5 ATP/FADH$_2$		7.5 ATP
	Total	80 ATP

12.87 Only oxaloacetate or pyruvate are starting points for gluconeogenesis (see Figure 12.18). Acetyl CoA can not be converted to pyruvate, so fatty acids cannot generate glucose.

Credits

Unless otherwise acknowledged, all photographs are the property of Benjamin Cummings Publishers, Pearson Education.

FM

p. iii Laura D. Frost. p. iii Todd Deal. p. iv Karen Timberlake.

Chapter 1

p. 4 Caroline von Tuempling/Getty Images. p. 5 Eric Schrader - Pearson Science. p. 9 JUPITERIMAGES/BananaStock/Alamy. p. 10 NASA. p. 12 iofoto/Shutterstock. p. 16 *left-right*: Eric Schrader - Pearson Science. p. 19 *left*: Josef Bosak/Shutterstock. p. 19 *middle, right*: titelio/Shutterstock. p. 20 *left*: Li Wa/Shutterstock. p. 20 *right*: Mark Evans/iStockphoto. p. 23 Richard Megna - Fundamental Photo.

Chapter 2

p. 36 emre ogan/iStockphoto. p. 37 *top*: Elen/Shutterstock. p. 37 *middle*: Pearson Science. p. 37 *bottom*: Colin Cuthbert/Photo Researchers, Inc. p. 51 SIMON FRASER/ SPL/Photo Researchers, Inc. p. 52 *middle*: Tomaz Levstek/iStockphoto. p. 52 *bottom*: Stanford Dosimetry, LLC. p. 57 *left*: Myo Han. p. 57 *right*: Dr. Michael Tobin.

Chapter 3

p. 70 Chepko Danil Vitalevich/Shutterstock. p. 75 Kurhan/Shutterstock. p. 81 *bottom*: Mark Kostich/iStockphoto. p. 81 *top*: Shutterstock. p. 87 Eric Schrader - Pearson Science. p. 90 Eric Schrader - Pearson Science.

Chapter 4

p. 108 REUTERS/Eric Gaillard. p. 110 Big Cheese/Photolibrary. p. 124 Eric Schrader - Pearson Science. p. 125 *top*: Eric Schrader - Pearson Science. p. 125 *middle*: Glen Jones/ iStockphoto. p. 125 *bottom*: Steffen Foerster Photography/Shutterstock. p. 126 Leonid Nyshko/iStockphoto. p. 138 *top, left; top, right; bottom, left; bottom, right*: Eric Schrader - Pearson Science. p. 140 *left*: Photos.com. p. 140 *right*: Vladimir Kolobov/ iStockphoto. p. 141 *left-right*: Eric Schrader - Pearson Science.

Chapter 5

p. 160 Carl Newedel/Gettyimages. p. 162 AP Photo/Seth Perlman. p. 163 *left*: Wojtek Kryczka/iStockphoto. p. 163 *right*: Baloncici/iStockphoto. p. 164 Marcelo Wain/iStockphoto. p. 169 *middle*: FreezeFrameStudio/iStockphoto. p. 169 *bottom, right*: Elena Elisseeva/iStockphoto. p. 169 *bottom, left*: Kelly Cline/iStockphoto. p. 172 stinkyt/iStockphoto. p. 173 Saturn Stills/Photo Researchers, Inc. p. 185 *top*: Jon Larson/iStockphoto. p. 191 *top, left*: Shutterstock. p. 191 *top, right*: David Toase/ Getty Images. p. 191 *middle*: Biophoto Associates/Photo Researchers. p. 192 Shutterstock. p. 197 Freddie Vargas/ iStockphoto.

Chapter 6

p. 210 Eric Schrader - Pearson Science. p. 212 Eric Schrader - Pearson Science. p. 219 Shutterstock. p. 225 *top*: Shutterstock. p. 225 *bottom*: iStockphoto. p. 233 Eric Schrader - Pearson Science.

Chapter 7

p. 244 Eric Schrader - Pearson Science. p. 246 Eric Schrader - Pearson Science. p. 248 *top*: Douglas Freer/iStockphoto. p. 248 *middle, top*: Shutterstock. p. 248 *middle, bottom*: Eric Schrader - Pearson Science. p. 248 *bottom*: Shutterstock. p. 249 *left*: Dorling Kindersley. p. 249 *right*: Klaus Guldbrandsen/Photo Researchers. p. 251 *top*: Dr. P. Marazzi/Photo Researchers. p. 251 *bottom*: Stephen J. Krasemann/Photo Researchers. p. 259 Eric Schrader - Pearson Science. p. 262 Eric Schrader - Pearson Science. p. 264 Eric Schrader - Pearson Science. p. 265 Paul Kline/iStockphoto. p. 270 Visuals Unlimited/Getty Images. p. 271 *bottom, left-right*: Sam Singer - Pearson Science. p. 273 Shutterstock.

Chapter 8

p. 286 REUTERS/David Gray. p. 290 *bottom*: Eric Schrader - Pearson Science. p. 292 Eric Schrader - Pearson Science. p. 304 *left*: Richard Megna - Fundamental Photo. p. 304 *right*: Alina Hart/iStockphoto. p. 317 Nina Shannon/iStockphoto.

Chapter 9

p. 324 Joselito Briones/iStochphoto. p. 338 *top*: Brucker, E.A., Olson, J.S., Phillips Jr., G.N., Dou, Y., Ikeda-Saito, M. (1996) High resolution crystal structures of the deoxy, oxy and aquomet forms of the cobalt myoglobin, J. Biol. Chem., 271: 25419–25422. p. 338 *bottom*: Shutterstock. p. 343 *left*: Liddington, R., Derewenda, Z., Dodson, E., Hubbard, R., Dodson, G. (1992) High resolution crystal structures and comparisons of T-state deoxy-haemoglobins: T(alpha-oxy)haemoglobin and T(met)haemoglobin, J. Mol. Biol., 228:551–579. p. 343 *right*: Silva, M.M., Rogers, P.H., Arnone, A., A third quaternary structure of human HaemoglonA at 1.7–A resolution. (1992), J. Biol. Chem., 267:17248–17256.

Chapter 10

p. 354 iStockphoto. p. 356 Coordinates for protein 1HKG (hexokinase from yeast) attributed to Steitz, T.A. et. al., Philos. Trans. R. Soc. London, Ser. B, 1981, 293: 43–52. p. 358 *left*: Steitz, T.A., Shoham, M., & Bennett, W.S., Jr. (1981), Structural dynamics of yeast hexokinase during catalysis. Philos. Trans. R. Soc. London, Ser. B, 293: 43–52. p. 358 *right*: Anderson, C.M., Stenkamp, R.E., Steitz, T.A. (1978). Sequencing a protein by x-ray crystallography. II. Refinement of yeast hexokinase B co-ordinates and sequence at 2.1A resolution. J. Mol. Biol. 123:15–33. p. 359 Eric Schrader - Pearson Science. p. 364 Eric Schrader - Pearson Science.

Chapter 11

p. 380 Monashee Frantz/OJO Images/Getty Images. p. 381 SPL/Photo Researchers. p. 404 Will & Deni McIntyre/Photo Researchers. p. 407 Stephen Ferreira. p. 408 *left*: AP Photo/PA/Files. p. 408 *right*: AP Photo/Texas A&M University.

Chapter 12

p. 420 Galina Barskaya/iStockphoto. p. 421 Ciaran Griffin /Getty Images. p. 427 iStockphoto. p. 435 Diego Cervo/ iStockphoto. p. 443 Laura Delong Frost.

TEXT CREDIT

Chapter 8

p. 299 "General Chemistry Online!" antoine.frostburg.edu/chem/senese/101/ index.shtml.

Glossary/Index

water molecules interacting with the ions and then surrounding each of the newly formed ions; the solvation process when water is the solvent. 219

Hydrogen bonding The strong dipole–dipole attraction of a hydrogen, covalently bonded to an O, N, or F (the hydrogen-bond donor), to a nonbonding pair of electrons on an O, N, or F (the hydrogen-bond acceptor). 213–215, 237

Hydrocarbons, 106
  molecule set, 106

Hydrogen:
  ionic name/formula, 75
  and octet rule, 83

Hydrogen peroxide, 342, 360
  oxidation by, 342

Hydrogenation The addition of hydrogen (H_2) to a carbon–carbon double bond to convert it to a single bond. 231

Hydrolysis reactions, 181, 196–197

Hydrolysis A reaction involving the breaking of one large organic molecule into two smaller ones. One of the reactants is H_2O. The reverse reaction is condensation. 181, 323

Hydronium ion (H_3O^+) The ion formed by the attraction of a proton (H^+) to an H_2O molecule. 287–288, 300

Hydrophilic Literally, "water-loving"; polar compounds or parts of compounds that readily dissolve in water. 220

Hydrophobic Literally, "water-fearing"; nonpolar compounds or parts of compound that do not dissolve in water. 220

Hydrophobic effect The movement of nonpolar amino acid side chains to the interior of a protein caused by an unfavorable interaction with the aqueous environment. 337

Hydroxyapaptite, 75

Hydroxyproline, 342–343

Hypertonic solution A solution outside of a cell having a higher concentration of solutes than the solution inside the cell. 271

Hyperventilation A condition of expelling too much CO_2 from the lungs, thereby upsetting the bicarbonate equilibrium. 313

Hyponatremia, 271

Hypotonic solution A solution outside of a cell having a lower concentration of solutes than the solution inside the cell. 271

Hypoventilation A condition of expelling too little CO_2 from the lungs, thereby upsetting the bicarbonate equilibrium. 285

I

Immunoglobulins, 344. *See also* Antibodies

Incomplete proteins, 329

Induced dipole. *See* London forces

Induced-fit model A model describing the initial interaction of an enzyme to its substrate; an enzyme's active site and its substrate have flexible shapes that are roughly complementary and adjust to allow the formation of ES. 357

Inhibitor A molecule that causes an enzyme's catalytic activity to decrease. 366–369, 371

Inner membrane space, 440

Inorganic compounds Compounds composed of elements other than carbon and hydrogen. 109

Inorganic phosphate, P_i, 355

Insulin, 407

Integral membrane protein A protein that is found within or spanning the phospholipid bilayer of a membrane; many serve as passages for polar compounds to move across the membrane. 344–345

Integrase, 404

Intermolecular forces Attractive forces between two or more molecules that are caused by uneven distributions of electrons within a molecule. 211–243
  and the cell membrane, 235–236
  and changes of state, 223–229, 237
  changing a liquid to a solid, 231
  dipole–dipole attractions, 213
  fats, 229–230
  golden rule of solubility, 208, 218
  and heat, 223
  hydrogen bonding, 213–215
  London forces, 211–213
  margarine, 231
  oils, 230–231
  and solubility, 218–222, 237
  trans fats, 233
  types of, 211–217, 237

International System of Units, 6

International Union of Pure and Applied Chemistry (IUPAC), 22

Inverse logarithms, 305

Invert sugar, relative sweetness of, 186

Iodine-131, 56
  and cancer treatment, 58

Ion An atom with a charge due to an unequal number of protons and electrons. 72
  biologically important, 75
  formation, 73–74
  found in fluids/cells of the body, 75
  naming, 74–75
  polyatomic, 74–75

Ion–dipole attraction An electrical attraction between an ion and a polar molecule. 216, 219–220, 237

Ionic bonds, 66, 77, 81

Ionic compounds, 66, 77–81, 97
  applying the golden rule of solubility to, 219
  distinguishing from covalent compounds, 81
  formulas for, 77–79
  naming, 79–81

Ionic interactions, 337

Ionic solutions and equivalents, 258

Ionize To dissociate forming ions when dissolved. 254

Ionizing radiation A general name for high-energy radiation of any kind (for example, nuclear, X-ray). 61
  properties of (table), 52

Iron-59, 57

Irreversible inhibition The loss of enzyme catalytic activity that is permanent. 368

Isobutane, 128

Isoelectric point (pI) The pH at which an amino acid has a net charge of zero. 311

Isoelectronic Having the same number of electrons. 72

Isopropyl alcohol, 104

Isotonic solution A solution outside of a cell having the same concentration of solutes as the solution on the inside of a cell. 270

Isotopes Atoms that have the same number of protons but different numbers of neutrons. 34
  and atomic mass, 41–42, 60
  and mass number, 60

J

Junk DNA, 399

K

K^+/H^+ ATPase, 276

Kelvin scale, 14–15

Kelvins, 14

Ketone bodies Ketone compounds formed from the condensation of excess acetyl CoA in the liver. 449–450, 455

Ketones A family of organic compounds whose functional group is a carbonyl ($C=O$) bonded to two alkyl groups. 123, 164

Ketose A monosaccharide that contains the ketone functional group. 164, 173

Ketosis A condition where ketone bodies are not all metabolized, leading to low blood pH. 449, 455

Kidney dialysis, 273

Kidney stones, 250–251

Kilo-, 7

Kilogram (kg), 6

L

L-dopa, 145–146

Lactase, 185

Lactose A disaccharide consisting of glucose and galactose found in milk and milk products. 185, 428
  relative sweetness of, 186

the general formula $C_n(H_2O)_n$. 162–172, 196–197

  functional groups in, 162–166

    alcohols, 162–163

    aldehydes, 163–164

    ketones, 164

    important, 168–170

    and redox, 173–175

    stereochemistry in, 166, 166–172

      diastereomers, 168

      Fischer projection, 166–168

      multiple chiral centers, 166

    structure, 176–178

Monounsaturated A term used to describe organic compounds that have one double bond in their structure. 118

Monounsaturated fatty acids, 118–120

mRNA. *See* messenger RNA (mRNA)

Muscular dystrophy (Duchenne), 401

Mutagen An environmental agent that causes a mutation. 400, 411

Mutation A change in DNA sequence. 399–400, 411

  sources of, 400

Myoglobin, 337–338

N

N-acetylglucosamine, 192–194, 197

NADH/NAD$^+$ and FADH$_2$/FAD, 425

Natural log (ln), 305

Negatively charged beta (β) particle, 51

Neutral The term that describes a solution with equal concentrations of H_3O^+ and OH^-. 303

Neutral gamma (γ) ray, 51

Neutral solution, 303

Neutralization The reaction between an acid and a base to form a salt and water. 290–291, 315

Neutrons A subatomic particle with no charge, symbol n. 37, 59

Nicotinamide adenine dinucleotide (NADH/NAD$^+$), 422

Nitrogen, ionic name/formula, 75

Nitrogenous bases, 376–378, 382

Nitrosamine, 400

Noble gases, 71

Nonbonded pair A pair of electrons belonging to only one atom. It is also referred to as a lone pair. 89–90, 98

Nonbonding clouds, 88

Noncompetitive inhibitor A molecule that inhibits an enzyme's catalytic activity by interacting with a site other than the active site and distorting the shape of the active site. 367

Nonelectrolyte A substance that does not conduct electricity because it does not ionize in aqueous solution. 244, 255–256, 277

Nonmetal atoms, 96–97

Nonpolar amino acids Amino acids whose side chains are composed almost entirely of carbon and hydrogen, resulting in an even distribution of electrons over the side chain portion of the molecule. 329, 346

Nonpolar compounds, applying the golden rule of solubility to, 219

Nonpolar covalent bond A bond in which the electrons are shared equally by the two atoms. 91, 98, 250

Nonpolar interactions The association of nonpolar amino acid side chains with each other; London forces. 337

Nonpolar molecule A molecule that has an even distribution of electrons over the entire molecule. 94–95

N-terminus (N-terminal amino acid or amino terminus) In a peptide, the amino acid with the free protonated amine, always written to the far left of the structure. 331, 346

Nuclear decay equations A chemical equation where the reactant is a radioactive isotope and the products are a radioactive particle and a second radioisotope. 53

  writing, 54

Nuclear equations, and radioactive decay, 53–55, 62

Nuclear radiation Energy emitted spontaneously from the nucleus of an atom. 50–51

  types/properties of (table), 51

Nuclear transplantation, 408

Nucleic acid sequence, 387–388

Nucleic acids, 170, 377–378, 381–417

  backbone of, 377

  components of, 381–387, 409

  condensation of the components, 382–383

  deoxyribose, 382, 385

  formation of, 387–390, 410

  primary structure, 387–388

  ribose, 382

Nucleoside A pentose sugar condensed with one of the five nitrogenous bases (adenine, thymine, guanine, cytosine, or uracil) at C1′. 383, 386, 409

Nucleotide A pentose sugar condensed with one of the five nitrogenous bases (adenine, thymine, guanine, cytosine, or uracil) at C1′ and up to three phosphates at C5′. 381, 382, 386, 409

  components of, 376–377, 381

  condensation of, 388

  metabolically relevant, 424–427

  naming, 385–386

Nucleus The central part of an atom containing protons and neutrons. 423

  atoms, 37–38

Nutrasweet®, 186

O

Octet, 82

Octet rule Atoms react with other atoms to obtain eight electrons in their outer shell. 71–72, 96

  and hydrogen, 83

Oil A lipid molecule composed of three fatty acids joined to a glycerol backbone and that exists as a liquid at room temperature; also known as a triglyceride. 230–231, 238

Oligosaccharides A carbohydrate composed of three to nine monosaccharide units joined through glycosidic bonds. 161–162

Optimum pH The pH at which an enzyme functions most efficiently. 365–366

Optimum temperature The temperature at which an enzyme functions most efficiently. 366, 371

Organelles, 423

Organic chemistry The field of chemistry dedicated to studying the structure, characteristics, and reactivity of carbon-containing compounds. 109, 112

Organic compounds Compounds composed primarily of carbon and hydrogen but that may also include oxygen, nitrogen, sulfur, phosphorus, and a few other elements in their structures. 109–157

  condensed structural formulas, 113–114

  defined, 109

  families of, 148–149

    table, 121–123

  functional groups, 121–127

  isomerism in, 127–147, 149–150

    conformational isomers, 128–129

    structural isomers, 127–128

  representing the structures of, 113–116

  simple, nomenclature of, 129–134

  skeletal structures, 114–115

Osmosis The passage of water across a semipermeable membrane in an effort to equalize the solution concentrations on either side. This passage does not require energy (it is a passive process). 244–245, 270–272, 278

Osmotic pressure The pressure that water exerts during osmosis. This amount of pressure applied to the more concentrated solution of the two separated solutions would stop osmosis. 271

Oxidation The loss of electrons during a chemical reaction; in organic reactions, often appears as a gain of oxygen or a loss of hydrogen. 170–171, 196

  by hydrogen peroxide, 342

METRIC UNITS AND SOME USEFUL CONVERSION FACTORS

Length meter (m)	Volume cubic meter (m³)	Mass kilogram (kg)
1 meter (m) = 100 centimeters (cm)	1 liter (L) = 1000 milliliters (mL)	1 kilogram (kg) = 1000 grams (g)
1 meter (m) = 1000 millimeters (mm)	1 mL = 1 cc = 1 cm³	1 g = 1000 milligrams (mg)
1 cm = 10 mm	1 tsp = 5 mL	1 kg = 2.20 lb
1 kilometer (km) = 0.621 mile (mi)	1 L = 1.06 quart (qt)	1 lb = 454 g
1 inch (in.) = 2.54 cm (exact)	1 qt = 946 mL	1 mole = 6.02×10^{23} particles
		Water
		density = 1.00 g/mL

Temperature kelvin (K)	Pressure pascal (Pa)	Energy calorie (cal)
$°F = 1.8(°C) + 32$	1 atmosphere = 760 mmHg	1 kcal = 1000 calories (cal)
$°C = \dfrac{(°F - 32)}{1.8}$	1 atm = 14.7 psi	1 nutritional calorie (Cal) = 1 kcal
$K = °C + 273$		

METRIC PREFIXES

Prefix	Abbreviation	Relationoship to Standard Unit
kilo-	k	$1000 \times$
Base unit (has no prefix)		$1 \times$ (gram, liter, meter)
deci-	d	$\div 10$
centi-	c	$\div 100$
milli-	m	$\div 1000$
micro-	$\propto$	$\div 1,000,000$

SOLUTION EQUATIONS

Concentration Equations	Dilution Equation
$\text{Concentration} = \dfrac{\text{amount of solute}}{\text{amount of solution}}$	$C_{\text{initial}} \times V_{\text{initial}} = C_{\text{final}} \times V_{\text{final}}$

Molarity, M $M = \dfrac{\text{Mole solute}}{\text{L Solution}}$

Percent concentration

$\% \text{ concentration} = \dfrac{\text{parts of solute}}{100 \text{ parts of solution}}$

$\% \text{ (m/v)} = \dfrac{\text{g solute}}{\text{mL solution}} \times 100\%$

$\% \text{ (m/m)} = \dfrac{\text{g solute}}{\text{g solution}} \times 100\%$

$\% \text{ (v/v)} = \dfrac{\text{mL solute}}{\text{mL solution}} \times 100\%$

$\text{ppm} = \dfrac{\text{g solute}}{\text{mL solution}} \times 1,000,000$

$\text{ppb} = \dfrac{\text{g solute}}{\text{mL solution}} \times 1,000,000,000$